ATHEMATICS

PLE GROUPS

TERISTIC 2 TYPE

AND

n to Their Classification

NTIAL

S

PERATORS

ATORS

HEORY

THE UNIVERSITY SERIES IN M

Series Editor: Joseph J. Kohn
Princeton University

THE CLASSIFICATION OF FINITE SIM
Daniel Gorenstein
VOLUME 1: GROUPS OF NONCHARAC

ELLIPTIC DIFFERENTIAL EQUATION
OBSTACLE PROBLEMS
Giovanni Maria Troianiello

FINITE SIMPLE GROUPS: An Introducti
Daniel Gorenstein

INTRODUCTION TO PSEUDODIFFERE
AND FOURIER INTEGRAL OPERATOR
François Treves
VOLUME 1: PSEUDODIFFERENTIAL O
VOLUME 2: FOURIER INTEGRAL OPER

MATRIX THEORY: A Second Course
James M. Ortega

A SCRAPBOOK OF COMPLEX CURVE T
C. Herbert Clemens

Elliptic Differential Equations and Obstacle Problems

Giovanni Maria Troianiello

Università degli Studi di Roma I
Rome, Italy

Plenum Press • New York and London

Library of Congress Cataloging in Publication Data

Troianiello, Giovanni Maria.
Elliptic differential equations and obstacle problems.

(The University series in mathematics)
Bibliography: p.
Includes index.
1. Differential equations, Elliptic. 2. Variational inequalities (Mathematics). 3. Calculus of variations. I. Title. II. Series: University series in mathematics (Plenum Press)
QA377.T75 1987 515.3′53 87-12370
ISBN 0-306-42448-7

A Division of Plenum Publishing Corporation
233 Spring Street, New York, N.Y. 10013

Printed in the United States of America

To the memory of
Guido Stampacchia

Preface

In the few years since their appearance in the mid-sixties, variational inequalities have developed to such an extent and so thoroughly that they may now be considered an "institutional" development of the theory of differential equations (with appreciable feedback as will be shown). This book was written in the light of these considerations both in regard to the choice of topics and to their treatment. In short, roughly speaking my intention was to write a book on second-order elliptic operators, with the first half of the book, as might be expected, dedicated to function spaces and to linear theory whereas the second, nonlinear half would deal with variational inequalities and nonvariational obstacle problems, rather than, for example, with quasilinear or fully nonlinear equations (with a few exceptions to which I shall return later). This approach has led me to omit any mention of "physical" motivations in the wide sense of the term, in spite of their historical and continuing importance in the development of variational inequalities. I here addressed myself to a potential reader more or less aware of the significant role of variational inequalities in numerous fields of applied mathematics who could use an analytic presentation of the fundamental theory, which would be as general and self-contained as possible.

Having said all of this, I cannot fail to point out the extent to which my treatment of the subject does not succeed in being general or self-contained. On the first point I hasten to indicate that, in order to avoid an overly technical presentation, I have chosen to make C^1 regularity assumptions on (portions of) boundaries even where $C^{0,1}$ would have been sufficient. But above all, I have bypassed "truly" mixed problems. In effect I do systematically consider Dirichlet conditions on one portion of the boundary, and Neumann (or, more generally, regular oblique derivative) conditions on the remaining part. However, the basic reason for this was to avoid introducing separate statements and demonstrations for each type of boundary value problem; therefore, I adopt the hypothesis that both boundary portions are closed whenever the absence of such a hypothesis might introduce excessively delicate problems of regularity.

Coming to the second point, background results in functional analysis and in the theory of Lebesgue spaces have been listed without proofs; I have furthermore assumed that the reader has a graduate level knowledge of the real variable, and omitted the technically more complex part of the John–Nirenberg lemma. Detailed demonstrations are provided for all other results in the book.

What do I consider the relevant features of my book? First of all, I must mention Chapter 3 in which I develop what are generally called the Schauder and L^p theories (here referred to globally as $H^{k,p}$ and $C^{k,\delta}$ theory). Usually the essential *a priori* estimates are obtained, for the former by means of the Schauder–Caccioppoli techniques in Hölder function spaces, for the latter by using singular integrals in the light of the Calderón–Zygmund theory, and in particular by applying the Agmon–Douglis–Nirenberg method for boundary estimates. But here I have chosen Campanato's approach, which is more unified and, to my mind, simpler: Schauder's Hölderian estimates are absorbed by others, of a basically variational type, in the spaces of Morrey, John–Nirenberg, and Campanato, whereas L^p estimates are obtained from the previous ones by interpolation without resorting to singular integrals. My presentation, designed to be as complete as is reasonably possible, covers both the variational and the nonvariational case, as well as Dirichlet, Neumann, and regular oblique derivative boundary conditions. (The reader familiar with Campanato's method may notice some minor improvements introduced in Chapter 3. For instance, in the problems at the end of the chapter the L^p theory is extended to the range $1 < p < 2$.) But the use of Campanato's techniques is not limited to Chapter 3. They are also used in Chapter 1, reformulating those of Morrey, to show part of Sobolev inequalities. Campanato's method is further used in Chapter 2 to extend the De Giorgi–Nash theorem to nonhomogeneous equations with lower order coefficients, and in Chapter 4 to show $C^{0,\delta}$ and $C^{1,\delta}$ regularity results for solutions of variational inequalities of obstacle type.

These remarks should in no way give the impression that any one method has been given systematic preference. Quite the contrary. For example, still on the subject of variational inequalities, the reader will find the "natural" utilization of Lewy–Stampacchia inequalities for $H^{2,p}$ regularity (p finite), of difference quotients for H^2 regularity in a more general case, and of the penalty method for $H^{1,\infty}$ and $H^{2,\infty}$ regularity.

As has already been said, one of my aims throughout the book has been to go beyond the Dirichlet type of boundary conditions, and consequently I have had to tackle the problem of trace spaces in detail. This has been done in Chapter 1, where I have defined Sobolev spaces for orders between 0 and 1 by using the rapid and to my mind handy method of quotient spaces. This does not mean that I have systematically avoided any intrinsic definition of function spaces on manifolds. In point of fact I have presented full details, however tedious they may be, in the Lebesgue case since it furnishes the concrete basis for later abstract constructions. Before moving on from the material dealt with in Chapter 1, I would like to add that it includes a detailed study of lattice properties, and that the study of Sobolev spaces is probably more extensive here than is usually the case in texts about partial differential equations.

Passing to Chapter 2 I want to mention, in addition to the standard topics (Lax–Milgram and De Giorgi–Nash theorems, method of difference quotients), L^s regularity results for solutions of linear equations, and a study of interior regularity for solutions of a class of quasilinear equations, up to the point where the De Giorgi–Nash theorem comes into play and makes possible the automatic application of the linear theory.

Abstract existence results for nonlinear equations are discussed in Chapter 4 as byproducts of the study of variational inequalities. The reason for this is that Brézis' very general existence theorem for pseudomonotone operators (and its consequent application to differential operators of the Leray–Lions type) fits naturally into this wider setting. One last observation on my treatment of variational inequalities: I have included new existence and uniqueness results for variational obstacle problems involving a class of noncoercive bilinear forms, as well as existence theorems concerning quasilinear operators under natural growth conditions. In the latter context I used Lewy–Stampacchia inequalities to bring the study of equations in the presence of lower and upper solutions quite naturally back to bilateral variational problems.

There is a correspondence between the last point above and the nonvariational case dealt with in Chapter 5, where, among other things, I redemonstrate (and extend) results of Amann–Crandall and Kazdan–Kramer for semilinear equations. Here again lower and upper solutions are treated as obstacles in a constrained problem. (The study of the nonlinear case utilizes prerequisites for linear operators which are demonstrated at the beginning of the chapter.) Chapter 5 also takes up the problem of providing a sufficiently weak notion of solutions to unilateral problems for some nonlinear operators when we cannot (at least *a priori*) be certain of the existence of an $H^{2,p}$ solution (nor even perhaps of an H^1 solution in the case of divergence form operators). I also show that the characteristics of these weak solutions make it possible, in certain circumstances, to work back to an optimal threshold of regularity: The case considered is that of implicit unilateral problems (nonvariational counterparts of quasi-variational inequalities), in particular that of stochastic impulse control.

The ground covered in this book should be more than sufficient as the basis of a two-semester graduate course on second-order elliptic operators. With this end in mind I have provided problems at the end of each chapter and hints to their solution in informal style similar to that of suggestions which might be given orally in a seminar. The problems should present no difficulties to anyone who has a sound grasp of the preceding theoretical matter.

This book would probably never have been written had I not had the privilege of studying with teachers such as P. D. Lax, L. Nirenberg, and G. Stampacchia, nor had the good fortune to work in daily contact with colleagues and friends in the Mathematics Department of the University of Rome, of whom I should single out M. G. Garroni, U. Mosco, and F. Scarpini. The constructive telephone conversations I occasionally had over the years with C. Baiocchi should also be mentioned here. However, for what regards specifically this endeavor, the help and encouragement given to me by J. J. Kohn have been of special importance. I am glad to be able here to acknowledge my indebtedness to all these persons and to express my gratitude.

Contents

Glossary of Basic Notations

$\mathbb{N}$ is the set of natural numbers, $\mathbb{R}$ the real line. For $N \in \mathbb{N}$ the typical point of the Euclidean N-space $\mathbb{R}^N$ is denoted by $x \equiv (x_1, \ldots, x_N)$ and also by (x', x_N) with $x' \equiv (x_1, \ldots, x_{N-1})$ if $N \geq 2$; for x, $y \in \mathbb{R}^N$, $x \cdot y \equiv \sum_{i=1}^N x_i y_i$ and $|x| \equiv (x \cdot x)^{1/2}$. For derivatives of a function $u(x)$ we shall often adopt the *multi-index notation*: $D^\alpha u \equiv \partial^{|\alpha|} u / \partial x_1^{\alpha_1} \ldots \partial x_N^{\alpha_N}$, where each α_i is in $\mathbb{N} \cup \{0\}$, $\alpha \equiv (\alpha_1, \ldots, \alpha_N)$, $|\alpha| \equiv \alpha_1 + \cdots + \alpha_N$. We shall, however, usually write u_{x_i} for $\partial u / \partial x_i$ and $u_{x_i x_j}$ for $\partial^2 u / \partial x_i \, \partial x_j$ if $N \geq 2$, u' for du/dx and u'' for $d^2 u / dx^2$ if $N = 1$.

If D is a subset of $\mathbb{R}^N$, its boundary is denoted by ∂D and its closure $D \cup \partial D$ by $\bar{D}$.

Ω is an open subset of $\mathbb{R}^N$; from Chapter 2 on we shall constantly assume that Ω is connected as well as bounded. (Openness and connectedness make Ω a *domain*.) The notation $\omega \subset\subset \Omega$ means that ω is an open subset of $\mathbb{R}^N$ with $\bar{\omega} \subset \Omega$.

Γ is a relatively open portion of $\partial \Omega$; the unit outward normal at a point $x \in \Gamma$, if existing, is denoted by $\nu : x \mapsto (\nu^1(x), \ldots, \nu^N(x))$.

For $x^0 \in \mathbb{R}^N$ and $0 < r < \infty$,

$$B_r(x^0) \equiv \{x \in \mathbb{R}^N \mid |x - x^0| < r\},$$

$$B_r^+(x^0) \equiv \{(x', x_N) \in B_r(x^0) \mid x_N > x_N^0\},$$

$$S_r(x^0) \equiv \partial B_r(x^0),$$

$$S_r^+(x^0) \equiv \{(x', x_N) \in S_r(x^0) \mid x_N > x_N^0\},$$

$$S_r^0(x^0) \equiv \{(x', x_N) \in \mathbb{R}^N \mid |x' - x^{0\prime}| < r, \ x_N = x_N^0\};$$

in these notations we shall usually depress the dependence on x^0 if $x^0 = 0$, on r if $r = 1$. The set $\Omega \cap B_r(x^0)$ is denoted by $\Omega[x^0, r]$.

We shall follow the practice of using the same symbol C for different constants depending on prescribed sets of arguments.

1

Function Spaces

In the modern approach to partial differential equations a pivotal role is played by various function spaces which are defined in terms of the existence of derivatives (either in the classical or in a generalized, weaker sense). In this chapter we develop the study of such spaces to the extent required for the investigation of second-order elliptic problems.

We begin by listing, without proofs, some fundamental background results of functional analysis (Section 1.1). We then pass to spaces of smooth functions, with a special emphasis on extensions and traces as well as on local representations of boundary portions (Section 1.2). In Section 1.3 we dwell on Lebesgue spaces. After recalling (without proofs) some basic properties, we illustrate the technique of approximation by convolution and introduce L^p spaces which are defined through surface integrals. Section 1.4 is devoted to $L^{2,\mu}$ spaces, which for certain values of μ are identifiable with Hölder spaces. We call the reader's attention especially to Lemma 1.18, which will be utilized on several occasions.

The rest of the chapter is centered on the theory of Sobolev spaces, which will play a fundamental role throughout. Sections 1.5 and 1.6 deal with such topics as density results, extensions, continuous or compact imbeddings into Lebesgue or Hölder spaces. In Section 1.7 traces of functions from Sobolev spaces are defined through a density argument. Finally, in Section 1.8 various notions of inequalities, which will be essential to the study of equations and especially of obstacle problems, are introduced and mutually compared.

1.1. Preliminaries from Functional Analysis

1.1.1. Banach and Hilbert Spaces

All linear spaces considered in this book are assumed to be defined over $\mathbb{R}$. If V, W are two such spaces and F is an operator $V \to W$, the notation $F(v)$ for the value of F at $v \in V$ is replaced by Fv when F is linear and by $\langle F, v\rangle$ when in addition $W = \mathbb{R}$, that is, when F is a linear functional.

A *seminorm* on a linear space V is a mapping $v \mapsto [v]_V$ from V into $[0, \infty[$ such that

$$[\lambda v]_V = |\lambda|\, [v]_V \qquad \text{for } \lambda \in \mathbb{R},\ v \in V,$$

$$[v + w]_V \le [v]_V + [w]_V \qquad \text{for } v,\ w \in V.$$

The following analytic formulation of the *Hahn–Banach theorem* guarantees the possibility of extending linear functionals dominated by seminorms.

THEOREM 1.A. *Let W be a proper subspace of a linear space V. Suppose that F is a linear functional on W and $[\cdot]_V$ a seminorm on V such that*

$$|\langle F, v\rangle| \le [v]_V \qquad \textit{for } v \in W.$$

Then there exists a linear functional $\tilde{F}$ on V such that

$$\langle \tilde{F}, v\rangle = \langle F, v\rangle \qquad \textit{for } v \in W,$$

$$|\langle \tilde{F}, v\rangle| \le [v]_V \qquad \textit{for } v \in V.$$

A *norm* $v \mapsto |v|_V$ on a linear space V is a seminorm that vanishes only for $v = 0$. Two norms $|\cdot|_V$ and $|\cdot|_V'$ on V are said to be *equivalent* if

$$C^{-1}|v|_V \le |v|_V' \le C|v|_V \qquad \text{for } v \in V,$$

C being some positive constant; we then write $|\cdot|_V \sim |\cdot|_V'$. When V is endowed with a norm we call it a *normed space*. Any (linear) subspace W of V is then a normed space with $|\cdot|_W \equiv |\cdot|_V$. Since the mapping

$$u, v \mapsto |u - v|_V$$

is a metric on V, we can freely utilize metric notions such as: convergence, also called *strong convergence* (of a sequence $\{v_n\}$ to v in V, denoted by the symbol

$$v_n \to v),$$

continuity (of a mapping from a subset of V into another normed space), density, compactness or relative compactness (of a subset of V), completeness or separability (of V). The (*topological*) *dual space* of V is the linear space V' of continuous, or *bounded*, linear functionals F on V, endowed with the norm

$$|F|_{V'} \equiv \sup_{v \in V,\ |v|_V \leq 1} |\langle F, v\rangle|;$$

in this context $\langle \cdot, \cdot \rangle$ is the *duality pairing* between V and V'. By *weak convergence* of a sequence $\{v_n\}$ to v in V, denoted by the symbol

$$v_n \rightharpoonup v,$$

we mean convergence of $\langle F, v_n\rangle$ to $\langle F, v\rangle$ in $\mathbb{R}$ whatever $F \in V'$. Strong convergence implies weak convergence, and viceversa if V is finite dimensional. Weakly convergent sequences are bounded, and $\langle F_n, v_n\rangle \to \langle F, v\rangle$ if $F_n \to F$ in V', $v_n \rightharpoonup v$ in V.

The Hahn–Banach theorem can be given a geometric formulation that assures an adequate supply of continuous linear functionals, as is stated in the following theorem.

THEOREM 1.B. *Let W be a subspace of a normed space V. If W is not dense in V there exists a nonzero element F of V' such that*

$$\langle F, v\rangle = 0 \qquad \textit{for } v \in W.$$

Two normed spaces V and W are (*topologically*) *isomorphic* if there exists an injective and surjective linear operator $T : W \to V$ such that both T and $T^{-1} : V \to W$ are continuous, i.e., satisfy

$$|Tw|_V \leq C|w|_W \qquad \text{for } w \in W$$

and

$$|T^{-1}v|_W \leq C|v|_V \qquad \text{for } v \in V$$

with some positive constant C. We then write $V \sim W$. V and W are *isometrically isomorphic* in the particular case when

$$|Tw|_V = |w|_W \qquad \text{for } w \in W.$$

If the linear operator T is only required to be injective and continuous (which can happen to be the case with $T =$ identity when W is a subspace of V as well as a normed space on its own), we say that W is *continuously imbedded*, or *injected*, in V and write

$$W \hookrightarrow V;$$

the particular choice of T is algebraically and topologically irrelevant because W and its image $T(W)$ are isometrically isomorphic when the latter is normed by

$$z \mapsto |\, w \,|_W \qquad \text{for } z = Tw, \quad w \in W.$$

If $T(W)$ is dense in V (so that

$$V' \hookrightarrow W')$$

we write

$$W \underset{\text{ds}}{\hookrightarrow} V.$$

When a normed space is complete we call it a *Banach space*. Simple considerations show that V' is always a Banach space whether the normed space V is complete or not (see Problem 1.3). Any closed subspace of a Banach space is a Banach space in its turn.

Let V be a Banach space.

LEMMA 1.C. *Let $\mathbb{K} \subseteq V$ be closed in (the metric of) V. If $\mathbb{K}$ is convex and $\{v_n\} \subset \mathbb{K}$ converges weakly to v in V, then $v \in \mathbb{K}$.*

The linear mapping I defined by

$$\langle Iv, F\rangle \equiv \langle F, v\rangle \qquad \text{for } F \in V'$$

is a continuous injection of V in the dual space V'' of V', and even more, namely, an isometric isomorphism between V and the image space $I(V)$, by the Hahn–Banach theorem (see Problem 1.1). If I is surjective, that is, $I(V) = V''$, we call V *reflexive*.

THEOREM 1.D. *A Banach space is reflexive if and only if its dual space is such. Any closed subspace of a reflexive Banach space is reflexive as well.*

An important property of reflexive Banach spaces is given by the following theorem.

THEOREM 1.E. *Every bounded sequence in a reflexive Banach space contains a weakly convergent subsequence.*

A special class of normed spaces is that of *pre-Hilbert spaces.* They are linear spaces V such that there exists a mapping

$$u, v \mapsto (u, v)_V$$

from the Cartesian product $V \times V$ into $\mathbb{R}$, called a *scalar product* on V, which is linear in each variable and satisfies

$$(u, v)_V = (v, u)_V \qquad \text{for } u, v \in V$$

as well as

$$(u, u)_V > 0 \qquad \text{for } u \in V, \qquad u \neq 0.$$

On pre-Hilbert spaces the *Cauchy–Schwarz inequality* holds:

THEOREM 1.F. *Let $(\cdot, \cdot)_V$ be a scalar product on V. Then,*

$$|(u, v)_V| \leq (u, u)_V^{1/2}(v, v)_V^{1/2} \qquad \textit{for } u, v \in V.$$

A norm on V is given by the mapping

$$u \mapsto (u, u)_V^{1/2} \equiv |u|_V.$$

When we say that a normed space is a pre-Hilbert one, we mean that $|\cdot|_V$ is associated with a scalar product on V as above.

Two scalar products on V are said to be *equivalent* if the corresponding norms are such.

When a pre-Hilbert space is complete (and is therefore a Banach space) we call it a *Hilbert space.*

THEOREM 1.G. *Hilbert spaces are reflexive.*

A Hilbert space is isometrically isomorphic to its image in the dual space V' under the mapping

$$u \mapsto (u, \cdot)_V.$$

As a matter of fact, the *Riesz representation theorem* (see the corollary of Theorem 2.1 below) asserts that the above mapping is surjective; its inverse, that is, the isometric isomorphism $\mathcal{I} : V' \to V$ defined by

$$(\mathcal{I}F, v)_V \equiv \langle F, v \rangle \qquad \text{for } v \in V,$$

$F \in V'$, is called the *Riesz isomorphism.*

We conclude this subsection with a few considerations about product and quotient spaces.

If $V_1, \ldots, V_k$ are normed spaces, so is their Cartesian product

$$V \equiv V_1 \times \cdots \times V_k,$$

with

$$|v|_V \equiv \left(\sum_{i=1}^{k} |v_i|_{V_i}^p \right)^{1/p} \qquad \text{for some } p \in [1, \infty[$$

or

$$|v|_V \equiv \max_{i=1,\ldots,k} |v_i|_{V_i}$$

[$v \equiv (v_1, \ldots, v_n)$ with $v_i \in V_i$], all these norms being equivalent; V is separable, or complete, or reflexive, if each V_i is such.

Somewhat more delicate is the question of quotient spaces. For the sequel all we need is the following theorem.

THEOREM 1.H. *Let W be a closed subspace of a normed space V, and let V/W denote the linear space of equivalence classes*

$$[v] \equiv \{v + w \mid w \in W\},$$

$v \in V$. Then the mapping

$$[v] \mapsto \inf_{w \in W} |v + w|_V$$

defines a norm on V/W. If V is a Banach (Hilbert) space, so is V/W.

1.1.2. Fixed Points and Compact Operators

It is well known that in a complete metric space (in particular, in a Banach space) a contraction has a unique fixed point. More sophisticated existence (not uniqueness) results for fixed points will now be listed.

For finite-dimensional Banach spaces we have at our disposal *Brouwer's fixed point theorem*:

THEOREM 1.I. *Let V be a finite-dimensional Banach space, let $\mathbb{K}$ be a closed convex subset of V, and let T be a continuous mapping of $\mathbb{K}$ into itself such that the image $T(\mathbb{K})$ is bounded. Then T has a fixed point*

$$u \in \mathbb{K}, \qquad u = Tu.$$

Brouwer's theorem utilizes the fact that in Euclidean spaces bounded sets are relatively compact. Its direct extension to infinite-dimensional spaces is *Schauder's theorem*:

THEOREM 1.J. *Theorem 1.I remains valid in any Banach space provided the image $T(\mathbb{K})$ is required to be relatively compact.*

For the next result, known as the *Leray–Schauder theorem*, we need the following important notion. A mapping T between two normed spaces is said to be *compact* if it is continuous and maps bounded sets into relatively compact sets; when T is linear the requirement of continuity, which then amounts to boundedness, is clearly redundant.

THEOREM 1.K. *Let V be a Banach space. Suppose $\mathscr{C}$ is a compact mapping of $V \times [0, 1]$ into V with the following properties:*

(i) $\mathscr{C}(u, 0) = 0$ *whenever* $u \in V$;

(ii) *there exists a constant C such that*

$$|u|_V \leq C \text{ whenever } u \in V \text{ with } u = \mathscr{C}(u, s) \text{ for some } s \in [0, 1].$$

Then the mapping $T \equiv \mathscr{C}(\cdot, 1) : V \to V$ has a fixed point.

A linear mapping T of a normed space into itself admits always 0 as a fixed point. If T is in addition supposed compact the question of the existence of fixed points different from 0 must be formulated in terms of the *Fredholm alternative* as follows.

THEOREM 1.L. *Let V be a normed space and suppose $T : V \to V$ is linear and compact. Then, either the homogeneous equation*

$$u \in V, \qquad u - Tu = 0$$

has a solution $u \neq 0$, or the inhomogeneous equation

$$u \in V, \qquad u - Tu = v$$

is uniquely solvable for any choice of $v \in V$, *in which case the inverse of the operator* $u \mapsto u - Tu$ *is also bounded.*

REMARK. If V in Theorem 1.L is assumed to be a Hilbert space, the content of the Fredholm alternative can be considerably enriched. To wit, the equation

$$u \in V, \qquad \lambda u - Tu = v$$

can be shown to be uniquely solvable for any choice of v in V if $0 \neq \lambda \in \mathbb{R} \setminus \Sigma$, Σ being a certain countable set of real numbers with no limit points except possibly $\lambda = 0$, whereas the range of the mapping $u \mapsto \lambda u - Tu$ when $\lambda \in \Sigma$ can be characterized in terms of the null space of the mapping $u \mapsto \lambda u - T^*u$, with $T^* : V \to V'$ defined by

$$(T^*u, v)_V \equiv (u, Tv)_V \qquad \text{for } u, v \in V.$$

The proofs of the results stated in this section can be found in monographs on functional analysis such as those by H. Brézis [19] and A. E. Taylor and D. C. Lay [144]; for what concerns in particular fixed point theorems, we refer to D. Gilbarg and N. S. Trudinger [67].

1.2. Various Spaces of Smooth Functions

1.2.1. C^k and $C^{k,\delta}$ Spaces

For $D \subseteq \mathbb{R}^N$, $C^0(D)$ is the linear space of continuous real functions on D. When $u = u(x)$, $x \in D$ is uniformly continuous on D, any nonnegative and nondecreasing function τ on $]0, \infty[$ such that $\tau(r) \to 0$ as $r \to 0^+$ and

$$| u(x) - u(y) | \leq \tau(| x - y |) \qquad \text{for } x, y \in D$$

is called a *modulus of uniform continuity* for u.

Let D be compact. It is known from calculus that functions from $C^0(D)$ are uniformly continuous on D. Moreover, $C^0(D)$ becomes a Banach space with the choice of the norm

$$| u |_{C^0(D)} \equiv \max_D | u |;$$

convergence in $C^0(D)$ is called *uniform convergence*. A necessary and sufficient condition for a subset of $C^0(D)$ to be relatively compact is given by the celebrated *Ascoli–Arzelà theorem*, which states the following.

THEOREM 1.M. *A subset of the Banach space $C^0(D)$ is relatively compact if and only if its elements are uniformly bounded in the norm of $C^0(D)$ and admit a common modulus of uniform continuity.*

For the proof see, for instance, A. Kufner, O. John, and S. Fučik [92].

$C^k(\Omega)$, with $k \in \mathbb{N}$, is the linear space of functions on Ω having all derivatives of order $\leq k$ in $C^0(\Omega)$, and $C^\infty(\Omega) \equiv \bigcap_{k\in N} C^k(\Omega)$.

Given a continuous function $u = u(x)$, $x \in \Omega$, let $\operatorname{supp} u$ denote its *support*, that is, the closure of the set $\{x \in \Omega \mid u(x) \neq 0\}$: $C_c^k(\Omega)$, with k a nonnegative integer or $k = \infty$, is the linear subspace of $C^k(\Omega)$ consisting of functions u such that $\operatorname{supp} u$ is a compact subset of Ω. In particular, an important subset of $C_c^\infty(\mathbb{R}^N)$ is introduced as follows: Let $\varrho \in C_c^\infty(\mathbb{R}^N)$, $\varrho \geq 0$, $\varnothing \neq \operatorname{supp} u \subseteq \bar{B}$ [an admissible choice being $\varrho(x) \equiv e^{1/(|x|^2-1)}$ if $|x| < 1$, $\varrho(x) \equiv 0$ otherwise]. Set $\varrho_n(x) \equiv n^N \varrho(nx)/\int_{\mathbb{R}^N} \varrho(y)\,dy$ for $x \in \mathbb{R}^N$, so that $\varrho_n \geq 0$, $\operatorname{supp} \varrho_n \subseteq \bar{B}_{1/n}$, and $\int_{\mathbb{R}^N} \varrho_n(x)\,dx = 1$ $(n \in N)$. Each function of the sequence $\{\varrho_n\}$ is called a *mollifier*.

For $k \in \mathbb{N}$, $C^k(\bar{\Omega})$ is the linear space of functions in $C^k(\Omega)$ which can be continuously extended to $\bar{\Omega}$ together with all their derivatives of order $\leq k$. It is clear that, if Ω is bounded, $C^k(\bar{\Omega})$ becomes a Banach space with the choice of the norm

$$| u |_{C^k(\bar{\Omega})} \equiv \sum_{i=0}^{k} \sum_{|\alpha|=i} | D^\alpha u |_{C^0(\bar{\Omega})},$$

where we have used the multi-index notation. $C^\infty(\bar{\Omega})$ is the linear space $\bigcap_{k\in N} C^k(\bar{\Omega})$.

$C_c^k(\Omega \cup \Gamma)$, with k a nonnegative integer or $k = \infty$, is the linear subspace of $C^k(\bar{\Omega})$ consisting of functions u such that $\operatorname{supp} u$ is a compact subset of $\Omega \cup \Gamma$. For k finite a norm on $C_c^k(\Omega \cup \Gamma)$ can be defined in the obvious way also when Ω is not bounded; however, $C_c^k(\Omega \cup \Gamma)$ is not complete unless Ω is bounded and $\Gamma = \partial\Omega$, in which case $C_c^k(\Omega \cup \Gamma) = C^k(\bar{\Omega})$.

For $0 < \delta \leq 1$ let

$$[u]_{\delta;D} \equiv \sup_{\substack{x,y\in D \\ x\neq y}} \frac{| u(x) - u(y) |}{| x - y |^\delta}$$

whenever u is a function defined on a closed subset D of $\mathbb{R}^N$. If $[u]_{\delta;D} < \infty$

{so that u is uniformly continuous on D with a modulus of continuity given by $\tau(r) \equiv [u]_{\delta;D} r^{\delta}$}, we write $u \in C^{0,\delta}(D)$ and say that u is *Hölder continuous* or *Hölderian* in D (with exponent δ) when $0 < \delta < 1$, *Lipschitz continuous* or *Lipschitzian* in D when $\delta = 1$. If D is compact, a norm in the linear space $C^{0,\delta}(D)$ is defined by

$$| u |_{C^{0,\delta}(D)} \equiv | u |_{C^0(D)} + [u]_{\delta;D}.$$

If $u = u(x)$, $x \in \Omega$, is such that its restriction $u|_D$ belongs to $C^{0,\delta}(D)$ whenever D is a closed subset of Ω, we write $u \in C^{0,\delta}(\Omega)$ and say that u is *Hölder continuous* or *Hölderian* in Ω (with exponent δ) when $0 < \delta < 1$, *Lipschitz continuous* or *Lipschitzian* in Ω when $\delta = 1$. (Note that these notations and terminology are consistent with the above ones for Ω both open and closed, i.e., $\Omega = \mathbb{R}^N$.) For $k \in \mathbb{N}$, $C^{k,\delta}(\bar{\Omega})$ $[C^{k,\delta}(\Omega)]$ is the linear space of functions $u \in C^k(\bar{\Omega})$ $[u \in C^k(\Omega)]$ such that $D^\alpha u \in C^{0,\delta}(\bar{\Omega})$ $[D^\alpha u \in C^{0,\delta}(\Omega)]$ whenever $|\alpha| = k$. When Ω is bounded, a norm on $C^{k,\delta}(\bar{\Omega})$ is defined by

$$| u |_{C^{k,\delta}(\bar{\Omega})} \equiv | u |_{C^k(\bar{\Omega})} + \sum_{|\alpha|=k} [D^\alpha u]_{\delta;\bar{\Omega}}.$$

LEMMA 1.1. *For $k = 0, 1, \ldots$ and $0 < \delta \leq 1$, $C^{k,\delta}(\bar{\Omega})$, with Ω bounded, is a Banach space. For $k = 0$ the result remains valid if $\bar{\Omega}$ is replaced by any compact subset of $\mathbb{R}^N$.*

PROOF. Let $\{u_n\}$ be a Cauchy sequence in $C^{k,\delta}(\bar{\Omega})$. Since $\{u_n\}$ is also a Cauchy sequence in the Banach space $C^k(\bar{\Omega})$, it converges in the latter space toward a function u. Let $\varepsilon > 0$ be arbitrarily fixed, and let n_ε be so large that

$$[D^\alpha u_{n+p} - D^\alpha u_n]_{\delta;\bar{\Omega}} < \varepsilon$$

whenever $n \geq n_\varepsilon$, $p \in \mathbb{N}$, $|\alpha| = k$. As $p \to \infty$ we obtain the inequality

$$[D^\alpha u - D^\alpha u_n]_{\delta;\bar{\Omega}} \leq \varepsilon,$$

which proves that $u \in C^{k,\delta}(\bar{\Omega})$ and that $u_n \to u$ in $C^{k,\delta}(\bar{\Omega})$.

The last statement of the lemma is at this point obvious. □

REMARK. When Ω is bounded, by the Weierstrass theorem (see M. A. Naimark [124]) the set of all polynomials in $x_1, \ldots, x_N$ with rational coefficients is dense in $C^0(\bar{\Omega})$. This shows that $C^0(\bar{\Omega})$ is separable. So also is $C^k(\bar{\Omega})$ for $k \in \mathbb{N}$, since it can be identified with a subspace of a suitable Cartesian power of $C^0(\bar{\Omega})$.

For $0 < \delta \leq 1$, instead, $C^{k,\delta}(\bar{\Omega})$ is not separable. It is convenient to stipulate the notational convention $C^{k,0} \equiv C^k$. For $k = 0, 1, \ldots$ and $0 \leq \delta \leq 1$, $C^{k,\delta}(\bar{\Omega})$ is not reflexive. See A. Kufner, O. John, and S. Fučik [92] and the references therein.

When Ω is bounded the following facts can be readily ascertained. For $0 \leq \delta \leq 1$ the product uv belongs to $C^{0,\delta}(\bar{\Omega})$ whenever u, v do, and

$$| uv |_{C^{0,\delta}(\bar{\Omega})} \leq | u |_{C^{0,\delta}(\bar{\Omega})} \, | v |_{C^{0,\delta}(\bar{\Omega})} .$$

For $k = 0, 1, \ldots$ and $0 \leq \gamma < \delta \leq 1$, $C^{k,\delta}(\bar{\Omega}) \subsetneq C^{k,\gamma}(\bar{\Omega})$. (As a matter of fact, this injection is compact: see Problem 1.4.) More delicate is the question whether for $k \in \mathbb{N}$

$$C^k(\bar{\Omega}) \subsetneq C^{k-1,\delta}(\bar{\Omega}) \tag{1.1}$$

if $\delta > 0$. When Ω is convex, (1.1) is an immediate consequence of the mean value theorem whatever $\delta \leq 1$; in particular, if Λ belongs to $[C^{k,\delta}(\bar{\Omega})]^N$ and $\Omega' \equiv \Lambda(\Omega)$ is open, each function $u \equiv u' \circ \Lambda$, $u' \in C^{k,\delta}(\bar{\Omega}')$, belongs to $C^{k,\delta}(\bar{\Omega})$ with norm estimate

$$| u |_{C^{k,\delta}(\bar{\Omega})} \leq C \, | u' |_{C^{k,\delta}(\bar{\Omega}')} .$$

Note that the convexity of Ω can be dispensed with when $\delta = 0$. However, (1.1) is not true in general.

EXAMPLE. Let $N = 2$, $\Omega = \{(x_1, x_2) \in \mathbb{R}^2 \mid -1 < x_1 < 1, -1 < x_2 < | x_1 |^{1/2}\}$. The function

$$u(x_1, x_2) \equiv \begin{cases} (\text{sign } x_1) x_2^{\alpha}, \ 1 < \alpha < 2, & \text{if } x_2 > 0, \\ 0 \quad \text{otherwise} & \end{cases}$$

belongs to $C^1(\bar{\Omega})$, but does not belong to $C^{0,\delta}(\bar{\Omega})$ if $\alpha/2 < \delta \leq 1$.

When $N \geq 2$ all the above definitions and properties can be automatically transferred from $\mathbb{R}^N$ to $\mathbb{R}^{N-1}$, or even to $\mathbb{R}^{N-1} \times \{0\}$, provided the latter is endowed with its relative topology.

1.2.2. Extensions

When dealing with a function $u = u(x)$, $x \in D \subset \mathbb{R}^N$, it may be desirable to know whether u can be extended as a function of $x \in D'$, $D \subset$

$D' \subseteq \mathbb{R}^N$, which keeps throughout D' certain properties u has in D. To clarify this requirement we consider the case when D is not closed, and suppose that $0 \in \bar{D} \setminus D$. Let $\{x^n\} \subset D$ be such that $|x^n|$ tends to 0 in $\mathbb{R}$ with $|x^n| > |x^{n+1}|$. If $g \in C^0(]0, \infty[)$ is such that $g(|x^n|) = 1/(n\pi)$, $g > 0$ in $]0, \infty[$, the continuous function $u = u(x) \equiv \sin[g(|x|)^{-1}]$, $x \in D$, admits no continuous extension to $\bar{D}$.

As the preceding example shows, in order that every $u \in C^0(D)$ admit a continuous extension to $\mathbb{R}^N$ it is necessary that D be closed. Remarkably enough, this condition is also sufficient. Indeed, call $\tilde{u} = \tilde{u}(x)$, $x \in \mathbb{R}^N$, a *controlled* C^0 *extension* of $u \in C^0(D)$ to $\mathbb{R}^N$ whenever $\tilde{u} \in C^0(\mathbb{R}^N)$, $\tilde{u}|_D = u$ and $\sup_{\mathbb{R}^N} |\tilde{u}| = \sup_D |u|$; then we have the following theorem.

THEOREM 1.N. *If D is a closed subset of $\mathbb{R}^N$ every function $u \in C^0(D)$ admits a controlled C^0 extension to $\mathbb{R}^N$.*

This is a formulation of the *Tietze extension theorem*, a result in general topology whose proof can for instance be found in K. Kuratowski [93].

Let us progressively increase the amount of regularity to be kept under the extension procedure.

Call $\tilde{u} = \tilde{u}(x)$, $x \in \mathbb{R}^N$, a *controlled* $C^{0,\delta}$ *extension* of $u \in C^{0,\delta}(D)$, $0 < \delta \leq 1$, to $\mathbb{R}^N$ if $\tilde{u}$ is a controlled C^0 extension of u such that $[\tilde{u}]_{\delta;\mathbb{R}^N} = [u]_{\delta;D}$.

THEOREM 1.2. *If D is a closed subset of $\mathbb{R}^N$ every function $u \in C^{0,\delta}(D)$, $0 < \delta \leq 1$, admits a controlled $C^{0,\delta}$ extension to $\mathbb{R}^N$.*

PROOF. For $x \in \mathbb{R}^N$ set

$$v(x) \equiv \sup_{\xi \in D} \{u(\xi) - [u]_{\delta;D} |x - \xi|^\delta\}.$$

Then $\sup_{\mathbb{R}^N} v \leq \sup_D u$ and $v(x) \geq u(x)$ for $x \in D$. If $v(x)$ were $> u(x)$ for some $x \in D$ there would exist $\xi \in D$ such that $u(\xi) > u(x) + [u]_{\delta;D} \times |x - \xi|^\delta$, which is impossible. Thus, $v = u$ on D.

The function

$$\tilde{u}(x) \equiv \max\{v(x), -\sup_D |u|\}, \qquad x \in \mathbb{R}^N,$$

satisfies $\tilde{u} = u$ on D and $\sup_{\mathbb{R}^N} |\tilde{u}| = \sup_D |u|$. Let $x, y \in \mathbb{R}^N$ be such that $\tilde{u}(x) > \tilde{u}(y)$. Then $\tilde{u}(x) = v(x)$ and

$$\begin{aligned} 0 < \tilde{u}(x) - \tilde{u}(y) \leq v(x) - v(y) &= \sup_{\xi \in D} \{u(\xi) - [u]_{\delta;D} \mid x - \xi \mid^{\delta}\} \\ &\quad - \sup_{\eta \in D} \{u(\eta) - [u]_{\delta;D} \mid y - \eta \mid^{\delta}\} \\ &\leq [u]_{\delta;D} \sup_{\xi \in D} (\mid y - \xi \mid^{\delta} - \mid x - \xi \mid^{\delta}) \\ &\leq [u]_{\delta;D} \mid x - y \mid^{\delta}. \end{aligned}$$

(For the last inequality see Problem 1.6). The proof is complete. □

To proceed further into extension techniques we specialize with $D = \bar{\Omega}$ and introduce a useful terminology for the description of open portions Γ of $\partial\Omega$, as follows.

When Ω' and Ω'' are bounded open subsets of $\mathbb{R}^N$ we say that a map $\Lambda : \bar{\Omega}' \to \bar{\Omega}''$ is a $C^{k,\delta}$ *diffeomorphism*, with $k \in \mathbb{N} \cup \{0\}$ and $0 \leq \delta \leq 1$, if it is one-to-one and onto with

$$\Lambda(\Omega') = \Omega'', \qquad \Lambda \in [C^{k,\delta}(\bar{\Omega}')]^N, \qquad \Lambda^{-1} \in [C^{k,\delta}(\bar{\Omega}'')]^N.$$

We say that Γ is *straightened* by a $C^{k,\delta}$ diffeomorphism $\Lambda : \bar{U} \to \bar{B}$ if U is a bounded domain of $\mathbb{R}^N$, $U \cap \partial\Omega = \Gamma$, $\Lambda(U \cap \Omega) = B^+$ and $\Lambda(\Gamma) = S^0$. More generally, we call Γ of *class* $C^{k,\delta}$ if it is the union of a family (which we can always assume discrete, and even finite if Γ is compact) of open sets Γ_j straightened by $C^{k,\delta}$ diffeomorphisms $\Lambda_j : \bar{U}_j \to \bar{B}$; we call the family $\{(\Gamma_j, \Lambda_j)\}$ a $C^{k,\delta}$ *atlas* on Γ. If $k \in \mathbb{N}$ the Jacobian matrices of Λ_j and Λ_j^{-1} are, for each j, (defined and) nonsingular, respectively, throughout $\bar{U}_j$ and $\bar{B}$. This means that, whenever $x^0 \in \Gamma$, there exist a positive real number r and a permutation $\xi_1 = x_{i_1}, \ldots, \xi_N = x_{i_N}$ of coordinate axes such that $\Gamma \cap Q_r$, with $Q_r \equiv]\xi_1{}^0 - r, \xi_1{}^0 + r[\times \cdots \times]\xi_N{}^0 - r, \xi_N{}^0 + r[$, is the graph of a $C^{k,\delta}$ function $\xi_N = \lambda(\xi')$, and $\Omega \cap Q_r = \{(\xi', \xi_N) \in Q_r \mid \xi_N > \lambda(\xi')\}$. Moreover, the unit outward normal ν is defined throughout Γ.

We can now return to the problem of extending a function $u(x)$, $x \in \bar{\Omega}$. If $\operatorname{supp} u \subset \Omega$ the *trivial extension* of u to $\mathbb{R}^N$, defined as 0 in $\mathbb{R}^N \setminus \bar{\Omega}$, obviously shares the same properties of regularity as u independently of the regularity of $\partial\Omega$. In more general situations, however, the latter plays a crucial role, as is illustrated by

THEOREM 1.3. *Let Ω be bounded with $\partial\Omega$ of class $C^{k,\delta}$ for some $k \in \mathbb{N}$, $\delta \in [0, 1]$. Whenever Ω' is a bounded open subset of $\mathbb{R}^N$ such that $\Omega \subset\subset \Omega'$,* *every function $u \in C^{k,\delta}(\bar{\Omega})$ admits an extension $\tilde{u}$ to Ω' with $\tilde{u} \in C^{k,\delta}(\bar{\Omega}')$,* supp $\tilde{u} \subset \Omega'$ *and*

$$| \tilde{u} |_{C^{k,\delta}(\bar{\Omega}')} \leq C \, | u |_{C^{k,\delta}(\bar{\Omega})},$$

C being independent of u.

PROOF. We begin with suitable changes of variables near boundary points. Precisely, we fix $x^0 \in \partial\Omega$ and denote by U a bounded domain of $\mathbb{R}^N$ such that $x^0 \in U$, $U \cap \partial\Omega$ being straightened by a $C^{k,\delta}$ diffeomorphism $\Lambda : \bar{U} \to \bar{B}$. The function $u'(x) \equiv (u \circ \Lambda^{-1})(x)$, $x \in \overline{B^+}$, belongs to $C^{k,\delta}(\overline{B^+})$.

We extend u' across S^0 by setting

$$\widetilde{u'}(x', x_N) \equiv \sum_{h=1}^{k+1} C_h u'(x', -x_N/h)$$

for $(x', -x_N) \in \overline{B^+}$, where the vector $(C_1, \ldots, C_{k+1})$ is determined as the unique solution to

$$\sum_{h=1}^{k+1} \left(-\frac{1}{h} \right)^{j-1} C_h = 1 \qquad \text{for } j = 1, \ldots, k+1.$$

{The coefficient matrix $[(-1/h)^{j-1}]_{h,j=1,\ldots,k+1}$ of the above system is nonsingular, since it coincides with the Vandermonde matrix of the numbers $-1, -1/2, \ldots, -1/(k+1)$.} Thus, $\widetilde{u'} \in C^{k,\delta}(\bar{B})$ with

$$D^\alpha \widetilde{u'}(x', x_N) = \sum_{h=1}^{k+1} \left(-\frac{1}{h} \right)^{\alpha_N} C_h (D^\alpha u')(x', -x_N/h)$$

for $(x', -x_N) \in \bar{B}^+$ whatever the multi-index $\alpha \equiv (\alpha_1, \ldots, \alpha_N)$, $| \alpha | \leq k$, hence

$$| \widetilde{u'} |_{C^{k,\delta}(\bar{B})} \leq C \, | u' |_{C^{k,\delta}(\bar{B}^+)} \leq C \, | u |_{C^{k,\delta}(\bar{\Omega})}.$$

We now go back to the original variables. Let r be so small that $\overline{B_r(x^0)} \subset U \cap \Omega'$. The function $w(x) \equiv (\widetilde{u'} \circ \Lambda)(x)$, $x \in \overline{B_r(x^0)}$, belongs to $C^{k,\delta}(\overline{B_r(x^0)})$ with

$$| w |_{C^{k,\delta}(\overline{B_r(x^0)})} \leq C \, | u |_{C^{k,\delta}(\bar{\Omega})}$$

[see the observation following (1.1)]; notice that $w = u$ on $\overline{B_r(x^0)} \cap \bar{\Omega}$.

By compactness $\partial\Omega = \bigcup_{i=1}^m \Gamma^i$ for some $m \in N$, where each Γ^i is the intersection of $\partial\Omega$ with some open sphere B^i constructed through the same procedure as the one above for $B_r(x^0)$.

At this point we need the following result, which is said to provide a *partition of unity.*

LEMMA 1.4. *Let* ω *be an open subset of* $\mathbb{R}^N$ *and for* $j = 0, 1 \ldots$ *let* $\omega_j \subset\subset \omega$ *be such that* $\omega = \bigcup_{j=0}^{\infty} \omega_j$, *any compact subset of* ω *intersecting only a finite number of the* ω_j*'s. Then there exists a sequence* $\{g_j\} \subset C^\infty(\mathbb{R}^N)$ *with* supp $g_j \subset \omega_j$, $0 \leq g_j \leq 1$, $\sum_{j=0}^{\infty} g_j = 1$ *throughout* ω.

For the proof see Problem 1.9. An open covering $\{\omega_j\}$ of the required type can, for instance, be obtained by setting $\omega_0 \equiv \omega_1'$ and $\omega_j \equiv \omega_{j+1}' \setminus \overline{\omega_{j-1}'}$ for $j \geq 1$ whenever $\omega_j' \subset\subset \omega_{j+1}'$, $\omega = \bigcup_{j=0}^{\infty} \omega_j'$.

As a matter of fact, here we need the following straightforward consequence of Lemma 1.4, corresponding to the case $\omega = \mathbb{R}^N$ for a suitable choice of $\omega_{m+p} \subset\subset \mathbb{R}^N \setminus D$, $p \in \mathbb{N}$:

COROLLARY. *Let D be a compact subset of* $\mathbb{R}^N$ *such that* $D \subset \bigcup_{j=0}^{m} \omega_j$, *each* ω_j *being a bounded open subset of* $\mathbb{R}^N$. *Then there exist* $g_0, \ldots, g_m \in C^\infty(\mathbb{R}^N)$ *with* supp $g_j \subset \omega_j$, $0 \leq g_j \leq 1$, $\sum_{j=0}^{m} g_j = 1$ *throughout D.*

We take $D = \bar{\Omega}$, $\omega_i = B^i$ for $i = 1, \ldots, m$, $\Omega \supset\supset \omega_0 \supset \Omega \setminus \bigcup_{i=1}^{m} B^i$, and set

$$\tilde{u}(x) \equiv (ug_0)(x) + \sum_{i=1}^{m} (w^i g_i)(x), \qquad x \in \bar{\Omega}'$$

where each w^i is the function from $C^{k,\delta}(\overline{B^i})$ constructed through the procedure previously illustrated for w [with the understanding that products $(fg_j)(x)$ are defined to be 0 for $x \in \mathbb{R}^N \setminus \operatorname{supp} g_j$]. The function $\tilde{u}$ is the sought-for extension of u. □

REMARK. Since Ω' can be chosen convex, an immediate consequence of Theorem 1.3 is that (1.1) holds for $k \in \mathbb{N}$ and $0 < \delta \leq 1$ if $\partial\Omega$ is of class C^1.

1.2.3. Traces

Let $0 \leq \delta \leq 1$ and assume Ω bounded, Γ not only open but also closed.

Every function $u \in C^{0,\delta}(\bar{\Omega})$ admits a trace $u|_\Gamma$ on Γ, $u|_\Gamma$ belonging to $C^{0,\delta}(\Gamma)$ with

$$| u|_\Gamma |_{C^{0,\delta}(\Gamma)} \leq | u |_{C^{0,\delta}(\bar{\Omega})}.$$

In fact $C^{0,\delta}(\Gamma)$ is exactly the space of traces of functions from $C^{0,\delta}(\bar{\Omega})$

since any function $\eta \in C^{0,\delta}(\Gamma)$ admits an extension $u \in C^{0,\delta}(\bar{\Omega})$ ($\eta = u|_\Gamma$) with

$$| u |_{C^{0,\delta}(\bar{\Omega})} = | \eta |_{C^{0,\delta}(\Gamma)}$$

(see Theorems 1.N and 1.2).

Now let $k \in N$ and assume Γ of class $C^{k,\delta}$. We define a linear space $C^{k,\delta}(\Gamma)$ as follows. If $\{(\Gamma_j, \Lambda_j)\}_{j=1,\ldots,n}$, $n \in N$, is a $C^{k,\delta}$ atlas on Γ, a function $\eta = \eta(x)$, $x \in \Gamma$, belongs to $C^{k,\delta}(\Gamma)$ if, for each j, $\eta \circ (\Lambda_j^{-1})|_{S^0}$ belongs to $C^{k,\delta}(S^0)$ (S^0 being endowed with its relative topology); note that the unit outward normal ν belongs to $[C^{k-1,\delta}(\Gamma)]^N$.

It is clear that the traces on Γ of functions from $C^{k,\delta}(\bar{\Omega})$ belong to $C^{k,\delta}(\Gamma)$. Vice versa, we have the following lemma.

Lemma 1.5. *If $\eta \in C^{k,\delta}(\Gamma)$ there exists $u \in C^{k,\delta}(\bar{\Omega})$ such that $u|_\Gamma = \eta$, $u = 0$ near $\partial\Omega \setminus \Gamma$ if the latter is not empty.*

Proof. Let $x^0 \in \Gamma_j$ for some j. The function

$$v_j(x', x_N) \equiv (\eta \circ \Lambda_j^{-1})(x', 0), \qquad (x', x_N) \in B$$

belongs to $C^{k,\delta}(B)$.

Let $B_r(x^0) \subset\subset U_j$, $B_r(x^0) \cap (\partial\Omega \setminus \Gamma) = \varnothing$. The function $w_j(x) \equiv (v_j \circ \Lambda_j)(x)$, $x \in B_r(x^0)$, belongs to $C^{k,\delta}(B_r(x^0))$, and $w_j|_{B_r(x^0)\cap\Gamma} = \eta|_{B_r(x^0)\cap\Gamma}$. As in the proof of Lemma 1.4 we can reach the sought-for conclusion by appropriately choosing an open covering of Γ and the corresponding partition of unity. □

By Lemma 1.5 the definition of $C^{k,\delta}(\Gamma)$ does not depend on the particular choice of the $C^{k,\delta}$ atlas on Γ. Indeed, $C^{k,\delta}(\Gamma)$ could equivalently be defined as the linear space of functions $\eta = u|_\Gamma$ with $u \in C^{k,\delta}(\bar{\Omega})$, $u = 0$ on $\partial\Omega \setminus \Gamma$. Moreover, a norm on $C^{k,\delta}(\Gamma)$ is provided by

$$| \eta |_{C^{k,\delta}(\Gamma)} \equiv \inf\{| u |_{C^{k,\delta}(\bar{\Omega})} \mid u|_\Gamma = \eta,\ u|_{\partial\Omega\setminus\Gamma} = 0\}.$$

Since $|\cdot|_{C^{k,\delta}(\Gamma)}$ is a norm on a Banach space quotient, $C^{k,\delta}(\Gamma)$ is a Banach space (see Theorem 1.H). We leave the details to the reader.

1.3. Lebesgue Spaces

We assume that the reader is familiar with the basic theory of Lebesgue measure and integral, such as can be found, for instance in W. Rudin [133]. We write a.a. and a.e., respectively, for "almost any" and "almost everywhere" with respect to Lebesgue measure meas_N on $\mathbb{R}^N$. (We shall often write $|E|$ for $\text{meas}_N E$.) Let u be a measurable real function on Ω. We identify u with the equivalence class of all functions on Ω which equal u a.e.; we attribute a pointwise property in Ω or $\bar{\Omega}$ to u if that property holds for a representative of the equivalence class of u. We shall mostly write $\int_\Omega u\,dx$ instead of $\int_\Omega u(x)\,dx$.

For u measurable in Ω denote by $\{\omega_\alpha\}_{\alpha\in A}$ the indexed family of all open sets $\omega_\alpha \subset \Omega$ such that $u = 0$ a.e. in ω_α. The open set $\omega \equiv \bigcup_{\alpha\in A} \omega_\alpha$ is the union of a countable family of compact sets, and each one of these can be covered by a finite number of the ω_α's: hence, $\omega = \bigcup_{n=1}^\infty \omega_{\alpha_n}$ for a suitable sequence $\{\alpha_n\} \subset A$, and $u = 0$ a.e. in ω. We call $\bar{\Omega}\setminus\omega$ the *support* of u and denote it by $\operatorname{supp} u$. For $u \in C^0(\Omega)$ the present definition of $\operatorname{supp} u$ is rapidly seen to coincide with that of Section 1.2.1.

1.3.1. L^p Spaces over Ω

For $1 \le p < \infty$ we denote by $L^p(\Omega)$ the linear space of measurable functions u on Ω such that $|u|^p$ is integrable over Ω, and set

$$|u|_{p;\Omega} \equiv \left(\int_\Omega |u|^p\,dx\right)^{1/p}.$$

Let

$$\operatorname*{ess\,sup}_\Omega u \equiv \inf\{C \in \mathbb{R} \mid u \le C \text{ a.e. in } \Omega\},$$

$$\operatorname*{ess\,inf}_\Omega u \equiv -\operatorname*{ess\,sup}_\Omega(-u).$$

We denote by $L^\infty(\Omega)$ the linear space of measurable functions u on Ω such that $\operatorname{ess\,sup}_\Omega |u| < \infty$, and set

$$|u|_{\infty;\Omega} \equiv \operatorname*{ess\,sup}_\Omega |u|.$$

Note that $|u|_{\infty;\Omega} = |u|_{C^0(\bar{\Omega})}$ if Ω is bounded and $u \in C^0(\bar{\Omega})$. For $1 \le p \le \infty$ we attribute the same meaning as above to the symbol $|u|_{p;\Omega}$ also when u is an $\mathbb{R}^N$-valued function from $[L^p(\Omega)]^N$. Moreover, for $x^0 \in \mathbb{R}^N$ and $0 < r < \infty$ we write $|u|_{p;x^0,r}$ instead of $|u|_{p;B_r(x^0)}$, $|u|_{p;x^0,r,+}$ instead of $|u|_{p;B_r^+(x^0)}$, and (usually) depress the dependence on x^0 if $x^0 = 0$, on r if $r = 1$.

Whatever $p \in [1, \infty]$, $u \mapsto |u|_{p;\Omega}$ defines a norm on $L^p(\Omega)$, whereas $u, v \mapsto \int_\Omega uv\,dx$ defines a scalar product in $L^2(\Omega)$.

If Ω is bounded and $u \in L^\infty(\Omega)$ we have

$$\left(\frac{1}{|\Omega|}\int_\Omega |u|^p\,dx\right)^{1/p} \leq |u|_{\infty;\Omega}$$

as well as

$$\left(\int_\Omega |u|^p\,dx\right)^{1/p} \geq (1-\varepsilon)|u|_{\infty;\Omega}[\operatorname{meas}_N\{x\in\Omega \mid |u(x)| \geq (1-\varepsilon)|u|_{\infty;\Omega}\}]^{1/p}$$

for $\varepsilon > 0$. Therefore,

$$|u|_{\infty;\Omega} = \lim_{p\to\infty}\left(\int_\Omega |u|^p\,dx\right)^{1/p}. \tag{1.2}$$

With every p we associate the *conjugate exponent* p' defined by

$$\begin{aligned} p' &\equiv p/(p-1) \quad &&\text{if } 1 < p < \infty,\\ p' &\equiv \infty &&\text{if } p = 1,\\ p' &\equiv 1 &&\text{if } p = \infty. \end{aligned}$$

The next result is *Hölder's inequality*; it contains the Cauchy–Schwarz inequality when $p = 2$.

THEOREM 1.O. *For $1 \leq p \leq \infty$ let $u \in L^p(\Omega)$ and $v \in L^{p'}(\Omega)$. Then uv belongs to $L^1(\Omega)$, and*

$$|uv|_{1;\Omega} \leq |u|_{p;\Omega}\,|v|_{p';\Omega}.$$

More generally: let $u_i \in L^{p_i}(\Omega)$, $1 \leq p_i \leq \infty$, for $i = 1, \ldots, n$, with $p^{-1} \equiv p_1^{-1} + \cdots + p_n^{-1} \leq 1$ ($p_i^{-1} \equiv 0$ if $p_i = \infty$). Then $u \equiv u_1 \cdots u_n$ belongs to $L^p(\Omega)$, and

$$|u|_{p;\Omega} \leq |u_1|_{p_1;\Omega} \cdots |u_n|_{p_n;\Omega}.$$

For what concerns structure properties of $L^p(\Omega)$ we have the following theorem.

THEOREM 1.P. *For $1 \leq p \leq \infty$, $L^p(\Omega)$ is a Banach space with respect to the norm $u \mapsto |u|_{p;\Omega}$; $L^2(\Omega)$ is a Hilbert space with respect to the scalar product $u, v \mapsto \int_\Omega uv\,dx$. $L^p(\Omega)$ is separable for $1 \leq p < \infty$, whereas $L^\infty(\Omega)$ is not. $L^p(\Omega)$ is reflexive for $1 < p < \infty$, whereas $L^1(\Omega)$ and $L^\infty(\Omega)$ are not.*

Note that $L^p(\Omega) \subsetneq L^q(\Omega)$ for $1 \leq q < p \leq \infty$ if Ω is bounded.

The next result relates convergence in $L^p(\Omega)$ to convergence a.e. in Ω.

THEOREM 1.Q. *Let* $1 \le p \le \infty$ *and suppose that* $\{u_n\}_n$ *converges toward* u *in* $L^p(\Omega)$. *Then there exists a subsequence* $\{u_{n_k}\}_k$ *which converges to* u *a.e. in* Ω; *moreover, there exists* $u^* \in L^p(\Omega)$ *such that* $|u_{n_k}(x)| \le u^*(x)$ *for a.a.* $x \in \Omega$, *any* $k \in \mathbb{N}$.

Passing to dual spaces we have the following theorem.

THEOREM 1.R. *Let* $F \in [L^p(\Omega)]'$, $1 \le p < \infty$. *Then there exists* $f \in L^{p'}(\Omega)$, *with* $|f|_{p';\Omega} = |F|_{[L^p(\Omega)]'}$, *such that*

$$\langle F, v\rangle = \int_\Omega fv\,dx \qquad \text{for } v \in L^p(\Omega). \tag{1.3}$$

For $1 \le p < \infty$ we identify $[L^p(\Omega)]'$ with $L^{p'}(\Omega)$ by reformulating (1.3) as $F = f$. Note that $L^1(\Omega) \subsetneq [L^\infty(\Omega)]'$, but there exist bounded linear functionals F on $L^\infty(\Omega)$ such that for no choice of f in $L^1(\Omega)$ (1.3) holds with $p = \infty$: take for instance

$$\langle F, v\rangle \equiv v(x^0) \qquad \text{for } v \in C_c^\infty(\Omega),$$

$x^0 \in \Omega$, and utilize the Hahn–Banach theorem to extend F as an element of $[L^\infty(\Omega)]'$ (see Problem 1.10).

Finally we have the following theorem.

THEOREM 1.S. $C_c^0(\Omega)$ *is dense in* $L^p(\Omega)$ *for* $1 \le p < \infty$.

For the proofs of Theorems 1.O–1.S we refer to H. Brézis [19].

We denote by $L^p_{\text{loc}}(\Omega)$ the linear space of measurable functions u on Ω such that $u|_\omega \in L^p(\omega)$ whenever $\omega \subset\subset \Omega$.

1.3.2. Approximation by Convolution in C^0, $C^{0,\delta}$, L^p

If $u \in L^1_{\text{loc}}(\Omega)$ and $\varrho \in C^\infty(\mathbb{R}^N)$ with supp $\varrho \subseteq \bar{B}_r$ for some r we denote by $\varrho * u$ the *convolution* of ϱ and u, that is, the function

$$\begin{aligned}(\varrho * u)(x) &\equiv \int_{B_r(x)} \varrho(x-y)u(y)\,dy \\ &= \int_{B_r} u(x-y)\varrho(y)\,dy \qquad \text{for } x \in \Omega, \qquad \operatorname{dist}(x, \partial\Omega) > r;\end{aligned}$$

the definition of $\varrho * u$ is trivially extended to the whole of $\bar{\Omega}$ if dist(supp u, $\partial\Omega) > 2r$. [For $S, S' \subset \mathbb{R}^N$ the symbol dist(S, S') denotes $\inf\{|x - y| \mid x \in S, y \in S'\}$.]

LEMMA 1.6. *Let* $u \in L^1_{\text{loc}}(\Omega)$, $\varrho \in C^\infty(\mathbb{R}^N)$ *with* supp $\varrho \subseteq B_r$. *Then* $(\varrho * u)|_{\bar{\omega}} \in C^\infty(\bar{\omega})$ *whenever* $\omega \subset\subset \Omega$ *with* dist$(\omega, \partial\Omega) > r$, *and*

$$D^\alpha(\varrho * u)(x) = [(D^\alpha\varrho) * u](x) \qquad \text{for } x \in \bar{\omega}$$

whatever the multi-index α. *Moreover,* $\varrho * u \in C_c^\infty(\Omega)$ *when* dist(supp u, $\partial\Omega$) $> 2r$.

PROOF. It clearly suffices to prove that, for $x \in \omega$, $(\varrho * u)(x)$ is differentiable and verifies

$$\frac{\partial}{\partial x_i}(\varrho * u)(x) = \left(\frac{\partial \varrho}{\partial x_i} * u\right)(x),$$

$i = 1, \ldots, N$. We arbitrarily fix $h \in \mathbb{R}^N$ with $|h|$ sufficiently small and apply the mean value theorem:

$$\begin{aligned} &|\varrho(x + h - y) - \varrho(x - y) - h \cdot \nabla\varrho(x - y)| \\ &\quad \leq |h|\, \varepsilon(|h|), \qquad \varepsilon(r) \to 0 \qquad \text{as } r \to 0^+. \end{aligned}$$

Thus,

$$\begin{aligned} &|(\varrho * u)(x + h) - (\varrho * u)(x) - [(h \cdot \nabla\varrho) * u](x)| \\ &\quad \leq |h|\varepsilon(|h|) \int_{B_r(x) \cup B_r(x+h)} |u(y)|\, dy \end{aligned}$$

and the conclusion is patent. □

We now specialize with convolutions $\varrho_n * u$, called *regularizations* of u, where $\{\varrho_n\}$ is a sequence of mollifiers (see Section 1.2.1.).

LEMMA 1.7. *Let* $u \in C^0(\Omega)$ *and* $\omega \subset\subset \Omega$. *Then for n large enough all functions* $(\varrho_n * u)|_{\bar{\omega}}$ *have the same modulus of uniform continuity as* $u|_{\overline{\omega'}}$ *and satisfy*

$$|\varrho_n * u|_{C^0(\bar{\omega})} \leq |u|_{C^0(\overline{\omega'})}$$

if $\omega \subset\subset \omega' \subset\subset \Omega$; *moreover,*

$$(\varrho_n * u)|_{\bar{\omega}} \to u|_{\bar{\omega}} \qquad \text{in } C^0(\bar{\omega}) \qquad \text{as } n \to \infty.$$

PROOF. The statement about the modulus of uniform continuity and the norm estimate are obvious.

Let $\varepsilon > 0$ be arbitrarily fixed and choose $\delta > 0$ in such a way that $B_\delta(x) \subset \omega'$ and

$$| u(x-y) - u(x) | < \varepsilon$$

for $x \in \bar{\omega}$, $y \in \overline{B_\delta}$. Since

$$(\varrho_n * u)(x) - u(x) = \int_{B_{1/n}} [u(x-y) - u(x)]\varrho_n(y)\, dy$$

and the right-hand side is majorized in absolute value by $\varepsilon \int_{B_{1/n}} \varrho_n(y)\, dy = \varepsilon$ for $1/n < \delta$, the conclusion follows. ∎

A simple consequence of the above is that $C_c^\infty(\Omega)$ is dense in $C_c^k(\Omega)$ for $k = 0, 1, \ldots$, hence also in $L^p(\Omega)$ for $1 \leq p < \infty$ by Theorem 1.S.

If u is Hölderian or Lipschitzian in Ω we can improve Lemma 1.7 as follows.

LEMMA 1.8. *Let $u \in C^{0,\delta}(\Omega)$, $0 < \delta \leq 1$, and $\omega \subset\subset \Omega$. Then, whenever $\omega \subset\subset \omega' \subset\subset \Omega$,*

$$[\varrho_n * u]_{\delta;\bar{\omega}} \leq [u]_{\delta;\overline{\omega'}}$$

for n large enough, and

$$(\varrho_n * u)|_{\bar{\omega}} \to u|_{\bar{\omega}} \quad \text{in } C^{0,\gamma}(\bar{\omega}) \quad \text{as } n \to \infty \quad \text{for } 0 \leq \gamma < \delta.$$

PROOF. The estimate is obvious. The last statement follows from the compactness of the imbedding $C^{0,\delta}(\bar{\omega}) \subsetneq C^{0,\gamma}(\bar{\omega})$ (see Problem 1.4). ∎

REMARK. In the setting of Lemma 1.8 we can estimate the rate of convergence of $(\varrho_n * u)|_{\bar{\omega}}$ to $u|_{\bar{\omega}}$ in $C^0(\bar{\omega})$. Namely, for n large enough

$$| (\varrho_n * u)(x) - u(x) | \leq \int_{B_{1/n}} | u(x-y) - u(x) | \varrho_n(y)\, dy$$
$$\leq \int_{B_{1/n}} [u]_{\delta;\overline{\omega'}} n^{-\delta} \varrho_n(y)\, dy = n^{-\delta}[u]_{\delta;\overline{\omega'}}$$

for $x \in \bar{\omega}$. We can also majorize the rate of divergence, for $\delta < 1$, of

$|\varrho_n * u|_{C^1(\bar\omega)}$: indeed (assuming $\int_{\mathbb{R}^N} \varrho\, dy = 1$),

$$
\begin{aligned}
|(\varrho_n * u)_{x_i}(x)| &= nn^N \left| \int_{B_{1/n}(x)} \varrho_{x_i}[n(x-y)]u(y)\, dy \right| \\
&= nn^N \left| \int_{B_{1/n}} u(x-y)\varrho_{x_i}(ny)\, dy \right| \\
&= nn^N \left| \int_{B_{1/n}} [u(x-y)-u(x)]\varrho_{x_i}(ny)\, dy \right| \\
&\le n \int_{B_{1/n}} [u]_{\delta;\overline{\omega'}} n^{-\delta} |\varrho_{x_i}(ny)| n^N\, dy
\end{aligned}
$$

for $x \in \bar\omega$, $i = 1, \ldots, N$, hence

$$|\varrho_n * u|_{C^1(\bar\omega)} \le Cn^{1-\delta} |u|_{C^{0,\delta}(\overline{\omega'})}.$$

Passing to Lebesgue spaces we have the following lemma.

LEMMA 1.9. *Let $u \in L^p_{\mathrm{loc}}(\Omega)$, $1 \le p < \infty$, and $\omega \subset\subset \Omega$. Then, whenever $\omega \subset\subset \omega' \subset\subset \Omega$,*

$$|\varrho_n * u|_{p;\omega} \le |u|_{p;\omega'}$$

for n large enough, and

$$(\varrho_n * u)|_\omega \to u|_\omega \qquad \text{in } L^p(\omega) \qquad \text{as } n \to \infty.$$

PROOF. When $x \in \omega$ and $1/n < \operatorname{dist}(\omega, \partial\omega')$ Hölder's inequality yields, for $p > 1$,

$$
\begin{aligned}
|(\varrho_n * u)(x)|^p &\le \left[\int_{B_{1/n}(x)} \varrho_n^{1/p}(x-y)\varrho_n^{1/p'}(x-y)\, |u(y)|\, dy \right]^p \\
&\le \int_{B_{1/n}(x)} \varrho_n(x-y)\, |u(y)|^p\, dy \left[\int_{B_{1/n}(x)} \varrho_n(x-y)\, dy \right]^{p/p'} \\
&= \int_{\omega'} \varrho_n(x-y)\, |u(y)|^p\, dy,
\end{aligned}
$$

so that

$$\int_\omega |(\varrho_n * u)(x)|^p\, dx \le \int_{\omega'} |u(y)|^p\, dy \int_\omega \varrho_n(x-y)\, dx \le \int_{\omega'} |u(y)|^p\, dy$$

whatever $p \in [1, \infty[$.

Fix $\varepsilon > 0$ and let $u_1 \in C_c^0(\omega')$ be such that $|u - u_1|_{p;\omega'} < \varepsilon$ (see Theorem 1.S). Then

$$\begin{aligned}|\varrho_n * u - u|_{p;\omega} &\leq |\varrho_n * (u - u_1)|_{p;\omega} + |u_1 - u|_{p;\omega} + |\varrho_n * u_1 - u_1|_{p;\omega}\\ &\leq 2|u_1 - u|_{p;\omega'} + \max_{\bar\omega} |\varrho_n * u_1 - u_1| \, |\operatorname{supp}(\varrho_n * u_1) \cup \operatorname{supp} u_1|^{1/p}\\ &\leq 2\varepsilon + \max_{\bar\omega} |\varrho_n * u_1 - u_1| \, |\operatorname{supp}(\varrho_n * u_1) \cup \operatorname{supp} u_1|^{1/p}\end{aligned}$$

by the preceding norm estimate, and

$$\limsup_{n\to\infty} |\varrho_n * u - u|_{p;\omega} \leq 2\varepsilon$$

by Lemma 1.7 with u replaced by u_1. □

REMARK. When Ω is bounded and $u \in C^{0,\delta}(\bar\Omega)$, $0 \leq \delta \leq 1$, the norm estimates and the convergence results of Lemmas 1.7 and 1.8 remain valid with ω and ω' replaced by Ω provided $\varrho_n * u$ is replaced by $\varrho_n * \tilde u$, $\tilde u$ being any controlled $C^{0,\delta}$ extension of u to $\mathbb{R}^N$ (see Theorems 1.N and 1.2). An analogous consideration can be made for Lemma 1.10 if $u \in L^p(\Omega)$, also if Ω is not bounded, provided $\varrho_n * u$ is replaced for $|\mathbb{R}^N \setminus \Omega| > 0$ by $\varrho_n * \tilde u$, where $\tilde u$ is the trivial extension of u to $\mathbb{R}^N$.

Lemma 1.9 enables us to give an L^p counterpart, known as the *Fréchet–Kolmogorov theorem*, to the Ascoli–Arzelà sufficient condition for relative compactness in $C^0(\bar\Omega)$.

THEOREM 1.10. *Let $\mathscr{F} \subset L^p_{\text{loc}}(\Omega)$, $1 \leq p < \infty$, be such that*

$$\sup_{u\in\mathscr{F}} |u|_{p;\omega'} < \infty \quad \textit{whenever } \omega' \subset\subset \Omega.$$

Fix $\omega \subset\subset \Omega$ and denote by $\mathscr{F}|_\omega$ the family of restrictions to ω of functions from $\mathscr{F}$. $\mathscr{F}|_\omega$ is relatively compact in $L^p(\omega)$ if for every $\varepsilon > 0$ there exists $\delta > 0$, $\delta <$ dist$(\omega, \partial\Omega)$, such that

$$\int_\omega |u(x-h) - u(x)|^p \, dx < \varepsilon^p \qquad \textit{for } u \in \mathscr{F},\ h \in \mathbb{R}^N \qquad \textit{with } |h| < \delta.$$

PROOF. For each n large enough the family $\mathscr{H}_n \equiv \{(\varrho_n * u)|_{\bar\omega} \mid u \in \mathscr{F}\}$ satisfies the assumptions of the Ascoli–Arzelà theorem, since

$$\sup_{u\in\mathscr{F}} |\varrho_n * u|_{C^0(\bar\omega)} \leq \sup_{u\in\mathscr{F}} |u|_{1;\omega'} \, |\varrho_n|_{C^0(\overline{B_{1/n}})}$$

and

$$\sup_{u\in\mathscr{F}} |(\varrho_n * u)(x) - (\varrho_n * u)(y)| \le |x-y| \max_{\overline{B}_{1/n}} |\nabla\varrho_n| \sup_{u\in\mathscr{F}} |u|_{1;\omega'}$$

for $x, y \in \bar\omega$, with $\omega' \equiv \bigcup_{x\in\bar\omega} B_{1/n}(x)$. This means that $\mathscr{H}_n$ is relatively compact in $C^0(\bar\omega)$, hence in $L^p(\omega)$. Let $\varepsilon > 0$ be fixed, and let $\{u_1^{(n)}, \ldots, u_{m_n}^{(n)}\} \subset L^p(\omega)$, $m_n \in N$, be such that, whenever $u \in \mathscr{F}$, $|\varrho_n * u - u_i^{(n)}|_{p;\omega} < \varepsilon$ for some i. Let $u \in \mathscr{F}$. If $x \in \omega$ and $n > \delta^{-1}$, $\delta = \delta(\varepsilon)$, we have (by using Hölder's inequality as in the proof of Lemma 1.9)

$$|(\varrho_n * u)(x) - u(x)|^p = \left|\int_{B_{1/n}} [u(x-y) - u(x)]\varrho_n(y)\, dy\right|^p$$
$$\le \int_{B_{1/n}} |u(x-y) - u(x)|^p \varrho_n(y)\, dy,$$

hence

$$|\varrho_n * u - u|_{p;\omega}^p \le \int_{B_{1/n}} \varrho_n(y)\, dy \int_\omega |u(x-y) - u(x)|^p\, dx < \varepsilon^p.$$

Thus, whenever $u \in \mathscr{F}$ there exists some $u_i^{(n)}$ such that $|u - u_i^{(n)}|_{p;\omega} < 2\varepsilon$. This proves that $\mathscr{F}|_\omega$ is relatively compact. ∎

1.3.3. L^p Spaces over Γ

Up until now we have considered L^p spaces only over open subsets of $\mathbb{R}^N$. When $N \ge 2$ we can turn to measurable functions defined on open subsets of $\mathbb{R}^{N-1}$, or even of $\mathbb{R}^{N-1}\times\{0\}$ if the latter is endowed with its relative topology as well as with the $(N-1)$-dimensional Lebesgue measure meas_{N-1}: we write a.e.$[N-1]$ for "almost everywhere with respect to meas_{N-1}." We can define the Banach space $L^p(S^0)$, $1 \le p < \infty$, of all measurable functions η on S^0 such that $|\eta|^p$ is integrable over S^0, and set

$$|\eta|_{p;S^0} \equiv \left(\int_{S^0} |\eta|^p\, dx'\right)^{1/p}.$$

We can also define the Banach space $L^\infty(S^0)$ of all measurable functions η on S^0 such that $|\eta| \le C$ a.e. $[N-1]$ in S^0 for some $C \in [0, \infty[$, and set

$$|\eta|_{\infty;S^0} \equiv \inf\{C \in [0,\infty[\ \big|\ |\eta| \le C \text{ a.e. } [N-1] \text{ in } S^0\}.$$

The matter becomes considerably more delicate when S^0 is replaced by a "curved surface" of $\mathbb{R}^N$. This is the situation we are now going to deal with. More precisely, we are going to consider the case of a C^1 (open and) compact portion Γ of $\partial\Omega$: we mention at the outset that, here and throughout most of the sequel, the class C^1 could be safely replaced by the class $C^{0,1}$, as in J. Nečas [127], at the price of a few additional technical difficulties.

Let $\{(\Gamma_j, \Lambda_j)\}_{j=1,\dots,m}$ be a C^1 atlas on Γ.

Suppose that $\Gamma' \subset \Gamma$ is such that $\text{meas}_{N-1}[\Lambda_j(\Gamma' \cap \Gamma_j)]$ (exists and) vanishes for every j. Γ' is then said to be a *zero subset of* Γ; a property that holds at all points of $\Gamma_0 \setminus \Gamma'$, Γ_0 being a subset of Γ, is said to hold a.e. $[N-1]$ in Γ_0.

For $1 \le p \le \infty$ we write $\eta \in L^p(\Gamma)$ if η is a function on Γ (to be identified with any other such function that equals it a.e. $[N-1]$ in Γ) with $\eta \circ (\Lambda_j^{-1})|_{S^0} \in L^p(S^0)$ for every j. For $\eta \in L^\infty(\Gamma)$ we set

$$| \eta |_{\infty;\Gamma} \equiv \inf\{C \in [0, \infty[\, \big| \, | \eta | \le C \text{ a.e. } [N-1] \text{ in } \Gamma\}.$$

We now define an integral over Γ through the following procedure. We first consider all $(N-1)\times(N-1)$ submatrices of the Jacobian matrix of $\Lambda_j^{-1}(x', 0)$, $| x' | \le 1$. The sum of the squares of their determinants is a strictly positive continuous function of x', whose square root we denote by $H_j(x')$. Next, we introduce a partition of unity $\{g_j\}$ relative to the open covering $\{U_j\}$ of Γ, $U_j = \Lambda_j^{-1}(B)$, and set

$$\int_\Gamma \eta \, d\sigma \equiv \sum_{j=1}^m \int_{S^0} (g_j\eta) \circ (\Lambda_j^{-1})|_{S^0} H_j \, dx' \tag{1.4}$$

for $\eta \in L^1(\Gamma)$.

Let $1 \le p < \infty$. For $\eta \in L^p(\Gamma)$ we set

$$| \eta |_{p;\Gamma} \equiv \left(\int_\Gamma | \eta |^p \, d\sigma \right)^{1/p}.$$

LEMMA 1.11. *There exist two positive constants C_1, C_2 such that*

$$C_1 | \eta |_{p;\Gamma}^p \le \sum_{j=1}^m \left| \eta \circ (\Lambda_j^{-1})|_{S^0} \right|_{p;S^0}^p \le C_2 | \eta |_{p;\Gamma}^p$$

whenever $\eta \in L^p(\Gamma)$, $1 \le p < \infty$.

PROOF. Since the nonnegative function $(g_i | \eta |^p) \circ (\Lambda_j^{-1})|_{S^0}$ vanishes outside $\Lambda_j(\Gamma_i \cap \Gamma_j)$, the change of variables formula for $(N-1)$-fold

Lebesgue integrals yields

$$\begin{aligned}\int_{S^0} (g_i \mid \eta \mid^p) \circ (\Lambda_j^{-1})|_{S^0}\, dx' &= \int_{\Lambda_j(\Gamma_i \cap \Gamma_j)} (g_i \mid \eta \mid^p) \circ (\Lambda_j^{-1})|_{\Lambda_j(\Gamma_i \cap \Gamma_j)}\, dx' \\ &\leq C \int_{\Lambda_i(\Gamma_i \cap \Gamma_j)} (g_i \mid \eta \mid^p) \circ (\Lambda_i^{-1})|_{\Lambda_i(\Gamma_i \cap \Gamma_j)}\, dx' \\ &\leq C \int_{\Lambda_i(\Gamma_i \cap \Gamma_j)} (g_i \mid \eta \mid^p) \circ (\Lambda_i^{-1})|_{\Lambda_i(\Gamma_i \cap \Gamma_j)} H_i\, dx' \\ &\leq C \int_{S^0} (g_i \mid \eta \mid^p) \circ (\Lambda_i^{-1})|_{S^0} H_i\, dx'.\end{aligned}$$

Thus,

$$\begin{aligned}\int_{S^0} \mid \eta \mid^p \circ (\Lambda_j^{-1})|_{S^0}\, dx' &\leq C \sum_{i=1}^{m} \int_{S^0} (g_i \mid \eta \mid^p) \circ (\Lambda_i^{-1})|_{S^0} H_i\, dx' \\ &= C \int_{\Gamma} \mid \eta \mid^p d\sigma,\end{aligned}$$

and the right-hand side inequality follows from summation over j. The left-hand side inequality is straightforward. □

It is clear that $|\cdot|_{p;\Gamma}$ is a norm when $p = \infty$; as for $1 \leq p < \infty$, it suffices to write $\mid \eta \mid_{p;\Gamma}^p$, $\eta \in L^p(\Gamma)$, as a sum $\sum_{j=1}^m \mid \eta_j' \mid_{p;S^0}^p$ with η_1', $\ldots, \eta_m' \in L^p(S^0)$. The same argument also shows that, by the $(N-1)$-dimensional version of Theorem 1.O, Hölder's inequality is still valid with Ω replaced by Γ.

We can now collect all the results about the structure of $L^p(\Gamma)$ that will be needed in the sequel.

THEOREM 1.12. *For $1 \leq p < \infty$ $L^p(\Gamma)$ is a Banach space with respect to the norm $\eta \mapsto \mid \eta \mid_{p;\Gamma}$; $L^2(\Gamma)$ is a Hilbert space with respect to the scalar product $\eta, \theta \mapsto \int_\Gamma \eta\theta\, d\sigma$. Convergence of a sequence in $L^p(\Gamma)$ implies convergence of a subsequence a.e. $[N-1]$ in Γ. Finally,*

$$L^{p'}(\Gamma) \subsetneq [L^p(\Gamma)]'$$

for $1 < p < \infty$.

PROOF. To prove completeness at the same time as convergence a.e. $[N-1]$ on Γ of suitable subsequences, consider a Cauchy sequence $\{\eta_n\}_n \subset L^p(\Gamma)$. Each sequence $\{\eta_n \circ (\Lambda_j^{-1})|_{S^0}\}_n$, $j = 1, \ldots, m$, is then a Cauchy sequence in $L^p(S^0)$ by Lemma 1.11 and converges in $L^p(S^0)$ toward a function η_j' by the $(N-1)$-dimensional version of Theorem 1.P. As a matter of fact, the $(N-1)$-dimensional version of Theorem 1.Q shows that we can find a subsequence of indices such that $\{\eta_{n_k} \circ (\Lambda_j^{-1})|_{S^0}\}_k$ converges to η_j' a.e. $[N-1]$ in S^0 for $j = 1, \ldots, m$. Let Γ_j' be the subset of points $x \in \Gamma_j$ such that $\eta_{n_k}(x)$ does not tend toward $(\eta_j' \circ \Lambda_j)(x)$: it is clear that $\text{meas}_{N-1} \Lambda_j(\Gamma_j') = 0$, hence also that $\text{meas}_{N-1} \Lambda_i(\Gamma_j' \cap \Gamma_i) = 0$ for every i since $\Lambda_i(\Gamma_j' \cap \Gamma_i) = \Lambda_i \circ \Lambda_j^{-1} \circ \Lambda_j(\Gamma_j' \cap \Gamma_i)$. This shows that $\eta_{n_k}|_{\Gamma_j}$ tends toward $\eta_j' \circ (\Lambda_j)|_{\Gamma_j}$ a.e. $[N-1]$ in Γ_j. But then $\eta_{n_k}|_{\Gamma_i \cap \Gamma_j}$ tends toward both functions $\eta_i' \circ (\Lambda_i)|_{\Gamma_i \cap \Gamma_j}$ and $\eta_j' \circ (\Lambda_j)|_{\Gamma_i \cap \Gamma_j}$ a.e. $[N-1]$ in $\Gamma_i \cap \Gamma_j$. This means that a function $\eta \in L^p(\Gamma)$ is defined a.e. $[N-1]$ in Γ by setting $\eta \equiv \eta_j' \circ (\Lambda_j)|_{\Gamma_j}$ on Γ_j. Again by Lemma 1.11, $\eta_n \to \eta$ in $L^p(\Gamma)$; moreover, $\eta_{n_k} \to \eta$ a.e. $[N-1]$ in Γ.

The statement concerning $p = 2$ is obvious.

The last statement of the theorem is proved as follows. The linear mapping

$$\theta \mapsto \int_\Gamma \eta\theta \, d\sigma \qquad \text{for } \theta \in L^p(\Gamma)$$

defines an element of $[L^p(\Gamma)]'$, whatever the choice of $\eta \in L^{p'}(\Gamma)$, by Hölder's inequality. If $\eta \in L^{p'}(\Gamma)$ is such that

$$\int_\Gamma \eta\theta \, d\sigma = 0 \qquad \text{for } \theta \in L^p(\Gamma),$$

then in particular

$$\int_\Gamma |\eta|^p \, d\sigma = 0$$

with the choice of $\theta \equiv |\eta|^{p-2}\eta$ where η does not vanish, $\theta \equiv 0$ elsewhere, so that $\eta = 0$. These considerations prove that $L^{p'}(\Gamma) \subsetneq [L^p(\Gamma)]'$. □

It is obvious that $C^0(\Gamma) \subsetneq L^p(\Gamma) \subsetneq L^q(\Gamma)$ if $1 \le q < p < \infty$.

Let $\eta \in L^p(\Gamma)$, $1 \le p < \infty$, and take $\varepsilon > 0$. For every $j = 1, \ldots, m$ let $\zeta_j' \in C^1(\overline{S^0})$ be such that

$$|\eta \circ (\Lambda_j^{-1})|_{S^0} - \zeta_j'|_{p;S^0} < \varepsilon.$$

We extend trivially each function $\zeta_j \equiv \zeta_j' \circ (\Lambda_j)|_{\Gamma_j}$ to the whole of Γ. Then $g_j|_\Gamma \zeta_j$ belongs to $C^1(\Gamma)$, and

$$|g_j(\eta - \zeta_j)|_{p;\Gamma}^p \le C\,|[g_j(\eta - \zeta_j)] \circ (\Lambda_j^{-1})|_{S^0}|_{p;S^0}^p \le C\varepsilon^p$$

(see the proof of Lemma 1.11). This shows that

$$\left| \eta - \sum_{j=1}^{m} g_j \zeta_j \right|_{p;\Gamma} \leq \sum_{j=1}^{m} | g_j(\eta - \zeta_j) |_{p;\Gamma} \leq C\varepsilon,$$

and we have the following lemma.

LEMMA 1.13. *For* $1 \leq p < \infty$, $L^p(\Gamma)$ *is the completion of* $C^1(\Gamma)$ *with respect to* $|\cdot|_{p;\Gamma}$.

REMARK. Lemma 1.13 can be utilized to prove that the definition of $\int_\Gamma \eta \, d\sigma$, hence also the definition of $L^p(\Gamma)$, does not depend on the particular choice of the atlas $\{(\Gamma_j, \Lambda_j)\}$—nor, *a fortiori*, on the partition of unity $\{g_j\}$.

Indeed, let $\eta = u|_\Gamma$ for $u \in C_c^1(\Omega \cup \Gamma)$: clearly, supp u lies in $\Omega' \cup (\partial\Omega' \cap \Gamma)$ for some bounded open set $\Omega' \subseteq \Omega$ with $\partial\Omega'$ of class C^1. Then the divergence theorem of advanced calculus yields

$$\int_\Gamma \eta \nu^i \, d\sigma = \int_\Omega u_{x_i} \, dx.$$

This demonstrates the required property of independence for $\int_\Gamma \eta\nu^i \, d\sigma$ when $\eta \in C^1(\Gamma)$, hence also when $\eta \in C^0(\Gamma)$ by density. Replace η by $\eta\nu^i$, which belongs to $C^0(\Gamma)$ if η does: then

$$\int_\Gamma \eta \, d\sigma = \sum_{i=1}^{N} \int_\Gamma \eta(\nu^i)^2 \, d\sigma,$$

and the required property holds for $\int_\Gamma \eta \, d\sigma$ when $\eta \in C^0(\Gamma)$ and finally when $\eta \in L^1(\Gamma)$.

The same observation applies to the notion of a zero subset Γ' of Γ: indeed, $\text{meas}_{N-1}[\Lambda_j(\Gamma' \cap \Gamma_j)] = 0$ for every j if and only if the characteristic function $\chi_{\Gamma'}$ of Γ' belongs to $L^1(\Gamma)$ with $\int_\Gamma \chi_{\Gamma'} \, d\sigma = 0$.

1.4. Morrey, John–Nirenberg, and Campanato Spaces

Throughout this section we assume Ω bounded.

1.4.1. Definition and Basic Properties

Let u be a function $\Omega \to \mathbb{R}(\Omega \to \mathbb{R}^N)$. If $u \in L^2(\Omega)$ ($u \in [L^2(\Omega)]^N$) and $\omega \neq \varnothing$ is an open subset of Ω, the scalar (the vector) $|\omega|^{-1} \int_\omega u \, dx$ is denoted by $(u)_\omega$. A straightforward computation shows that the function $\int_\omega |u - \lambda|^2 \, dx$ of $\lambda \in \mathbb{R}$ ($\lambda \in \mathbb{R}^N$) attains its minimum at $\lambda = (u)_\omega$. Therefore $\int_\omega |u - (u)_\omega|^2 \, dx$ increases with ω. We write $(u)_{x^0,\varrho}$ instead of $(u)_{\Omega[x^0,\varrho]}$, depressing the dependence on x^0 if $x^0 = 0$. For $0 \le \mu \le N + 2$ we set

$$[u]_{2,\mu;\Omega} \equiv \left(\sup_{\substack{x^0 \in \bar{\Omega} \\ 0<\varrho<\infty}} \varrho^{-\mu} \int_{\Omega[x^0,\varrho]} | u - (u)_{x^0,\varrho} |^2 \, dx \right)^{1/2}.$$

Note that, whenever $0 < r < \varrho < \infty$,

$$\begin{aligned} \varrho^{-\mu} \int_{\Omega[x^0,\varrho]} | u - (u)_{x^0,\varrho} |^2 \, dx &\le r^{-\mu} \int_{\Omega[x^0,\varrho]} | u |^2 \, dx \\ &\le r^{-\mu} \int_{\Omega} | u |^2 \, dx, \end{aligned}$$

hence

$$\begin{aligned} [u]^2_{2,\mu;\Omega} &\le r^{-\mu} | u |^2_{2;\Omega} \\ &\quad + \sup_{\substack{x^0 \in \bar{\Omega} \\ 0<\varrho\le r}} \varrho^{-\mu} \int_{\Omega[x^0,\varrho]} | u - (u)_{x^0,\varrho} |^2 \, dx. \end{aligned}$$

This circumstance will often be tacitly utilized. For instance, if $u : \Omega \to \mathbb{R}$, $\operatorname{supp} u \subset \Omega$ and $\Omega' \supset \Omega$, then the trivial extension $\tilde{u}$ of u to Ω' satisfies

$$[\tilde{u}]_{2,\mu;\Omega'} \le C(| u |^2_{2;\Omega} + [u]^2_{2,\mu;\Omega})^{1/2}$$

with C dependent on u only through $\operatorname{dist}(\operatorname{supp} u, \partial\Omega)$.

If $[u]_{2,\mu;\Omega}$ is finite we set

$$| u |_{2,\mu;\Omega} \equiv (| u |^2_{2;\Omega} + [u]^2_{2,\mu;\Omega})^{1/2}.$$

We now specialize with scalar functions and denote by $L^{2,\mu}(\Omega)$ the linear space of function $u \in L^2(\Omega)$ such that $[u]_{2,\mu;\Omega} < \infty$.

LEMMA 1.14. *For $0 \le \mu \le N + 2$, $L^{2,\mu}(\Omega)$ is a Banach space with respect to the norm $u \mapsto | u |_{2,\mu;\Omega}$.*

PROOF. We need only prove completeness. Let $\{u_n\}$ be a Cauchy sequence in $L^{2,\mu}(\Omega)$. Since $\{u_n\}$ is a Cauchy sequence in $L^2(\Omega)$ as well, it converges in the latter space toward a function u. Let $\varepsilon > 0$ be arbitrarily fixed and choose $n_\varepsilon \in N$ in such a way that

$$[u_{n+p} - u_n]_{2,\mu;\Omega} < \varepsilon$$

for $n \geq n_\varepsilon$ and $p \in N$. Whenever $x^0 \in \bar{\Omega}$ and $0 < \varrho < \infty$ we have

$$\begin{aligned}\varrho^{-\mu} \int_{\Omega[x^0,\varrho]} &| u - u_n - (u - u_n)_{x^0,\varrho} |^2 \, dx \\ &\leq 2\varrho^{-\mu} \int_{\Omega[x^0,\varrho]} | u - u_{n+p} - (u - u_{n+p})_{x^0,\varrho} |^2 \, dx \\ &\quad + 2\varrho^{-\mu} \int_{\Omega[x^0,\varrho]} | u_n - u_{n+p} - (u_n - u_{n+p})_{x^0,\varrho} |^2 \, dx \\ &< 2\varrho^{-\mu} \int_{\Omega} | u - u_{n+p} |^2 \, dx + 2\varepsilon^2,\end{aligned}$$

hence also

$$\varrho^{-\mu} \int_{\Omega[x^0,\varrho]} | u - u_n - (u - u_n)_{x^0,\varrho} |^2 \, dx \leq 2\varepsilon^2$$

after letting $p \to \infty$. By a passage to the supremum over x^0 and ϱ we obtain

$$[u - u_n]_{2,\mu;\Omega} \leq \sqrt{2}\, \varepsilon,$$

so that $u \in L^{2,\mu}(\Omega)$ and $u_n \to u$ in $L^{2,\mu}(\Omega)$ as $n \to \infty$. □

For what concerns the behavior of functions in $L^{2,\mu}$ spaces under suitable changes of variables and under extensions we have the following lemmas.

LEMMA 1.15. *Let Ω be convex and let $\Lambda : \bar{\Omega} \to \bar{\Omega}'$ be a C^1 diffeomorphism. Then each function $u \equiv u' \circ \Lambda$, $u' \in L^{2,\mu}(\Omega')$, belongs to $L^{2,\mu}(\Omega)$. Moreover,*

$$| u |_{2,\mu;\Omega} \leq C \, | u' |_{2,\mu;\Omega'}$$

with C independent of u'.

PROOF. Since Ω is convex there exists a constant K such that

$$| \Lambda(x) - \Lambda(x^0) | \leq K \, | x - x^0 | \qquad \text{for } x, x^0 \in \bar{\Omega}.$$

Therefore,

$$\Lambda(\Omega[x^0, \varrho]) \subseteq \Omega'[y^0, K\varrho]$$

for $x^0 \in \bar{\Omega}$, $y^0 = \Lambda(x^0)$, $0 < \varrho < \infty$. Let $u' \in L^{2,\mu}(\Omega')$. Then $u \in L^2(\Omega)$,

and

$$\int_{\Omega[x^0,\varrho]} |u(x) - (u)_{x^0,\varrho}|^2\,dx \le \int_{\Omega[x^0,\varrho]} |u(x) - (u')_{y^0,K\varrho}|^2\,dx$$
$$= \int_{\Lambda(\Omega[x^0,\varrho])} |u'(y) - (u')_{y^0,K\varrho}|^2 J(y)\,dy$$
$$\le \int_{\Omega'[y^0,K\varrho]} |u'(y) - (u')_{y^0,K\varrho}|^2 J(y)\,dy$$
$$\le (\max_{\bar\Omega'} J) K^\mu \varrho^\mu [u']^2_{2,\mu;\Omega'},$$

where J denotes the absolute value of the Jacobian determinant of Λ^{-1} and $(u')_{y^0,K\varrho} \equiv (u')_{\Omega'[y^0,K\varrho]}$. ◻

Lemma 1.16. *Let $\Omega = B_R{}^+$, $0 < R < \infty$. If $u \in L^{2,\mu}(B_R{}^+)$ its extension $\tilde u$ to B_R defined by*

$$\tilde u(x', x_N) \equiv u(x', -x_N) \qquad \textit{for } (x', -x_N) \in B_R{}^+$$

belongs to $L^{2,\mu}(B_R)$ with

$$|\tilde u|_{2,\mu;B_R} \le \sqrt{2}\,|u|_{2,\mu;B_R{}^+}.$$

Proof. It suffices to utilize the inequalities

$$\int_{B_R[x^0,\varrho]} |\tilde u - (\tilde u)_{B_R[x^0,\varrho]}|^2\,dx \le \int_{B_R[x^0,\varrho]} |\tilde u - (u)_{x^0,\varrho}|^2\,dx$$
$$\le 2\int_{B_R{}^+[x^0,\varrho]} |u - (u)_{x^0,\varrho}|^2\,dx$$

for $x^0 \in \overline{B_R{}^+}$, $0 < \varrho < \infty$. {Note that, when $(x', x_N) \in B_R[x^0,\varrho] \setminus B_R{}^+$, then $(x', -x_N) \in B_R{}^+[x^0,\varrho]$ since

$$|-x_N - x_N{}^0| \le |x_N - x_N{}^0|$$

if $x_N{}^0 \ge 0$, $x_N < 0$.} ◻

The above function $\tilde u$ is called the *extension by reflection* of u.

1.4.2. Equivalent Norms and Multipliers

It is obvious that $u|_\omega \in L^{2,\mu}(\omega)$ if $u \in L^{2,\mu}(\Omega)$, ω being an open subset of Ω, and that $L^{2,\lambda}(\Omega) \subsetneq L^{2,\mu}(\Omega)$ if $0 \le \mu < \lambda \le N+2$. Moreover,

$L^{2,0}(\Omega)$ is isomorphic to $L^2(\Omega)$, and $L^p(\Omega) \subsetneq L^{2,\mu}(\Omega)$ for $p > 2$, $\mu = N(p-2)/p$ by Hölder's inequality. To obtain deeper properties of functions from $L^{2,\mu}(\Omega)$ we introduce the following definitions. We say that Ω has the *property* (A) if there exists a positive constant, which we denote by A, such that $|\Omega[x^0, \varrho]| \geq A\varrho^N$ whenever $x^0 \in \bar{\Omega}$ and $0 < \varrho \leq \operatorname{diam} \Omega$ or, equivalently, $0 < \varrho \leq r$ for some $r < \operatorname{diam} \Omega$. (For $S \subset \mathbb{R}^N$ the symbol diam S denotes $\sup\{|x-y| \mid x, y \in S\}$.) When $N \geq 2$ we say that Ω has the *cone property* if there exists a bounded open cone $\mathscr{K}$ such that each $x^0 \in \partial\Omega$ is the vertex of a cone $\mathscr{K}(x^0) \subset \Omega$ congruent to $\mathscr{K}$. The cone property clearly implies property (A). On the other hand, Ω has the cone property if $\partial\Omega$ is of class C^1.

To see this we fix $x^0 \in \partial\Omega$ and operate a permutation $\xi_1 = x_{i_1}, \ldots, \xi_N = x_{i_N}$ of coordinate axes in such a way that, for some positive constant r which can be assumed independent of x^0, the set $\partial\Omega \cap Q_r$, with

$$Q_r \equiv]\xi_1{}^0 - r, \xi_1{}^0 + r[\times \cdots \times]\xi_N{}^0 - r, \xi_N{}^0 + r[\quad (\xi_k{}^0 \equiv x_{i_k}^0)$$

is the graph of a C^1 function $\xi_N = \lambda(\xi')$ and $\Omega \cap Q_r = \{(\xi', \xi_N) \in Q_r \mid \xi_N > \lambda(\xi')\}$. Then for some constant $C \geq 1$ independent of x^0 the cone

$$\{(\xi', \xi_N) \in \mathbb{R}^N \mid |\xi' - \xi^{0\prime}| < rC^{-1},\ C|\xi' - \xi^{0\prime}| < \xi_N - \xi_N{}^0 < r\}$$

lies in Ω.

Note that a hemisphere has the cone property.

THEOREM 1.17. *Let Ω have the property (A).*

(i) *If $0 \leq \mu < N$ the mapping*

$$u \mapsto \Bigg(\sup_{\substack{x^0 \in \bar{\Omega} \\ 0<\varrho<\infty}} \varrho^{-\mu} \int_{\Omega[x^0, \varrho]} u^2\, dx \Bigg)^{1/2} \tag{1.5}$$

defines a norm on $L^{2,\mu}(\Omega)$ which is equivalent to $|\cdot|_{2,\mu;\Omega}$.

(ii) *If $N < \mu \leq N + 2$, $L^{2,\mu}(\Omega)$ is isomorphic to $C^{0,\delta}(\bar{\Omega})$ for $\delta = (\mu - N)/2$.*

The proof of Theorem 1.17(i) relies on the following results, which will again be useful on several future occasions.

LEMMA 1.18. *Let φ and Φ be nonnegative functions, the first one defined on some interval $]0, R]$, the second one on the half-line $]1, \infty[$. Let β, γ, K*

be real numbers satisfying $\beta, \gamma > 0$, $K > 1$. *Suppose that there exists a positive function* $H(s)$, $s > 1$, *such that* $H(s) \leq R$ *and*

$$\varphi(\varrho) \leq K \frac{\varrho^\beta}{r^\beta} \varphi(r) + \varrho^\gamma \Phi(s) \tag{1.6}$$

whenever

$$0 < \varrho < r \leq H(s), \qquad r/\varrho \leq s.$$

Then, given any $\varepsilon \in]0, \beta - \gamma[$, *the inequality*

$$\varphi(\varrho) \leq K \frac{\varrho^{\beta-\varepsilon}}{r^{\beta-\varepsilon}} \varphi(r) + \varrho^\gamma \Phi(K^{1/\varepsilon}) \frac{K^{(\beta-\gamma)/\varepsilon}}{K^{(\beta-\gamma)/\varepsilon} - K}$$

is satisfied for $0 < \varrho < r \leq H(K^{1/\varepsilon})$ *without any further restriction on* r/ϱ.

Proof. Fix $\varepsilon \in]0, \beta - \gamma[$, $r \in]0, H(K^{1/\varepsilon})]$, $\varrho \in]0, r[$. Setting $s \equiv K^{1/\varepsilon} > 1$ we denote by k the first nonnegative integer such that $s^{k+1} \geq r/\varrho$ and put

$$\varrho_i \equiv \varrho s^i \qquad \text{for } i = 0, 1, \ldots, k.$$

We write (1.6) with ϱ replaced by ϱ_i, r by ϱ_{i+1}, and obtain

$$\varphi(\varrho_i) \leq K s^{-\beta} \varphi(\varrho_{i+1}) + \varrho_i^\gamma \Phi(s),$$

hence also

$$\varphi(\varrho) \leq (K s^{-\beta})^k \varphi(\varrho_k) + \varrho^\gamma \sum_{i=0}^{k-1} (K s^{\gamma-\beta})^i \Phi(s)$$

by iteration. To estimate $\varphi(\varrho_k)$ we apply (1.6) with ϱ replaced by ϱ_k and obtain

$$\varphi(\varrho_k) \leq K s^{\beta k} \frac{\varrho^\beta}{r^\beta} \varphi(r) + s^{\gamma k} \varrho^\gamma \Phi(s).$$

Summing up,

$$\varphi(\varrho) \leq K^{k+1} \frac{\varrho^\beta}{r^\beta} \varphi(r) + \varrho^\gamma \sum_{i=0}^{k} (K s^{\gamma-\beta})^i \Phi(s).$$

Since

$$K^k = s^{\varepsilon k} < r^\varepsilon / \varrho^\varepsilon$$

and

$$\sum_{i=0}^{k} (K s^{\gamma-\beta})^i \leq (1 - K s^{\gamma-\beta})^{-1} = (1 - K^{1+(\gamma-\beta)/\varepsilon})^{-1},$$

the conclusion is patent. □

Proof of Theorem 1.17. *Step* 1: *Proof of* (i). Let $x^0 \in \bar{\Omega}$, $0 < \varrho < r \leq \operatorname{diam} \Omega$. Whenever $u \in L^{2,\mu}(\Omega)$ the inequalities

$$\int_{\Omega[x^0,\varrho]} u^2\,dx \leq 2\,|\Omega[x^0,\varrho]|\,|(u)_{x^0,r}|^2 + 2\int_{\Omega[x^0,r]} |u-(u)_{x^0,r}|^2\,dx$$

$$\leq 2\,|\Omega[x^0,\varrho]|\,|\Omega[x^0,r]|^{-1}\int_{\Omega[x^0,r]} u^2\,dx + r^\mu 2[u]_{2,\mu;\Omega}$$

show that the function $\varphi(\varrho) \equiv \int_{\Omega[x^0,\varrho]} u^2\,dx$ satisfies (1.6) with $\Phi(s) = 2[u]^2_{2,\mu;\Omega}s^\mu$, $\beta = N$, $\gamma = \mu$ and $K \geq 2\,|B|/A$ [where property (A) has been utilized together with the inequality $|\Omega[x^0,\varrho]| \leq |B_\varrho(x^0)| = |B|\,\varrho^N$]. If we simply divide by ϱ^μ at this point we do not arrive at the required bound on $\varrho^{-\mu}\varphi(\varrho)$ as ϱ varies near 0, because of the restriction $r/\varrho \leq s$. But Lemma 1.18 applies: by choosing $\varepsilon = N - \mu$ and $r = \operatorname{diam} \Omega$ we obtain

$$\varrho^{-\mu}\varphi(\varrho) \leq C(|u|^2_{2;\Omega} + [u]^2_{2,\mu;\Omega}),$$

hence

$$\sup_{\substack{x^0\in\bar{\Omega}\\ 0<\varrho<\infty}} \varrho^{-\mu}\int_{\Omega[x^0,\varrho]} u^2\,dx \leq C\,|u|^2_{2,\mu;\Omega}.$$

On the other hand,

$$\int_{\Omega[x^0,\varrho]} |u-(u)_{x^0,\varrho}|^2\,dx \leq \int_{\Omega[x^0,\varrho]} u^2\,dx,$$

hence

$$|u|^2_{2,\mu;\Omega} \leq C \sup_{\substack{x^0\in\bar{\Omega}\\ 0<\varrho<\infty}} \varrho^{-\mu}\int_{\Omega[x^0,\varrho]} u^2\,dx.$$

Note that the proof of the last inequality did not utilize property (A).

Step 2: *Proof of* (ii). Let $u \in C^{0,\delta}(\bar{\Omega})$. Whenever $x^0 \in \bar{\Omega}$ and $0 < \varrho < \infty$, the integral mean value theorem yields

$$|u(x) - (u)_{x^0,\varrho}| \leq [u]_{\delta;\bar{\Omega}} 2^\delta \varrho^\delta$$

for $x \in \Omega[x^0,\varrho]$. Thus,

$$\int_{\Omega[x^0,\varrho]} |u-(u)_{x^0,\varrho}|^2\,dx \leq [u]^2_{\delta;\bar{\Omega}} 2^{2\delta}\varrho^{2\delta}\,|\Omega[x^0,\varrho]|,$$

and $C^{0,\delta}(\bar{\Omega}) \subsetneq L^{2,\mu}(\Omega)$; here property (A) has played no role.

Vice versa, let $u \in L^{2,\mu}(\Omega)$. Whenever $x^0 \in \bar{\Omega}$ and $0 < r_1 < r_2 \leq \operatorname{diam} \Omega$ we have

$$| (u)_{x^0,r_1} - (u)_{x^0,r_2} |^2 \leq 2(| u(x) - (u)_{x^0,r_1} |^2 + | u(x) - (u)_{x^0,r_2} |^2),$$

hence

$$\begin{aligned} | (u)_{x^0,r_1} - (u)_{x^0,r_2} |^2 & \\ \leq \frac{2}{A} r_1^{-N}\Big(\int_{\Omega[x^0,r_1]} | u - (u)_{x^0,r_1} |^2 \, dx + \int_{\Omega[x^0,r_2]} | u - (u)_{x^0,r_2} |^2 \, dx\Big) & \\ \leq C r_1^{-N}(r_1^{\mu} + r_2^{\mu})[u]^2_{2,\mu;\Omega} & \qquad (1.7) \end{aligned}$$

after integration over $\Omega[x^0, r_1]$.

Now let $\varrho \in]0, 2 \operatorname{diam} \Omega]$ be arbitrarily fixed. From (1.7) with $r_1 = \varrho 2^{-(i+1)}$ and $r_2 = \varrho 2^{-i}$, $i \in \mathbb{N}$, we obtain

$$| (u)_{x^0,\varrho 2^{-(i+1)}} - (u)_{x^0,\varrho 2^{-i}} | \leq C 2^{i(N-\mu)/2} \varrho^{(\mu-N)/2}[u]_{2,\mu;\Omega}$$

and therefore

$$\begin{aligned} | (u)_{x^0,\varrho 2^{-h}} - (u)_{x^0,\varrho 2^{-k}} | &\leq C 2^{h(N-\mu)/2} \sum_{i=0}^{k-h-1} 2^{i(N-\mu)/2} \varrho^{(\mu-N)/2}[u]_{2,\mu;\Omega} \\ &\leq C 2^{h(N-\mu)/2} \varrho^{(\mu-N)/2}[u]_{2,\mu;\Omega}. \end{aligned}$$

for $h < k$. This shows that $\{(u)_{x^0,\varrho 2^{-i}}\}_i \subset \mathbb{R}$ is a Cauchy sequence, hence a convergent one: its limit $\hat{u}(x^0)$ is clearly independent from the choice of ϱ, since (1.7) can be applied with $r_1 = \bar{\varrho} 2^{-i}$ and $r_2 = \varrho 2^{-i}$ whenever $0 < \bar{\varrho} < \varrho$. Thus,

$$\hat{u}(x^0) = \lim_{r \to 0^+} (u)_{x^0,r}$$

with

$$| (u)_{x^0,r} - \hat{u}(x^0) | \leq C r^{(\mu-N)/2}[u]_{2,\mu;\Omega} \qquad (1.8)$$

for $0 < r \leq \operatorname{diam} \Omega$.

By the above procedure we have defined a function $\hat{u}$ on $\bar{\Omega}$. From (1.8) we first of all deduce that

$$| \hat{u}(x) | \leq C[u]_{2,\mu;\Omega} + | (u)_{x,\operatorname{diam} \Omega} |,$$

hence that $\hat{u}$ is bounded in Ω with

$$\sup_{\Omega} | \hat{u} | \leq C \, | u |_{2,\mu;\Omega}. \qquad (1.9)$$

Next, let us prove that $\hat{u}$ is Hölderian. Let $x, y \in \bar{\Omega}$ with $R \equiv |x - y| < (\operatorname{diam} \Omega)/2$. Then (1.8) yields

$$|\hat{u}(x) - \hat{u}(y)| \leq |(u)_{x,2R} - \hat{u}(x)| + |(u)_{x,2R} - (u)_{y;2R}| + |(u)_{y,2R} - \hat{u}(y)|$$
$$\leq C[u]_{2,\mu;\Omega} R^{(\mu-N)/2} + |(u)_{x;2R} - (u)_{y;2R}|.$$

On the other hand, the inequality

$$|(u)_{x,2R} - (u)_{y,2R}| \leq |(u)_{x;2R} - u(\xi)| + |u(\xi) - (u)_{y,2R}|$$

yields

$$|(u)_{x,2R} - (u)_{y,2R}|^2 \leq 2^{\mu+1} R^\mu |E|^{-1} \Big[(2R)^{-\mu} \int_{\Omega[x,2R]} |u - (u)_{x,2R}|^2 \, d\xi + (2R)^{-\mu} \int_{\Omega[y,2R]} |u - (u)_{y,2R}|^2 \, d\xi \Big]$$

after integration over the set $E \equiv \Omega[x, 2R] \cap \Omega[y, 2R]$ which contains $\Omega[x, R]$ and therefore satisfies $|E| \geq AR^N$. Thus,

$$|(u)_{x,2R} - (u)_{y,2R}| \leq C[u]_{2,\mu;\Omega} R^{(\mu-N)/2},$$

and finally

$$|\hat{u}(x) - \hat{u}(y)| \leq C[u]_{2,\mu;\Omega} |x - y|^\delta.$$

For $|x - y| > (\operatorname{diam} \Omega)/2$ we obtain instead

$$|\hat{u}(x) - \hat{u}(y)| \leq C |u|_{2,\mu;\Omega} \leq C \left(\frac{\operatorname{diam} \Omega}{2} \right)^{-\delta} |u|_{2,\mu;\Omega} |x - y|^\delta$$

by (1.9).

We have thus proved that $\hat{u} \in C^{0,\delta}(\bar{\Omega})$ with $|\hat{u}|_{C^{0,\delta}(\bar{\Omega})} \leq C |u|_{2,\mu;\Omega}$. To complete the present proof we need only take into account the Lebesgue theorem, which ensures the convergence a.e. in Ω of the function $x \mapsto (u)_{x,r}$, as $r \to 0^+$, toward (a representative of) u: thus, $\hat{u}$ is nothing but (a representative of) u. □

REMARK. By Theorem 1.17, if $\mu < N$ Lemma 1.16 remains valid for the extension $\tilde{u}(x', x_N) \equiv -u(x', -x_N)$ for $(x', -x_N) \in B_R^+$.

Theorem 1.17 indicates that the role of $L^{2,\mu}(\Omega)$ varies according to whether $0 \leq \mu < N$, $N < \mu \leq N + 2$ or $\mu = N$.

For $0 < \mu < N$, $L^{2,\mu}(\Omega)$ is called a *Morrey space*. The norm (1.5) [or, equivalently, any other norm obtained from (1.5) by replacing the range $0 < \varrho < \infty$ with a range $0 < \varrho \leq r$, where $0 < r < \infty$] is clearly more convenient to deal with than $|\cdot|_{2,\mu;\Omega}$.

For $N < \mu \leq N + 2$, $L^{2,\mu}(\Omega)$ is called a *Campanato space*.

$L^{2,N}(\Omega)$ is called a *John–Nirenberg space*; its elements are also said to have the *bounded mean oscillation* property. From the inequalities

$$\int_{\Omega[x^0,\varrho]} | u - (u)_{x^0,\varrho} |^2 \, dx \leq \int_{\Omega[x^0,\varrho]} u^2 \, dx$$
$$\leq | \Omega[x_0, \varrho] | \operatorname*{ess\,sup}_{\Omega} u^2$$

it follows that $L^\infty(\Omega) \subsetneq L^{2,N}(\Omega)$. On the other hand, $L^\infty(\Omega)$ is a proper subset of $L^{2,N}(\Omega)$: in the case $N = 1$, $\Omega =]0, 1[$, for instance, the latter space contains the unbounded function $u(x) \equiv \log x$.

Let Ω have property (A). If we agree to qualify as a *space of multipliers* for $L^{2,\mu}(\Omega)$ a Banach space X of functions v defined on Ω such that

$$u \in L^{2,\mu}(\Omega) \text{ implies } uv \in L^{2,\mu}(\Omega),$$

with

$$| uv |_{2,\mu;\Omega} \leq C \, | u |_{2,\mu;\Omega} \, | v |_X ,$$

it is not difficult to ascertain that $L^\infty(\Omega)$ is one such space when $0 \leq \mu < N$, $C^{0,\delta}(\bar{\Omega})$ when $\mu = N + 2\delta$ with $0 < \delta \leq 1$ (see Lemma 1.18). When $\mu = N$ we proceed as follows: We fix any δ in $]0, 1]$ and multiply $u \in L^{2,N}(\Omega)$ by $v \in C^{0,\delta}(\bar{\Omega})$. Then we fix $x^0 \in \bar{\Omega}$, $\varrho \in]0, 1]$, and obtain

$$\int_{\Omega[x^0,\varrho]} | uv - (uv)_{x^0,\varrho} |^2 \, dx$$
$$= \int_{\Omega[x^0,\varrho]} | [u - (u)_{x^0,\varrho}]v + (u)_{x^0,\varrho} v - (uv)_{x^0,\varrho} |^2 \, dx$$
$$\leq C\Big(| v |^2_{C^0[\bar{\Omega}]} \int_{\Omega[x^0,\varrho]} | u - (u)_{x^0,\varrho} |^2 \, dx + \varrho^{2\delta} [v]^2_{\delta;\bar{\Omega}} \int_{\Omega[x^0,\varrho]} u^2 \, dy \Big),$$

where the estimate

$$\left| \frac{1}{| \Omega[x^0, \varrho] |} \int_{\Omega[x^0,\varrho]} u(y)[v(x) - v(y)] \, dy \right|^2$$
$$\leq \frac{1}{| \Omega[x^0, \varrho] |} (2\varrho)^{2\delta} [v]^2_{\delta;\bar{\Omega}} \int_{\Omega[x^0,\varrho]} u^2 \, dy,$$

$x \in \Omega[x^0, \varrho]$, has been utilized. Since $L^{2,N}(\Omega) \subsetneq L^{2,N-2\delta}(\Omega)$ if $N \geq 2\delta$, we also have

$$\int_{\Omega[x^0,\varrho]} u^2\, dy \leq C\varrho^{(N-2\delta)} |u|^2_{2,N;\Omega}$$

by Theorem 1.17(i), hence

$$\sup_{\substack{x^0\in\bar\Omega \\ 0<\varrho\leq 1}} \varrho^{-N} \int_{\Omega[x^0,\varrho]} |uv - (uv)_{x^0,\varrho}|^2\, dx$$
$$\leq C(|v|^2_{C^0(\bar\Omega)}[u]^2_{2,N;\Omega} + [v]^2_{\delta;\bar\Omega}|u|^2_{2,N;\Omega}).$$

Summing up, we have proved the following

LEMMA 1.19. *Let Ω have property* (A). *If $0 \leq \mu < N$, $L^\infty(\Omega)$ is a space of multipliers for $L^{2,\mu}(\Omega)$. If $N < \mu \leq N + 2$, $C^{0,\delta}(\bar\Omega)$ is a space of multipliers for $L^{2,\mu}(\Omega)$ provided $\delta = (\mu - N)/2$. Finally, $C^{0,\delta}(\bar\Omega)$ is a space of multipliers for $L^{2,N}(\Omega)$ whenever $0 < \delta \leq 1$.*

REMARK. The results of the present section can be extended to the class of spaces $L^{p,\mu}(\Omega)$, with $1 \leq p < \infty$ and $0 \leq \mu \leq N + p$, constructed via the obvious definitions of $[\cdot]_{p,\mu;\Omega}$ and $|\cdot|_{p,\mu;\Omega}$. See S. Campanato [32].

1.5. Sobolev Spaces

Suppose Γ of class C^1 and compact. Whenever $w \in C_c^1(\Omega \cup \Gamma)$,

$$\int_\Omega w_{x_i}\, dx = \int_\Gamma w|_\Gamma \nu^i\, d\sigma \qquad \text{for } i = 1, \ldots, N$$

by the divergence theorem. This is the motivation for the generalized notion of derivative which will be provided below, and even, indeed, the starting point of the whole variational theory of elliptic equations. Note that, regardless of Γ,

$$\int_\Omega u_{x_i} v\, dx = -\int_\Omega u v_{x_i}\, dx \qquad \text{for } i = 1, \ldots, N \tag{1.10}$$

whenever $u \in C^1(\Omega)$, $v \in C_c^1(\Omega)$.

1.5.1. Distributional Derivatives

A sequence $\{v_n\} \subset C_c^\infty(\Omega)$ is said to converge *in (the sense of)* $\mathscr{D}(\Omega)$ toward 0 if $\operatorname{supp} v_n \subseteq E$, E being some compact subset of Ω independent of n, and $D^\alpha v_n \to 0$ uniformly on E for every multi-index α. If $T : v \mapsto \langle T, v\rangle$ is a linear functional on $C_c^\infty(\Omega)$ which satisfies $\langle T, v_n\rangle \to 0$ whenever $v_n \to 0$ in $\mathscr{D}(\Omega)$, we call it a *distribution* (on Ω), and write $T \in \mathscr{D}'(\Omega)$. If T is a distribution, so is the linear functional $v \mapsto -\langle T, v_{x_i}\rangle$ on $C_c^\infty(\Omega)$: we denote it by $\partial T/\partial x_i$ or T_{x_i} and call it the *derivative in the sense of* $\mathscr{D}'(\Omega)$, or *distributional derivative*, of T with respect to x_i. More generally, if α is any multi-index the αth distributional derivative of T is the distribution

$$D^\alpha T : v \mapsto (-1)^{|\alpha|}\langle T, D^\alpha v\rangle \qquad \text{for } v \in C_c^\infty(\Omega).$$

Note that $T_{x_i x_j} = T_{x_j x_i}$.

The single most important example of a distribution may be considered to be the *Dirac measure* δ_{x^0} concentrated at any given point $x^0 \in \Omega$, which is defined by

$$\langle \delta_{x^0}, v\rangle \equiv v(x^0) \qquad \text{for } v \in C_c^\infty(\Omega).$$

A whole class of distributions is introduced instead by setting

$$\langle T^u, v\rangle \equiv \int_\Omega uv\,dx \qquad \text{for } v \in C_c^\infty(\Omega)$$

whenever $u \in L^1_{\text{loc}}(\Omega)$; note that the identity $u_1 = u_2$ holds in $L^1_{\text{loc}}(\Omega)$ if and only if $T^{u_1} = T^{u_2}$ (see Problem 1.11). Let $1 \le p < \infty$, and let $T \in \mathscr{D}'(\Omega)$ with

$$|\langle T, v\rangle| \le C\,|v|_{p;\Omega} \qquad \text{for } v \in C_c^\infty(\Omega):$$

since T can be extended as an element of $[L^p(\Omega)]'$, there exists a function $u \in L^{p'}(\Omega) \subset L^1_{\text{loc}}(\Omega)$ such that $T = T^u$. On the other hand, no function $u \in L^1_{\text{loc}}(\Omega)$ can be found with the property $T^u = \delta_{x^0}$ (Problem 1.10).

As a distribution, T^u admits derivatives of all orders, regardless of the (lack of) regularity of the function u. Even if u admits a classical derivative $\partial u(x)/\partial x_i$ at a.a. $x \in \Omega$, with $\partial u/\partial x_i \in L^1_{\text{loc}}(\Omega)$, the identity $T^{\partial u/\partial x_i} = \partial(T^u)/\partial x_i$ need not hold: in the one-dimensional case $\Omega = \mathbb{R}$, for instance, the classical derivative of the so-called *Heaviside function*

$$H(x) \equiv \begin{cases} 1 & \text{for } x \ge 0, \\ 0 & \text{for } x < 0 \end{cases}$$

exists and equals 0 throughout $\mathbb{R}\setminus\{0\}$, whereas $d(T^H)/dx = \delta_0$. However, things change if we require that the distributional derivative $\partial(T^u)/\partial x_i$ equal T^{w_i} for some $w_i \in L^1_{\text{loc}}(\Omega)$.

To illustrate this case we need some additional terminology. If P denotes a straight line of $\mathbb{R}^N$ intersecting Ω on a nonvoid set, a function on Ω that is absolutely continuous on every compact interval $\subset P \cap \Omega$ is said to be *absolutely continuous* on P. When $N \geq 2$, a property which holds on all parallel straight lines from a given family, except those of a subfamily intersecting an orthogonal hyperplane on a set S with $\text{meas}_{N-1} S = 0$, is said to hold on *almost all straight lines* of the family. When $N = 1$ the same espression means that the property in question holds on $\mathbb{R}$.

THEOREM 1.20. *Let $u \in L^1_{\text{loc}}(\Omega)$, $i = 1, \ldots, N$. In order that $\partial(T^u)/\partial x_i = T^{w_i}$ for some $w_i \in L^1_{\text{loc}}(\Omega)$ it is necessary and sufficient that u admits a representative $u^* = u^*(x)$, $x \in \Omega$, which is absolutely continuous on almost all straight lines that are parallel to the ith coordinate axis and intersect Ω on a nonvoid set, and has the classical derivative $\partial u^*/\partial x_i$ in $L^1_{\text{loc}}(\Omega)$; if this is the case, $\partial(T^u)/\partial x_i = T^{\partial u^*/\partial x_i}$.*

PROOF. We shall repeatedly utilize Fubini's theorem, both in N and in $N + 1$ dimensions. For the sake of notational simplicity we shall consider only the index $i = N$.

Step 1: *Necessity*. If $\mathbb{R}^N\setminus\Omega \neq \emptyset$ we shall consider u and $w \equiv w_N$ as measurable functions on $\mathbb{R}^N$, $u = w = 0$ in $\mathbb{R}^N\setminus\Omega$.

Let $\Omega = \bigcup_{j=0}^{\infty} \omega_j$ with $\omega_j \subset\subset \Omega$, each compact subset of Ω intersecting only a finite number of the ω_j's. Let $\{g_j\}$ be the partition of unity relative to this open covering (see Lemma 1.4). We fix j and set $g \equiv g_j$, $z \equiv gu$. It is evident that $z \in L^1(\mathbb{R}^N)$ and

$$-\int_{\mathbb{R}^N} z v_{x_N}\, dx = \int_{\mathbb{R}^N} (g_{x_N} u + gw) v\, dx \qquad \text{for } v \in C_c^\infty(\mathbb{R}^N). \tag{1.11}$$

Let $z^{(N)}$ be any representative of $g_{x_N} u + gw$. The function z^* defined as

$$z^*(x) \equiv \int_{-\infty}^{x_N} z^{(N)}(x', t)\, dt, \qquad x_N \in \mathbb{R},$$

if

$$x' \in S' \equiv \left\{ y' \in \mathbb{R}^{N-1} \,\middle|\, \int_{-\infty}^{\infty} |z^{(N)}(y', t)|\, dt < \infty \right\},$$

$z^*(x) \equiv 0$ otherwise, is (a representative of) an element of $L^1_{loc}(\mathbb{R}^N)$, and

$$\int_{\mathbb{R}^N} z^* v\,dx = \int_{\mathbb{R}^N} \left[\int_{x_N}^{\infty} v(x', t)\,dt\right] z^{(N)}(x)\,dx$$
$$= \int_{\mathbb{R}^N} \left[\int_{x_N}^{\infty} v(x', t)\,dt\right] \tilde{g}(x) z^{(N)}(x)\,dx$$
$$= \int_{\mathbb{R}^N} zv\,dx \qquad \text{for } v \in C_c^\infty(\mathbb{R}^N)$$

by (1.11) with $v(x)$ replaced by $[\int_{x_N}^\infty v(x', t)dt]\tilde{g}(x)$, provided $\tilde{g} \in C_c^\infty(\mathbb{R}^N)$ with $\tilde{g} = 1$ on $\operatorname{supp} g$. This shows that z^* is a representative of z.

Since $\operatorname{meas}_{N-1}(\mathbb{R}^{N-1} \setminus S') = 0$, z^* is absolutely continuous on almost all straight lines parallel to the Nth coordinate axis; a.e. in $\mathbb{R}^N$ its classical derivative $z^*_{x_N}$ exists and equals $z^{(N)}$: hence,

$$-\int_{\mathbb{R}^N} z v_{x_N}\,dx = \int_{\mathbb{R}^N} z^*_{x_N} v\,dx \qquad \text{for } v \in C_c(\mathbb{R}^N).$$

The necessity of the condition has thus been proven with u replaced by $g_j u$. At this point we need only utilize the identity $u = \sum_{j=0}^\infty g_j u$ and the fact that $(\operatorname{supp} g_j) \cap E = \varnothing$ for all but a finite number of the g_j's if E is a compact subset of Ω.

Step 2: Sufficiency. If $\mathbb{R}^N \setminus \Omega \neq \varnothing$ we shall consider u^* and $u^*_{x_N}$ as measurable functions on $\mathbb{R}^N$, $u^* = u^*_{x_N} = 0$ in $\mathbb{R}^N \setminus \Omega$.

Let $v \in C_c^\infty(\mathbb{R}^N)$ with $\operatorname{supp} v \subset \Omega$. Then,

$$\left\langle \frac{\partial(T^u)}{\partial x_N}, v \right\rangle = -\int_\Omega u v_{x_N}\,dx$$
$$= -\int_{\mathbb{R}^N} u^* v_{x_N}\,dx$$
$$= -\int_{\mathbb{R}^{N-1}} dx' \int_{-\infty}^{\infty} u^*(x', x_N) v_{x_N}(x', x_N)\,dx_N$$
$$= \int_{\mathbb{R}^{N-1}} dx' \int_{-\infty}^{\infty} u^*_{x_N}(x', x_N) v(x', x_N)\,dx_N$$
$$= \int_{\mathbb{R}^N} u^*_{x_N} v\,dx$$

since u^* and v are absolutely continuous on almost all straight lines parallel to the Nth coordinate axis, and $(u^* v)(x', x_N)$, $x' \in \mathbb{R}^{N-1}$, vanishes identically for $|x_N|$ large enough. □

REMARK. Inspection of Step 1 above shows that the classical derivative u_{x_i} of a function $u \in C^0(\Omega)$ exists and is continuous at all points of Ω, with $T^{\partial u/\partial x_i} = \partial(T^u)/\partial x_i$, if $\partial(T^u)/\partial x_i = T^{w_i}$ with $w_i \in C^0(\Omega)$. In particular, a function $u \in C^0(\Omega)$ belongs to $C^1(\Omega)$ if and only if $\partial(T^u)/\partial x_i = T^{w_i}$ with $w_i \in C^0(\Omega)$ for $i = 1, \ldots, N$ [see (1.10)].

We identify the distribution T^u associated to $u \in L^1_{\mathrm{loc}}(\Omega)$ with the function u itself. This creates no ambiguity for what concerns the present meaning of the symbol $\langle u, v\rangle$, which is perfectly consistent with that of the pairing between $u \in L^{p'}(\Omega)$ and $v \in L^p(\Omega)$ for $1 \leq p \leq \infty$. For what concerns the notation $\partial u/\partial x_i$ or u_{x_i} for derivatives, no ambiguity arises (at least, up to the equivalence relation for measurable functions) whenever distributional derivatives are (distributions associated with) functions from $L^1_{\mathrm{loc}}(\Omega)$. Note that, in such a case, $\partial(u|_\omega)/\partial x_i = (\partial u/\partial x_i)|_\omega$ whenever ω is an open subset of Ω, and $\operatorname{supp} u_{x_i} \subseteq \operatorname{supp} u$.

1.5.2. Difference Quotients

Let e^i denote the ith unit coordinate vector. For $x \in \Omega$ and $h \in \mathbb{R}$ with $x + he^i \in \Omega$ we set, whenever u is defined on Ω,

$$\tau_h{}^i u(x) \equiv u(x + he^i),$$

$$\delta_h{}^i u \equiv (\tau_h{}^i u - u)/h \qquad \text{if } h \neq 0;$$

$\delta_h{}^i$ is the classical difference quotient. If $\operatorname{dist}(\operatorname{supp} u, \partial\Omega) > |h|$, the definitions of $\tau_h{}^i u$, $\delta_h{}^i u$ can be trivially extended to the whole of Ω. We shall often depress the dependence on i. For $\omega \subset\subset \Omega$ and $|h|$ sufficiently small it is evident that

$$\delta_h(uv) = (\tau_h u)\delta_h v + (\delta_h u)v \qquad \text{in } \omega,$$

and that

$$\int_\omega (\delta_h u)v\, dx = -\int_\Omega u\delta_{-h} v\, dx$$

if $u, v \in L^2_{\mathrm{loc}}(\Omega)$, $\operatorname{supp} v \subset \omega$.

It can also be readily ascertained that the membership in $L^p_{\mathrm{loc}}(\Omega)$ of u together with a distributional derivative u_{x_j} implies

$$\frac{\partial}{\partial x_j}[(\delta_h u)|_\omega] = (\delta_h u_{x_j})|_\omega \in L^p(\omega).$$

THEOREM 1.21. *Let $1 < p \leq \infty$, $i = 1, \ldots, N$. In order that the distributional derivative u_{x_i} of $u \in L^p(\Omega)$ belong to $L^p(\Omega)$ it is necessary and sufficient that*

$$| \delta_h{}^i u |_{p;\omega} \leq C \tag{1.12}$$

for all bounded open sets $\omega \subset \Omega$ and real numbers h such that $x \in \omega$ implies $x + h'e^i \in \Omega$ for $0 < | h' | \leq | h |$. If this is the case, (1.12) holds with $C = | u_{x_i} |_{p;\Omega}$.

PROOF. We shall only consider $i = N$, $\delta_h \equiv \delta_h{}^N$.

Step 1: *Necessity*. By Theorem 1.20 there exists $S' \subset \mathbb{R}^{N-1}$ such that $\text{meas}_{N-1}(\mathbb{R}^{N-1} \setminus S') = 0$ and

$$\delta_h u(x) = \frac{1}{h} \int_{x_N}^{x_N+h} u_{x_N}(x', t)\, dt$$

at all points $x = (x', x_N) \in \omega$ with $x' \in S'$. We introduce the trivial extensions z and $(\widetilde{u_{x_N}})$ to $\mathbb{R}^N$ of $(\delta_h u)(x)$, $x \in \omega$, and $u_{x_N}(x)$, $x \in \Omega$, respectively. Let $1 < p < \infty$. For $x' \in S'$ we have

$$\begin{aligned}
\int_{-\infty}^{\infty} | z(x', x_N) |^p\, dx_N &\leq \int_{-\infty}^{\infty} \left| \frac{1}{h} \int_{x_N}^{x_N+h} (\widetilde{u_{x_N}})(x', t)\, dt \right|^p dx_N \\
&= \int_{-\infty}^{\infty} \left| \frac{1}{h} \int_0^h (\widetilde{u_{x_N}})(x', x_N + t)\, dt \right|^p dx_N \\
&\leq \frac{1}{| h |} \int_{-\infty}^{\infty} \left(\int_0^{|h|} | (\widetilde{u_{x_N}})(x', x_N + t) |^p\, dt \right) dx_N
\end{aligned}$$

by Hölder's inequality, hence

$$\begin{aligned}
\int_{-\infty}^{\infty} | z(x', x_N) |^p\, dx_N &\leq \frac{1}{| h |} \int_0^{|h|} \left(\int_{-\infty}^{\infty} | (\widetilde{u_{x_N}})(x', x_N + t) |^p\, dx_N \right) dt \\
&= \int_{-\infty}^{\infty} | (\widetilde{u_{x_N}})(x', x_N) |^p\, dx_N .
\end{aligned}$$

Therefore,

$$\begin{aligned}
\int_{\omega} | \delta_h u |^p\, dx &= \int_{\mathbb{R}^{N-1}} dx' \int_{-\infty}^{\infty} | z(x', x_N) |^p\, dx_N \\
&\leq \int_{\mathbb{R}^{N-1}} dx' \int_{-\infty}^{\infty} | (\widetilde{u_{x_N}})(x', x_N) |^p\, dx_N \\
&= \int_{\Omega} | u_{x_N} |^p\, dx.
\end{aligned}$$

For $i = N$ this amounts to (1.12) with $C = | u_{x_N} |_{p;\Omega}$ if $1 < p < \infty$.

If $p = \infty$ it suffices to consider the case when Ω is bounded and let $q \to \infty$ in the inequality

$$\left(\int_\omega |\delta_h u|^q \, dx\right)^{1/q} \leq \left(\int_\Omega |u_{x_N}|^q \, dx\right)^{1/q}$$

[see (1.2) in Section 1.3.1].

Step 2: *Sufficiency*. Take v in $C_c^\infty(\Omega)$ and let $\omega \subset\subset \Omega$ be such that $\operatorname{supp} v \subset \omega$. Then,

$$\left|\int_\Omega u\delta_{-h} v \, dx\right| = \left|\int_\omega (\delta_h u) v \, dx\right| \leq C \, |v|_{p';\omega},$$

hence

$$\left|\int_\Omega u v_{x_N} \, dx\right| \leq C \, |v|_{p';\Omega} \tag{1.13}$$

after a passage to the limit as $h \to 0$. By (1.13) the distribution $u_{x_N}: v \mapsto -\int_\Omega u v_{x_N} \, dx$ on $C_c^\infty(\Omega)$ is an element of $L^p(\Omega)$. □

REMARK. For what concerns the necessary part of Theorem 1.21 see also Problem 1.14.

1.5.3. $H^{k,p}$ Spaces: Definitions and First Properties

Let $1 \leq p \leq \infty$. We define $H^{1,p}(\Omega)$ as the linear space of functions $u \in L^p(\Omega)$ whose distributional derivatives $u_{x_1}, \ldots, u_{x_N}$ belong to $L^p(\Omega)$, and, by recurrence, $H^{k,p}(\Omega)$ as the linear space of functions $u \in H^{k-1,p}(\Omega)$ with $u_{x_1}, \ldots, u_{x_N} \in H^{k-1,p}(\Omega)$ $(k \in N,\ k \geq 2)$. For $k \in N$, $H^{k,p}(\Omega)$ is called a *Sobolev space*. Local Sobolev spaces are introduced by writing $u \in H^{k,p}_{\text{loc}}(\Omega)$ if $u|_\omega \in H^{k,p}(\omega)$ whenever $\omega \subset\subset \Omega$. For the sake of notational uniformity we set $H^{0,p}(\Omega) \equiv L^p(\Omega)$, $H^{0,p}_{\text{loc}}(\Omega) \equiv L^p_{\text{loc}}(\Omega)$. In the symbol $H^{k,p}$ we depress the dependence on p if $p = 2$.

THEOREM 1.22. *For $k \in N$ and $1 \leq p \leq \infty$, $H^{k,p}(\Omega)$ is a Banach space with respect to the norm*

$$|u|_{H^{k,p}(\Omega)} \equiv \left(\sum_{|\alpha| \leq k} |D^\alpha u|^p_{p;\Omega}\right)^{1/p} \qquad \text{if } p < \infty,$$

$$|u|_{H^{k,\infty}(\Omega)} \equiv \max_{|\alpha| \leq k} |D^\alpha u|_{\infty;\Omega} \, .$$

$H^k(\Omega)$ is a Hilbert space with respect to the scalar product

$$(u, v)_{H^k(\Omega)} \equiv \sum_{|\alpha| \leq k} (D^\alpha u, D^\alpha v)_{L^2(\Omega)} .$$

$H^{k,p}(\Omega)$ is separable for $1 \leq p < \infty$, reflexive for $1 < p < \infty$.

PROOF. It is clear that $H^{k,p}(\Omega)$ is a normed space, $H^k(\Omega)$ a pre-Hilbert one.

Let

$$X^{(p)} \equiv \prod_{|\alpha| \leq k} X_\alpha^{(p)},$$

where $X_\alpha^{(p)} \equiv L^p(\Omega)$ whatever the multi-index α with $|\alpha| \leq k$. For $\mathbf{u} \equiv \{u_\alpha\}_{|\alpha| \leq k} \in X^{(p)}$ set

$$|\mathbf{u}|_{X^{(p)}} \equiv \sum_{|\alpha| \leq k} |u_\alpha|_{p;\Omega}$$

and define I: $H^{k,p}(\Omega) \to X^{(p)}$ by

$$Iu \equiv \{D^\alpha u\}_{|\alpha| \leq k} \qquad \text{for } u \in H^{k,p}(\Omega).$$

$I(H^{k,p}(\Omega))$ is a closed subspace of $X^{(p)}$ by definition of distributional derivatives. At this point completeness, separability for $1 \leq p < \infty$, and reflexivity for $1 < p < \infty$ are easily transferred from $L^p(\Omega)$ to ($X^{(p)}$ and from $X^{(p)}$ to) $H^{k,p}(\Omega)$, the latter space being isomorphic to its image under I. ∎

REMARK. $H^{k,\infty}(\Omega)$ is not separable; neither $H^{k,1}(\Omega)$ nor $H^{k,\infty}(\Omega)$ is reflexive (see A. Kufner, O. John, and S. Fučik [92]).

If Ω is bounded the following inclusions are obvious:

$$H^{k,p}(\Omega) \subsetneq H^{k,q}(\Omega) \qquad \text{if } 1 \leq q < p \leq \infty,$$
$$C^k(\bar{\Omega}) \subsetneq H^{k,\infty}(\Omega).$$

From Theorem 1.21 it follows also that

$$C^{k-1,1}(\bar{\Omega}) \subsetneq H^{k,\infty}(\Omega);$$

the example of Section 1.2.1 shows that the inclusion

$$H^{k,\infty}(\Omega) \subsetneq C^{k-1,1}(\bar{\Omega})$$

is not true in general.

Membership in $H^{k,p}(\Omega)$ of a function from $L^p(\Omega)$ is a local property, in the sense clarified by the next result.

LEMMA 1.23. *Let $k \in N$, $1 \leq p \leq \infty$. If $\Omega = \bigcup_{j=0}^{m} \Omega_j$, each Ω_j being an open subset of $\mathbb{R}^N$, and $u \in L^p(\Omega)$ with $u|_{\Omega_j} \in H^{k,p}(\Omega_j)$ for $j = 0, 1, \ldots, m$, then $u \in H^{k,p}(\Omega)$.*

PROOF. Let α be any multi-index with $|\alpha| \leq k$, $u_j^\alpha \equiv D^\alpha(u|_{\Omega_j})$. Since $u_i^\alpha = u_j^\alpha$ in $\Omega_i \cap \Omega_j$, the function $u^\alpha \equiv u_j^\alpha$ on Ω_j, $j = 0, 1, \ldots, m$, is a well-defined element of $L^p(\Omega)$. Let $v \in C_c^\infty(\Omega)$ and denote by $\{g_j\}$ a partition of unity relative to the open covering $\{\Omega_j\}$ of supp v (see the corollary of Lemma 1.4). Then,

$$\begin{aligned}
\int_\Omega u^\alpha v \, dx &= \sum_{j=0}^{m} \int_\Omega u^\alpha g_j v \, dx \\
&= \sum_{j=0}^{m} \int_{\Omega_j} u_j^\alpha g_j v \, dx = (-1)^{|\alpha|} \sum_{j=0}^{m} \int_{\Omega_j} u D^\alpha(g_j v) \, dx \\
&= (-1)^{|\alpha|} \int_\Omega u D^\alpha v \, dx,
\end{aligned}$$

hence $u^\alpha = D^\alpha u$. □

For what concerns dual spaces the following considerations will suffice to our purposes. Let $1 \leq p < \infty$. Since $H^{k,p}(\Omega)$ is densely injected in $L^p(\Omega)$, $L^{p'}(\Omega)$ is continuously injected in $[H^{k,p}(\Omega)]'$.

We can therefore safely utilize the same symbol $\langle F, v\rangle$ for the pairing between $F \in [H^{k,p}(\Omega)]'$ and $v \in H^{k,p}(\Omega)$ as for the one between $F \in [L^p(\Omega)]'$ and $v \in L^p(\Omega)$, after identifying $F \in [H^{k,p}(\Omega)]'$ with $u \in L^{p'}(\Omega)$ when

$$F : v \mapsto \int_\Omega uv \, dx \qquad \text{for } v \in H^{k,p}(\Omega).$$

Note that, when u and v belong to $H^k(\Omega)$, $\langle u, v\rangle$ equals their scalar product in $L^2(\Omega)$, not in $H^k(\Omega)$. An element F of $[H^{k,p}(\Omega)]'$ is defined by

$$\langle F, v\rangle \equiv \sum_{|\alpha| \leq k} \langle u_\alpha, D^\alpha v\rangle \qquad \text{for } v \in H^{k,p}(\Omega)$$

if $u_\alpha \in L^{p'}(\Omega)$ for any multi-index α satisfying $|\alpha| \leq k$. Vice versa, it can be proven that every element of $[H^{k,p}(\Omega)]'$ admits the above representation: see R. Adams [1].

1.5.4. Density Results

When $\Omega = \mathbb{R}^N$ we have at our disposal the following lemma.

LEMMA 1.24. *$C_c^\infty(\mathbb{R}^N)$ is dense in $H^{k,p}(\mathbb{R}^N)$ whatever $k \in N$, $1 \leq p < \infty$.*

PROOF. We operate a preliminary reduction by the so-called *cutoff method.* For $1 < r < \infty$ let $g_r(x) \equiv g(x/r)$, where $g \in C_c^\infty(\mathbb{R}^N)$, $g = 1$ on $\bar{B}$. If u belongs to $H^{k,p}(\Omega)$, so does $g_r u$ with

$$|D^\alpha(g_r u)| \leq C \sum_{|\beta| \leq |\alpha|} |D^\beta u| \qquad \text{in } \mathbb{R}^N \qquad \text{for } 0 \leq |\alpha| \leq k.$$

[Compare with (1.11).] Straightforward arguments prove that

$$|u - g_r u|_{H^{k,p}(\mathbb{R}^N)} \to 0 \qquad \text{as } r \to \infty,$$

so that we can restrict our considerations to functions u with compact supports. Let $\{\varrho_n\}$ be a sequence of mollifiers and denote by $\varrho_n{}^x$, $x \in \mathbb{R}^N$, the function $y \mapsto \varrho_n(x - y)$. Since $\varrho_n{}^x \in C_c^\infty(\mathbb{R}^N)$ we have

$$\begin{aligned}
[D^\alpha(\varrho_n * u)](x) &= [(D^\alpha \varrho_n) * u](x) \\
&= (-1)^{|\alpha|}\langle D^\alpha \varrho_n{}^x, u\rangle = \langle \varrho_n{}^x, D^\alpha u\rangle \\
&= (\varrho_n * D^\alpha u)(x) \qquad \text{for } x \in \mathbb{R}^N.
\end{aligned}$$

We can therefore apply Lemma 1.9 with $\omega \supset \operatorname{supp} u$: we obtain

$$D^\alpha(\varrho_n * u)|_\omega \to D^\alpha u|_\omega \qquad \text{in } L^p(\omega) \qquad \text{for } |\alpha| \leq k,$$

hence

$$\varrho_n * u \to u \qquad \text{in } H^{k,p}(\mathbb{R}^N).$$

Note that for n large enough $\varrho_n * u \in C_c^\infty(\mathbb{R}^N)$ because supp u is compact. ▯

Approximation in $H^{k,p}(\Omega)$ becomes considerably more delicate if $R^N \setminus \Omega \neq \emptyset$ since we cannot take much advantage of the cutoff method (see the beginning of the proof of Theorem 1.27 below). The same argument as in the preceding proof does however show that $D^\alpha(\varrho_n * u)|_\omega = (\varrho_n * D^\alpha u)|_\omega$ for n large enough, $|\alpha| \leq k$, if $u \in H^{k,p}(\Omega)$ [or even $u \in H^{k,p}_{\text{loc}}(\Omega)$ only] and $\omega \subset\subset \Omega$. Thus, Lemma 1.9 yields

$$(\varrho_n * u)|_\omega \to u|_\omega \qquad \text{in } H^{k,p}(\omega).$$

If supp u is a compact subset of Ω we are of course in the same situation as in the proof of Lemma 1.24, so that

$$\varrho_n * u \to u \qquad \text{in } H^{k,p}(\Omega),$$

$\varrho_n * u$ being an element of $C_c^\infty(\Omega)$ for n large enough.

Note that $(\varrho_n * u)_{x_i}(x)$ vanishes identically for $x \in \omega \subset\subset \Omega$ if $u_{x_i} = 0$ a.e. in Ω. Thus, if $u_{x_1} = \cdots = u_{x_N} = 0$ a.e. in Ω and ω is connected, each smooth function $(\varrho_n * u)|_\omega$ is a constant, and finally $u|_\omega$ is a constant by Theorem 1.Q.

Summing up, we have the following lemma and corollary.

LEMMA 1.25. *Let Ω be any open subset of $\mathbb{R}^N$. If $u \in H^{k,p}(\Omega)$ with $k \in \mathbb{N}$ and $1 \leq p < \infty$, the function $u|_\omega$ is the limit of $\{(\varrho_n * u)|_\omega\}_n$ in $H^{k,p}(\omega)$ for any $\omega \subset\subset \Omega$, and even for $\omega = \Omega$ if* supp u *is a compact subset of Ω* [*in which case $\{\varrho_n * u\}_{n \geq n_0} \subset C_c^\infty(\Omega)$ if n_0 is large enough*].

COROLLARY. *Let Ω be connected. If $u \in H^{1,p}(\Omega)$, $u_{x_1} = \cdots = u_{x_N} = 0$ in Ω, then u is a constant.*

From Lemma 1.25 it is easy to deduce that $uv \in H^{1,p}(\Omega)$, with $(uv)_{x_i} = u_{x_i}v + uv_{x_i}$, if $u \in C^{0,1}(\bar{\Omega}) \cap L^\infty(\Omega)$ and $v \in H^{1,p}(\Omega)$. See also Problem 1.21.

Lemma 1.25 has a local character. The most general global result in approximation is the *Meyers–Serrin theorem*:

THEOREM 1.26. *$C^\infty(\Omega) \cap H^{k,p}(\Omega)$ is dense in $H^{k,p}(\Omega)$ whatever $k \in \mathbb{N}$, $1 \leq p < \infty$.*

PROOF. Let $\Omega = \bigcup_{j=0}^\infty \omega_j$ with $\omega_j \subset\subset \Omega$, $\omega_j \cap E = \varnothing$ for all but a finite number of indices j whenever E is a compact subset of Ω. Denote by $\{g_j\}$ a partition of unity relative to the above open covering of Ω. If $u \in H^{k,p}(\Omega)$ we can find, for any $\varepsilon > 0$ and $j = 0, 1, \ldots$, a natural number n_j such that supp$[\varrho_{n_j} * (g_j u)] \subset \omega_j$ and

$$|\, g_j u - \varrho_{n_j} * (g_j u)\,|_{H^{k,p}(\Omega)} < \varepsilon/2^{j+1} \tag{1.14}$$

(see Lemma 1.25). The function $w \equiv \sum_{j=0}^\infty \varrho_{n_j} * (g_j u)$ belongs to $C^\infty(\Omega)$; (1.14) implies that $w \in H^{k,p}(\Omega)$ with

$$|\, u - w\,|_{H^{k,p}(\Omega)} < \varepsilon. \qquad \square$$

The approximating functions provided by the Meyers–Serrin theorem need not be smooth up to $\partial\Omega$. As a matter of fact, there exist bounded domains Ω for which $C^1(\bar{\Omega})$ is not dense in $H^{1,p}(\Omega)$.

EXAMPLE. Let $N = 2$ and take

$$\Omega = (]0, 2[\times]-1, 1[)\setminus(]0, 1[\times\{0\}),$$

$\Omega_1 \equiv]0, 2[\times]-1, 1[$. Since $\bar{\Omega} = \bar{\Omega}_1$ and $\int_\Omega = \int_{\Omega_1}$, the limit of any sequence $\{u_n\} \subset C^1(\bar{\Omega})$ that converges in $H^{1,p}(\Omega)$ must be an element of $H^{1,p}(\Omega_1)$. But the two spaces do not coincide: for instance, the function

$$u(x) \equiv \begin{cases} e^{1/(4x_1^2-1)} & \text{for } 0 < x_1 < 1/2, \quad 0 < x_2 < 1, \\ 0 & \text{otherwise} \end{cases}$$

belongs to $H^{1,p}(\Omega)$, whereas its derivative u_{x_2} in the sense of $\mathscr{D}'(\Omega_1)$ is not a measurable function on Ω_1.

We now introduce a class of open sets $\Omega \subset \mathbb{R}^N$ for which Theorem 1.26 can be improved by replacing functions from $C^\infty(\Omega) \cap H^{k,p}(\Omega)$ with functions that are smooth up to $\partial\Omega$. Ω is said to have the *segment property* if, given any $x^0 \in \partial\Omega$, there exist an open neighborhood U of x^0 and a nonzero vector $\xi \in \mathbb{R}^N$ such that $x + t\xi \in \Omega$ whenever $x \in \bar{\Omega} \cap U$ and $0 < t < 1$. For $N \geq 2$ this property is certainly satisfied if every point $x^0 \in \partial\Omega$ has an open neighborhood $\tilde{U}$ such that $\tilde{U} \cap \Omega$ lies on one side of $\tilde{U} \cap \partial\Omega$, the latter being the graph of a continuous function of $N - 1$ among the coordinates $x_1, \ldots, x_N$.

THEOREM 1.27. *If $\Omega \subset \mathbb{R}^N$ has the segment property, the set of restriction to $\bar{\Omega}$ of functions from $C_c^\infty(\mathbb{R}^N)$ is dense in $H^{k,p}(\Omega)$ whatever $k \in \mathbb{N}$, $1 \leq p < \infty$.*

PROOF. If Ω is unbounded we can apply the cutoff method of the proof of Lemma 1.23 and show that for our present purposes it suffices to approximate functions from $H^{k,p}(\Omega)$ whose supports are compact subsets of $\bar{\Omega}$ (not of Ω, though). Let u be one such function. We can find finitely many open neighborhoods $U_1, \ldots, U_m$ of the type required by the segment property, and open sets $\omega_1, \ldots, \omega_m$ with $\bar{\omega}_i \subset U_i$, $\operatorname{supp} u \cap \partial\Omega \subset \bigcup_{i=1}^m \omega_i$. (Compare with Problem 1.9.) Let ω_0 be such that $\operatorname{supp} u \setminus \bigcup_{i=1}^m \omega_i \subset \omega_0 \subset\subset \Omega$ and denote by $\{g_j\}_{j=0,\ldots,m}$ a partition of unity relative to the open covering $\{\omega_j\}_{j=0,\ldots,m}$ of supp u (see the corollary of Lemma 1.4). We shall prove the theorem by showing that each function

$g_j u$ is approximated in $H^{k,p}(\Omega)$ by restrictions to Ω of functions from $C_c^\infty(\mathbb{R}^N)$.

The above is true for $j = 0$ by Lemma 1.25, since $\operatorname{supp}(g_0 u)$ is a compact subset of Ω. For $j = 1, \ldots, m$ set $u_j \equiv$ trivial extension of $g_j u$ to $\mathbb{R}^N$: we have $u_j|_\Omega \in H^{k,p}(\Omega)$, $u_j|_{\mathbb{R}^N \setminus \operatorname{supp} u_j} \in H^{k,p}(\mathbb{R}^N \setminus \operatorname{supp} u_j)$, and Lemma 1.23 implies $u_j|_{\mathbb{R}^N \setminus K_j} \in H^{k,p}(\mathbb{R}^N \setminus K_j)$, where $K_j \equiv \bar{\omega}_j \cap \partial\Omega$. Let $K_{j;t}$ denote the set of points $y = x - t\xi$, $x \in K_j$, where ξ is the vector associated with U_j by the segment property. If

$$0 < t < \min\{1, |\xi|^{-1}\operatorname{dist}(\bar{\omega}_j, \mathbb{R}^N \setminus U_j)\}$$

we have both $K_{j;t} \subset U_j$ and $K_{j;t} \cap \bar{\Omega} = \varnothing$. Then $u_{j;t} \mapsto u_j(x + t\xi)$ is an element of $H^{k,p}(\mathbb{R}^N \setminus K_{j;t})$, with $(D^\alpha u_{j;t})(x) = (D^\alpha u_j)(x + t\xi)$ by definition of distributional derivatives, and $\operatorname{supp} u_{j;t} \subset U_j$. Thus, $u_{j;t}|_\Omega \to u_j|_\Omega$ in $H^{k,p}(\Omega)$ as $t \to 0^+$ (see Problem 1.8). Let $\Omega_j' \equiv \Omega \cap U_j$. Since $\Omega_j' \subset\subset \mathbb{R}^N \setminus K_{j;t}$ we have $(\varrho_n * u_{j;t})|_{\Omega_j'} \to u_{j;t}|_{\Omega_j'}$ in $H^{k,p}(\Omega_j')$ by Lemma 1.25 with Ω replaced by $\mathbb{R}^N \setminus K_{j;t}$, and this concludes the proof because $\varrho_n * u_{j;t} \in C_c^\infty(\mathbb{R}^N)$ and $\varrho_n * u_{j;t} = 0$ in $\bar{\Omega} \setminus \Omega_j'$ for n large enough.

(The above procedure can be greatly illuminated by sketching the graphs of functions such as u_j and $u_{j;t}$ in the one-dimensional case.) □

1.5.5. Changes of Variables and Extensions

LEMMA 1.28. *Let Ω be bounded and let $\Lambda\colon \bar{\Omega} \to \bar{\Omega}'$ be a C^k diffeomorphism for some $k \in \mathbb{N}$. Then for $1 \le p \le \infty$ the mapping $u \mapsto u' \equiv u \circ \Lambda^{-1}$ defines an isomorphism of $H^{k,p}(\Omega)$ onto $H^{k,p}(\Omega')$, all distributional derivatives $D^\alpha u'$ with $|\alpha| \le k$ obeying the classical chain rule almost everywhere. When $k \ge 2$ the same conclusion remains valid if Λ is a $C^{k-1,1}$ diffeomorphism.*

PROOF. Let $u \in H^{k,p}(\Omega)$, $1 \le p < \infty$, $\omega' \subset\subset \Omega'$, $\omega \equiv \Lambda^{-1}(\omega')$. We apply the chain rule for derivatives to each function $u_n' \equiv u_n \circ (\Lambda^{-1})|_{\omega'}$, where $u_n \equiv (\varrho_n * u)|_\omega$: for $|\alpha| \le k$ we obtain

$$D^\alpha u_n'(y) = \sum_{|\beta| \le |\alpha|} P_{\alpha\beta}(y)(D^\beta u_n)[x(y)] \tag{1.15}$$

for $y \in \omega$, $x(y) \equiv \Lambda^{-1}(y)$, where $P_{\alpha\beta}$ is a suitable polynomial in derivatives, of order $\le |\alpha|$, of the components of Λ^{-1}. Since $u_n \to u|_\omega$ in $H^{k,p}(\omega)$ by Lemma 1.25, we have

$$(D^\beta u_n) \circ (\Lambda^{-1})|_{\omega'} \to (D^\beta u) \circ (\Lambda^{-1})|_{\omega'} \qquad \text{in } L^p(\omega') \qquad \text{for } |\beta| \le k,$$

so that $\{D^\alpha u_n'\}$ is a Cauchy sequence in $L^p(\omega')$ by (1.15). Let u'^α be the limit of $\{D^\alpha u_n'\}$ in $L^p(\omega')$: a passage to the limit in (1.15) for a suitable subsequence of indices yields

$$u'^\alpha(y) = \sum_{|\beta| \leq |\alpha|} P_{\alpha\beta}(y)(D^\beta u)[x(y)] \tag{1.16}$$

for a.a. $y \in \omega'$. Since

$$\begin{aligned}(-1)^{|\alpha|} \int_{\omega'} u' D^\alpha v' \, dy &= \lim_{n\to\infty} (-1)^{|\alpha|} \int_{\omega'} u_n' D^\alpha v' \, dy \\ &= \lim_{n\to\infty} \int_{\omega'} (D^\alpha u_n') v' \, dy \\ &= \int_{\omega'} u'^\alpha v' \, dy \qquad \text{for } v' \in C_c^\infty(\omega'),\end{aligned}$$

(1.16) is valid with $u'^\alpha(y) = D^\alpha u'(y)$ for a.a. $y \in \omega'$, hence for a.a. $y \in \Omega'$ by the arbitrariness of ω'. We have thus proved that all distributional derivatives, of order $\leq k$, of u' belong to $L^p(\Omega')$ and are obtained from those of u by the classical chain rule, which yields

$$\begin{aligned}| u' |_{H^{k,p}(\Omega')} &\leq C \sum_{|\alpha| \leq k} | (D^\alpha u) \circ \Lambda^{-1} |_{p;\Omega'} \\ &\leq C | u |_{H^{k,p}(\Omega)},\end{aligned}$$

since each function $P_{\alpha\beta}$ is at least continuous on the compact set $\bar{\Omega}'$.

When $k \geq 2$ and Λ is a $C^{k-1,1}$ diffeomorphism, (1.16) remains valid for a.a. $y \in \Omega'$, with $u'^\alpha = D^\alpha u'$, if $|\alpha| \leq k - 1$. Let $|\alpha| = k - 1$, $|\beta| \leq |\alpha|$: then each function $y \mapsto (D^\beta u)[x(y)]$ belongs to $H^{1,p}(\Omega')$, and all its first derivatives can be computed through the chain rule, by the first part of the proof with k replaced by 1; moreover, each polynomial $P_{\alpha\beta}$ belongs at least to $C^{0,1}(\bar{\Omega}')$. Hence each function $y \mapsto P_{\alpha\beta}(y)(D^\beta u)[x(y)]$ belongs to $H^{1,p}(\Omega')$, and all first derivatives of the function $y \mapsto D^\alpha u'(y)$ can be obtained through the chain rule.

When $p = \infty$ we replace p by any $q < \infty$ and arrive again at the expression (1.16) for $u'^\alpha = D^\alpha u'$.

The roles of u and u', Ω and Ω', Λ and Λ^{-1} can obviously be simultaneously interchanged. □

REMARK. If Λ is a $C^{0,1}$ diffeomorphism Lemma 1.28 is valid for $k = 1$: see C. B. Morrey, Jr. [118] or J. Nečas [127].

LEMMA 1.29. *Let* $u \in H^{k,p}(B^+)$ *for some* $k \in \mathbb{N}$, $p \in [1, \infty]$, *and denote by* $\tilde{u}$ *its extension to* B *defined by*

$$\tilde{u}(x', x_N) \equiv \sum_{h=1}^{k} C_h u(x', -x_N/h) \qquad \text{for } (x', -x_N) \in B^+, \tag{1.17}$$

where the vector $(C_1, \ldots, C_k)$ *is the unique solution to the linear system*

$$\sum_{h=1}^{k} \left(-\frac{1}{h}\right)^{j-1} C_h = 1 \qquad \text{for } j = 1, \ldots, k.$$

Then $\tilde{u} \in H^{k,p}(B)$ *with*

$$| D^\alpha \tilde{u} |_{p;1} \leq C \,| D^\alpha u |_{p;+} \qquad \text{for } |\alpha| \leq k, \tag{1.18}$$

C being independent of u.

PROOF. Let us first assume $u \in C^{k-1,1}(\overline{B^+})$. Then $\tilde{u} \in C^{k-1,1}(\bar{B})$ with

$$D^\alpha \tilde{u}(x', x_N) = \sum_{h=1}^{k} \left(-\frac{1}{h}\right)^{\alpha_N} C_h (D^\alpha u)(x', -x_N/h)$$

for $(x', -x_N) \in \overline{B^+}$ whenever $|\alpha| \leq k - 1$. (Compare with the proof of Theorem 1.3.)

Let $|\alpha| = k - 1$, so that all first distributional derivatives of $D^\alpha \tilde{u}$ belong to $L^\infty(B)$ by Theorem 1.21 and

$$\left(\frac{\partial}{\partial x_i} D^\alpha \tilde{u}\right)\bigg|_{B^+} = \frac{\partial}{\partial x_i} [(D^\alpha \tilde{u})|_{B^+}],$$

$$\left(\frac{\partial}{\partial x_i} D^\alpha \tilde{u}\right)\bigg|_{B\setminus\overline{B^+}} = \frac{\partial}{\partial x_i} [(D^\alpha \tilde{u})|_{B\setminus\overline{B^+}}].$$

We utilize Lemma 1.28 to compute $(\partial/\partial x_i)[(D^\alpha u)(x', -x_N/h)]$ whenever $(x', -x_N) \in B^+$, $h = 1, \ldots, k$. Thus,

$$\left(\frac{\partial}{\partial x_i} D^\alpha \tilde{u}\right)(x', x_N) = \begin{cases} (D^\alpha u_{x_i})(x', x_N) & \text{for a.a. } (x', x_N) \in B^+, \\ \displaystyle\sum_{h=1}^{k} c_{ih}\left(-\frac{1}{h}\right)^{\alpha_N} C_h D^\alpha u_{x_i}(x', -x_N/h) & \\ \quad \text{for a.a. } (x', x_N) \in B\setminus\overline{B^+} \text{ with } c_{ih} \equiv 1 & \\ \quad \text{if } i = 1, \ldots, N-1,\ c_{Nh} \equiv -1/h. & \end{cases}$$

In the general case $u \in H^{k,p}(B^+)$ we avail ourselves of Theorem 1.27 and approximate $D^\alpha u$ for $|\alpha| \leq k$ with $\{D^\alpha u_n\}$, $u_n \in C^\infty(\overline{B^+})$, in $L^p(B^+)$ if $p < \infty$, in $L^q(B^+)$ for any $q < \infty$ if $p = \infty$, thus obtaining

$$\begin{aligned}
\int_B \tilde{u} D^\alpha v \, dx &= \lim_{n\to\infty} \int_B \tilde{u}_n D^\alpha v \, dx \\
&= \lim_{n\to\infty} (-1)^{|\alpha|} \Bigg[\int_{B^+} (D^\alpha u_n) v \, dx \\
&\quad + \int_{B \setminus B^+} \sum_{h=1}^{k} \left(-\frac{1}{h}\right)^{\alpha_N} C_h (D^\alpha u_n)(x', -x_N/h) v(x', x_N) \, dx \Bigg] \\
&= (-1)^{|\alpha|} \Bigg[\int_{B^+} (D^\alpha u) v \, dx \\
&\quad + \int_{B \setminus B^+} \sum_{h=1}^{k} \left(-\frac{1}{h}\right)^{\alpha_N} C_h (D^\alpha u)(x', -x_N/h) v(x', x_N) \, dx \Bigg] \\
&\quad \text{for } v \in C_c^\infty(B).
\end{aligned}$$

This shows that $u \in H^{k,p}(B)$ with the norm estimate (1.18). ☐

If $k = 1$, (1.17) is the extension by reflection of u.

If Ω is bounded we say that it has the *extension property* (k, p) if, whenever Ω' is another open subset of $\mathbb{R}^N$ with $\Omega \subset\subset \Omega'$, every $u \in H^{k,p}(\Omega)$ admits an extension $\tilde{u} \in H^{k,p}(\Omega')$ with $\operatorname{supp} \tilde{u} \subset \Omega'$ and

$$|\tilde{u}|_{H^{k,p}(\Omega')} \leq C |u|_{H^{k,p}(\Omega)},$$

C being independent of u. Note that, by Lemma 1.25, Ω cannot have the extension property (k, p) if $C^\infty(\bar{\Omega})$ is not dense in $H^{k,p}(\Omega)$. Thanks to Lemmas 1.28 and 1.29, a procedure analogous to that for Theorem 1.3 demonstrates the following.

THEOREM 1.30. *When Ω is bounded, it has the extension property* (k, p), $1 \leq p \leq \infty$, *if $\partial\Omega$ is of class C^1 for $k = 1$, of class $C^{k-1,1}$ for $k \geq 2$.*

REMARK 1. For $k = 1$ Theorem 1.30 admits a generalization which requires only that Ω has a strengthened cone property: see R.A. Adams [1]. However, the extension property need not be valid if Ω is only assumed to have the segment property: see the example following Theorem 1.33 below.

REMARK 2. Theorem 1.30 can easily be generalized as follows: Let $u \in H^{k,p}(\Omega)$ with compact support $\subset \Omega \cup \Gamma$, Γ being of class C^1 for $k = 1$, of class $C^{k-1,1}$ for $k \geq 2$. If U is an open subset of $\mathbb{R}^N$ such that $U \cap \partial\Omega \subseteq \Gamma$ and $U \cap \bar{\Omega} \supset \operatorname{supp} u$, then u admits an extension $\tilde{u} \in H^{k,p}(\Omega \cup U)$, supp $\tilde{u}$ being a compact subset of U. However, the constant of the norm estimate depends on dist(supp u, $\partial\Omega \setminus \Gamma$) unless Γ is closed.

REMARK 3. Lemma 1.29 and Theorem 1.30 imply the validity of any extension property $(1, p)$ if Ω is a hemisphere.

REMARK 4. By Theorem 1.21, $H^{k,\infty}(\Omega) \subsetneq C^{k-1,1}(\bar{\Omega})$ for $k \in N$ if Ω has the extension property $(1, \infty)$. See also Theorem 1.41 below.

1.6. Continuous and Compact Imbeddings of Sobolev Spaces

1.6.1. Sobolev Inequalities I

LEMMA 1.31. *Let $N \geq 2$, $f_1, \ldots, f_N \in L^{N-1}(\mathbb{R}^{N-1})$. The function*

$$f(x) \equiv f_1(\hat{x}_1) \cdots f_N(\hat{x}_N),$$

where $\hat{x}_i \equiv (x_1, \ldots, x_{i-1}, x_{i+1}, \ldots, x_N)$ for $i = 1, \ldots, N$, belongs to $L^1(\mathbb{R}^N)$ and satisfies

$$|f|_{1;\mathbb{R}^N} \leq \prod_{i=1}^{N} |f_i|_{N-1;\mathbb{R}^{N-1}}.$$

PROOF. The result is obviously true when $N = 2$. We assume its validity for some value of N and proceed to prove it for $N + 1$. Let $\hat{x}_i \equiv (x_1, \ldots, x_{i-1}, x_{i+1}, \ldots, x_N, x_{N+1})$ for $i = 1, \ldots, N + 1$. For a.a. $x_{N+1} \in \mathbb{R}$,

$$\int_{\mathbb{R}^N} \prod_{i=1}^{N} |f_i(\hat{x}_i)|^{N'} dx_1 \cdots dx_N$$
$$\leq \prod_{i=1}^{N} \left(\int_{\mathbb{R}^{N-1}} |f_i(\hat{x}_i)|^N dx_1 \cdots dx_{i-1}\, dx_{i+1} \cdots dx_N \right)^{N'/N}$$

by the inductive assumption applied to the functions

$$(x_1, \ldots, x_{i-1}, x_{i+1}, \ldots, x_N) \mapsto |f_i(\hat{x}_i)|^{N'},$$

$N' = N/(N-1)$, which belong to $L^{N-1}(\mathbb{R}^{N-1})$; thus, Hölder's inequality yields

$$\begin{aligned}\int_{\mathbb{R}^N} \prod_{i=1}^{N+1} |f_i(\hat{x}_i)|\, dx_1 \cdots dx_N &\leq |f_{N+1}|_{N;\mathbb{R}^N} \left(\int_{\mathbb{R}^N} \prod_{i=1}^{N} |f_i(\hat{x}_i)|^{N'}\, dx_1 \cdots dx_N\right)^{1/N'} \\ &\leq |f_{N+1}|_{N;\mathbb{R}^N} \prod_{i=1}^{N} \left(\int_{\mathbb{R}^{N-1}} |f_i(\hat{x}_i)|^N\, dx_1 \cdots dx_{i-1}\, dx_{i+1} \cdots dx_N\right)^{1/N}. \qquad (1.19)\end{aligned}$$

Again by Hölder's inequality, the membership of all functions

$$x_{N+1} \mapsto \left(\int_{\mathbb{R}^{N-1}} |f_i(\hat{x}_i)|^N\, dx_1 \cdots dx_{i-1}\, dx_{i+1} \cdots dx_N\right)^{1/N},$$

$i = 1, \ldots, N$, in $L^N(\mathbb{R})$ implies the membership of their product in $L^1(\mathbb{R})$ with

$$\begin{aligned}\int_{-\infty}^{\infty} \prod_{i=1}^{N} &\left(\int_{\mathbb{R}^{N-1}} |f_i(\hat{x}_i)|^N\, dx_1 \cdots dx_{i-1}\, dx_{i+1} \cdots dx_N\right)^{1/N} dx_{N+1} \\ &\leq \prod_{i=1}^{N} \left[\int_{-\infty}^{\infty} \left(\int_{\mathbb{R}^{N-1}} |f_i(\hat{x}_i)|^N\, dx_1 \cdots dx_{i-1}\, dx_{i+1} \cdots dx_N\right) dx_{N+1}\right]^{1/N} \\ &= \prod_{i=1}^{N} |f_i|_{N;\mathbb{R}^N}.\end{aligned}$$

The desided result for $N+1$ then follows from (1.19) after integration in dx_{N+1}. ∎

Lemma 1.31 will be utilized in the proof of the next result.

THEOREM 1.32. *Let $u \in H^{1,p}(\mathbb{R}^N)$ with $1 \leq p < N$. Then $u \in L^{p^*}(\mathbb{R}^N)$ where $p^* \equiv Np/(N-p)$; moreover,*

$$|u|_{p^*;\mathbb{R}^N} \leq C\, |\nabla u|_{p;\mathbb{R}^N} \qquad (1.20)$$

with C independent of u.

PROOF. Without loss of generality we assume $u \in C_c^\infty(\mathbb{R}^N)$ (see Lemma 1.24).

Step 1: The case $p = 1$. Let $x \in \mathbb{R}^N$. Since

$$\begin{aligned}|u(x)| &= \left|\int_{-\infty}^{x_i} u_{x_i}(x_1, \ldots, x_{i-1}, t, x_{i+1}, \ldots, x_N)\, dt\right| \\ &\leq \int_{-\infty}^{\infty} |u_{x_i}(x_1, \ldots, x_{i-1}, t, x_{i+1}, \ldots, x_N)|\, dt \equiv [f_i(\hat{x}_i)]^{N-1}\end{aligned}$$

for $i = 1, \ldots, N$, we have

$$| u(x) |^{N/(N-1)} \leq \prod_{i=1}^{N} f_i(\hat{x}_i),$$

hence

$$\begin{aligned}\int_{\mathbb{R}^N} | u(x) |^{N/(N-1)} \, dx &\leq \prod_{i=1}^{N} | f_i |_{N-1;\mathbb{R}^{N-1}} \\ &= \prod_{i=1}^{N} \Big(\int_{\mathbb{R}^N} | u_{x_i}(x) | \, dx \Big)^{1/(N-1)} \leq \Big(\int_{\mathbb{R}^N} | \nabla u | \, dx \Big)^{N/(N-1)}\end{aligned}$$

by Lemma 1.31.

Step 2: *The case* $p > 1$. For $t > 0$ set $v \equiv v^{(t)} \equiv | u |^{1+t}$. Then $v \in C_c^1(\mathbb{R}^N)$ with $| \nabla v | = (1 + t) | u |^t | \nabla u |$. Step 1 with u replaced by v yields

$$\begin{aligned}\Big(\int_{\mathbb{R}^N} | u |^{(1+t)N/(N-1)} \, dx \Big)^{(N-1)/N} &\leq (1 + t) \int_{\mathbb{R}^N} | u |^t | \nabla u | \, dx \\ &\leq (1 + t) \Big(\int_{\mathbb{R}^N} | \nabla u |^p \, dx \Big)^{1/p} \Big(\int_{\mathbb{R}^N} | u |^{tp'} \, dx \Big)^{1/p'}.\end{aligned}$$

Let $t \equiv N(p - 1)/(N - p)$: then,

$$\frac{(1 + t)N}{N - 1} = tp' = \frac{Np}{N - p} = p^*,$$

and (1.20) holds with $C = p(N - 1)/(N - p)$. □

Passing to bounded open sets we have the following theorem.

THEOREM 1.33. *Let $k \in \mathbb{N}$, $p \in [1, \infty[$ with $kp \leq N$. If Ω is bounded and has the extension property* $(1, r)$ *for $p \leq r \leq Np/[N - (k - 1)p]$, then*

$$H^{k,p}(\Omega) \subsetneq L^{Np/(N-kp)}(\Omega) \tag{1.21}$$

if $kp < N$,

$$H^{k,p}(\Omega) \subsetneq L^q(\Omega) \qquad \text{for any } q < \infty \tag{1.22}$$

if $kp = N$, and even, if Ω has the extension property $(N, 1)$,

$$H^{N,1}(\Omega) \subsetneq C^0(\bar{\Omega}).$$

PROOF. Take $u \in H^{1,p}(\Omega)$, $p < N$, and let $\tilde{u} \in H^{1,p}(\mathbb{R}^N)$ be an extension of u,

$$| \tilde{u} |_{H^{1,p}(\mathbb{R}^N)} \leq C | u |_{H^{1,p}(\Omega)}$$

with C independent of u. By Theorem 1.32 $\tilde{u}$ belongs to $L^{p^*}(\mathbb{R}^N)$, so that $u \in L^{p^*}(\Omega)$ and

$$| u |_{p^*;\Omega} \leq | \tilde{u} |_{p^*;\mathbb{R}^N} \leq C \, | \nabla\tilde{u} |_{p;\mathbb{R}^N} \leq C \, | u |_{H^{1,p}(\Omega)}.$$

This proves the theorem if $k = 1$, $p < N$. As a consequence, a function $u \in L^1(\Omega)$ with $u_{x_1}, \ldots, u_{x_N} \subset L^q(\Omega)$, $1 < q < \infty$, must belong to $H^{1,q}(\Omega)$.

If $k > 1$, $kp < N$, we apply the above result to all derivatives of order $k - 1$, then to all those of order $k - 2$, and so on. Thus,

$$H^{k,p}(\Omega) \subsetneq H^{k-1,p^*}(\Omega) \subsetneq H^{k-2,p^{**}}(\Omega),$$

and so on for k steps. Note that

$$\overbrace{p^{*\cdots*}}^{h \text{ times}} = Np/(N - hp).$$

If $kp = N$, (1.22) follows from (1.21) with p replaced by $p - \varepsilon$ for any $\varepsilon \in]0, p[$.

Now take $u \in H^{N,1}(\Omega)$ and let $\tilde{u} \in H^{N,1}(\mathbb{R}^N)$ be an extension of u, $| \tilde{u} |_{H^{N,1}(\mathbb{R}^N)} \leq C \, | u |_{H^{N,1}(\Omega)}$. Assume $\tilde{u} \in C_c^\infty(\mathbb{R}^N)$ [hence also $u \in C^\infty(\bar{\Omega})$]; then,

$$\tilde{u}(x) = \int_{-\infty}^{x_1} \cdots \int_{-\infty}^{x_N} \tilde{u}_{x_1 \cdots x_N}(t_1, \ldots, t_N) \, dt_1 \cdots dt_N,$$

hence

$$| u |_{\infty;\Omega} \leq | \tilde{u} |_{\infty;\mathbb{R}^N} \leq C \, | u |_{H^{N,1}(\Omega)}.$$

If $\tilde{u} \notin C_c^\infty(\mathbb{R}^N)$ it suffices to proceed by density. □

We know that $\partial\Omega$ must have some regularity for Ω to have the extension property $(1, p)$. The next example shows that (1.21) need not hold if no regularity restriction is imposed on $\partial\Omega$.

Example. Let $N = 2$,

$$\Omega = \{x \in \mathbb{R}^2 \mid 0 < x_1 < 1, \ | x_2 | < e^{-1/x_1^2}\}.$$

The function $u(x) \equiv x_1^3 \exp(1/x_1^2)$ belongs to the Sobolev space $H^{1,1}(\Omega)$ but to no Lebesgue space $L^q(\Omega)$ if $q > 1$. This is also an indirect way of showing that the segment property (which holds for Ω chosen as above) is not sufficient for extension property (1,1) even though it guarantees the density of $C^\infty(\bar{\Omega})$ in $H^{1,1}(\Omega)$.

If $kp = N$ with $p > 1$ a function $u \in H^{k,p}(\Omega)$ need not be bounded.

EXAMPLE. Let $N \geq 2$,

$$u(x) \equiv \left(\ln \frac{1}{|x|}\right)^{\delta} \qquad \text{for } x \in \Omega = B_{1/2}, \qquad x \neq 0,$$

with $0 < \delta < 1 - 1/N$. Theorem 1.20 shows that the unbounded function u belongs to $H^{1,N}(\Omega)$.

1.6.2. Rellich's Theorem with Some Applications

The next result is *Rellich's theorem*.

THEOREM 1.34. *Let* $1 \leq p < \infty$. *Whenever* Ω *is bounded and has the extension property* $(1, p)$, *the space* $H^{1,p}(\Omega)$ *is compactly injected into any* $L^q(\Omega)$ *with* $1 \leq q < p^* \equiv Np/(N-p)$ *if* $p < N$, $1 \leq q < \infty$ *otherwise. In particular,* $H^{1,p}(\Omega)$ *is always compactly injected into* $L^p(\Omega)$.

PROOF. Let $1 \leq p < N$, $1 \leq q < p^*$, and choose $\lambda \in]0, 1]$ so that

$$\lambda q + \frac{1-\lambda}{p^*} q = 1.$$

Let $\omega \subset\subset \Omega$ be arbitrarily fixed and set

$$\tau_h u(x) \equiv u(x+h) \qquad \text{for } x \in \omega,$$

with $h \in \mathbb{R}^N$, $|h| < \operatorname{dist}(\omega, \partial\Omega)$. If $\mathscr{F}$ is a bounded subset of $H^{1,p}(\Omega)$, Theorem 1.21 yields

$$|\tau_h u - u|_{1;\omega} \leq C|h| \qquad \text{for } u \in \mathscr{F};$$

thanks to Theorem 1.33,

$$\begin{aligned} |\tau_h u - u|_{q;\omega} &\leq |\tau_h u - u|_{1;\omega}^{\lambda} \, |\tau_h u - u|_{p^*;\omega}^{1-\lambda} \\ &\leq C|h|^{\lambda}(2|u|_{p^*;\Omega})^{1-\lambda} \\ &\leq C|h|^{\lambda} \qquad \text{for } u \in \mathscr{F}, \end{aligned}$$

by Hölder's inequality for the product of $|\tau_h u - u|^{\lambda q}$ and $|\tau_h u - u|^{(1-\lambda)q}$ in ω. From Theorem 1.10 it follows that $\mathscr{F}|_{\omega}$ is relatively compact in $L^q(\omega)$ whatever $\omega \subset\subset \Omega$.

Let $\varepsilon > 0$ be arbitrarily fixed. By Hölder's inequality,

$$| u |_{q;\Omega\setminus\bar\omega} \leq | u |_{p^*;\Omega\setminus\bar\omega} \, | \Omega\setminus\bar\omega |^{1/q-1/p^*},$$

hence

$$| u |_{q;\Omega\setminus\bar\omega} < \varepsilon \qquad \text{for } u \in \mathscr{F}$$

if $\omega \subset\subset \Omega$ is suitably chosen. By the relative compactness of $\mathscr{F}|_\omega$, there exist $u_1, \ldots, u_n \in L^q(\omega)$ such that, whenever $u \in \mathscr{F}$,

$$| u - u_i |_{q;\omega} < \varepsilon \qquad \text{for some } i;$$

but then

$$| u - \tilde{u}_i |_{q;\Omega} < 2\varepsilon \qquad \text{for some } i,$$

where $\tilde{u}_i$ denotes the trivial extension of u_i to Ω. This shows that $\mathscr{F}$ is relatively compact in $L^q(\Omega)$.

If $p = N = 1$ the same procedure as above can be repeated with p^* replaced by ∞, and $1/p^*$ by 0 (see Theorem 1.33).

If p is $\geq N > 1$ or $> N = 1$ it suffices to replace it by any $r < N$ for $N > 1$, by 1 for $N = 1$. □

In the sequel we shall often make a crucial use of Rellich's theorem. In Chapter 2, for instance, we shall exploit it in the study of linear partial differential operators defined on bounded open sets.

For the time being we shall need Theorem 1.34 for the next three results.

LEMMA 1.35. *Let $1 \leq p < \infty$. Suppose that Ω is a bounded domain that has the extension property $(1, p)$. Then there exists a constant $C(\Omega)$ such that*

$$| u - (u)_\Omega |_{p;\Omega} \leq C(\Omega) \, | \nabla u |_{p;\Omega} \tag{1.23}$$

whenever $u \in H^{1,p}(\Omega)$. The same inequality holds with $C(\Omega) = Cr$, C independent of $x^0 \in \mathbb{R}^N$ as well as of $r \in]0, \infty[$, if $\Omega = B_r(x^0)$ or $\Omega = B_r^+(x^0)$.

PROOF. Without loss of generality we consider only the case $(u)_\Omega = 0$. If the theorem were false there would exist a sequence $\{u_n\} \subset H^{1,p}(\Omega)$ with $(u_n)_\Omega = 0$, $| u_n |_{p;\Omega} = 1$, $| \nabla u_n |_{p;\Omega} \leq 1/n$. But then, for a suitable subsequence of indices,

$$u_{n_k} \to u \qquad \text{in } L^p(\Omega) \qquad \text{as } k \to \infty$$

by Rellich's theorem and $(u)_\Omega = 0$, $| u |_{p;\Omega} = 1$; besides,

$$| \nabla u_n |_{p;\Omega} \to 0 \qquad \text{as } n \to \infty$$

so that $u \in H^{1,p}(\Omega)$ with $\nabla u = 0$ a.e. in Ω. By the connectedness of Ω, u would be a constant (see the corollary of Lemma 1.25), thus contradicting either $(u)_\Omega = 0$ or $| u |_{p;\Omega} = 1$.

Now let $\Omega = B_r(x^0)$; it is not restrictive to take $x^0 = 0$. Writing $C(r)$ for $C(B_r)$ we obtain the inequality

$$\int_B | u' - (u')_B |^p \, dx \leq C(1) \int_B | \nabla u' |^p \, dx$$

for the function $u' \in H^{1,p}(B)$ defined as $u'(x) \equiv u(rx)$ for $x \in B$. The conclusion in the case at hand follows from the identities

$$(u')_B = (u)_{B_r}, \qquad \nabla u'(x) = r(\nabla u)(rx) \qquad \text{for } x \in B.$$

Finally let $\Omega = B_r^+(x^0)$, or, more specifically, $\Omega = B_r^+$. Let $u \in H^{1,p}(B_r^+)$ with $(u)_{B_r^+} = 0$ and denote by $\tilde{u}$ the extension by reflection of u to B_r (see Lemma 1.29): then, $(\tilde{u})_{B_r} = 0$ and

$$\int_{B_r} | \tilde{u} |^p \, dx \leq C r^p \int_{B_r} | \nabla \tilde{u} |^p \, dx.$$

The conclusion follows from the identities

$$\int_{B_r} | \tilde{u} |^p \, dx = 2 \int_{B_r^+} | u |^p \, dx,$$

$$\int_{B_r} | \nabla \tilde{u} |^p \, dx = 2 \int_{B_r^+} | \nabla u |^p \, dx. \qquad \square$$

Inequality (1.23) is called *Poincaré's inequality.*

A proof very close to the above yields another inequality of the Poincaré type:

LEMMA 1.36. *Let p and Ω be as in Lemma 1.35 and take $\delta \in]0, 1[$. Then there exists a constant $C(\Omega, \delta)$ such that*

$$| u |_{p;\Omega} \leq C(\Omega, \delta) \, | \nabla u |_{p;\Omega}$$

whenever the function $u \in H^{1,p}(\Omega)$ vanishes a.e. in a subset of Ω whose *measure is* $\geq \delta \mid \Omega \mid$. *The same inequality holds with* $C(\Omega, \delta) = C(\delta)r$, $C(\delta)$ *being independent of* $x^0 \in \mathbb{R}^N$ *as well as of* $r \in]0, \infty[$, *if* $\Omega = B_r(x^0)$ *or* $\Omega = B_r^+(x^0)$.

REMARK. The connectedness of Ω is necessary for the validity of the above lemma. To see this, take Ω as the union of two disjoint open sets Ω_0 and Ω_1, $u = 0$ on Ω_0, $u = 1$ on Ω_1.

LEMMA 1.37. *Let* p *and* Ω *be as in Lemma* 1.35 *and take* $\varepsilon > 0$, $h, k \in N$ *with* $h < k$. *Then there exists a constant* C *such that*

$$\sum_{|\alpha|=h} | D^\alpha u |_{p;\Omega} \leq \varepsilon \sum_{|\beta|=k} | D^\beta u |_{p;\Omega} + C | u |_{p;\Omega}$$

whenever $u \in H^{k,p}(\Omega)$.

PROOF. Suppose that there exist $\varepsilon > 0$ and $u_n \in H^{k,p}(\Omega)$, $n \in N$, with $| u_n |_{H^{k,p}(\Omega)} = 1$ and

$$\sum_{|\alpha|=h} | D^\alpha u_n |_{p;\Omega} > \varepsilon \sum_{|\beta|=k} | D^\beta u_n |_{p;\Omega} + n | u_n |_{p;\Omega}. \tag{1.24}$$

By Rellich's theorem we may assume that $\{D^\gamma u_n\}$ converges in $L^p(\Omega)$ whatever the multi-index γ, $| \gamma | \leq k - 1$, hence that $u_n \to u$ in $H^{k-1,p}(\Omega)$. Since all norms $| D^\alpha u_n |_{p;\Omega}$ are uniformly bounded it follows from (1.24) that $u_n \to 0$ in $L^p(\Omega)$, hence that $u = 0$. But then $| D^\alpha u_n |_{p;\Omega} \to 0$ for $| \alpha | = h$, and (1.24) can again be applied to yield

$$\sum_{|\beta|=k} | D^\beta u_n |_{p;\Omega} \to 0,$$

hence

$$| u_n |_{H^{k,p}(\Omega)} \to 0,$$

a contradiction. □

The above lemma is said to provide an *interpolation inequality* for intermediate derivatives. More results of this sort will be given in Section 5.2.1.

1.6.3. Sobolev Inequalities II

For $\mu > N - 2$ the next three results are, respectively, interior, boundary, and global formulations of *Morrey's theorem.*

THEOREM 1.38. *Let Ω be bounded and assume $u \in H^1(\Omega)$ with $u_{x_1}, \ldots, u_{x_N} \in L^{2,\mu}(\Omega)$ for some $\mu \in [0, N[$. Then, whenever $\omega \subset\subset \Omega$, $u|_\omega$ belongs to $L^{2,\mu+2}(\omega)$ [so that $u \in C^{0,\delta}(\Omega)$ with $\delta = (\mu + 2 - N)/2$ if $\mu > N - 2$]; moreover,*

$$[u]_{2,\mu+2;\omega} \leq C \mid \nabla u \mid_{2,\mu;\Omega},$$

C being independent of u.

PROOF. Let $0 < \varrho < \operatorname{dist}(\omega, \partial\Omega)$. Whenever $x^0 \in \bar{\omega}$ Poincaré's inequality in $H^1(B_\varrho(x^0))$ yields

$$\begin{aligned}\int_{\omega[x^0,\varrho]} |u - (u)_{\omega[x^0,\varrho]}|^2\,dx &\leq \int_{B_\varrho(x^0)} |u - (u)_{x^0,\varrho}|^2\,dx \\ &\leq C\varrho^2 \int_{B_\varrho(x^0)} |\nabla u|^2\,dx \\ &\leq C\varrho^{\mu+2} \mid \nabla u \mid^2_{2,\mu;\Omega}.\end{aligned}$$ □

THEOREM 1.39. *Let $\Omega = B^+$ and assume $u \in H^1(B^+)$ with $u_{x_1}, \ldots, u_{x_N} \in L^{2,\mu}(B^+)$ for some $\mu \in [0, N[$. Then, whenever $0 < r < 1$, $u|_{B_r^+}$ belongs to $L^{2,\mu+2}(B_r^+)$, therefore to $C^{0,\delta}(\overline{B_r^+})$ with $\delta = (\mu + 2 - N)/2$ if $\mu > N - 2$; moreover,*

$$[u]_{2,\mu+2;B_r^+} \leq C \mid \nabla u \mid_{2,\mu;B^+},$$

C being independent of u.

PROOF. Set $r_h \equiv (1 - r)2^{-h}$, $h \in \mathbb{N}$. If $x \in \overline{B_r^+}$ with $x_N > r_1$, then

$$\begin{aligned}B_r^+[x, \sigma] &= B_r^+ \cap B_\sigma(x) \\ &\subseteq B_\sigma(x) \subset \omega \equiv \{y \in B^+_{r+r_2} \mid y_N > r_2\}\end{aligned}$$

for $0 < \sigma \leq r_2$. Since $\omega \subset\subset B^+$ Theorem 1.38 applies: hence,

$$\begin{aligned}\sigma^{-(\mu+2)} \mid u - (u)_{B_r^+[x,\sigma]} \mid^2_{2;B_r^+[x,\sigma]} &\leq \sigma^{-(\mu+2)} \mid u - (u)_{x,\sigma} \mid^2_{2;x,\sigma} \\ &\leq [u]^2_{2,\mu+2;\omega} \leq C \mid \nabla u \mid^2_{2,\mu;B^+}.\end{aligned}$$

This means that there remains to bound $\sigma^{-(\mu+2)} \mid u - (u)_{B_r^+[x,\sigma]} \mid^2_{2;B_r^+[x,\sigma]}$, $0 < \sigma \leq r_2$, only when $x \in \overline{B_r^+}$ with $x_N \leq r_1$. But then,

$$B_r^+[x, \sigma] \subset B_\varrho^+(x^0) \subset B^+,$$

where x^0 is the projection of x on S_r^0 and $\varrho \equiv 3\sigma$. In such a case Poincaré's inequality yields

$$\sigma^{-(\mu+2)} \mid u - (u)_{B_r^+[x,\sigma]} \mid^2_{2;B_r^+[x,\sigma]} \leq 3^{\mu+2}\varrho^{-(\mu+2)} \mid u - (u)_{x^0,\varrho} \mid^2_{2;x^0,\varrho,+}$$
$$\leq C\varrho^{-\mu} \mid \nabla u \mid^2_{2;x^0,\varrho,+} \leq C \mid \nabla u \mid^2_{2,\mu;B^+}. \qquad \square$$

THEOREM 1.40. *Let Ω be bounded with $\partial\Omega$ of class C^1. Assume $u \in H^1(\Omega)$ with $u_{x_1}, \ldots, u_{x_N} \in L^{2,\mu}(\Omega)$ for some $\mu \in [0, N[$. Then u belongs to $L^{2,\mu+2}(\Omega)$, or equivalently to $C^{0,\delta}(\bar{\Omega})$ with $\delta = (\mu + 2 - N)/2$ if $\mu > N - 2$; moreover,*

$$\mid u \mid_{2,\mu+2;\Omega} \leq C(\mid u \mid_{2;\Omega} + \mid \nabla u \mid_{2,\mu;\Omega})$$

with C independent of u.

PROOF. Fix $x^0 \in \partial\Omega$ and denote by U a bounded domain of $\mathbb{R}^N$, $U \ni x^0$, such that $\partial\Omega \cap U$ is straightened by a C^1 diffeomorphism Λ: $\bar{U} \to \bar{B}$. Moreover, let $R > 0$ be so small that $B_R(x^0) \subset \Lambda^{-1}(B_r)$ for some $r \in]0, 1[$. By Lemma 1.15, Theorem 1.39 applies to the function $u' \equiv u \circ (\Lambda^{-1})|_{B^+}$. We extend $u'|_{B_r^+}$ to B_r by reflection (see Lemma 1.16): again by Lemma 1.15, $\widetilde{u'} \circ \Lambda|_{B_R(x^0)}$ belongs to $L^{2,\mu+2}(B_R(x^0))$, hence $u|_{\Omega \cap B_R(x^0)} = \widetilde{u'} \circ \Lambda|_{\Omega\cap B_R(x^0)}$ to $L^{2,\mu+2}(\Omega \cap B_R(x^0))$, with norm estimate.

At this point we cover $\bar{\Omega}$ with a finite number of open sets ω_j, $j = 0, 1, \ldots, m$, where $\omega_0 \subset\subset \Omega$ and $\omega_1, \ldots, \omega_m$ are spheres constructed through the same procedure illustrated above for $B_R(x^0)$. Letting $\{g_j\}$ denote the partition of unity relative to $\{\omega_j\}$ we obtain the desired conclusion by writing u as $\sum_{j=0}^m g_j u$ and applying Theorem 1.38 to $g_0 u$. $\square$

We now fill up the gap in the range of k, p left over by Theorem 1.33.

THEOREM 1.41. *Let $k \in \mathbb{N}$, $p \geq 1$ with $kp > N$. If Ω is bounded and has the extension property $(1, r)$ for every finite r, then*

$$H^{k,p}(\Omega) \subsetneq C^{[k-N/p],k-N/p-[k-N/p]}(\bar{\Omega})$$

(where $[a] \equiv$ integer part of $a \in \mathbb{R}$) if $N/p \notin \mathbb{N}$, and

$$H^{k,p}(\Omega) \subsetneq C^{k-N/p-1,\delta}(\bar{\Omega}) \qquad \text{for any } \delta \in]0, 1[$$

if $N/p \in \mathbb{N}$. In particular, the set-theoretical inclusion

$$\bigcap_{k\in N} H^{k,p}(\Omega) \subset C^\infty(\bar{\Omega})$$

holds for any $p \geq 1$.

PROOF. Let $k = 1$. If $N < p$ it cannot be $N/p \in \mathbb{N}$. We fix a bounded open subset Ω' of $\mathbb{R}^N$, $\Omega \subset\subset \Omega'$ and extend every $u \in H^{1,p}(\Omega)$ to $\tilde{u} \in H^{1,p}(\Omega')$ with

$$| \tilde{u} |_{H^{1,p}(\Omega')} \leq C \, | u |_{H^{1,p}(\Omega)} .$$

If $N = 1$, $\tilde{u}$ has the absolute continuity property provided by Theorem 1.20, and the membership of u in $C^{0,1-1/p}(\bar{\Omega})$ follows from the fundamental theorem of calculus together with Hölder's inequality. Let $N \geq 2$. Since $L^p(\Omega') \subsetneq L^{2,\mu}(\Omega')$ for $\mu = N(p-2)/p$ Theorem 1.38 applies with Ω replaced by Ω'. Hence, $H^{1,p}(\Omega) \subsetneq C^{0,\delta}(\bar{\Omega})$ with $\delta = (\mu + 2 - N)/2 = 1 - N/p$.

Let $k > 1$, $N/p \notin \mathbb{N}$, and set $h \equiv [k - N/p]$. Each derivative $D^\alpha u$, $| \alpha | = h + 1$, belongs to $H^{k-h-1,p}(\Omega)$ and therefore to $L^q(\Omega)$ with

$$q = \frac{Np}{N - (k - h - 1)p} = \frac{N}{h + 1 - (k - N/p)} > N$$

by Theorem 1.33. The above considerations about the case $k = 1$ yield

$$D^\beta u \in C^{0,1-N/q}(\bar{\Omega}) \qquad \text{for } | \beta | = h$$

with norm estimate, and $1 - N/q = k - N/p - h$.

If $N/p \in \mathbb{N}$ replace p by $p - \varepsilon$, where ε is any positive number such that $k(p - \varepsilon) > N$. ☐

The norm estimates corresponding to the continuous injections in Theorems 1.33 and 1.41 are called *Sobolev inequalities.*

REMARK 1. Theorem 1.41 can also be given a proof which does not necessitate Theorem 1.38 (see H. Brézis [19]); in the sequel, however, we shall repeatedly need the latter result anyway.

REMARK 2. Theorems 1.33, 1.34, and 1.41 are valid for $\Omega = B_r^+(x^0)$: see Remark 3 after Theorem 1.30.

1.7. $H_0^{k,p}$ Spaces and Trace Spaces

1.7.1. $H_0^{k,p}(\Omega)$ Spaces

For $k \in \mathbb{N}$ and $1 \leq p < \infty$ we denote by $H_0^{k,p}(\Omega)$ the closure of $C_c^\infty(\Omega)$ in $H^{k,p}(\Omega)$, depressing the dependence on p when $p = 2$.

From Lemma 1.24 we know that $H_0^{k,p}(\mathbb{R}^N) = H^{k,p}(\mathbb{R}^N)$. When Ω is a proper subset of $\mathbb{R}^N$, $H_0^{k,p}(\Omega)$ certainly contains all functions $u \in H^{k,p}(\Omega)$ such that supp u is a compact subset of Ω (see Lemma 1.25), but whether the identity $H_0^{k,p}(\Omega) = H^{k,p}(\Omega)$ is valid or not depends on Ω as well as on the values of k and p.

EXAMPLE. Let $\Omega = \mathbb{R}^N \setminus \{0\}$.

From Theorem 1.41 it is easy to deduce that, when $kp > N$, every function from $H_0^{k,p}(\Omega)$ has a representative in $C^0(\mathbb{R}^N)$ which vanishes at 0. This shows that $H_0^{k,p}(\Omega)$ does not contain, for instance, any function from $H^{k,p}(\Omega)$ which takes on a nonzero constant value in $B_r \setminus \{0\}$ for some $r > 0$.

Now take $k = 1$, $1 \leq p < N$. Note that $H^{1,p}(\Omega) = H^{1,p}(\mathbb{R}^N)$ by Theorem 1.20. Let $g \in C^\infty(\mathbb{R})$ satisfy $g(t) = 0$ for $|t| < 1/2$, $g(t) = 1$ for $|t| > 1$, and set $g_n(t) \equiv g(nt)$ for $n \in \mathbb{N}$, so that $g_n(t) = 0$ for $|t| < 1/2n$, $g_n(t) = 1$ for $|t| > 1/n$, and $\sup_{\mathbb{R}} |g_n'| \leq n \sup_{\mathbb{R}} |g'|$ (where the prime denotes d/dx). If $u \in H^{1,p}(\Omega)$ with supp $u \subset B_R$ for some $R > 0$, each function $x \mapsto u_n(x) \equiv g_n(|x|)u(x)$ belongs to $H_0^{1,p}(\Omega)$. We claim that

$$u_n \to u \qquad \text{in } H^{1,p}(\Omega).$$

To verify this claim we majorize $\int_\Omega |\nabla(u_n - u)|^p \, dx$ with a quantity

$$C \int_\Omega \left|1 - g_n(|x|)\right|^p |\nabla u(x)|^p \, dx + C \int_\Omega \left|g_n'(|x|)\right|^p |u(x)|^p \, dx.$$

Since

$$\begin{aligned}\int_\Omega \left|g_n'(|x|)\right|^p |u(x)|^p \, dx &\leq n^p |g'|^p_{\infty;\mathbb{R}} \int_{B_{1/n}} |u|^p \, dx \\ &\leq Cn^p \left(\int_{B_{1/n}} |u|^{p^*} \, dx\right)^{p/p^*} |B_{1/n}|^{1-p/p^*} \\ &\leq Cn^p n^{-N(1-p/p^*)} \left(\int_{B_{1/n}} |u|^{p^*} \, dx\right)^{p/p^*}\end{aligned}$$

with $1/p^* \equiv 1/p - 1/N$, where Theorem 1.32 has been taken into account, we have

$$\int_\Omega |\nabla(u_n - u)|^p \, dx \to 0 \qquad \text{as } n \to \infty,$$

hence the claimed property. At this point we need only apply the cutoff method to approximate every element in $H^{1,p}(\Omega)$ with functions having compact supports, thus obtaining $H_0^{1,p}(\Omega) = H^{1,p}(\Omega)$.

An exhaustive treatment of the above matter would require the theory of "polar sets," as in R. Adams [1]. We can, however, rather easily prove that $H_0^{1,p}(\Omega)$ does not equal $H^{1,p}(\Omega)$ if $|\mathbb{R}^N\setminus\Omega| > 0$. This we shall do in a minute. First we prove the following lemma.

LEMMA 1.42. *Let $u \in H_0^{1,p}(\Omega)$ for some $p \in [1,\infty[$. If Ω' is another open subset of $\mathbb{R}^N$, $\Omega' \supset \Omega$, the trivial extension $\tilde{u}$ of u to Ω' belongs to $H_0^{1,p}(\Omega')$ with $\partial\tilde{u}/\partial x_i = \widetilde{(\partial u/\partial x_i)} \equiv$ trivial extension of $\partial u/\partial x_i$ to Ω' ($i = 1, \ldots, N$).*

PROOF. Let u be the limit in $H^{1,p}(\Omega)$ of $\{u_n\} \subset C_c^\infty(\Omega)$: the trivial extension $\tilde{u}_n$ of each function u_n to Ω' belongs to $C_c^\infty(\Omega')$, $\{\tilde{u}_n\}$ is a Cauchy sequence in $H^{1,p}(\Omega')$, $\tilde{u}_n \to \tilde{u}$ in $L^p(\Omega')$, and $\tilde{u} \in H_0^{1,p}(\Omega')$ with $\partial\tilde{u}/\partial x_i = 0$ in $\Omega'\setminus\bar{\Omega}$,

$$\left(\frac{\partial \tilde{u}}{\partial x_i}\right)\Big|_{\Omega} = \frac{\partial}{\partial x_i}(\tilde{u}|_{\Omega}) = \frac{\partial u}{\partial x_i}. \qquad \square$$

THEOREM 1.43. *Let $|\mathbb{R}^N\setminus\Omega| > 0$. Then $H_0^{1,p}(\Omega)$ is a proper subspace of $H^{1,p}(\Omega)$; moreover, if $x^0 \in \mathbb{R}^N$ and $R > 0$ are such that $B_R(x^0) \cap \Omega \neq \varnothing$ and $|B_R(x^0)\setminus\Omega| > 0$, there exists a constant $C(\delta)$, where $\delta \equiv |B_R(x^0)\setminus\Omega|/R^N$, such that the inequality*

$$|u|_{p;\Omega\cap B_R(x^0)} \leq C(\delta)R\,|\nabla u|_{p;\Omega\cap B_R(x^0)} \tag{1.25}$$

holds for every function $u \in H_0^{1,p}(\Omega)$.

PROOF. Let u be arbitrarily fixed in $H_0^{1,p}(\Omega)$ and denote by $\tilde{u}$ the trivial extension of u to $\mathbb{R}^N$: $\tilde{u}|_{B_R(x^0)}$ is a function from $H^{1,p}(B_R(x^0))$ which vanishes on $B_R(x^0)\setminus\bar{\Omega}$. Since

$$\nabla(\tilde{u}|_{B_R(x^0)}) = (\nabla\tilde{u})|_{B_R(x^0)} = (\widetilde{\nabla u})|_{B_R(x^0)}$$

Lemma 1.36 yields

$$\begin{aligned} |u|_{p;\Omega\cap B_R(x^0)} &= |\tilde{u}|_{p;x^0,R} \\ &\leq C(\delta)R\,|\nabla\tilde{u}|_{p;x^0,R} = C(\delta)R\,|\nabla u|_{p;\Omega\cap B_R(x^0)}. \end{aligned}$$

Now let $u \in H^{1,p}(\Omega)$ equal a constant $\neq 0$ in $\Omega \cap B_R(x^0)$. Then each first derivative of $u|_{\Omega\cap B_R}$ vanishes identically, and (1.25) cannot hold. Thus, $u \notin H_0^{1,p}(\Omega)$. $\square$

With no difficulty we arrive at the following corollary.

COROLLARY. *Let Ω be bounded. Then there exists a constant $C(\Omega)$, which equals Cr with C independent both of $x^0 \in \mathbb{R}^N$ and of $r \in]0, \infty[$ if $\Omega = B_r(x^0)$ or $\Omega = B_r^+(x^0)$, such that Poincaré's inequality*

$$| u |_{p;\Omega} \leq C(\Omega) \, | \nabla u |_{p;\Omega} \tag{1.26}$$

holds for $u \in H_0^{1,p}(\Omega)$. A norm on $H_0^{1,p}(\Omega)$ equivalent to $u \mapsto | u |_{H^{1,p}(\Omega)}$ is defined by $u \mapsto | \nabla u |_{p;\Omega}$.

[Take $R = 2r$ if $\Omega = B_r(x^0)$, $R = r$ if $\Omega = B_r^+(x^0)$.]

$[H_0^{1,p}(\Omega)]'$ is denoted by $H^{-1,p'}(\Omega)$, $p' \equiv p/(p-1)$ if $p > 1$ and $1' \equiv \infty$, the dependence on p' being depressed if $p' = p = 2$. If $f \in L^{p'}(\Omega)$ the distribution f_{x_i} is (identifiable with) an element of $H^{-1,p'}(\Omega)$.

1.7.2. $H_0^{1,p}(\Omega \cup \Gamma)$ Spaces

We now assume Γ of class C^1 and denote by $H_0^{1,p}(\Omega \cup \Gamma)$ the closure of $C_c^\infty(\Omega \cup \Gamma)$ in $H^{1,p}(\Omega)$, $1 \leq p < \infty$; we write $H_0^1(\Omega \cup \Gamma)$ instead of $H_0^{1,2}(\Omega \cup \Gamma)$. If the support of $u \in H^{1,p}(\Omega)$ is a compact subset of $\Omega \cup \Gamma$, Remark 2 after Theorem 1.30 and Lemma 1.25 easily show that $u \in H_0^{1,p}(\Omega \cup \Gamma)$. In particular, $H_0^{1,p}(\Omega \cup \Gamma)$ could equivalently be defined as the closure of $C_c^1(\Omega \cup \Gamma)$ in $H^{1,p}(\Omega)$; consequently, the mapping $u \mapsto u' \equiv u \circ \Lambda^{-1}$ defines an isomorphism between $H_0^{1,p}(\Omega \cup \Gamma)$ and $H_0^{1,p}(\Omega' \cup \Gamma')$ if Ω is bounded and $\Lambda: \bar{\Omega} \to \bar{\Omega}'$ is a C^1 diffeomorphism, $\Gamma' = \Lambda(\Gamma)$. As for the problem whether $H_0^{1,p}(\Omega \cup \Gamma)$ equals $H^{1,p}(\Omega)$ or not when $\partial\Omega \setminus \Gamma \neq \emptyset$, consider $\Omega = B(x^0)$ with $x^0 = (0, \ldots, 0, 1)$ and $\Gamma = S(x^0) \setminus \{0\}$ in the light of the example at the beginning of this section: $H_0^{1,p}(\Omega \cup \Gamma) = H^{1,p}(\Omega)$ if $p < N$, whereas nonzero constants do not belong to $H_0^{1,p}(\Omega \cup \Gamma)$ if $p > N$.

Lemma 1.42 admits the following straightforward generalization.

LEMMA 1.44. *Let Γ be of class C^1 and suppose there exists an open subset U of $\mathbb{R}^N$ such that $U \cap \Omega \neq \emptyset$, $U \cap \Gamma = \emptyset$. Let $u \in H_0^{1,p}(\Omega \cup \Gamma)$ for some $p \in [1, \infty[$. The trivial extension $\tilde{u}$ of u to $\Omega' \equiv \Omega \cup U$ belongs to $H_0^{1,p}(\Omega' \cup \Gamma)$ with $\partial\tilde{u}/\partial x_i = \widetilde{(\partial u/\partial x_i)}$ for $i = 1, \ldots, N$.*

In the same vein of Theorem 1.43 we therefore arrive at the following theorem.

THEOREM 1.45. *Let Γ be of class C^1 and suppose there exists a bounded domain U of $\mathbb{R}^N$ such that $U \cap \Omega \neq \emptyset$, $U \cap \Gamma = \emptyset$ and $| U \setminus \Omega | > 0$.*

Moreover, suppose that U has the extension property $(1, p)$. *Then there exists a constant* $C(\Omega, U)$, *which has the form* $C(\delta)R$ *if U is a sphere* $B_R(x^0)$ *or a hemisphere* $B_R{}^+(x^0)$ *and* $\delta \equiv |U\setminus\Omega|/R^N$, *such that the inequality*

$$|u|_{p;\Omega\cap U} \leq C(\Omega, U) |\nabla u|_{p;\Omega\cap U} \tag{1.27}$$

holds for $u \in H_0{}^{1,p}(\Omega \cup \Gamma)$. *Consequently,* $H^{1,p}(\Omega)\setminus H_0{}^{1,p}(\Omega \cup \Gamma) \neq \emptyset$.

If Ω is a hemisphere $B_r{}^+(x^0)$ we can take $U = B_{2r}^+(x^0)$ if $\Gamma = S_r{}^0(x^0)$, $U = B_r(x^0)$ if $\Gamma = S_r{}^+(x^0)$, and arrive at the following corollary of Theorem 1.45.

COROLLARY. *There exists a constant C independent of* $x^0 \in \mathbb{R}^N$ *and of* $r \in]0, \infty[$ *such that Poincaré's inequality*

$$|u|_{p;x^0,r,+} \leq Cr |\nabla u|_{p;x^0,r,+}$$

holds for $u \in H_0{}^{1,p}\big(B_r{}^+(x^0) \cup S_r{}^0(x^0)\big)$ *or* $u \in H_0{}^{1,p}\big(B_r{}^+(x^0) \cup S_r{}^+(x^0)\big)$.

REMARK. The formulation of Poincaré's inequality in the above corollary can also be directly proven as well as sharpened. To wit, consider $(x', x_N) \in B_r{}^+$. (We take $x^0 = 0$ for the sake of notational simplicity). Then, both quantities

$$\left|\int_{x_N}^{(r^2-|x'|^2)^{1/2}} u_{x_N}(x', t)\, dt\right|^p = |u(x)|^p$$

[when $u \in C_c{}^1(B_r{}^+ \cup S_r{}^0)$] and

$$\left|\int_0^{x_N} u_{x_N}(x', t)\, dt\right|^p = |u(x)|^p$$

[when $u \in C_c{}^1(B_r{}^+ \cup S_r{}^+)$] are bounded by

$$r^{p-1}\int_0^{(r^2-|x'|^2)^{1/2}} |u_{x_N}(x', t)|^p\, dt,$$

so that the double integration

$$\int_{|x'|\leq r} dx' \int_0^{(r^2-|x'|^2)^{1/2}} [\cdots]\, dx_N$$

and a density argument yield the inequality

$$| u |_{p;r,+}^{p} \leq r^p | u_{x_N} |_{p;r,+}^{p}$$

both for $u \in H_0^1(B_r^+ \cup S_r^0)$ and for $u \in H_0^1(B_r^+ \cup S_r^+)$.

In general, if Ω is a bounded domain and $\Omega \cup U$ has the same regularity as U in Theorem 1.45, we can obtain (1.27) with U replaced by $\Omega \cup U$, hence with $\Omega \cap U$ replaced by Ω. (The connectedness of Ω cannot be dispensed with: see the example in the remark after Lemma 1.36, with $\Gamma = \partial\Omega_1$ of class C^1 so that $\bar{\Omega}_1 \cap \bar{\Omega}_2 = \varnothing$.) This amounts to (1.26). For the validity of the latter the setting of Theorem 1.45, although very simple, is, however, too restricted. The same technique as for Lemma 1.36 does indeed yield the following lemma.

LEMMA 1.46. *Let Ω be a bounded domain. Suppose Γ is of class C^1 and such that $H_0^{1,p}(\Omega \cup \Gamma)$, where $1 \leq p < \infty$, does not contain any nonzero constant. Then* (1.26) *holds whenever $u \in H_0^{1,p}(\Omega \cup \Gamma)$, and a norm on $H_0^{1,p}(\Omega \cup \Gamma)$ equivalent to $u \mapsto | u |_{H^{1,p}(\Omega)}$ is defined by $u \mapsto | \nabla u |_{p;\Omega}$.*

We shall return to the (rather indirect) requirement that nonzero constants do not belong to $H_0^{1,p}(\Omega \cup \Gamma)$ in Lemma 1.58.

REMARK. Let Ω be bounded with Γ of class C^1 and closed (possibly empty). Let Ω' be a bounded open set with $\partial\Omega'$ of class C^1, $\Omega' \cap \partial\Omega = \partial\Omega \setminus \Gamma$, $\Omega' \supset \Omega$, and consider trivial extensions to Ω' of functions from $H_0^{1,p}(\Omega \cup \Gamma)(1 \leq p < \infty)$. From Theorems 1.33, 1.34, and 1.41 it follows that, for $p < N$, $H_0^{1,p}(\Omega \cup \Gamma)$ is continuously injected into $L^{p^*}(\Omega)$ and compactly injected into $L^q(\Omega)$ for $q < p^*$, whereas it is continuously injected into $C^{0,1-N/p}(\bar{\Omega})$ for $p > N$.

The above statements do not remain valid if Γ is not closed. Consider for instance the example of Section 1.2.1, with the right angles of Ω conveniently smoothed in order that $\Gamma \equiv \partial\Omega \setminus \{0\}$ be of class C^1: u belongs to $H_0^{1,p}(\Omega \cup \Gamma)$ because it is the limit in $C^1(\bar{\Omega})$ of the sequence $\{u_n\} \subset C_c^1(\Omega \cup \Gamma)$ defined by $u_n(x_1, x_2) \equiv u(x_1, x_2 - 1/n)$, and this proves that, whatever p, $H_0^{1,p}(\Omega \cup \Gamma)$ is not injected into $C^{0,\delta}(\bar{\Omega})$ if $\delta > \alpha/2$.

1.7.3. Boundary Values and $H^{1/p',p}(\Gamma)$ Spaces

Assume Ω bounded, Γ of class C^1 and closed, and fix p in $[1, \infty[$. For functions from $H^{1,p}(\Omega)$ we want to define the space of traces on Γ, as we did in Section 1.2.3 for functions from $C^{k,\delta}(\bar{\Omega})$.

In the present situation, however, the preliminary necessity of giving sense to $u|_\Gamma$ when $u \in H^{1,p}(\Omega)$ is already a relevant problem, in that u is only defined up to equivalence in $L^p(\Omega)$, and $|\Gamma| = 0$. When $p > N$ this difficulty can be overcome by defining $u|_\Gamma$ as the trace on Γ of the continuous representative of u (see the final remark of Section 1.7.2). In the general case we need instead more elaborate considerations.

We now proceed to illustrate them for $N \geq 2$: the corresponding study for $N = 1$ is left to the reader as an easy exercise.

We begin with two lemmas concerning regular functions in the space $H_0^{1,p}(\Omega \cup \Gamma)$.

LEMMA 1.47. *Let $u \in C_c^1(\Omega \cup \Gamma)$. Its trace $u|_\Gamma$ on Γ vanishes identically if and only if $u \in H_0^{1,p}(\Omega)$.*

PROOF. If u vanishes on Γ each function $u_n \equiv (1/n)G(nu)$ with $G \in C^1(\mathbb{R})$, $G(t) = 0$ for $|t| \leq 1$, $G(t) = t$ for $|t| \geq 2$ belongs to $C_c^1(\Omega)$. By the dominated convergence theorem $u_n \to u$ in $H^{1,p}(\Omega)$, so that $u \in H_0^{1,p}(\Omega)$.

Vice versa, suppose that $u \in H_0^{1,p}(\Omega)$. By Lemma 1.42 the trivial extension $\tilde{u}$ of u to $\mathbb{R}^N$ belongs to $H^{1,p}(\mathbb{R}^N)$ with

$$\int_\Omega u_{x_i} v \, dx = \int_{\mathbb{R}^N} \tilde{u}_{x_i} v \, dx = -\int_{\mathbb{R}^N} \tilde{u} v_{x_i} \, dx = -\int_\Omega u v_{x_i} \, dx$$

hence

$$\int_\Gamma (uv)|_\Gamma \, \nu^i \, d\sigma = 0$$

by the divergence theorem, for $v \in C_c^\infty(\mathbb{R}^N)$, $i = 1, \ldots, N$. Since $\sum_{i=1}^N (\nu^i)^2 = 1$ the above implies $u|_\Gamma = 0$. $\square$

LEMMA 1.48. *Let $q = (Np - p)/(N - p)$ if $1 \leq p < N$, $q \in [1, \infty[$ arbitrary if $p \geq N$. There exists a constant C such that*

$$|u|_\Gamma|_{q;\Gamma} \leq C |u|_{H^{1,p}(\Omega)}$$

whenever $u \in C_c^1(\Omega \cup \Gamma)$.

PROOF. Through a partition of unity and a change of coordinates we see that the only thing to prove is the existence of a constant C such that

$$|u|_{S^0}|_{q;S^0} \leq C |u|_{H^{1,p}(B^+)} \tag{1.28}$$

whenever $u \in C_c^1(B^+ \cup S^0)$. We write

$$u(x', 0) = -\int_0^{(1-|x'|^2)^{1/2}} u_{x_N}(x', t)\, dt \qquad \text{for } |x'| < 1.$$

If $p = 1$, then $q = 1$ and (1.28) follows from the inequality

$$\int_{|x'|<1} |u(x', 0)|\, dx' \le \int_{B^+} |u_{x_N}|\, dx.$$

Let $1 < p < N$. The function $w \equiv |u|^{(Np-p)/(N-p)}$ belongs to $C_c^1(B^+ \cup S^0)$, and

$$\begin{aligned} w(x', 0) &= -\int_0^{(1-|x'|^2)^{1/2}} w_{x_N}(x', t)\, dt \\ &\le \frac{Np-p}{N-p}\int_0^{(1-|x'|^2)^{1/2}} |u(x', t)|^{N(p-1)/(N-p)}\, |u_{x_N}(x', t)|\, dt. \end{aligned}$$

By Hölder's inequality,

$$\begin{aligned} \big|u|_{S^0}\big|_{q;S^0}^q &= \int_{|x'|<1} |u(x', 0)|^{(Np-p)/(N-p)}\, dx' = \int_{|x'|<1} w(x', 0)\, dt \\ &\le \frac{Np-p}{N-p}\left(\int_{B^+} |u|^{Np/(N-p)}\, dx\right)^{1-1/p}\left(\int_{B^+} |u_{x_N}|^p\, dx\right)^{1/p}. \end{aligned} \tag{1.29}$$

Since

$$H^{1,p}(B^+) \subsetneq L^{Np/(N-p)}(B^+)$$

by Theorem 1.33 (see Remark 3 after Theorem 1.30),

$$\int_{B^+} |u|^{Np/(N-p)}\, dx \le C\, |u|_{H^{1,p}(B^+)}^{Np/(N-p)},$$

and (1.28) follows from (1.29) for the present choice of p.

Finally, if $p \ge N$ (1.28) is valid with $q = (Nr-r)/(N-r)$ whenever $1 \le r < N$, hence with any $q \in]1, \infty[$. □

To $u \in H_0^{1,p}(\Omega \cup \Gamma)$ we now associate the equivalence class $[u]$ of all functions $z \in H_0^{1,p}(\Omega \cup \Gamma)$ satisfying $z - u \in H_0^{1,p}(\Omega)$. This means that $[u]$ is an element of the Banach (Hilbert, if $p = 2$) quotient space $H_0^{1,p}(\Omega \cup \Gamma)/H_0^{1,p}(\Omega)$, normed by

$$|[u]|_{H_0^{1,p}(\Omega\cup\Gamma)/H_0^{1,p}(\Omega)} \equiv \inf\{|z|_{H^{1,p}(\Omega)} \mid z \in H_0^{1,p}(\Omega \cup \Gamma),\ z - u \in H_0^{1,p}(\Omega)\}$$

(see Theorem 1.H). Note that the linear space of equivalence classes $[u]$ with $u \in C_c^1(\Omega \cup \Gamma)$ is dense in $H_0^{1,p}(\Omega \cup \Gamma)/H_0^{1,p}(\Omega)$. If $u \in C_c^1(\Omega \cup \Gamma)$ verifies $[u] = 0$, then $u \in H_0^{1,p}(\Omega)$, and $u|_\Gamma = 0$ by Lemma 1.47. This means that the mapping

$$u|_\Gamma \mapsto |\,[u]\,|_{H_0^{1,p}(\Omega \cup \Gamma)/H_0^{1,p}(\Omega)}, \qquad u \in C_c^1(\Omega \cup \Gamma), \tag{1.30}$$

defines a norm on $C^1(\Gamma)$ (see Lemma 1.5): we denote the completion of $C^1(\Gamma)$ with respect to (1.30) by $H^{1/p',p}(\Gamma)$, by $H^{1/2}(\Gamma)$ if $p = 2$. Obviously, the following holds.

LEMMA 1.49. *$H^{1/p',p}(\Gamma)$ is a Banach space (a Hilbert space, if $p = 2$) isometrically isomorphic to $H_0^{1,p}(\Omega \cup \Gamma)/H_0^{1,p}(\Omega)$.*

It is clear that $H^{1/r',r}(\Gamma) \subsetneq H^{1/p',p}(\Gamma)$ for $1 \le p < r < \infty$.

We define a continuous linear mapping T from $H_0^{1,p}(\Omega \cup \Gamma)$ onto $H^{1/p',p}(\Gamma)$ by density after setting

$$Tu \equiv u|_\Gamma \qquad \text{for } u \in C_c^1(\Omega \cup \Gamma);$$

of course $Tu = 0 \Leftrightarrow u \in H_0^{1,p}(\Omega)$.

THEOREM 1.50. *Let $q = (Np - p)/(N - p)$ if $1 \le p < N$, $q < \infty$ arbitrary if $p \ge N$. For $u \in H_0^{1,p}(\Omega \cup \Gamma)$, Tu is a function from $L^q(\Gamma)$ with*

$$|\,Tu\,|_{q;\Gamma} \le C\,|\,Tu\,|_{H^{1/p',p}(\Gamma)}, \tag{1.31}$$

C independent of u, and $Tu = 0$ a.e. $[N-1]$ on Γ if and only if $u \in H_0^1(\Omega)$; equivalently,

$$H^{1/p',p}(\Gamma) \subsetneq L^q(\Gamma).$$

Moreover, $Tu = u|_\Gamma$ if $u \in H_0^{1,p}(\Omega \cup \Gamma) \cap C^0(\Omega \cup \Gamma)$.

PROOF. By Lemma 1.48 a bound

$$|\,Tu\,|_{q;\Gamma} \le C\,|\,u + v\,|_{H^{1,p}(\Omega)}$$

is valid whenever $u \in C_c^1(\Omega \cup \Gamma)$ if $v \in C_c^1(\Omega)$, hence also if $v \in H_0^1(\Omega)$. This shows that (1.31) holds for $u \in C_c^1(\Omega \cup \Gamma)$, and therefore

$$|\,\eta\,|_{q;\Gamma} \le C\,|\,Tu\,|_{H^{1/p',p}(\Gamma)}$$

for $u \in H_0^{1,p}(\Omega \cup \Gamma)$, η denoting the limit in $L^q(\Gamma)$ of $\{Tu_n\}$ if $\{u_n\} \subset C_c^1(\Omega \cup \Gamma)$, $u_n \to u$ in $H^{1,p}(\Omega)$. We will be allowed to identify η with Tu

after proving that Tu is the zero element of $H^{1/p',p}(\Gamma)$ if $\eta = 0$ a.e. $[N-1]$ on Γ. If this is the case, then,

$$\sum_{i=1}^{N} \int_{\Omega} (u_{nx_i} v + u_n v_{x_i})\, dx = \sum_{i=1}^{N} \int_{\Gamma} u_n|_\Gamma\, v|_\Gamma\, \nu^i\, d\sigma \to 0$$

for $v \in C_c^\infty(\mathbb{R}^N)$. This shows that the trivial extension $\tilde{u}$ of u to $\mathbb{R}^N$ satisfies

$$\int_{\mathbb{R}^N} \tilde{u} v_{x_i}\, dx = -\int_{\Omega} u_{x_i} v\, dx \qquad \text{for } v \in C_c^\infty(\mathbb{R}^N)$$

and consequently belongs to $H^{1,p}(\mathbb{R}^N)$. We claim that the above guarantees the membership of u in $H_0{}^{1,p}(\Omega)$, so that $Tu = 0$. Indeed, through a partition of unity and a change of coordinates the problem at hand is reduced to the case

$$\Omega \cap B = B^+, \qquad \partial\Omega \cap B = \Gamma \cap B = S^0, \qquad \operatorname{supp} u \subset B^+ \cup S^0. \tag{1.32}$$

Then each function

$$\hat{u}_n(x', x_N) \equiv \begin{cases} u(x', x_N - 1/n) & \text{if } (x', x_N - 1/n) \in \operatorname{supp} u = \operatorname{supp} \tilde{u}, \\ 0 & \text{otherwise} \end{cases}$$

satisfies $\hat{u}_n \in H^{1,p}(\mathbb{R}^N)$ as well as $\operatorname{supp} \hat{u}_n \subset \Omega$ and therefore $\hat{u}_n|_\Omega \in H_0{}^{1,p}(\Omega)$ for n large enough, so that $u = \lim_{n\to\infty} \hat{u}_n|_\Omega$ in $H^{1,p}(\Omega)$ belongs to $H_0{}^{1,p}(\Omega)$.

Rather simple considerations show that also the last statement of the theorem need only be proven in the particular case (1.32). This time we denote by $\check{u}$ the extension by reflection of $u|_{\overline{B^+}}$ to $\bar{B}$: thus,

$$\check{u} \in H^{1,p}(B) \cap C^0(\bar{B}) \qquad \text{with } \operatorname{supp} \check{u} \subset B.$$

The sequence $\{\varrho_n * \check{u}\}$ of regularizations of $\check{u}$ verifies

$$(\varrho_n * \check{u})|_{\bar{B}} \to \check{u} \qquad \text{in } C^0(\bar{B})$$

by Lemma 1.7, hence also

$$(\varrho_n * \check{u})|_{\overline{S^0}} \to u|_{\overline{S^0}} \qquad \text{in } C^0(\overline{S^0});$$

on the other hand,

$$(\varrho_n * \check{u})|_B \to \check{u} \qquad \text{in } H^{1,p}(B)$$

by Lemma 1.25, hence also

$$(\varrho_n * \check{u})|_{B^+} \to u|_{B^+} \qquad \text{in } H^{1,p}(B^+)$$

and finally,

$$(\varrho_n * \check{u})|_{S^0} \to (Tu)|_{S^0} \qquad \text{in } L^q(S^0).$$

This shows that the continuous function $u|_{\overline{S^0}}$ is a representative of the function $(Tu)|_{S^0} \in L^q(S^0)$. □

Theorem 1.50 allows us to introduce the notations $u|_\Gamma \equiv Tu$ for $u \in H_0^{1,p}(\Omega \cup \Gamma)$, $u|_\Gamma \equiv (gu)|_\Gamma$ for $u \in H^{1/p}(\Omega)$ if $\Gamma \not\equiv \partial\Omega$, where g is any function from $C_c^\infty(\Omega \cup \Gamma)$ with $g = 1$ near Γ [a definition that is clearly independent of the particular choice of g), $(uv)|_\Gamma \equiv u|_\Gamma v|_\Gamma$ [an element of $L^q(\Gamma)$, not necessarily of $H^{1/p',p}(\Gamma)$] for $u \in H^{1,p}(\Omega)$ and $v \in C^0(\bar{\Omega})$.

By construction, $H^{1/p',p}(\Gamma)$ is exactly the space of traces $u|_\Gamma$ on Γ (in the sense of the above definition) of functions $u \in H^{1,p}(\Omega)$. On the other hand, an "intrinsic definition" of $H^{1/p',p}(\Gamma)$, which underlies our choice of the symbol (of a "Sobolev space of fractionary order $1/p'$"), can also be given: see J. Nečas [127].

Note that $C^{0,1}(\Gamma) \subsetneq H^{1/p',p}(\Gamma)$ since $C^{0,1}(\bar{\Omega}) \subsetneq H^{1,p}(\Omega)$ (see Theorem 1.21).

1.7.4. Supplementary Results

The next two results are not necessary for the sequel but cast more light on the structure of $H^{1/p',p}(\Gamma)$ when $p > 1$.

LEMMA 1.51. *For $p > 1$ the injection of $H^{1/p',p}(\Gamma)$ into $L^s(\Gamma)$, with $1 \le s < (Np - p)/(N - p)$ if $p < N$ and $s < \infty$ arbitrary if $p \ge N$, is compact.*

PROOF. We may safely restrict ourselves to the range $1 < p < N$. Let $\{\eta_n\}$ be a bounded sequence in $H^{1/p',p}(\Gamma)$, hence $\eta_n = u_n|_\Gamma$ with $\{u_n\} \subset H_0^{1,p}(\Omega \cup \Gamma)$ bounded. Without loss of generality we assume (1.32) with u replaced by u_n for every n. Let $r \in [1, p[$ be such that $s = (Nr - r)/(N - r)$. By density we may apply (1.29) with u replaced by $u_n - u_m$, p by r, q by s, and obtain

$$\begin{aligned} |\eta_n - \eta_m|_{s;\Gamma}^s &= \left| (u_n - u_m)|_{S^0} \right|_{s;S^0} \\ &\le C\left(\int_{B^+} |u_n - u_m|^{Nr/(N-r)}\, dx\right)^{1-1/r}\left(\int_{B^+} |(u_n - u_m)_{x_N}|^r\, dx\right)^{1/r} \\ &\le C\left(\int_\Omega |u_n - u_m|^{Nr/(N-r)}\, dx\right)^{1-1/r} |u_n - u_m|_{H^{1,p}|\Omega|}. \end{aligned}$$

Since we have $Nr/(N - r) < p^* \equiv Np/(N - p)$, $H_0^{1,p}(\Omega \cup \Gamma)$ is compactly injected into $L^{Nr/(N-r)}(\Omega)$ (see the final remark of Section 1.7.1). Thus, a subsequence of $\{u_n\}$ converges in $L^{Nr/(N-r)}(\Omega)$, and so does the corresponding subsequence of $\{\eta_n\}$ in $L^s(\Gamma)$ by the above estimate. □

LEMMA 1.52. *For $p > 1$ the continuous injection $H^{1/p',p}(\Gamma) \subsetneq L^q(\Gamma)$ of Theorem 1.50 is not onto.*

PROOF. If $H^{1/p',p}(\Gamma)$ were continuously injected onto $L^q(\Gamma)$, $L^q(\Gamma)$ would be continuously injected onto $H^{1/p',p}(\Gamma)$ by the open mapping theorem (see H. Brézis [19]), and $L^q(\Gamma)$ would be compactly injected in $L^1(\Gamma)$ by the previous result. But we can easily construct (on the basis of the sequence $\{\sin nx\}$, $0 < x < \pi$) a bounded sequence $\{\eta_n\}_n \subset L^q(\Gamma)$ with $|\eta_n|_{1;\Gamma} = 1$ which converges weakly to 0 in $L^1(\Gamma)$, so that no subsequence $\{\eta_{n_k}\}_k$ can converge strongly (to 0) in $L^1(\Gamma)$. □

By density, the divergence theorem can be extended as follows.

THEOREM 1.53. *If $u \in H_0^{1,p}(\Omega \cup \Gamma)$ and $v \in H^{1,p}(\Omega)$, the identity*

$$\int_\Omega u_{x_i} v\, dx = -\int_\Omega u v_{x_i}\, dx + \int_\Gamma u|_\Gamma v|_\Gamma \nu^i\, d\sigma$$

holds for $i = 1, \ldots, N$. In particular, let $\Gamma = \emptyset$: if $u \in H_0^{1,p}(\Omega)$,

$$\int_\Omega u_{x_i}\, dx = 0$$

for $i = 1, \ldots, N$.

REMARK. The regularity of v required in the above statement can be weakened in the light of the final remark of Section 1.7.2.

A few extensions of the above notions of boundary values will now be given.

If $\partial\Omega$ is of class C^1, $\partial\Omega \setminus \Gamma \neq \emptyset$, $H_0^{1,p}(\Omega \cup \Gamma)$ is the space of functions $u \in H^{1,p}(\Omega)$ such that $u|_{\partial\Omega\setminus\Gamma}$ is the zero element of $H^{1/p',p}(\partial\Omega \setminus \Gamma)$. This circumstance leads us to the following definition for the case when no regularity is assumed about $\partial\Omega\setminus\Gamma$. We say that $u \in H^{1,p}(\Omega)$ equals 0 on $\partial\Omega\setminus\Gamma$ *in the sense of* $H^{1,p}(\Omega)$ if $u \in H_0^{1,p}(\Omega \cup \Gamma)$.

Boundary values $u|_\Gamma$ can be given an unambiguous meaning also when Γ is no longer assumed closed. Take for instance $\Omega = B^+$, $\Gamma = S^0$. If u is a measurable function on B^+ such that $u|_{B_R^+} \in H^{1,p}(B_R^+)$ whenever $0 < R < 1$, we define its trace $u|_{S^0}$ a.e. $[N-1]$ by setting $u|_{S^0_{1-1/n}} \equiv$

$(g_n u)|_{S^0_{1-1/n}}$ for $n = 2, 3, \ldots$, where $g_n \in C^\infty(\mathbb{R}^N)$, $g_n = 1$ on $\bar{B}_{1-1/n}$, with supp $g_n \subset B$.

We finally observe that, since $H^{1/p',p}(\Gamma)$ has been identified with a dense subspace of $L^q(\Gamma)$ (see Lemma 1.13), $L^{q'}(\Gamma)$ is automatically identified to a subspace of $[H^{1/p',p}(\Gamma)]'$ with unambiguous meaning of the symbol $\langle \Phi, \eta \rangle$ when $\Phi \in L^{q'}(\Gamma)$, $\eta \in H^{1/p',p}(\Gamma)$; note that each mapping $v \mapsto \langle \Phi, v|_\Gamma \rangle$ with Φ given in $[H^{1/p',p}(\Gamma)]'$ is a bounded linear functional on $H_0{}^{1,p}(\Omega \cup \Gamma)$, hence an element of $[H_0{}^{1,p}(\Omega \cup \Gamma)]'$.

1.8. Inequalities and Lattice Properties

1.8.1. Some Notions from the Abstract Theory of Ordered Linear Spaces

Let a linear space V be endowed with an order structure defined by a reflexive, transitive and antisymmetric binary relation $\leq$: V is called an *ordered linear space* with respect to $\leq$ if $u \leq v$ implies $u + w \leq v + w$ and $\lambda u \leq \lambda v$ for $u, v, w \in V$ and $0 \leq \lambda < \infty$. If E is a subset of V and there exists $z \in V$ satisfying $z \geq u (z \leq u)$ whenever $u \in E$, E is said to be *order bounded from above* (*from below*), or *majorized* (*minorized*), and z is called an *upper* (a *lower*) *bound*, or a *majorant* (a *minorant*) of E. If, moreover, E has an upper (a lower) bound z_0 satisfying $z_0 \leq z (z_0 \geq z)$ whenever z is an upper (a lower) bound of E, then z_0 is unique and is called the *least upper bound* (*greatest lower bound*), or *supremum* (*infimum*) of E. We denote the supremum (infimum) of E by sup E (inf E), or by $\bigvee_{i\in I} u_i (\bigwedge_{i\in I} u_i)$ if $E = \{u_i \in V \mid i \in I\}$ for some index family I; $u \vee 0$ $[-(u \wedge 0)]$ is also denoted by u^+ (u^-), $u^+ + u^-$ by $|u|$. If V is an ordered linear space such that $u \vee v$ and $u \wedge v$ exist whenever $u, v \in V$, we say that V is a *linear lattice*. If F is a linear functional on an ordered linear space V, we say that F is *nonnegative* (*nonpositive*), in symbol $F \geq 0$ $(F \leq 0)$, if

$$\langle F, v \rangle \geq 0 \quad (\langle F, v \rangle \leq 0) \qquad \text{for } v \in V, \qquad v \geq 0.$$

A Banach space V which is an ordered linear space (a linear lattice) is called an *ordered Banach space* (a *Banach lattice* if in addition $|v^\pm|_V \leq |v|_V$ for $v \in V$).

Many properties of ordered Banach spaces and Banach lattices can be proven exactly as in the case $V = \mathbb{R}$. If V is an ordered Banach space, so is correspondingly V'; sometimes we will find it useful to refer to inequalities between elements of V' as inequalities *in the sense of* V'. V' need not be a Banach lattice, even if V is; however, V' has a property that V need not have if the index family I in the following lemma is not finite.

LEMMA 1.54. *Let V be a Banach lattice. If $\{F_i\}_{i\in I} \subset V'$ is order bounded from above, then $\bigvee_{i\in I} F_i$ exists.*

PROOF. We begin with the particular case $I = \{1, 2\}$, $F_1 = 0$, and write F for F_2. Let $G \in V'$ satisfy $G \geq 0$, $G \geq F$. Whenever $v \in V$, $v \geq 0$, we have

$$H(v) \equiv \sup_{\substack{w\in V\\ 0\leq w\leq v}} \langle F, w\rangle \leq \langle G, v\rangle,$$

so that $0 \leq H(v) < \infty$. It is clear that $H(\lambda v) = \lambda H(v)$ for $\lambda \geq 0$. For $k = 1, 2$ let $v_k \in V$, $v_k \geq 0$: if $w_k \in V$, $0 \leq w_k \leq v_k$, then $0 \leq w_1 + w_2 \leq v_1 + v_2$ and therefore

$$H(v_1) + H(v_2) \leq H(v_1 + v_2).$$

On the other hand, let $w \in V$ with $0 \leq w \leq v_1 + v_2$: then,

$$w = w \wedge v_1 + (w - v_1)^+$$

and

$$0 \leq w_1 \equiv w \wedge v_1 \leq v_1,$$

$$0 \leq w_2 \equiv (w - v_1)^+ \leq v_2,$$

hence

$$\langle F, w\rangle \leq H(v_1) + H(v_2)$$

and finally

$$H(v_1 + v_2) \leq H(v_1) + H(v_2).$$

This shows that a nonnegative linear functional H on V is defined by

$$\langle H, v\rangle \equiv H(v^+) - H(v^-) \qquad \text{for } v \in V.$$

(For linearity: if $v = v_1 + v_2$ write $v_1^+ + v_2^+$ as $v^+ + w$, $v_1^- + v_2^-$ as $v^- + w$, $w \geq 0$.)

At this point we deduce that H is continuous on V, hence $H \in V'$, from the following theorem that we shall prove shortly.

THEOREM 1.55. *Let H be a nonnegative linear functional on a Banach lattice V. Then H belongs to V'.*

Since

$$\langle F, v\rangle \leq \langle H, v\rangle \leq \langle G, v\rangle$$

for any $v \in V$, $v \geq 0$, G being an arbitrary upper bound of F and 0, we have $H = 0 \vee F$.

If neither F_1 nor F_2 vanishes, both $F_1 - F_2$ and 0 are order bounded from above by $G - F_2$ if $G \in V'$ is such that $G \geq F_1$ and $G \geq F_2$, so that there exists $F_1 \vee F_2 = 0 \vee (F_1 - F_2) + F_2$ by the above.

Passing to a general index family I we remark that $\bigvee_{i\in I} F_i$ exists if and only if $\bigvee_{i,j\in I} (F_i \vee F_j)$ does, in which case the two coincide. Without loss of generality we therefore assume that the family $\{F_i\}_{i\in I}$ contains the supremum between any two of its elements. For $v \in V$, $v \geq 0$ we set

$$H(v) \equiv \sup_{i\in I} \langle F_i, v\rangle.$$

It is obvious that $H(\lambda v) = \lambda H(v)$ for $\lambda \geq 0$. Let $v_1, v_2 \in V$ with $v_1 \geq 0$, $v_2 \geq 0$. Clearly,

$$H(v_1 + v_2) \leq H(v_1) + H(v_2).$$

On the other hand, if $\varepsilon > 0$ is arbitrarily fixed and $i, j \in I$ are such that

$$\langle F_i, v_1\rangle > H(v_1) - \varepsilon, \qquad \langle F_j, v_2\rangle > H(v_2) - \varepsilon,$$

then

$$\langle F_i \vee F_j, v_1 + v_2\rangle > H(v_1) + H(v_2) - 2\varepsilon.$$

This shows that $H(v_1 + v_2) = H(v_1) + H(v_2)$. At this point we can proceed as in the first part of the proof and conclude that the linear functional

$$\langle H, v\rangle \equiv H(v^+) - H(v^-), \qquad v \in V,$$

belongs to V' and equals $\bigvee_{i\in I} F_i$. □

PROOF OF THEOREM 1.55. Since V is a Banach lattice we need only prove that

$$\sup_{v\in V,\, v\geq 0,\, |v|_V \leq 1} \langle H, v\rangle < \infty.$$

If the above were not true, for every $n \in N$ there would exist $u_n \in V$ with $u_n \geq 0$, $|u_n|_V \leq 1/n$ such that $\langle H, u_n\rangle > 1$. But then $\{v_m\} \subset V$ defined by

$$v_m \equiv \sum_{n=1}^{m} u_{n^2}$$

would be a Cauchy sequence with $v_m \leq v_{m+1}$, hence $v_m \to v$ in V with

$$\langle H, v \rangle \geq \langle H, v_m \rangle > m \qquad \text{for every } m,$$

which is absurd. □

1.8.2. Inequalities and Lattice Properties in Function Spaces over Ω

All function spaces over Ω of interest to us are linear subspaces of $L^1_{\text{loc}}(\Omega)$. The latter is an ordered linear space and even a linear lattice with respect to the relation $\leq$ defined by

$$u \leq 0 \qquad \text{if } u(x) \leq 0 \qquad \text{for a.a. } x \in \Omega.$$

All spaces $H^{k,p}(\Omega)$ are then automatically endowed with the structure of ordered linear spaces, and so are all spaces $C^{k,\delta}(\Omega)$ after the obvious passage to continuous representatives so that

$$u(x) \leq 0 \qquad \text{for } x \in \Omega \qquad \text{if } u \in C^0(\Omega), \qquad u \leq 0.$$

Note that a function u of $L^1_{\text{loc}}(\Omega)$ is ≤ 0 if and only if

$$\int_\Omega uv\, dx \leq 0 \qquad \text{for } v \in C_c^\infty(\Omega), \qquad v \geq 0.$$

The passage to lattice properties is more delicate. All spaces $L^p_{\text{loc}}(\Omega)$ $[L^p(\Omega)]$ and $C^{0,\delta}(\Omega)$ $[C^{0,\delta}(\bar{\Omega})$ with Ω bounded$]$ are linear (Banach) lattices. $C^1(\Omega)$ is not a linear lattice. For what concerns Sobolev spaces we make use of *Stampacchia's theorem*:

THEOREM 1.56. *For* $1 \leq p < \infty$, $H^{1,p}(\Omega)$ *and* $H_0^{1,p}(\Omega \cup \Gamma)$ *with* Γ *of class* C^1 *are Banach lattices, and*

$$\nabla u^+ = \chi_{u>0} \nabla u, \qquad \nabla u^- = \chi_{u<0} \nabla u$$

where $\chi_{u>0}$ $(\chi_{u<0})$ *denotes the characteristic function of the subset of* Ω *where (an arbitrarily fixed representative of)* u *is* >0 (<0); *hence,* $\nabla u = 0$ *a.e. in the subset of* Ω *where* $u = 0$. *Moreover, the mappings* $u \mapsto u^+$ *and* $u \mapsto u^-$ *are continuous in* $H^{1,p}(\Omega)$.

For the proof of this theorem we need the following lemma.

LEMMA 1.57. *Let $G \in C^1(\mathbb{R})$ with $G' \in L^\infty(\mathbb{R})$ and in addition $G(0) = 0$ if Ω is not bounded. Whenever $u \in H^{1,p}(\Omega)$, $1 \leq p < \infty$, $G \circ u$ belongs to $H^{1,p}(\Omega)$ with $\nabla(G \circ u) = G'(u)\nabla u$. If Γ is of class C^1, $G(0) = 0$ and $u \in H_0^{1,p}(\Omega \cup \Gamma)$, then $G \circ u$ belongs to $H_0^{1,p}(\Omega \cup \Gamma)$. Finally, the above remains valid without the requirement $G' \in L^\infty(\mathbb{R})$ if in addition $u \in L^\infty(\Omega)$.*

PROOF. If $u \in H^{1,p}(\Omega)$ the functions $G \circ u$, $(G' \circ u)u_{x_i}$ belong to $L^p(\Omega)$. (Note $|G(u(x))| \leq |G(0)| + |G'|_{\infty;\mathbb{R}} |u(x)|$.) Take $\omega \subset\subset \Omega$.

If $\{u_n\} \subset C^1(\bar\omega)$ is such that $u_n \to u|_\omega$ and $u_{nx_i} \to u_{x_i}|_\omega$ in $L^p(\omega)$ as well as a.e. in ω (see Lemma 1.25), we have $\nabla(G \circ u_n) = G'(u_n)\nabla u_n$ and $G'(u_n) \to G'(u)$ a.e. in ω. Thus, from the inequalities

$$\int_\omega |G(u_n) - G(u)|^p \, dx \leq |G'|^p_{\infty;\mathbb{R}} \int_\omega |u_n - u|^p \, dx,$$

$$\int_\omega |G'(u_n)\nabla u_n - G'(u)\nabla u|^p \, dx \leq C |G'|^p_{\infty;\mathbb{R}} \int_\omega |\nabla u_n - \nabla u|^p \, dx + C \int_\omega |G'(u_n) - G'(u)|^p |\nabla u|^p \, dx$$

and with the help of the dominated convergence theorem for the last integral, we deduce that

$$G \circ u_n \to (G \circ u)|_\omega,$$

$$(G \circ u_n)_{x_i} \to [(G' \circ u)u_{x_i}]|_\omega \qquad \text{for } i = 1, \ldots, N$$

in $L^p(\omega)$. This shows that $(G \circ u)|_\omega \in H^{1,p}(\omega)$ with $\nabla(G \circ u)|_\omega = (G' \circ u)|_\omega \times \nabla u|_\omega$. The first conclusion of the lemma is valid by definition of distributional derivatives. The statement concerning functions $u \in H_0^{1,p}(\Omega \cup \Gamma)$ is obviously valid under the stronger assumption $u \in C_c^1(\Omega \cup \Gamma)$ and obtains in general through inequalities in $L^p(\Omega)$ analogous to those above in $L^p(\omega)$. Finally, $G \circ u$ equals $\hat{G} \circ u$, $\hat{G}$ being a suitable function from $C_c^1(\mathbb{R})$, if $u \in L^\infty(\Omega)$. ☐

PROOF OF THEOREM 1.56. *Step 1: Proof of the lattice property.* For $\varepsilon > 0$ set

$$G_\varepsilon(t) \equiv \begin{cases} (t^2 + \varepsilon^2)^{1/2} - \varepsilon & \text{for } t > 0, \\ 0 & \text{for } t \leq 0. \end{cases}$$

We are in a position to apply Lemma 1.57, which yields $G_\varepsilon \circ u \in H^{1,p}(\Omega)$ for $u \in H^{1,p}(\Omega)$ and $G_\varepsilon \circ u \in H_0^{1,p}(\Omega \cup \Gamma)$ for $u \in H_0^{1,p}(\Omega \cup \Gamma)$ if Γ is of class C^1, as well as

$$\int_\Omega (G_\varepsilon \circ u) v_{x_i}\, dx = - \int_{\Omega_+} \frac{u u_{x_i}}{(u^2 + \varepsilon^2)^{1/2}} v\, dx \qquad \text{for } v \in C_c^\infty(\Omega)$$

with $\Omega_+ \equiv \{x \in \Omega \mid u(x) > 0\}$ (for an arbitrarily fixed representative of u). By letting $\varepsilon \to 0$ we obtain

$$\int_\Omega u^+ v_{x_i}\, dx = - \int_{\Omega_+} u_{x_i} v\, dx \qquad \text{for } v \in C_c^\infty(\Omega),$$

hence $u^+ \in H^{1,p}(\Omega)$ if $u \in H^{1,p}(\Omega)$, $u^+ \in H_0^{1,p}(\Omega \cup \Gamma)$ if $u \in H_0^{1,p}(\Omega \cup \Gamma)$, with

$$(u^+)_{x_i} = u_{x_i} \chi_{u>0}.$$

This proves the required property of u^+. As for u^-, it suffices to utilize the identity $u^- = (-u)^+$. Finally, a.e. in the subset of Ω where $u = 0$ both functions $\chi_{u>0}$ and $\chi_{u<0}$ vanish, so that

$$\nabla u = \nabla u^+ - \nabla u^- = 0.$$

Step 2: Continuity of $u \mapsto u^\pm$. Let $\{u_n\}$ converge to u in $H^{1,p}(\Omega)$ and set $\chi_+ \equiv \chi_{u>0}$, $\chi_{n+} \equiv \chi_{u_n>0}$ after fixing representatives. We have $u_n \to u$ and therefore $|u_n| \to |u|$, $u_n^\pm \to u^\pm$ in $L^p(\Omega)$ as well as (after passing to a subsequence still denoted by the same symbol as the whole sequence) a.e. in Ω. Let Ω_+ be defined as in Step 1,

$$\Omega_- \equiv \{x \in \Omega \mid u(x) < 0\}, \qquad \Omega_0 \equiv \Omega \setminus (\Omega_+ \cup \Omega_-):$$

we have

$$u^+ = u\chi_+ = \lim_{n\to\infty} u_n^+ = \lim_{n\to\infty} u_n \chi_{n+},$$

hence $\chi_{n+} \to 1$ a.e. in Ω_+ because $u = u^+ > 0$ and $\chi_+ = 1$ there, whereas $\chi_{n+} \to 0$ a.e. in Ω_- and $\nabla u = 0$ a.e. in Ω_0. But,

$$\begin{aligned}
\int_\Omega |\nabla(u_n^+ - u^+)|^p\, dx &= \int_\Omega |\chi_{n+} \nabla u_n - \chi_+ \nabla u|^p\, dx \\
&\le C\left(\int_\Omega \chi_{n+} |\nabla(u_n - u)|^p\, dx + \int_\Omega |\nabla u|^p |\chi_{n+} - \chi_+|\, dx\right) \\
&\le C\left(\int_\Omega |\nabla(u_n - u)|^p\, dx + \int_{\Omega_+} |\nabla u|^p |\chi_{n+} - \chi_+|\, dx \right. \\
&\qquad \left. + \int_{\Omega_-} |\nabla u|^p \chi_{n+}\, dx + \int_{\Omega_0} |\nabla u|^p\, dx\right).
\end{aligned}$$

By the dominated convergence theorem, the integrals over Ω_+ and Ω_- tend to 0, and so does the integral over Ω because $u_n \to u$ in $H^{1,p}(\Omega)$, whereas the integral over Ω_0 vanishes. This shows that $u_n^+ \to u^+$ in $H^{1,p}(\Omega)$ for a subsequence of indices, hence also for the whole sequence by uniqueness of the limit. The statement about the continuity of $u \to u^+$, therefore also of $u \to u^-$, has thus been proven. □

From Theorem 1.56 we deduce the following.

LEMMA 1.58. *Let Γ be of class C^1, $1 \leq p < \infty$. If $u \in H^{1,p}(\Omega)$ and there exists $v \in H_0^{1,p}(\Omega \cup \Gamma)$ with $0 \leq u \leq v$, then $u \in H_0^{1,p}(\Omega \cup \Gamma)$. In particular, if Ω is bounded a nonzero constant cannot belong to $H_0^{1,p}(\Omega \cup \Gamma)$ if the latter is a proper subspace of $H^{1,p}(\Omega)$.*

PROOF. Let $\{v_n\} \subset C_c^1(\Omega \cup \Gamma)$ converge to v in $H^{1,p}(\Omega)$. Then u is the limit in $H^{1,p}(\Omega)$ of $\{v_n \wedge u\}$, and each function $v_n \wedge u$ belongs to $H_0^{1,p}(\Omega \cup \Gamma)$ because its support lies in $\Omega \cup \Gamma$. This proves the membership of u in $H_0^{1,p}(\Omega \cup \Gamma)$.

Now let Ω be bounded and assume $u \in H^{1,p}(\Omega) \setminus H_0^{1,p}(\Omega \cup \Gamma)$, say $u^+ \notin H_0^{1,p}(\Omega \cup \Gamma)$. If nonzero constants belonged to $H_0^{1,p}(\Omega \cup \Gamma)$ each function $u^+ \wedge n$, $n \in N$, would belong to $H_0^{1,p}(\Omega \cup \Gamma)$ by the previous conclusion, and so would u^+ since $u^+ \wedge n \to u^+$ in $H^{1,p}(\Omega)$. □

1.8.3. Boundary Inequalities

Passing to boundary inequalities we assume Ω bounded, Γ of class C^1, and say that $u \in H^{1,p}(\Omega)$, $1 \leq p < \infty$, satisfies $u \leq 0$ ($u \geq 0$) on $\partial\Omega \setminus \Gamma$ *in the sense of* $H^{1,p}(\Omega)$ if $u^+ \in H_0^{1,p}(\Omega \cup \Gamma)$ [$u^- \in H_0^{1,p}(\Omega \cup \Gamma)$]; $u \in H_0^{1,p}(\Omega \cup \Gamma)$, $1 \leq p < \infty$, satisfies $u \leq 0$ ($u \geq 0$) on Γ *in the sense of* $H^{1,p}(\Omega)$ if $u^+ \in H_0^{1,p}(\Omega)$ [$u^- \in H_0^{1,p}(\Omega)$].

Note that u satisfies both $u \leq 0$ and $u \geq 0$ on $\partial\Omega \setminus \Gamma$ in the sense of $H^{1,p}(\Omega)$ if and only if $u \in H_0^{1,p}(\Omega \cup \Gamma)$, that is, $u = 0$ on $\partial\Omega \setminus \Gamma$ in the sense of $H^{1,p}(\Omega)$.

LEMMA 1.59. *Let Ω have the segment property and Γ be of class C^1. Then $u \in H^{1,p}(\Omega)$, $1 \leq p < \infty$, satisfies $u \leq 0$ on $\partial\Omega \setminus \Gamma$ in the sense of $H^{1,p}(\Omega)$ if and only if it is the limit in $H^{1,p}(\Omega)$ of a sequence $\{u_n\} \subset C^\infty(\bar{\Omega})$ with $u_n|_{\partial\Omega \setminus \Gamma} \leq 0$.*

PROOF. *Step 1: The "if" part.* For $n \in N$ the support of the Lipschitzian function $(u_n - 1/n)^+$ lies in $\Omega \cup \Gamma$, hence $(u_n - 1/n)^+ \in H_0^{1,p}(\Omega \cup \Gamma)$.

Since u^+ is the limit in $H^{1,p}(\Omega)$ of $(u_n - 1/n)^+$ by Theorem 1.55, u^+ belongs to $H_0^{1,p}(\Omega \cup \Gamma)$.

Step 2: The "only if" part. Let $\{u_n^{(+)}\} \subset C_c^\infty(\Omega \cup \Gamma)$ converge to u^+ in $H^{1,p}(\Omega)$. We approximate u^- in $H^{1,p}(\Omega)$ with a sequence $\{u_n^{(-)}\} \subset C^\infty(\bar{\Omega})$ constructed, as in Theorem 1.27, through a partition of unity, translations and convolutions with nonnegative mollifiers; hence, $u_n^{(-)} \geq 0$. Then, u is the limit in $H^{1,p}(\Omega)$ of $\{u_n^{(+)} - u_n^{(-)}\}$, and $u_n^{(+)} - u_n^{(-)} \in C^\infty(\bar{\Omega})$ with $(u_n^{(+)} - u_n^{(-)})|_{\partial\Omega \setminus \Gamma} \leq 0$. ∎

We now add the assumption that Γ is compact. $L^1(\Gamma)$ is an ordered linear space with respect to the relation $\leq$ defined by

$$\eta \leq 0 \qquad \text{if } \eta \leq 0 \qquad \text{a.e. } [N-1] \text{ on } \Gamma.$$

All linear subspaces of $L^1(\Gamma)$ are ordered linear spaces with respect to $\leq$ in $L^1(\Gamma)$, with the obvious pointwise meaning in the continuous case.

LEMMA 1.60. *Let $\eta \in L^1(\Gamma)$. In order that $\eta \leq 0$ it is necessary and sufficient that*

$$\int_\Gamma \eta\theta \, d\sigma \leq 0 \qquad \textit{for } \theta \in C^1(\Gamma), \qquad \theta \geq 0.$$

PROOF. Necessity is obvious. Passing to sufficiency, we consider a covering $\Gamma = \bigcup_{i=1}^n \Gamma_i$, each Γ_i being straightened by a C^1 diffeomorphism $\Lambda_i\colon \bar{U}_i \to \bar{B}$, and suppose there exists $E \subset S^0$, with $\operatorname{meas}_{N-1} E > 0$, such that $\eta \circ \Lambda_i^{-1} > 0$ on E for some i. We can always assume that E lies inside S_r^0 for some $r \in]0, 1[$, and find a partition of unity $\{g_i\}$ relative to the above covering of Γ with the property that $g_i = 1$ on $\Lambda_i^{-1}(S_R^0)$ for some $R \in]r, 1[$, hence $g_j = 0$ on $[\Lambda_j^{-1}(S_R^0)] \cap [\Lambda_i^{-1}(S_R^0)]$ for $j \neq i$. With the symbols of Section 1.33 we have

$$\int_E (g_i\eta) \circ (\Lambda_i^{-1})|_E H_i \, dx' > 0:$$

by approximating a.e. $[N-1]$ in S^0 the characteristic function of E with a sequence $\{\theta_n'\} \subset C^1(S^0)$ such that $0 \leq \theta_n' \leq 1$, $\operatorname{supp} \theta_n' \subset S_R^0$, we find an index n_0 such that $\theta_0' \equiv \theta_{n_0}'$ satisfies

$$\int_{S^0} (g_i\eta) \circ (\Lambda_i^{-1})|_{S^0} \theta_0' H_i \, dx' > 0,$$

hence

$$\int_\Gamma \eta\theta_0 \, d\sigma > 0$$

with $\theta_0 \equiv \theta_0' \circ \Lambda_i$ on $\Lambda_i^{-1}(S_R^0)$, $\theta_0 \equiv 0$ elsewhere. □

THEOREM 1.61. *Let Γ be of class C^1 and compact, $1 \le p < \infty$. $H^{1/p',p}(\Gamma)$ is a Banach lattice with respect to the order relation in $L^1(\Gamma)$, and $\eta^\pm = u^\pm|_\Gamma$ if $\eta = u|_\Gamma$ with $u \in H_0^{1,p}(\Omega \cup \Gamma)$.*

PROOF. We know that $\eta = u|_\Gamma = u^+|_\Gamma - u^-|_\Gamma$. We need to show that $u^\pm|_\Gamma \ge 0$, $u^+|_\Gamma u^-|_\Gamma = 0$ a.e. $[N-1]$. But this is true with $u^\pm|_\Gamma$ replaced by $u_n^\pm|_\Gamma$ if $\{u_n\} \subset C_c^\infty(\Omega \cup \Gamma)$, $u_n \to u$ in $H^{1,p}(\Omega)$, and we only need pass to the limit a.e. $[N-1]$ on Γ (see Theorem 1.12). For what concerns the norm estimate it suffices to note that

$$|\eta^\pm|_{H^{1/p',p}(\Gamma)} \le |u^\pm|_{H^{1,p}(\Omega)} \le |u|_{H^{1,p}(\Omega)}$$

whenever $\eta = u|_\Gamma$. □

In the setting of the above theorem the following mutual implications

$$u^+ \in H_0^{1,p}(\Omega) \Leftrightarrow u^+|_\Gamma = 0 \Leftrightarrow (u|_\Gamma)^+ = 0 \Leftrightarrow u|_\Gamma = -(u|_\Gamma)^-$$

lead to the following corollary.

COROLLARY. *Let Γ be of class C^1 and compact, $1 \le p < \infty$. Then $u \in H_0^{1,p}(\Omega \cup \Gamma)$ is ≤ 0 on Γ in the sense of $H^{1,p}(\Omega)$ if and only if $u|_\Gamma \le 0$.*

The notion of inequalities in the sense of $H^{1,p}(\Omega)$, $1 \le p < \infty$, can be enlarged as follows. Let Ω have the segment property and let $E \subseteq \bar{\Omega}$, $\hat{\psi} \in C^0(E)$. We say that $u \in H^{1,p}(\Omega)$ satisfies $u \le \hat{\psi}$ on E *in the sense of* $H^{1,p}(\Omega)$ if u is the limit in $H^{1,p}(\Omega)$ of a sequence $\{u_n\} \subset C^\infty(\bar{\Omega})$ with $u_n \le \hat{\psi}$ on E. If $E = \partial\Omega \setminus \Gamma$ with Γ of class C^1 and $\hat{\psi} = 0$, Lemma 1.59 leads us back to the previous definition; as for the case $E = \Omega$, $\hat{\psi} = 0$, we can adapt an argument utilized in Step 2 of the proof of Lemma 1.59 and verify that $u \le 0$ on Ω in the sense of $H^{1,p}(\Omega)$ if and only if $u(x) \le 0$ for a.a. $x \in \Omega$. Finally, if $\bar{E} \subset \Omega \cup \Gamma$ the above definition can be given without any hypothesis of regularity about $\partial\Omega \setminus \Gamma$ for functions $u \in H_0^{1,p}(\Omega \cup \Gamma)$, the sequence $\{u_n\}$ being taken in $C_c^\infty(\Omega \cup \Gamma)$.

Problems

For brevity's sake, problems will often be phrased in the form of assertions that must be proven, possibly following a basic outline.

1.1. Let V be a normed space and let $v_0 \in V$ be arbitrarily fixed. Apply the Hahn–Banach theorem to the linear functional $v \mapsto t \mid v_0 \mid_V$ for $v = tv_0$, $t \in \mathbb{R}$, and prove that there exists $F \in V'$ with $\mid F \mid_{V'} \leq 1$, $\langle F, v_0 \rangle = \mid v_0 \mid_V$.

1.2. Prove that

$$\mid v \mid_V \leq \liminf_{n \to \infty} \mid v_n \mid_V$$

if $v_n \rightharpoonup v$ in a normed space V.

1.3. Let V be a normed space. Any Cauchy sequence in V' is weakly convergent and even strongly convergent.

1.4. Let Ω be bounded and fix δ in $]0, 1]$, γ in $[0, \delta[$. To prove that the injection $C^{k,\delta}(\bar{\Omega}) \subsetneq C^{k,\gamma}(\bar{\Omega})$ is compact for any $k = 0, 1, \ldots,$ utilize the inequality

$$\mid u(x) - v(x) - [u(y) - v(y)] \mid / \mid x - y \mid^{\gamma} \leq 2 \mid u - v \mid_{C^0(\bar{\Omega})} / \varepsilon^{\gamma}$$

for $x, y \in \bar{\Omega}$, $\mid x - y \mid \geq \varepsilon > 0$.

1.5. If Ω is connected and u is a function on Ω with $[u]_{\delta;\Omega} < \infty$ for some $\delta > 1$, then u is a constant.

1.6. Let $0 < \delta \leq 1$. Utilize the inequality

$$(1 + y)^{\delta} \leq 1 + y^{\delta} \qquad \text{for } y \geq 0$$

to show that the function $u(x) \equiv \mid x \mid^{\delta}$, $x \in \mathbb{R}^N$, verifies $[u]_{\delta;\mathbb{R}^N} = 1$.

1.7. Find a function $u \in C^0(\bar{B})$ with $[u]_{\delta;\bar{B}} = \infty$ for any $\delta > 0$.

1.8. Denote by $\tilde{u}$ the trivial extension to $\mathbb{R}^N$ of a function $u \in L^p(\Omega)$, $1 \leq p < \infty$, and set $u_h(x) \equiv \tilde{u}(x + h)$ for $x \in \Omega$, $h \in \mathbb{R}^N$. Utilize Theorem 1.S to prove that $u_h \to u$ in $L^p(\Omega)$ as $\mid h \mid \to 0$.

1.9. Lemma 1.4 is proven as follows. For a suitable $t > 0$ the set $\omega_0' \equiv \{x \in \omega_0 \mid \operatorname{dist}(x, \partial\omega_0) > t\}$ is such that $\omega = \omega_0' \cup (\bigcup_{i=1}^{\infty} \omega_i)$. An open covering $\{\omega_j'\}$ of ω, with $\omega_j' \subset\subset \omega_j$, is constructed by recurrence. For each j there exists $g_j' \in C^{\infty}(\mathbb{R}^N)$ with $g_j' = 1$ on $\overline{\omega_j'}$, supp $g_j' \subset \omega_j$. The required partition of unity is obtained by setting

$$g_j \equiv g_j' \Big/ \sum_{k=0}^{\infty} g_k'.$$

1.10. Let $x^0 \in \Omega$. There exists no function $u \in L^1_{\text{loc}}(\Omega)$ with the property

$$\int_{\Omega} uv \, dx = v(x^0) \qquad \text{for } v \in C_c^{\infty}(\Omega).$$

1.11. Let $u \in L^1_{\text{loc}}(\Omega)$ verify

$$\int_\Omega uv\,dx = 0 \qquad \text{for } v \in C_c^\infty(\Omega).$$

By considering the functions $(\varrho_n * u)|_\omega$, $\omega \subset\subset \Omega$, with n sufficiently large, show that $u = 0$ a.e. in Ω.

1.12. Let Ω be bounded and take $p \in]1, \infty[$. If $\{u_n\}$ is a bounded sequence from $L^p(\Omega)$, $u_n \to u$ a.e. in Ω, the sets $E_m \equiv \{x \in \Omega \mid |u_n(x) - u(x)| \le 1$ whatever $n \ge m\}$ satisfy $|E_m| \to |\Omega|$ as $m \to \infty$. The family Φ of functions $v \in L^{p'}(\Omega)$ such that $\operatorname{supp} v \subseteq E_m$ for some m is dense in $L^{p'}(\Omega)$, and

$$\int_\Omega (u_n - u)v\,dx \to 0$$

whatever $v \in \Phi$. Hence, $u_n \rightharpoonup u$ in $L^p(\Omega)$. See J. L. Lions [103].

1.13. Let $\Omega = \{x \in \mathbb{R}^N \mid |x'| < 1, 0 < x_N < 1\}$. Utilize Theorem 1.20 to prove the following: if u belongs to $L^\infty(\Omega)$ and its distributional derivative u_{x_N} to $L^1(\Omega)$, then the mapping $x_N \mapsto \int_{|x'|<1} u^2(x', x_N)\,dx'$ is continuous on $]0, 1[$. Next, utilize Problem 1.12 to prove that the mapping $x_N \mapsto u(\cdot, x_N)$ is continuous from $]0, 1[$ into $L^2(S^0)$.

1.14. For $\omega \subset\subset \Omega$ and $|h| < \operatorname{dist}(\omega, \partial\Omega)$ prove the necessary part of Theorem 1.21 with the help of Lemma 1.25 and Hölder's inequality [which yields

$$|\delta_h{}^i u(x)|^p \le \int_0^1 |\nabla u(x + the^i)|^p\,dt \qquad \text{for } x \in \omega$$

whenever $u \in C^1(\overline{\omega'})$, $\omega' \equiv \{x + the^i \mid x \in \omega, 0 \le t \le 1\}$]. Proceed analogously for $\Omega = B^+$, $\bar{\omega} \subset B^+ \cup S^0$, $i = 1, \ldots, N-1$.

1.15. Given $\alpha \in \mathbb{R}$ and $p \in [1, \infty]$, find the largest value of $k \in \mathbb{N}$ for which the function $|x|^\alpha$ belongs to $H^{k,p}(B)$.

1.16. Let $\partial\Omega$ be of class C^1. Find the smallest value of the natural number k (depending on N) such that uv belongs to $H^k(\Omega)$ whenever both u and v do.

1.17. Formula (1.17) can be given with B^+ replaced by a Cartesian product $]-a_1, a_1[\times\cdots\times]-a_{N-1}, a_{N-1}[\times]0, a_N[$, $a_i > 0$. Utilize this observation a convenient number of times (for instance, 4 times when $N = 2$) to prove that every Cartesian product $]b_1, c_1[\times\cdots\times]b_N, c_N[$, $b_i < c_i$, has the extension property (k, p) for $k \in \mathbb{N}$ and $1 \le p \le \infty$.

1.18. Lemma 1.37 can be given a different proof which yields the sharper estimate

$$\sum_{|\alpha|=h} |D^\alpha u|_{p;\Omega}^p \le \delta \sum_{|\beta|=k} |D^\beta u|_{p;\Omega}^p + C\delta^{h/(h-k)} |u|_{p;\Omega}^p$$

with C independent of $\delta > 0$, provided the latter is sufficiently small. For $h = 1$ and $k = 2$ such a proof is particularly simple if Ω is a cube, the constant C being then independent of the size of Ω. Indeed, begin with the case $N = 1$ and divide the interval Ω into subintervals Ω_i of length between $\delta_0^{1/p}/2$ and $\delta_0^{1/p}$, with $0 < \delta_0 < |\Omega|^p$. Assume $u \in C^2(\bar{\Omega})$. For $\Omega_i =]a, b[$ apply the mean value theorem to u between two points $\xi \in]a, a + \alpha[$ and $\eta \in]a + 3\alpha, b[$ with $\alpha \equiv (b - a)/4$, and obtain

$$| u'(x) | \leq \frac{| u(\xi) | + | u(\eta) |}{2\alpha} + \int_a^b | u''(t) | \, dt$$

for $x \in]a, b[$. After integrating in ξ over $]a, a + \alpha[$, in η over $]a + 3\alpha, b[$, take the pth power, apply Hölder's inequality, and integrate in x over the interval $]a, b[$: the result is

$$\int_a^b | u' |^p \, dx \leq \frac{C_0}{\alpha^p} \int_a^b | u |^p \, dx + C_0 \alpha^p \int_a^b | u'' |^p \, dx$$
$$\leq 2^{3p} \frac{C_0}{\delta_0} \int_a^b | u |^p \, dx + \frac{C_0 \delta_0}{2^{2p}} \int_a^b | u'' |^p \, dx.$$

For $0 < \delta < 2^{-2p} C_0 | \Omega |^p$ the result in the case at hand follows easily. The passage to the case $N > 1$ is an immediate consequence of the above for $u \in C^2(\bar{\Omega})$. (See A. Friedman [54].)

1.19. Let $1 \leq p < \infty$. If $T \in H^{-1}(\Omega)$ with $f \leq T \leq g$, where both f and g belong to $L^p(\Omega)$, then T is an element of $L^p(\Omega)$.

1.20. If Ω is bounded and $\psi \in C^0(\bar{\Omega})$, $\psi|_{\partial\Omega} = 0$, $\psi > 0$ in Ω, then for every $k \in N$ and $p \in [1, \infty]$ there exists $v \in H_0^{k,p}(\Omega) \cap C^0(\Omega)$ such that $0 < v < \psi$ in Ω. To see this, utilize Lemma 1.4 and define $v \equiv \sum_{j=0}^{\infty} \varepsilon_j g_j / | g_j |_{H^{k,p}(\Omega)}$ with $\varepsilon_j > 0$ suitably chosen.

1.21. Utilize Lemma 1.25 to prove that $uv \in H^{1,p}(\Omega) \cap L^\infty(\Omega)$ $(1 \leq p \leq \infty)$ with $(uv)_{x_i} = u_{x_i} v + u v_{x_i}$ for $i = 1, \ldots, N$ if $u, v \in H^{1,p}(\Omega) \cap L^\infty(\Omega)$. Proceed analogously for $H_0^{1,p}(\Omega \cup \Gamma)$, with Γ of class C^1 and $1 \leq p < \infty$, instead of $H^{1,p}(\Omega)$.

1.22. Let Ω be bounded (no regularity being assumed on $\partial\Omega$). Take $p \in [1, \infty[$, $u \in H^{1,p}(\Omega) \cap C^0(\bar{\Omega})$ with $u|_{\partial\Omega} = 0$, and prove that $u \in H_0^{1,p}(\Omega)$ by considering $(u - 1/n)^+$, $(u - 1/n)^-$. See also the proof of Lemma 1.47.

1.23. Let $\{u_n\}$ converge toward u in $H^1(\Omega)$. Then $u_n^\pm \rightharpoonup u^\pm$ in $H^1(\Omega)$; moreover, $| u_n^\pm |_{H^1(\Omega)} \to | u^\pm |_{H^1(\Omega)}$. Hence, $u_n^\pm \to u^\pm$ in $H^1(\Omega)$. Compare with Step 2 of the proof of Theorem 1.56.

1.24. Let $u \in H^{2,p}(\Omega)$ $(1 \leq p < \infty)$. Then $| \nabla u | \in H^{1,p}(\Omega)$ with $(\partial/\partial x_j) | \nabla u | = u_{x_i} u_{x_i x_j} / | \nabla u |$.

1.25. Let Ω be bounded, Γ of class C^1 and closed. For $1 < p < \infty$ utilize the reflexivity of $H_0^{1,p}(\Omega \cup \Gamma)$ to prove that, if $\eta \in H^{1/p',p}(\Gamma)$, there exists $u \in H_0^{1,p}(\Omega \cup \Gamma)$ with $u|_\Gamma = \eta$ and

$$| u |_{H^{1,p}(\Omega)} = | \eta |_{H^{1/p',p}(\Gamma)}.$$

1.26. Let $v \in H_0^{1,p}(B^+ \cup S^0)$ $(1 \le p < \infty)$ and set $\eta(x') \equiv v(x', 0)$, $|x'| < 1$. Prove that for $i = 1, \ldots, N-1$ the derivative η_{x_i} in the sense of distributions over S^0 (the latter being endowed with the relative topology) equals $v_{x_i}|_{S^0}$. Utilize this fact to prove that, if $\Gamma = \partial\Omega$ is of class C^1 and $u \in H_0^{1,p}(\Omega) \cap H^{2,p}(\Omega)$, then $\nabla u = u_{x_i}\nu^i\nu$.

1.27. If Γ is compact and of class C^1, $u_n \rightharpoonup u$ in $H^1(\Omega)$ implies $u_n|_\Gamma \rightharpoonup u|_\Gamma$ in $H^{1/2}(\Gamma)$.

2

The Variational Theory of Elliptic Boundary Value Problems

Consider the following "model problem":

$$
\begin{gathered}
-\Delta u + u = f \qquad \text{in } \Omega, \\
u = 0 \qquad \text{on } \partial\Omega \setminus \Gamma, \qquad (\nabla u)\cdot \nu = 0 \qquad \text{on } \Gamma,
\end{gathered}
\tag{2.1}
$$

where Δ denotes, as is usual in the literature, the Laplacian $\sum_{i=1}^{N} \partial^2/\partial x_i^2$, and f is an arbitrarily fixed function from $L^2(\Omega)$. (As stipulated in the Glossary of Basic Notations, Ω is from now on supposed to be a bounded domain.) Let $\partial\Omega$ be of class C^1 and let its open portion Γ be closed as well. With the help of Section 1.7.3 for what concerns boundary values, we see that (2.1) certainly makes sense in the function space $H^2(\Omega)$ and implies

$$
\begin{gathered}
u \in H_0^1(\Omega \cup \Gamma), \\
a(u, v) \equiv \int_\Omega (u_{x_i} v_{x_i} + uv)\, dx = \int_\Omega fv\, dx \qquad \text{for } v \in H_0^1(\Omega \cup \Gamma)
\end{gathered}
\tag{2.2}
$$

by the divergence theorem: see Theorem 1.53. (From now on we adopt the *summation convention*: repeated dummy indices indicate summation from 1 to N.)

Vice versa, any function $u \in H^2(\Omega)$ satisfying (2.2) is rapidly seen to satisfy (2.1) as well (see Theorem 2.6 below). The second formulation of the model problem does, however, have a great advantage over the first one. Indeed, from the Riesz representation theorem it immediately follows that (2.2) admits a unique solution, since $u, v \mapsto a(u, v)$ is the scalar product in the Hilbert space $H_0^1(\Omega \cup \Gamma)$ and $v \mapsto \int_\Omega fv\,dx$ is an element of $[H_0^1(\Omega \cup \Gamma)]'$.

It is worth mentioning that a function u minimizing the functional

$$\mathscr{J}(v) \equiv \frac{1}{2}\int_\Omega (|\nabla v|^2 + v^2)\,dx - \int_\Omega fv\,dx$$

over $H_0^1(\Omega \cup \Gamma)$ must satisfy the condition $(d/d\lambda)\mathscr{J}(u + \lambda v)|_{\lambda=0} = 0$ for $v \in H_0^1(\Omega \cup \Gamma)$, which, by the fact that $a(u, v) = a(v, u)$, clearly amounts to (2.2): the latter is called the *Euler–Lagrange equation* of the minimum problem. Note that the converse of the above is also true, since $\mathscr{J}(u) \leq \mathscr{J}(u + \lambda v)$ whenever u solves (2.2), $\lambda \in \mathbb{R}$, $v \in H_0^1(\Omega \cup \Gamma)$. In the present chapter we shall not amplify this point; we shall instead return to it in Chapter 4.

The solution of (2.2) is of course not *a priori* required to be an element of $H^2(\Omega)$. Thus, in order to go back to the initial setting of problem (2.1), one has to tackle the nontrivial task of proving that (2.2), at least under convenient regularity assumptions about the data $\partial\Omega$, Γ, and f, ensures the additional regularity $u \in H^2(\Omega)$.

These considerations are behind the approach of the present chapter to differential problems such as (2.1).

We first generalize the Riesz representation theorem, passing from scalar products to wider classes of functionals $u, v \mapsto a(u, v)$, not necessarily satisfying $a(u, v) = a(v, u)$, on Hilbert spaces (Section 2.1). We then specialize with the space $H_0^1(\Omega \cup \Gamma)$ and study the applicability of previous abstract results to a class of problems that includes (2.2) (Section 2.2). Next we investigate various types of conditions on the data which guarantee greater regularity of solutions than mere membership in $H_0^1(\Omega \cup \Gamma)$. More specifically we set conditions in order that u belong to some space $L^s(\Omega)$ (Section 2.3), that $u \in C^{0,\delta}(\Omega)$ or $u \in C^{0,\delta}(\bar{\Omega})$ for some $\delta \in]0, 1[$ (Section 2.4), that $u \in H^k_{\mathrm{loc}}(\Omega)$ or $u \in H^k(\Omega)$, $k \geq 2$ (Section 2.5). Section 2.4 can be read independently of Section 2.3; Section 2.5, independently of Sections 2.3 and 2.4, except for Theorem 2.24, whose proof is omitted because it is similar to that for Theorem 2.19.

In Section 2.6 we take up nonlinear equations, proving some interior regularity results for their solutions.

2.1. Abstract Existence and Uniqueness Results

Let V be a Hilbert space. A *bilinear form* on V is a functional $u, v \mapsto a(u, v)$ on $V \times V$ which is linear in each variable; we call it

- *bounded* if

$$| a(u, v) | \leq M \, | u |_V \, | v |_V \qquad \text{for } u, v \in V \qquad (M > 0), \tag{2.3}$$

- *coercive* if

$$a(u, u) \geq \alpha_0 \, | u |_V{}^2 \qquad \text{for } u \in V \qquad (\alpha_0 > 0), \tag{2.4}$$

- *nonnegative* if

$$a(u, u) \geq 0 \qquad \text{for } u \in V,$$

- *symmetric* if

$$a(u, v) = a(v, u) \qquad \text{for } u, v \in V.$$

If a bilinear form $a(u, v)$ is bounded, all linear functionals

$$u \mapsto a(u, v) \qquad \text{with } v \text{ fixed in } V$$

and

$$v \mapsto a(u, v) \qquad \text{with } u \text{ fixed in } V$$

are elements of V'. Moreover, it is obvious that

$$a(u, v) = \lim_{n \to \infty} a(u_n, v_n)$$

whenever $u_n \to u$ and $v_n \to v$ in V. The same conclusion remains valid if either $u_n \to u$ and $v_n \rightharpoonup v$ in V, or $u_n \rightharpoonup u$ and $v_n \to v$ in V, since weakly convergent sequences are bounded and

$$| a(u_n, v_n) - a(u, v) | \leq M \, | u_n - u |_V \, | v_n |_V + | a(u, v_n - v) |,$$
$$| a(u_n, v_n) - a(u, v) | \leq | a(u_n - u, v) | + M \, | u_n |_V \, | v_n - v |_V .$$

If $a(u, v)$ is also supposed nonnegative, the inequalities

$$a(u_n - u, \; u_n - u) \geq 0 \qquad \text{for } n \in \mathbb{N}$$

clearly imply

$$a(u, u) \leq \liminf_{n \to \infty} a(u_n, u_n)$$

whenever $u_n \rightharpoonup u$ in V.

Under assumption (2.3) a linear operator $A: V \to V'$ with $| Au |_{V'} \le M | u |_V$ is defined by

$$\langle Au, v \rangle \equiv a(u, v) \qquad \text{for } v \in V$$

as u varies in V. Notice that $\{Au_n\}$ converges weakly in V' toward Au if $\{u_n\}$ converges weakly in V toward u.

Whenever $F \in V'$ the equation

$$u \in V, \qquad Au = F \tag{2.5}$$

can also be written as

$$u \in V, \qquad a(u, v) = \langle F, v \rangle \qquad \text{for } v \in V; \tag{2.6}$$

in the sequel we shall refer indifferently to either formulation (2.5) or (2.6), whichever is notationally more convenient.

The fundamental tool for the investigation of (2.6) is the *Lax–Milgram theorem*:

THEOREM 2.1. *Let $a(u, v)$ be a bounded and coercive bilinear form on V and let $F \in V'$. Then there exists a unique solution u of* (2.6); *moreover, u depends linearly on F and verifies*

$$| u |_V \le \alpha_0^{-1} | F |_{V'} \tag{2.7}$$

with α_0 from (2.4).

PROOF. We obtain (2.7) by choosing $v = u$ in (2.6) and taking (2.4) into account.

Uniqueness is a straightforward consequence of (2.7), since the difference of two solutions of (2.5) is a solution of the same equation with F replaced by 0.

Another consequence of (2.7), rewritten as

$$| u |_V \le \alpha_0^{-1} | Au |_{V'},$$

is that a sequence $\{u_n\} \subset V$ satisfies the Cauchy condition if $\{Au_n\}$ is a Cauchy sequence in V'. Suppose that $Au_n \to F$ in V' and set $u \equiv \lim_{n\to\infty} u_n$. Then $Au = F$ by the continuity of A. Thus, the image $A(V)$ of V under the linear map A is a closed subspace of V'. The proof will be complete if we show that $A(V)$ is dense in V' (the linearity of the map $F \mapsto u$ being obvious). To this end we fix any vector z in the dual space of V', which equals V by the reflexivity of Hilbert spaces. If $\langle F, z\rangle = 0$ whenever $F \in A(V)$, then in particular $\alpha_0 \mid z \mid_V^2 \leq \langle Az, z\rangle = 0$, that is, $z = 0$, and the Hahn–Banach theorem (see Theorem 1.B) yields the desired conclusion. □

From Theorem 2.1 with the particular choice of $a(u, v) = (u, v)_V$ [which implies (2.3) and (2.4) with $M = \alpha_0 = 1$] we obtain the Riesz representation theorem as a corollary.

COROLLARY. *For any choice of $F \in V'$ there exists a unique vector $u \in V$ satisfying*

$$(u, v)_V = \langle F, v\rangle \qquad \text{for } v \in V;$$

moreover, the isomorphism $\mathscr{I}$ from V' onto V defined by $\mathscr{I}F \equiv u$ verifies

$$\mid \mathscr{I}F \mid_V = \mid F \mid_{V'}.$$

We now suppose that V is continuously and densely injected into another Hilbert space H, so that H' is continuously and densely injected into V'. Upon identification of H' with H via the corresponding Riesz isomorphism, we obtain the scheme

$$V \underset{\text{ds}}{\hookrightarrow} H = H' \underset{\text{ds}}{\hookrightarrow} V', \tag{2.8}$$

which is referred to by saying that (V, H, V') is a *Hilbert triplet*. Notice that $(u, v)_H = \langle u, v\rangle$ for $u, v \in H$, in particular for $u, v \in V$, whereas $(u, v)_V = \langle \mathscr{I}^{-1}u, v\rangle$ for $u, v \in V$. Notice also that, if the original injection of V into H is compact, so is the injection (2.8) of V into V'.

Returning to bilinear forms on V, we weaken the notion of coerciveness as follows: we say that $a(u, v)$ is *coercive relative to H* if there exists some $\lambda > 0$ such that $a_\lambda(u, v) \equiv a(u, v) + \lambda\langle u, v\rangle$ is coercive, i.e.,

$$a(u, u) + \lambda \mid u \mid_H^2 \geq \alpha_0 \mid u \mid_V^2 \qquad \text{for } u \in V \qquad (\alpha_0 > 0). \tag{2.9}$$

Let A_λ: $u \mapsto Au + \lambda u$, $u \in V$. If (2.9) holds, A_λ has a bounded inverse A_λ^{-1}: $V' \to V$ by Theorem 2.1, and (2.5) can be rewritten as

$$u \in V, \qquad u - \lambda A_\lambda^{-1}u = z \tag{2.10}$$

with $z \equiv A_\lambda^{-1}F$. Let the injection $V \hookrightarrow H$ be compact, so that A_λ^{-1} is compact when considered as an operator $V \to V$. By the Fredholm alternative (see Theorem 1.L) (2.10) is uniquely solvable for any choice of $z \in V$ if and only if $u = 0$ is the unique vector of V satisfying $u - \lambda A_\lambda^{-1}u$

constant α of uniform ellipticity (see for instance G. Stampacchia [141]). Rather than enlarging on this approach, we proceed to investigate (2.6) in the light of Theorem 2.2 instead of Theorem 2.1. Note that the inequalities

$$\left| \int_\Omega d^j u u_{x_j}\, dx \right| \le \frac{\alpha}{4} |\nabla u|_{2;\Omega}^2 + \alpha^{-1} \sum_{j=1}^N |d^j|_{\infty;\Omega}^2 |u|_{2;\Omega}^2$$

and

$$\left| \int_\Omega b^i u_{x_i} u\, dx \right| \le \frac{\alpha}{4} |\nabla u|_{2;\Omega}^2 + \alpha^{-1} \sum_{i=1}^N |b^i|_{\infty;\Omega}^2 |u|_{2;\Omega}^2$$

yield (2.9) with

$$\lambda = \alpha^{-1} \sum_{i=1}^N (|d^i|_{\infty;\Omega}^2 + |b^i|_{\infty;\Omega}^2) + |c|_{\infty;\Omega} + \frac{\alpha}{2}, \qquad \alpha_0 = \frac{\alpha}{2}.$$

Thus, whatever the choice of V as above, the bilinear form (2.11) is coercive on V relative to $L^2(\Omega)$.

$(V, L^2(\Omega), V)$ is a Hilbert triplet. If the injection $V \subsetneq L^2(\Omega)$ is compact (see Theorem 1.34 and the remark following Lemma 1.46), the unique solvability of (2.6) for any choice of $F \in V'$ is an immediate consequence of Theorem 2.2 whenever it can be shown that (2.6) with $F = 0$ implies $u = 0$. In Section 2.2.2 we shall provide sufficient conditions for this.

2.2.2. The Weak Maximum Principle

Throughout the rest of this chapter we shall take as V the space $H_0^1(\Omega \cup \Gamma)$ with Γ at least of class C^1, neither case $\Gamma = \varnothing$ nor $\Gamma = \partial\Omega$ being excluded.

We say that the *weak maximum principle* holds for $A\colon H^1(\Omega) \to V'$ if any function $u \in H^1(\Omega)$ satisfying

$$\begin{aligned} u &\le 0 \quad \text{on } \partial\Omega \setminus \Gamma \text{ in the sense of } H^1(\Omega), \\ Au &\le 0 \quad [\text{i.e., } a(u, v) \le 0 \quad \text{for } v \in V, \quad v \ge 0] \end{aligned} \tag{2.13}$$

is ≤ 0.

The validity of the weak maximum principle implies naturally that in the present situation (2.6) has only the trivial solution when $F = 0$. The notion we have just introduced is, however, of extreme importance also when the unique solvability of (2.6) can be directly deduced from the Lax–Milgram theorem. We therefore explicitly state the following result, concerning the coercive case, which immediately follows from (2.13) with the choice $v = u^+$.

THEOREM 2.3. *If the bilinear form* (2.11) *on* $V = H_0^1(\Omega \cup \Gamma)$ *is coercive, the weak maximum principle holds for* $A\colon H^1(\Omega) \to V'$.

Things become more difficult when the coerciveness assumption is dropped. The result we have is the following theorem.

THEOREM 2.4. *Let* Ω *be such that* $H_0^1(\Omega \cup \Gamma) \subsetneq L^q(\Omega)$ *with* $q > 2$. *Let the operator* $A\colon H^1(\Omega) \to V'$ *from* (2.11) *satisfy* $A1 \geq 0$, *and in addition* $A1 \neq 0$ *if* $V = H_0^1(\Omega \cup \Gamma)$ *equals* $H^1(\Omega)$. *Then the weak maximum principle holds for* A.

PROOF. Let $u \in H^1(\Omega)$ satisfy (2.13), and suppose that $K \equiv \operatorname{ess\,sup}_\Omega u > 0$.

If u is the positive constant K, it coincides with $u^+ \in H_0^1(\Omega \cup \Gamma)$. But then all constants belong to the space $H_0^1(\Omega \cup \Gamma)$, which must coincide with $H^1(\Omega)$ by Lemma 1.58. Our assumptions imply the existence of some function $v \in V$, $v \geq 0$, such that

$$K\langle A1, v\rangle = a(u, v) > 0,$$

and this contradicts (2.13).

Since the possibility that u equals K throughout Ω has been ruled out, there exists $K_0 \in]0, K[$ such that the measure of the set $\Omega^* \equiv \{x \in \Omega \mid u(x) \leq K_0\}$ is positive (and of course independent of the choice of the representative of u).

Take any number k in the interval $[K_0, K[$, so that $u > k$ on a set of positive measure. Since the nonnegative function $v_k \equiv (u - k)^+ \leq u^+$ belongs to V, $v = v_k$ is admissible in (2.13), and the assumption $A1 \geq 0$ yields

$$\begin{aligned} 0 \geq a(u, v_k) - a(u - k, v_k) + k\langle A1, v_k\rangle &\geq a(u - k, v_k) \\ &= a(v_k, v_k). \end{aligned}$$

Hence

$$\begin{aligned} \int_\Omega a^{ij} v_{kx_i} v_{kx_j}\, dx &\leq -\int_\Omega [d^j v_k v_{kx_j} + (b^i v_{kx_i} + c v_k) v_k]\, dx \\ &\leq \frac{\alpha}{2} \int_\Omega |\nabla v_k|^2\, dx + C(\alpha) \int_\Omega v_k^2\, dx, \end{aligned}$$

and finally

$$| \nabla v_k |_{2;\Omega} \leq C \, | v_k |_{2;\Omega} \tag{2.14}$$

by uniform ellipticity. Thanks to Hölder's inequality (2.14) yields

$$| v_k |_{q;\Omega} \leq C \, | v_k |_{H^1(\Omega)} \leq C \, | v_k |_{2;\Omega} \leq C \, | \operatorname{supp} v_k |^{1/2-1/q} \, | v_k |_{q;\Omega},$$

hence

$$| \operatorname{supp} v_k | \geq C > 0$$

with C independent of k, since $| v_k |_{q;\Omega} \neq 0$. Letting $k \to K$ we deduce that the measure of the subset of Ω where $u < K$ is $< | \Omega |$: therefore, K is finite, and $u = K$ on a subset of Ω having positive measure.

Now denote by v_0 the bounded function v_{K_0}, and set $v^{(\varepsilon)} \equiv v_0/(K - K_0 + \varepsilon - v_0)$ for $\varepsilon > 0$. If G is any function from $C_c^1(\mathbb{R})$ such that $G(t) = t/(K - K_0 + \varepsilon - t)$ for $0 \leq t \leq K - K_0$, Lemma 1.57 applies and yields $v^{(\varepsilon)} \in V$. Since $v^{(\varepsilon)}$ vanishes wherever $(u - K_0)^+$ does, the same procedure followed for v_k yields the inequality

$$a(v_0, v^{(\varepsilon)}) \leq 0,$$

which can be rewritten as

$$\int_\Omega [a^{ij} v_{0x_i} v^{(\varepsilon)}_{x_j} - (d^i - b^i) v_{0x_i} v^{(\varepsilon)}] \, dx \leq - \int_\Omega [d^i (v_0 v^{(\varepsilon)})_{x_i} + c v_0 v^{(\varepsilon)}] \, dx$$
$$= - \langle A1, v_0 v^{(\varepsilon)} \rangle \leq 0$$

since the nonnegative function $v_0 v^{(\varepsilon)}$ belongs to V by the boundedness of v_0 and $v^{(\varepsilon)}$ (see Problem 1.21). Computation shows that

$$v^{(\varepsilon)}_{x_j} = (K - K_0 + \varepsilon) v_{0x_j} / (K - K_0 + \varepsilon - v_0)^2,$$

hence that

$$(K - K_0 + \varepsilon) \int_\Omega \frac{a^{ij} v_{0x_i} v_{0x_j}}{(K - K_0 + \varepsilon - v_0)^2} \, dx \leq \int_\Omega (d^i - b^i) v_{0x_i} v^{(\varepsilon)} \, dx$$
$$\leq C \int_\Omega v_0 \frac{| \nabla v_0 |}{K - K_0 + \varepsilon - v_0} \, dx,$$

and finally that

$$\alpha \int_\Omega | \nabla w^{(\varepsilon)} |^2 \, dx \leq \int_\Omega a^{ij} w^{(\varepsilon)}_{x_i} w^{(\varepsilon)}_{x_j} \, dx$$
$$\leq C \int_\Omega | \nabla w^{(\varepsilon)} | \, dx \leq C \, | \Omega |^{1/2} \left(\int_\Omega | \nabla w^{(\varepsilon)} |^2 \, dx \right)^{1/2} \tag{2.15}$$

with

$$w^{(\varepsilon)} \equiv \ln \frac{K - K_0 + \varepsilon}{K - K_0 + \varepsilon - v_0}.$$

Again by Lemma 1.57, $w^{(\varepsilon)}$ belongs to V, and $w^{(\varepsilon)} = 0$ on Ω^* because v_0 vanishes there. By (2.15)

$$| \nabla w^{(\varepsilon)} |_{2;\Omega} \leq C \qquad \text{for } \varepsilon > 0,$$

so that Lemma 1.36 yields a uniform bound on $| w^{(\varepsilon)} |_{2;\Omega}$, $\varepsilon > 0$. But then the monotone convergence theorem shows that

$$\ln \frac{K - K_0}{K - K_0 + v_0},$$

the limit as $\varepsilon \to 0^+$ of $\{w^{(\varepsilon)}\}$, is integrable over Ω. This implies that u cannot equal K on a set of positive measure, thus contradicting our previous conclusion based on the assumption $K > 0$. Hence, $K \leq 0$. □

COROLLARY. *In addition to the assumptions of Theorem* 2.4 *suppose that* $H_0^1(\Omega \cup \Gamma)$ *injects compactly into* $L^2(\Omega)$. *Then* (2.6) *admits a unique solution for any choice of* $F \in [H_0^1(\Omega \cup \Gamma)]'$.

The scope of the considerations developed up until now can be appreciated more fully with the help of the following example.

EXAMPLE. Let $N = 1$, $\Omega =]0, R[$ with $0 < R < \infty$, $\Gamma = \varnothing$. On $H_0^1(\Omega)$ consider the bilinear form

$$a(u, v) = \int_0^R (u'v' + \lambda uv)\, dx.$$

If $\lambda \geq 0$, $a(u, v)$ is coercive. But if λ takes on a value $-\pi^2 n^2/R^2$, $n \in \mathbb{N}$, the function $u(x) \equiv \sin(\pi n x/R)$ is an element of $H_0^1(\Omega)$ satisfying $Au = 0$, so that the weak maximum principle does not hold.

2.2.3. Interpretation of Solutions

By choosing the space $V = H_0^1(\Omega \cup \Gamma)$ and the bilinear form (2.11), the following properties of a solution u (if it exists) to (2.6) are immediately ascertained.

First of all, u is an element of $H^1(\Omega)$ satisfying

$$a(u, v) = \langle F, v\rangle = \langle f, v\rangle \qquad \text{for } v \in H_0^1(\Omega),$$

where $f \in H^{-1}(\Omega)$ is the restriction of $F \in V'$ to $H_0^1(\Omega)$ ($f = F$ if $\Gamma = \emptyset$). This is expressed by saying that u is a (*variational*) *solution* of the *ordinary* (for $N = 1$) or *partial* (for $N > 1$) *differential equation*

$$Lu = f \qquad \text{in } \Omega$$

with *free term* f.

Secondly, the membership of u in V contains, if $\partial\Omega \setminus \Gamma \neq \emptyset$, the condition

$$u = 0 \qquad \text{on } \partial\Omega\setminus\Gamma \text{ in the sense of } H^1(\Omega),$$

which amounts to

$$u|_{\partial\Omega\setminus\Gamma} = 0$$

if $\partial\Omega$ is of class C^1 and Γ is closed. We express the above by saying that u satisfies a (*homogeneous*) *Dirichlet condition* on $\partial\Omega\setminus\Gamma$.

In order to investigate the behavior of u on Γ, supposed $\neq \emptyset$, it is convenient to deal with the following assumptions: Γ is closed and, for some $p \in]1, \infty[$, u is a function of $H^{1,p}(\Omega)$ which verifies $Lu \in L^p(\Omega)$. The linear functional $v \mapsto a(u, v) - \int_\Omega (Lu)v\, dx$ is then bounded on $H_0^{1,p'}(\Omega \cup \Gamma)$ and vanishes identically on $H_0^{1,p'}(\Omega)$. A bounded linear functional Bu on $H^{1/p,p'}(\Gamma)$ is therefore defined by the expression

$$\langle Bu, v|_\Gamma\rangle \equiv a(u, v) - \int_\Omega (Lu)v\, dx \qquad \text{for } v \in H_0^{1,p'}(\Omega \cup \Gamma) \tag{2.16}$$

which we refer to as *Green's formula*; Bu is said to be the *conormal derivative* of u on Γ, relative to $a(u, v)$. It is important to remark that, under suitable hypotheses, Bu can be given a more explicit expression than its mere definition (2.16). For, suppose that $a^{ij}, d^j \in C^{0,1}(\bar\Omega)$ and $u \in H^{2,p}(\Omega)$: then, $a^{ij}u_{x_i} + d^ju$ belongs to $H^{1,p}(\Omega)$ for $j = 1, \ldots, N$, and the divergence theorem (see Theorem 1.53) yields

$$\begin{aligned} a(u, v) - \int_\Omega (Lu)v\, dx &= \int_\Omega [(a^{ij}u_{x_i} + d^ju)v]_{x_j}\, dx \\ &= \int_\Gamma (a^{ij}u_{x_i} + d^ju)|_\Gamma\, v|_\Gamma\, \nu^j\, d\sigma \end{aligned}$$

for $v \in H_0^{1,p'}(\Omega \cup \Gamma)$, hence

$$Bu = (a^{ij}u_{x_i} + d^j u)|_\Gamma \nu^j. \tag{2.17}$$

[It would in fact be appropriate to say that (2.17) is Green's formula, Bu being defined by (2.16).]

Summing up, we have the following lemma.

LEMMA 2.5. *Let Γ be nonempty and closed. Let $1 < p < \infty$. Then any $u \in H^{1,p}(\Omega)$ with $Lu \in L^p(\Omega)$ admits a conormal derivative $Bu \in [H^{1/p,p'}(\Gamma)]'$ defined by* (2.16). *If, moreover, $u \in H^{2,p}(\Omega)$ and $a^{ij}, d^j \in C^{0,1}(\bar{\Omega})$, then Bu satisfies* (2.17).

At this point we can return to the interpretation of solutions u to (2.6). If F has the expression

$$\langle F, v\rangle = \int_\Omega fv\,dx + \langle \zeta, v|_\Gamma\rangle \qquad \text{for } v \in V,$$
$$\text{with } f \in L^2(\Omega) \quad \text{and} \quad \zeta \in [H^{1/2}(\Gamma)]', \tag{2.18}$$

then u verifies $Lu \in L^2(\Omega)$, and its conormal derivative satisfies the so-called *Neumann condition*

$$Bu = \zeta \qquad \text{on } \Gamma$$

as an identity in $[H^{1/2}(\Gamma)]'$ {or even a.e. $[N-1]$, if for instance $\zeta \in L^2(\Gamma)$}. Since this procedure can be inverted with no difficulty, we have proved

LEMMA 2.6. *Let Γ be closed and assume* (2.18). *Then a function $u \in H^1(\Omega)$ satisfies* (2.6) *with $V = H_0^1(\Omega \cup \Gamma)$ and $a(u, v)$ given by* (2.11) *if and only if it satisfies*

$$\begin{aligned} Lu &= f \quad && \text{in } \Omega, \\ u &= 0 && \text{on } \partial\Omega\setminus\Gamma \text{ in the sense of } H^1(\Omega), \\ Bu &= \zeta && \text{on } \Gamma. \end{aligned} \tag{2.19}$$

We call (2.19) a (*variational*) *boundary value problem* (henceforth b.v.p.); we say that it is of the *mixed* type, if neither Γ nor $\partial\Omega\setminus\Gamma$ is empty, of the *Dirichlet* type if $\Gamma = \emptyset$, of the *Neumann* type if $\partial\Omega\setminus\Gamma = \emptyset$.

REMARK. Assume $\partial\Omega\setminus\Gamma \neq \emptyset$ and let $g \in H^1(\Omega)$. If the hypotheses of Theorem 2.6 are satisfied and the function z solves

$$z \in V,$$
$$a(z, v) = \langle F - Ag, v\rangle \qquad \text{for } v \in V,$$

the function $u = z + g$ solves the b.v.p.

$$\begin{aligned} Lu &= f && \text{in } \Omega, \\ u &= g && \text{on } \partial\Omega\setminus\Gamma \text{ in the sense of } H^1(\Omega), \\ Bu &= \zeta && \text{on } \Gamma, \end{aligned}$$

the condition on $\partial\Omega\setminus\Gamma$ being a *nonhomogeneous Dirichlet condition.*

2.3. L^s Regularity of Solutions

Throughout the rest of this chapter and the first five sections of the next we shall investigate the regularity of solutions to problems such as (2.6) with $V = H_0^1(\Omega \cup \Gamma)$ and $a(u, v)$ given by (2.11). For the purposes of clarification we begin with a few simple observations.

The form (2.11) may or may not be coercive; in any case, however, there exists $\lambda \geq 0$ such that $u, v \mapsto a(u, v) + \lambda\langle u, v\rangle$ is coercive (see Section 2.2.1). We rewrite (2.6) as

$$u \in V, \qquad a(u, v) + \lambda\langle u, v\rangle = \langle F + \lambda u, v\rangle \qquad \text{for } v \in V, \tag{2.20}$$

and deduce from (2.7) the norm estimate

$$|u|_{H^1(\Omega)} \leq \alpha_0^{-1}(|F|_{V'} + \lambda |u|_{2;\Omega})$$

with α_0 from (2.9).

When $N = 1$, $H^1(\Omega)$ is continuously injected into $C^{0,1/2}(\bar{\Omega})$, so that

$$|u|_{C^{0,1/2}(\bar{\Omega})} \leq C(|F|_{V'} + |u|_{2;\Omega}).$$

When $N \geq 2$ we can give sufficient conditions on Γ in order that $V \hookrightarrow L^r(\Omega)$, hence

$$|u|_{r;\Omega} \leq C(|F|_{V'} + |u|_{2;\Omega}), \tag{2.21}$$

with r arbitrarily fixed in $]2, \infty[$ if $N = 2$, $r = 2^* = 2N/(N-2)$ otherwise (see Theorem 1.33 and the remark at the end of Section 1.7.2). For what concerns $|F|_{V'}$ we note that, whenever $q^{-1} - p^{-1} + N^{-1}$ with $p > 2$ if $N = 2$ and $p \geq 2$ if $N \geq 3$, $L^q(\Omega)$ is continuously injected into V'; we define a bounded linear functional F on V by setting

$$\begin{aligned} \langle F, v\rangle &\equiv \int_\Omega (f^0 v + f^i v_{x_i})\, dx \qquad \text{for } v \in V \\ \text{with } f^0 &\in L^q(\Omega), \qquad f^i \in L^p(\Omega) \qquad \text{for } i = 1, \ldots, N, \end{aligned} \tag{2.22}$$

and (2.21) becomes

$$|u|_{r;\Omega} \leq C\left(|f^0|_{q;\Omega} + \sum_{i=1}^N |f^i|_{p;\Omega} + |u|_{2;\Omega}\right).$$

Up to now the fact that u solves (2.6) has played a role only in the norm estimate, whereas the regularity of u has been deduced from general properties of V. In the rest of this section we shall give sufficient conditions in order that the validity of (2.6) imply $u \in L^s(\Omega)$, with norm estimate, for some $s > r$.

First we have the following theorem.

Theorem 2.7. *Let $N \geq 2$, and suppose Γ is such that $H_0^1(\Omega \cup \Gamma) \subsetneq L^r(\Omega)$ for $r = 2^*$ if $N > 2$, $r \in]2, \infty[$ arbitrary if $N = 2$. Let the bounded linear functional F on $V = H_0^1(\Omega \cup \Gamma)$ be defined by* (2.22) *with $p > N$, $q = pN/(N+p)$ if $N > 2$, $q > p2/(2+p)$ if $N = 2$, and let $a(u, v)$ be given by* (2.11). *Then any solution u of* (2.6) *belongs to $L^\infty(\Omega)$; moreover, there exists a constant C (independent of u, F) such that*

$$|u|_{\infty;\Omega} \leq C\left(|f^0|_{q;\Omega} + \sum_{i=1}^N |f^i|_{p;\Omega} + |u|_{2;\Omega}\right).$$

We shall obtain Theorem 2.7 as a straightforward consequence of the following lemma.

Lemma 2.8. *Under the same assumptions of Theorem 2.7, any function $u \in H^1(\Omega)$ such that*

$$\begin{aligned} u &\leq 0 \qquad \text{on } \partial\Omega \setminus \Gamma \text{ in the sense of } H^1(\Omega), \\ a(u, v) &\leq \langle F, v\rangle \qquad \text{for } v \in V, \qquad v \geq 0 \end{aligned} \tag{2.23}$$

satisfies

$$\operatorname*{ess\,sup}_{\Omega} u \leq C\left(| f^0 |_{q;\Omega} + \sum_{i=1}^{N} | f^i |_{p;\Omega} + | u |_{2;\Omega} \right),$$

C being independent of u, F.

PROOF. Since (2.23) remains valid with u and F replaced respectively by $u/| u |_{2;\Omega}$ and $F/| u |_{2;\Omega}$ if $| u |_{2;\Omega} \neq 0$, we need consider only the case $| u |_{2;\Omega} = 1$.

For $0 < k < \infty$ we set $v_k \equiv (u - k)^+$, $\Omega_k \equiv \{x \in \Omega \mid v_k > 0\}$; notice that

$$| \Omega_k | \leq | u |_{1;\Omega}/k \leq | \Omega |^{1/2}/k$$

by Hölder's inequality. We restrict ourselves to the values of k for which $| \Omega_k | < 1$.

Since $v_k \in V$ and $v_k \geq 0$, (2.23) yields

$$a(v_k, v_k) = a(u - k, v_k) \leq \langle F - Ak, v_k \rangle.$$

Set $\theta \equiv | \nabla v_k |_{2;\Omega}$. By uniform ellipticity,

$$\begin{aligned} \alpha\theta^2 \leq & -\int_{\Omega_k} [d^j v_k v_{kx_j} + (b^i v_{kx_i} + c v_k) v_k] \, dx \\ & + \int_{\Omega_k} [(f^0 - kc) v_k + (f^i - kd^i) v_{kx_i}] \, dx, \end{aligned}$$

so that Hölder's inequality yields

$$\begin{aligned} \theta^2 \leq C\Big[& | v_k |_{2;\Omega} (\theta + | v_k |_{2;\Omega} + | \Omega_k |^{1/2-1/q} | f^0 |_{q;\Omega} + k | \Omega_k |^{1/2}) \\ & + \theta\Big(| \Omega_k |^{1/2-1/p} \sum_{i=1}^{N} | f^i |_{p;\Omega} + k | \Omega_k |^{1/2} \Big) \Big]. \end{aligned} \tag{2.24}$$

We now utilize the continuous imbedding of V into $L^r(\Omega)$, with r so large that $1/r \leq 1/2 + 1/p - 1/q$ and $(1/2 - 1/p)r > 1$ if $N = 2$ (the same inequalities being obviously satisfied by $r = 2^*$ if $N \geq 3$), to obtain

$$\begin{aligned} | v_k |_{r;\Omega} & \leq C(| v_k |_{2;\Omega} + \theta) \\ & \leq C(| \Omega_k |^{1/2-1/r} | v_k |_{r;\Omega} + \theta), \end{aligned}$$

hence also

$$| v_k |_{r;\Omega} \leq C\theta \qquad \text{for } k \geq k_0 \tag{2.25}$$

if $k_0 > 0$ is large enough. Since $1/2 - 1/p \leq 1 - 1/r - 1/q$ and $|\Omega_k| < 1$ we deduce from (2.24) that

$$\theta^2 \leq C\Big[|\Omega_k|^{1/2-1/r}|v_k|_{r;\Omega}(\theta + |\Omega_k|^{1/2-1/r}|v_k|_{r;\Omega} + |\Omega_k|^{1/2-1/q}|f^0|_{q;\Omega} + k|\Omega_k|^{1/2}) + \theta\Big(|\Omega_k|^{1/2-1/p}\sum_{i=1}^N |f^i|_{p;\Omega} + k|\Omega_k|^{1/2}\Big)\Big]$$
$$\leq C[\theta^2|\Omega_k|^{1/2-1/r} + \theta(|\Omega_k|^{1/2-1/p}\tau + k|\Omega_k|^{1/2})],$$

where

$$\tau \equiv |f^0|_{q;\Omega} + \sum_{i=1}^N |f^i|_{p;\Omega};$$

if $k_1 \geq k_0$ is large enough, we have

$$\theta \leq C|\Omega_k|^{1/2-1/p}(\tau + k) \qquad \text{for } k \geq k_1. \tag{2.26}$$

At this point we utilize (2.25), (2.26), and the inequalities

$$|v_k|_{r;\Omega}^r = \int_{\Omega_k}(u-k)^r\,dx \geq \int_{\Omega_h}(u-k)^r\,dx \geq |\Omega_h|(h-k)^r,$$

valid for $k < h < \infty$, to arrive at

$$|\Omega_h|(h-k)^r \leq \hat{C}|\Omega_k|^{(1/2-1/p)r}(\tau+k)^r,$$

which we rewrite as

$$|\Omega_h| \leq K(h-k)^{-r}|\Omega_k|^\beta$$

with $K \equiv \hat{C}(\tau + h_1)^r$ and $\beta \equiv (1/2 - 1/p)r > 1$, for $k_1 < k < h < h_1$, $h_1 < \infty$.

Let us assume the validity of the following lemma.

Lemma 2.9. *Let φ be a nonnegative, nonincreasing function of $k \in]k_1, h_1[$, where $k_1 < h_1 \leq \infty$. Suppose that there exist positive constants K, r, β, with $\beta > 1$, such that*

$$\varphi(h) \leq K(h-k)^{-r}\varphi(k)^\beta$$

for $k_1 < k < h < h_1$. If the number

$$\hat{k} \equiv K^{1/r}2^{\beta/(\beta-1)}\varphi(k_1)^{(\beta-1)/r}.$$

is such that $k_1 + \hat{k} < h_1$, then $\varphi(k_1 + \hat{k}) = 0$.

We can apply Lemma 2.9 to $\varphi(k) = |\Omega_k|$, provided we find k_1 and h_1 with the required properties.

We let k_1 be so large that not only (2.26) holds but also

$$\hat{C}^{1/r}2^{\beta/(\beta-1)}\,|\,\Omega_{k_1}|^{(\beta-1)/r} \leq \hat{C}^{1/r}2^{\beta/(\beta-1)}\,|\,\Omega\,|^{(\beta-1)/2r}k_1{}^{(1-\beta)/r} < 1/2.$$

We then choose $h_1 = 2k_1 + \tau$ and obtain

$$k_1 + \hat{k} = k_1 + \hat{C}^{1/r}2^{\beta/(\beta-1)}\,|\,\Omega_{k_1}|^{(\beta-1)/r}(\tau + h_1) < k_1 + \tfrac{1}{2}(\tau + h_1) = h_1.$$

We can conclude that $|\,\Omega_{k_1+\hat{k}}\,| = 0$, that is, $u \leq k_1 + \hat{k}$ in Ω. This is nothing but the required bound on $\operatorname{ess\,sup}_\Omega u$ for $|\,u\,|_{2;\Omega} = 1$. □

Remark. In the above proof we did not utilize (2.23) in its full strength. Indeed, we exploited only the fact that

$$a\big(u, (u-k)^+\big) \leq \langle F, (u-k)^+\rangle$$

for all k sufficiently large.

Note that the constant of the estimate does not depend on the coefficient c if the latter is ≥ 0.

At this point we need only proceed to the proof of Lemma 2.9.

Proof of Lemma 2.9. For $n \in N$ let

$$k_n \equiv k_1 + \hat{k}(1 - 2^{-(n-1)}),$$

so that $\varphi(k_1 + \hat{k}) \leq \varphi(k_n)$. We shall prove Lemma 2.9 by showing that $\varphi(k_n) \to 0$ as $n \to \infty$. More precisely, we shall show by induction that

$$\varphi(k_n) \leq \varphi(k_1)/2^{\mu(n-1)} \tag{2.27}$$

where μ is the positive number $r/(\beta - 1)$.

For $n = 1$, (2.27) is obviously satisfied. If (2.27) holds for some value of n, the assumption of the lemma yields

$$\begin{aligned}\varphi(k_{n+1}) &\leq K2^{rn}\hat{k}^{-r}\varphi(k_n)^\beta\\ &\leq K2^{rn}\hat{k}^{-r}\varphi(k_1)^\beta 2^{-\beta\mu(n-1)}.\end{aligned}$$

But since

$$\hat{k}^r = K2^{r\beta/(\beta-1)}\varphi(k_1)^{\beta-1},$$

we can conclude that

$$\varphi(k_{n+1}) \leq 2^{rn-\beta\mu(n-1)}2^{-\beta\mu}\varphi(k_1) = \varphi(k_1)/2^{\mu n}. \qquad \square$$

We now prove the following theorem.

THEOREM 2.10. *Same assumptions as in Theorem 2.7, except that here we take $N > p \geq 2$. Then any solution of (2.6) belongs to $L^{p^*}(\Omega)$, $1/p^* \equiv 1/p - 1/N$, with norm estimate*

$$|u|_{p^*;\Omega} \leq C\left(|f^0|_{q;\Omega} + \sum_{i=1}^{N} |f^i|_{p;\Omega} + |u|_{2;\Omega}\right), \tag{2.28}$$

the constant C being independent of u, F.

PROOF. We proceed in two steps. Note that for $p = 2$ the theorem is an immediate consequence of the assumption $H_0^1(\Omega \cup \Gamma) \hookrightarrow L^{2^*}(\Omega)$.

Step 1: *A preliminary reduction.* Let us momentarily assume the validity of a norm estimate

$$|u|_{p^*;\Omega} \leq C\left(|f^0|_{q;\Omega} + \sum_{i=1}^{N} |f^i|_{p;\Omega}\right) \tag{2.29}$$

for all functions $u \in L^\infty(\Omega)$ that satisfy

$$u \in V, \qquad a(u, v) + \lambda\langle u, v\rangle = \langle F, v\rangle \qquad \text{for } v \in V \tag{2.30}$$

with λ sufficiently large; we can suppose that the bilinear form $u, v \mapsto a(u, v) + \lambda\langle u, v\rangle$ is coercive. We claim that, as a consequence, (2.28) holds for solutions of (2.6). To substantiate our claim, we first prove that (2.29) remains valid even if the solution of (2.30) does not belong to $L^\infty(\Omega)$. We approximate f^0 in $L^q(\Omega)$, $f^1, \ldots, f^N$ in $L^p(\Omega)$ with sequences $\{f_n^0\}$, $\{f_n^1\}, \ldots, \{f_n^N\}$ of bounded functions and denote by u_n the solution of (2.30) with F replaced by F_n: $v \mapsto \int_\Omega (f_n^0 v + f_n^i v_{x_i})\,dx$. Each u_n belongs to $L^\infty(\Omega)$ by Theorem 2.7, so that (2.29) yields

$$|u_n|_{p^*;\Omega} \leq C\left(|f_n^0|_{q;\Omega} + \sum_{i=1}^{N} |f_n^i|_{p;\Omega}\right) \qquad \text{for } n \in \mathbb{N}. \tag{2.31}$$

By the uniform boundedness of $|u_n|_V$ [see (2.7)], a subsequence of $\{u_n\}$ converges weakly in V toward a function $\hat{u}$: it is clear that $\hat{u}$ solves (2.30), hence $\hat{u} = u$ by uniqueness. We deduce that u is the weak limit in $L^{p^*}(\Omega)$ of $\{u_n\}$, with

$$| u |_{p^*;\Omega} \leq \liminf_{n\to\infty} | u_n |_{p^*;\Omega} \leq C\Big(| f^0 |_{q;\Omega} + \sum_{i=1}^{N} | f^i |_{p;\Omega}\Big)$$

by (2.31) (see Problem 1.2).

Let us turn to solutions of (2.6), and consequently of (2.20), instead of (2.30). The function $\hat{f}^0 \equiv f^0 + \lambda u$ belongs to $L^{q_1}(\Omega)$ with

$$| \hat{f}^0 |_{q_1;\Omega} \leq C(| f^0 |_{q;\Omega} + | u |_{2;\Omega}),$$

where $q_1 \equiv q \wedge 2$. Let p_1 be defined by $p_1^{-1} \equiv q_1^{-1} - N^{-1}$: from our previous considerations about solutions of (2.30) it follows that $u \in L^{p_1}(\Omega)$ with norm estimate. If $p_1 = p$ we have obtained (2.28); if not, we repeat the above procedure, with 2 replaced by p_1^*, and so on through a finite number of stages, until we reach the sought-for conclusion. (This procedure, called a *bootstrap argument*, will be met with again.) The claim is thus substantiated.

Step 2: *Proof of* (2.29) *for solutions* $u \in L^\infty(\Omega)$ *of* (2.30). If u belongs to $V \cap L^\infty(\Omega)$, so does $| u |^\delta u$, with

$$(| u |^\delta u)_{x_i} = (\delta + 1) | u |^\delta u_{x_i}$$

whenever $\delta \geq 0$. [For $\delta > 0$ apply Lemma 1.57, with $G \in C_c^1(\mathbb{R})$, $G(t) = | t |^\delta t$ if $| t | \leq | u |_{\infty;\Omega}$.] We fix δ through the requirement $2^*(\delta + 2)/2 = p^*$ and estimate

$$\begin{aligned}\alpha(\delta + 1) &\int_\Omega | u |^\delta | \nabla u |^2 \, dx \\ &\leq \int_\Omega a^{ij} u_{x_i} (\delta + 1) | u |^\delta u_{x_j} \, dx \\ &= a(u, | u |^\delta u) - \int_\Omega [d^j u(\delta + 1) | u |^\delta u_{x_j} + (b^i u_{x_i} + cu) | u |^\delta u] \, dx \\ &\leq a(u, | u |^\delta u) + \varepsilon \int_\Omega | u |^\delta | \nabla u |^2 \, dx + C(\varepsilon) \int_\Omega u^2 | u |^\delta \, dx.\end{aligned}$$

Since $(\delta/2 + 1)^2 | u |^\delta | \nabla u |^2 = \big| \nabla(| u |^{\delta/2} u)\big|^2$, we can choose $\varepsilon > 0$ and $\lambda = \lambda(\varepsilon)$ in such a way that

$$\big\| | u |^{\delta/2} u \big\|_{H^1(\Omega)}^2 \leq C[a(u, | u |^\delta u) + \lambda\langle u, | u |^\delta u\rangle].$$

If u solves (2.30) we repeatedly make use of Hölder's inequality and obtain, for $\varepsilon > 0$,

$$\begin{aligned}
\big\| \, |u|^{\delta/2}u \big\|^2_{H^1(\Omega)} &\le C \int_\Omega [f^0 \, |u|^\delta u + f^i(\delta+1) \, |u|^\delta u_{x_i}] \, dx \\
&\le C \int_\Omega \Big(|f^0| \, |u|^{\delta+1} + |f^i| \, |u|^{\delta/2} \big| (|u|^{\delta/2}u)_{x_i} \big| \Big) \, dx \\
&\le C \bigg[|f^0|_{q;\Omega} \, |u|^{\delta+1}_{p^*;\Omega} \\
&\quad + \bigg(\int_\Omega \sum_{i=1}^N |f^i|^2 \, |u|^\delta \, dx \bigg)^{1/2} \bigg(\int_\Omega \big| \nabla(|u|^{\delta/2}u) \big|^2 \, dx \bigg)^{1/2} \bigg] \\
&\le C \bigg(|f^0|_{q;\Omega} \, |u|^{\delta+1}_{p^*;\Omega} \\
&\quad + \frac{1}{2\varepsilon} \sum_{i=1}^N |f^i|^2_{p;\Omega} \, |u|^\delta_{p^*;\Omega} + \frac{\varepsilon}{2} \big\| \, |u|^{\delta/2}u \big\|^2_{H^1(\Omega)} \bigg),
\end{aligned}$$

since

$$\delta = \frac{N(p-2)}{N-p}$$

and therefore

$$1 - \frac{1}{q} = \frac{\delta+1}{p^*}, \qquad 1 - \frac{2}{p} = \frac{\delta}{p^*}.$$

At this point we set a suitable value of ε and majorize the quantity

$$\begin{aligned}
|u|^{\delta+2}_{p^*;\Omega} &= \bigg(\int_\Omega |u|^{p^*} \, dx \bigg)^{(\delta+2)/p^*} \\
&= \bigg(\int_\Omega |u|^{[(\delta+2)/2]2^*} \, dx \bigg)^{2/2^*} = \big| \, |u|^{\delta/2}u \big|^2_{2^*;\Omega}
\end{aligned}$$

with $C \big\| \, |u|^{\delta/2}u \big\|^2_{H^1(\Omega)}$, thus obtaining

$$|u|^{\delta+2}_{p^*;\Omega} \le C \bigg(|f^0|_{q;\Omega} \, |u|^{\delta+1}_{p^*;\Omega} + \sum_{i=1}^N |f^i|^2_{p;\Omega} \, |u|^\delta_{p^*;\Omega} \bigg).$$

After dividing by $|u|^\delta_{p^*;\Omega} \ne 0$ we arrive at

$$|u|^2_{p^*;\Omega} \le C \bigg(\frac{\varepsilon}{2} |u|^2_{p^*;\Omega} + \frac{1}{2\varepsilon} |f^0|^2_{q;\Omega} + \sum_{i=1}^N |f^i|^2_{p;\Omega} \bigg) \qquad \text{for } \varepsilon > 0,$$

hence (2.29) after another suitable choice of ε. □

2.4. The De Giorgi–Nash Theorem

Throughout this section we shall take $N \geq 3$. This restriction will be briefly commented upon in the remark following the proof of Lemma 2.15.

2.4.1. Pointwise Bounds on Subsolutions

A function $W \in H^1(\Omega)$ is a (*variational*) *subsolution* of the equation $-(a^{ij}w_{x_i})_{x_j} = 0$ in Ω, where the a^{ij}'s are the leading coefficients of the bilinear form (2.11), if the distribution $-(a^{ij}W_{x_i})_{x_j}$ is a nonpositive element of $H^{-1}(\Omega)$, that is

$$\int_\Omega a^{ij}W_{x_i}v_{x_j}\,dx \leq 0 \qquad \text{for } v \in H_0{}^1(\Omega), \qquad v \geq 0.$$

An important property of subsolutions, whose proof we postpone until later (see Lemma 4.28 below), is that the supremum of two of them is still a subsolution.

LEMMA 2.11. *Let W be a nonnegative subsolution and assume $x^0 \in \Omega$, $B_{2r}(x^0) \subseteq \Omega$ $(r > 0)$. Then a bound*

$$W(x) \leq C\varrho^{-N/2}\,|\,W\,|_{2;x^0,r+\varrho} \qquad \textit{for a.a. } x \in B_r(x^0),$$

$0 < \varrho \leq r$, *is valid; the constant C (independent of W, x^0, and r) depends on the a^{ij}'s only through α and the bound imposed on $|\,a^{ij}\,|_{\infty;\Omega}$.*

In the proof of Lemma 2.11 we shall utilize the following result.

LEMMA 2.12. *Let W be a nonnegative subsolution and set $W_k \equiv W \wedge k$ for $0 < k < \infty$. Assume $x^0 \in \Omega$, $B_{2r}(x^0) \subseteq \Omega$ $(r > 0)$. Then a bound*

$$\left(\int_{B_\varrho(x^0)} W^2 W_k{}^{p\lambda+2(\lambda-1)}\,dx\right)^{1/\lambda} \leq \frac{C(1+p)}{(R-\varrho)^2}\int_{B_R(x^0)} W^2 W_k{}^p\,dx, \qquad (2.32)$$

where $\lambda \equiv N/(N-2)$, holds whenever $0 \leq p < \infty$ and $0 < \varrho < R \leq 2r$; the constant C is independent of W, x^0, r, and k.

PROOF. Without loss of generality we assume $x^0 = 0$. Let $g \in C^1(\bar{\Omega})$ with supp $g \subset B_R$, $0 \leq g \leq 1$, $g = 1$ on $\bar{B}_\varrho$, $|\,\nabla g\,| \leq 2(R-\varrho)^{-1}$. Since W_k belongs to $H^1(\Omega) \cap L^\infty(\Omega)$ with $W_{kx_j} = W_{x_j}\chi_{W<k}$, $\chi_{W<k}$ being the

characteristic function of the subset of Ω where $W < k$ (see Theorem 1.56), the nonnegative function $v = g^2 W_k^p W$ belongs to $H_0^1(\Omega)$ with

$$v_{x_j} = g^2 W_k^p (p W_{kx_j} + W_{x_j}) + 2 g g_{x_j} W_k^p W.$$

[Write $W_k = (W - k) \wedge 0 + k$, $W = (W - k) \vee 0 + (W - k) \wedge 0 + k$.] Thus,

$$\int_\Omega a^{ij} W_{x_i} g^2 W_k^p (p W_{kx_j} + W_{x_j})\, dx \le -2 \int_\Omega a^{ij} W_{x_i} g g_{x_j} W_k^p W\, dx,$$

and

$$\begin{aligned}
\alpha \int_\Omega & g^2 W_k^p (p \,|\, \nabla W_k \,|^2 + |\, \nabla W \,|^2)\, dx \\
&\le C \int_\Omega g \,|\, \nabla W \,|\, W_k^p W \,|\, \nabla g \,|\, dx \\
&\le \varepsilon \int_\Omega g^2 \,|\, \nabla W \,|^2 W_k^p\, dx + C(\varepsilon) \int_\Omega W_k^p W^2 \,|\, \nabla g \,|^2\, dx,
\end{aligned}$$

$\varepsilon > 0$. Take $\varepsilon = \alpha/2$: then the gradient of the function $\hat{W} = W_k^{p/2} W$ satisfies

$$\begin{aligned}
\int_\Omega g^2 \,|\, \nabla \hat{W} \,|^2\, dx &\le 2 \int_\Omega g^2 W_k^p \Big(\frac{p^2}{4} \,|\, \nabla W_k \,|^2 + |\, \nabla W \,|^2 \Big)\, dx \\
&\le \Big(\frac{p}{2} + 2 \Big) \int_\Omega g^2 W_k^p (p \,|\, \nabla W_k \,|^2 + |\, \nabla W \,|^2)\, dx \\
&\le C(1 + p) \int_\Omega \hat{W}^2 \,|\, \nabla g \,|^2\, dx,
\end{aligned}$$

so that

$$\int_{B_R} |\, \nabla (g \hat{W}) \,|^2\, dx \le C(1 + p) \int_{B_R} \hat{W}^2 \,|\, \nabla g \,|^2\, dx. \tag{2.33}$$

Since

$$|\, g \hat{W} \,|_{2\lambda; R} \le C \,|\, \nabla (g \hat{W}) \,|_{2; R}$$

by Theorem 1.33 and the corollary of Theorem 1.43, (2.33) yields

$$\Big(\int_{B_\varrho} \hat{W}^{2\lambda}\, dx \Big)^{1/\lambda} \le \frac{C(1 + p)}{(R - \varrho)^2} \int_{B_R} \hat{W}^2\, dx.$$

The conclusion now follows from the inequality $W_k \leq W$, which implies $W^2 W_k^{2(\lambda-1)} \leq W^{2\lambda}$ and therefore

$$W^2 W_k^{p\lambda+2(\lambda-1)} \leq W_k^{p\lambda} W^{2\lambda} = \hat{W}^{2\lambda}. \qquad \square$$

PROOF OF LEMMA 2.11. Take $x^0 = 0$. For $m = 0, 1, 2, \ldots$ we set

$$r_m \equiv r\left(1 + \frac{1}{2^m}\right), \qquad p_m \equiv 2(\lambda^m - 1),$$

$$A_m \equiv \left(\int_{B_{r_m}} W^2 W_k^{p_m}\, dx\right)^{1/\lambda^m}.$$

By choosing $R = r_m$, $\varrho = r_{m+1}$ and $p = p_m$ in (2.32) we obtain

$$A_{m+1} \leq A_m \left[\frac{C(2\lambda^m - 1)^{1/2}}{r2^{-m-1}}\right]^{2/\lambda^m}$$

hence

$$A_{m+1} \leq A_0 \prod_{i=0}^{m} \left(\frac{C2\lambda^i}{r2^{-i-1}}\right)^{2/\lambda^i}.$$

Let $A_0 > 0$. For m large enough the logarithm of the right-hand side of the above inequality is bounded by

$$\ln A_0 + 2 \sum_{i=0}^{\infty} \frac{i+1}{\lambda^i} \ln\left[(C2\lambda^i)^{1/(i+1)}2\right] - 2 \ln r \sum_{i=0}^{\infty} \frac{1}{\lambda^i}$$
$$= \ln A_0 + C - N \ln r$$

since

$$\sum_{i=0}^{\infty} \left(\frac{N-2}{N}\right)^i = \frac{N}{2},$$

and therefore

$$A_{m+1} \leq Cr^{-N} \int_{B_{2r}} W^2\, dx. \tag{2.34}$$

Because of the inequality

$$\left(\int_{B_r} W_k^{2\lambda^{m+1}}\, dx\right)^{1/\lambda^{m+1}} \leq A_{m+1}$$

(2.34) yields

$$\operatorname*{ess\,sup}_{B_r} W_k^2 \leq Cr^{-N} \int_{B_{2r}} W^2\, dx \tag{2.35}$$

after a passage to the limit as $m \to \infty$. For $\varrho = r$ the sought-for conclusion follows from (2.35), whose right-hand side is independent of k. For $0 < \varrho < r$ we cover B_r by a finite number of spheres $B_{\varrho/2}(y^i)$, $y^i \in B_r$. Then $B_\varrho(y^i) \subset B_{r+\varrho}$, and from the preceding conclusion we deduce

$$W(x) \leq C\varrho^{-N/2} \mid W \mid_{2;\varrho,y^i} \leq C\varrho^{-N/2} \mid W \mid_{2;r+\varrho}$$

for a.a. $x \in B_{\varrho/2}(y^i)$, hence the desired inequality a.e. in B_r. □

2.4.2. Hölder Continuity of Solutions

We now turn from subsolutions to solutions of the equation $-(a^{ij}w_{x_i})_{x_j} = 0$ in Ω, that is,

$$\begin{gathered} w \in H^1(\Omega), \\ \int_\Omega a^{ij} w_{x_i} v_{x_j}\,dx = 0 \qquad \text{for } v \in H_0^1(\Omega), \end{gathered} \tag{2.36}$$

and prove a *Harnack type inequality* (see J. Moser [123] for a sharper result).

LEMMA 2.13. *Let w satisfy* (2.36) *and assume that $w \geq 0$ on $B_R(x^0) \subseteq \Omega$, the set $E \equiv \{x \in B_R(x^0) \mid w(x) \geq 1\}$ having measure $\geq K \mid B_R(x^0) \mid$ for some $K \in]0, 1[$. Then, $w(x) \geq c(K)$ for a.a. $x \in B_{R/2}(x^0)$, where $c(K) \in]0, 1[$ is independent of x^0 and R, but depends on w through K, on the a^{ij}'s through α and the bound imposed on $\mid a^{ij} \mid_{\infty;\Omega}$.*

PROOF. Take $x^0 = 0$, and let $k \in]1/2, 1[$ (independent of R) satisfy $\mid B_R \setminus B_{kR} \mid = K \mid B_R \mid/2$. Then,

$$\begin{aligned} K \mid B_R \mid \leq \mid E \mid = \mid E \cap (B_R \setminus B_{kR}) \mid \\ + \mid E \cap B_{kR} \mid \leq \frac{K}{2} \mid B_R \mid + \mid E \cap B_{kR} \mid, \end{aligned}$$

and therefore $\mid E \cap B_{kR} \mid \geq K \mid B_R \mid/2$.

The idea of the proof is to provide a bound independent of $\varepsilon > 0$ on $[-\ln(w + \varepsilon)]^+$ or, equivalently, on $-\ln[(w + \varepsilon) \wedge 1]$ throughout $B_{R/2}$. For $0 < \varepsilon < 1$ and $t \geq 0$ we set $H_\varepsilon(t) \equiv (t + \varepsilon) \wedge 1$ but, instead of dealing immediately with $-\ln H_\varepsilon(w)$, we first approximate H_ε uniformly from below with a monotone sequence of positive concave functions $H_{\varepsilon,n} \in C^2([0, \infty[)$, with $H_{\varepsilon,n}(t) = H_\varepsilon(t)$ except in a small neighborhood of

$t = 1 - \varepsilon$. Let $G_{\varepsilon,n}(t) = -\ln[H_{\varepsilon,n}(t)]$, so that $G'_{\varepsilon,n} = -H'_{\varepsilon,n}/H_{\varepsilon,n}$, $G''_{\varepsilon,n} = -[H''_{\varepsilon,n}H_{\varepsilon,n} - (H'_{\varepsilon,n})^2]/H^2_{\varepsilon,n} \geq (G'_{\varepsilon,n})^2$. If $g \in C^1(\bar{\Omega})$ with supp $g \subset B_R$, $0 \leq g \leq 1$, $g = 1$ on $\bar{B}_{kR}$ and $|\nabla g| \leq 2(R - kR)^{-1}$, the function $v = g^2 G'_{\varepsilon,n}(w)$ is admissible in (2.36) by an easy adaptation of Lemma 1.57. Thus,

$$\begin{aligned} 0 &= \int_\Omega a^{ij} w_{x_i}[G'_{\varepsilon,n}(w)2gg_{x_j} + g^2 G''_{\varepsilon,n}(w)w_{x_j}]\,dx \\ &\geq \int_\Omega \{a^{ij}[G_{\varepsilon,n}(w)]_{x_i} 2gg_{x_j} + \alpha g^2 G''_{\varepsilon,n}(w)\,|\nabla w|^2\}\,dx \\ &\geq \frac{\alpha}{2}\int_{B_R} g^2\,|\nabla G_\varepsilon(w)|^2\,dx - C\int_{B_R} |\nabla g|^2\,dx \end{aligned}$$

by standard arguments: note that the inequality $G''_{\varepsilon,n} \geq (G'_{\varepsilon,n})^2$ has played a fundamental role here. Summing up,

$$\int_{B_{kR}} |\nabla G_{\varepsilon,n}(w)|^2\,dx \leq CR^{N-2} \tag{2.37}$$

whenever $0 < \varepsilon < 1$, $n \in N$. As $n \to \infty$, $G_{\varepsilon,n}(w) \to W_\varepsilon^+$ a.e. in B_R, with $W_\varepsilon \equiv -\ln(w + \varepsilon)$, and hence also, by monotonicity, in $L^2(B_R)$, whereas from (2.37) we deduce that

$$\frac{\partial}{\partial x_i} G_{\varepsilon,n}(w) \rightharpoonup \frac{\partial}{\partial x_i} W_\varepsilon^+$$

in $L^2(B_{kR})$ and

$$\int_{B_{kR}} |\nabla W_\varepsilon^+|^2\,dx \leq CR^{N-2} \tag{2.38}$$

(see Problem 1.2).

Since (again by an adaptation of Lemma 1.57)

$$\begin{aligned} \int_\Omega a^{ij} W_{\varepsilon x_i} v_{x_j}\,dx &= -\int_{B_R} a^{ij} w_{x_i} \frac{v_{x_j}}{w + \varepsilon}\,dx \\ &= -\int_{B_R} a^{ij} w_{x_i}\left[\left(\frac{v}{w+\varepsilon}\right)_{x_j} + w_{x_j}\frac{v}{(w+\varepsilon)^2}\right] dx \\ &= -\int_{B_R} a^{ij} w_{x_i} w_{x_j}\frac{v}{(w+\varepsilon)^2}\,dx \leq 0 \end{aligned}$$

whenever $v \in C_c^1(B_R)$, $v \geq 0$, W_ε is a subsolution of the equation $-(a^{ij}w_{x_i})_{x_j}$

$= 0$ in B_R, and so too is W_ε^+, being the supremum of two subsolutions. Lemma 2.11 therefore yields

$$\operatorname*{ess\,sup}_{B_{R/2}} W_\varepsilon^+ \leq C(kR - R/2)^{-N/2} \mid W_\varepsilon^+ \mid_{2;kR}. \tag{2.39}$$

On the other hand, all functions W_ε^+, $\varepsilon > 0$, vanish on the set $E \cap B_{kR}$, whose measure is $\geq K \mid B_R \mid/2$ by our initial considerations. From Lemma 1.36 and (2.38) we deduce that

$$\mid W_\varepsilon^+ \mid_{2;kR} \leq C(K)kR \mid \nabla W_\varepsilon^+ \mid_{2;kR} \leq C(K)R^{N/2},$$

so that (2.39) yields a uniform bound on $W_\varepsilon^+(x)$, $0 < \varepsilon < 1$, for a.a. $x \in B_{R/2}$. The conclusion follows after letting $\varepsilon \to 0$, since

$$-\ln w(x) \leq C(K)$$

at a.a. point $x \in B_{R/2}$ where $w(x) < 1$. □

At this point we are in a position to prove the celebrated *De Giorgi–Nash theorem.*

THEOREM 2.14. *If w satisfies* (2.36), *then it belongs to $C^{0,\delta_0}(\Omega)$ for some $\delta_0 \in]0, 1[$; more precisely,*

$$\max_{\overline{B_\varrho(x^0)}} w - \min_{\overline{B_\varrho(x^0)}} w \leq CR^{-N/2}(\varrho/R)^{\delta_0} \mid w \mid_{2;2R,x^0} \tag{2.40}$$

whenever $0 < \varrho \leq R$, $B_{2R}(x^0) \subseteq \Omega$, where C and δ_0 (both independent of w) depend on the a^{ij}'s only through α and the bound imposed on $\mid a^{ij} \mid_{\infty;\Omega}$.

PROOF. Since both w and $-w$ are subsolutions, Lemma 2.11 yields

$$\operatorname*{ess\,sup}_{B_R(x^0)} \mid w \mid \leq CR^{-N/2} \mid w \mid_{2;2R,x^0}.$$

We set

$$m_0 \equiv \operatorname*{ess\,inf}_{B_R(x^0)} w, \qquad M_0 \equiv \operatorname*{ess\,sup}_{B_R(x^0)} w,$$

so that $M_0 - m_0 \leq KR^{-N/2} \mid w \mid_{2;2R,x^0}$ for some $K > 0$.

Next we fix the unique number $\hat{m}$ such that $\mid E^\pm \mid \leq \mid B_R(x^0) \mid/2$, E^+ (E^-) being the subset of $B_R(x^0)$ where $w > \hat{m}$ ($w < \hat{m}$); precisely, $\hat{m}$ is the supremum of all values $m \geq m_0$ such that $\operatorname{meas}_N\{x \in B_R(x^0) \mid w(x) < m\} \leq \mid B_R(x^0) \mid/2$.

Suppose $m_0 < \hat{m} < M_0$. Both nonnegative functions $[M_0 - w(x)]/(M_0 - \hat{m})$ and $[w(x) - m_0]/(\hat{m} - m_0)$ satisfy (2.36) with Ω replaced by $B_R(x^0)$, and are ≥ 1 on subsets of $B_R(x^0)$ having measure $\geq |B_R(x^0)|/2$. We can therefore apply Lemma 2.13 and obtain

$$\frac{M_0 - w(x)}{M_0 - \hat{m}} \geq c(1/2), \qquad \frac{w(x) - m_0}{\hat{m} - m_0} \geq c(1/2)$$

that is,

$$m_1 \leq w(x) \leq M_1$$

with

$$m_1 \equiv \hat{m} - h(\hat{m} - m_0), \qquad M_1 \equiv \hat{m} + h(M_0 - \hat{m}), \qquad h \equiv 1 - c(1/2),$$

for a.a. $x \in B_{R/2}(x^0)$. The same result holds if $\hat{m} = m_0$ or $\hat{m} = M_0$. Thus, $M_1 - m_1 \leq h(M_0 - m_0)$, and by iteration

$$\operatorname*{osc}_{2^{-n}R} w \leq h^n K R^{-N/2} | w |_{2;2R,x^0} \equiv h^n K'$$

where

$$\operatorname*{osc}_{\varrho} w \equiv \operatorname*{ess\,sup}_{B_\varrho(x^0)} w - \operatorname*{ess\,inf}_{B_\varrho(x^0)} w.$$

For $2^{-(n+1)}R < \varrho \leq 2^{-n}R$ we have

$$\begin{aligned}\ln(\operatorname*{osc}_{\varrho} w) &\leq \ln K' - \ln h + (n+1)\ln h \\ &< \ln(K'/h) + [(\ln h)/\ln 2] \ln(R/\varrho),\end{aligned}$$

hence

$$\operatorname*{osc}_{\varrho} w \leq h^{-1}K'(\varrho/R)^{\delta_0} \qquad \text{with } \delta_0 = -(\ln h)/\ln 2, \tag{2.41}$$

since the positive number h is < 1.

From (2.41) we immediately arrive at (2.40) provided we show that w has a pointwise representative from $C^0(\Omega)$. To do this we arbitrarily fix $\omega \subset\subset \Omega$ and denote by $\{y^i\}$ an everywhere dense sequence of points of ω. If $R > 0$ is $< \frac{1}{2}\operatorname{dist}(\omega, \partial\Omega)$ and $n \in N$ is $> 1/R$, there exists $S_n \subset \Omega$, with $|S_n| = 0$, such that (a representative of) w satisfies

$$| w(x) - w(y) | \leq h^{-1} K R^{-N/2} (1/nR)^{\delta_0} | w |_{2;\Omega}$$

whenever $x, y \in B_{1/n}(y^i) \setminus S_n$ for some $i \in N$ [see (2.41)], hence also whenever $x, y \in \omega \setminus S_n$ with $|x - y| < 1/n$. Set $S = \bigcup_n S_n$. The function w

is uniformly continuous when restricted to the dense subset $\omega \setminus S$ of $\bar{\omega}$, and therefore has an extension to $\bar{\omega}$ that belongs to $C^0(\bar{\omega})$. The conclusion is patent. □

2.4.3. $L^{2,\mu}$ Regularity of First Derivatives

Assume the validity of the next result, whose proof follows later.

LEMMA 2.15. *Let $\Omega = B_r$ for some $r > 0$. There exists a constant C such that for any $\varrho \in]0, r]$*

$$| \nabla u |_{2;\varrho}^2 \leq C\left(\frac{\varrho^{\mu_0}}{r^{\mu_0}} | \nabla u |_{2;r}^2 + r^2 | f^0 |_{2;r}^2 + \sum_{i=1}^N | f^i |_{2;r}^2\right) \tag{2.42}$$

with $\mu_0 = N - 2 + 2\delta_0$, δ_0 being the Hölder exponent of Theorem 2.14, whenever u satisfies

$$u \in H^1(B_r),$$

$$\int_{B_r} a^{ij} u_{x_i} v_{x_j} \, dx = \langle F, v \rangle \equiv \int_{B_r} (f^0 v + f^i v_{x_i}) \, dx \qquad \text{for } v \in H_0^1(B_r)$$

with $f^0, \ldots, f^N \in L^2(B_r)$. C is independent of r; it depends on the a^{ij}'s only through the bound imposed on their $L^\infty(B_r)$ norms as well as through α.

We can then pass from (2.36) to a complete equation such as

$$\begin{gathered} u \in H^1(\Omega), \\ a(u, v) = \langle F; v \rangle \equiv \int_\Omega (f^0 v + f^i v_{x_i}) \, dx \qquad \text{for } v \in H_0^1(\Omega) \end{gathered} \tag{2.43}$$

with $a(u, v)$ given by (2.11) and $f^0 \in L^{2,(\mu-2)^+}(\Omega)$, $f^1, \ldots, f^N \in L^{2,\mu}(\Omega)$, $0 < \mu < \mu_0$, and investigate interior regularity of solutions as below.

Set

$$\varkappa_\mu(F; u) \equiv | f^0 |_{2,(\mu-2)^+;\Omega}^2 + \sum_{i=1}^N | f^i |_{2,\mu;\Omega}^2 + | u |_{H^1(\Omega)}^2$$

and suppose μ is such that, whenever $\omega_1 \subset\subset \Omega$, $u|_{\omega_1}$ belongs to $L^{2,\mu}(\omega_1)$ with $| u |_{2,\mu;\omega_1}^2 \leq C\varkappa_\mu(F; u)$; note that $H^1(\omega_1) \subsetneq L^{2,2}(\omega_1)$ if $\partial\omega_1$ is of class C^1 (see Theorem 1.40).

Let $\omega \subset\subset \omega_1$, $x^0 \in \bar{\omega}$, $0 < r \le d \equiv [\text{dist}(\omega, \partial\omega_1)] \wedge 1$. The function u satisfies

$$\int_{B_r(x^0)} a^{ij} u_{x_i} v_{x_j}\, dx = \int_{B_r(x^0)} (\hat{f}_0 v + \hat{f}^i v_{x_i})\, dx$$
$$\equiv \int_{B_r(x^0)} [(f^0 - b^i u_{x_i} - cu)v + (f^i - d^i u)v_{x_i}]\, dx$$
$$\text{for } v \in H_0^1(B_r(x^0)).$$

Therefore, after an inessential translation of the origin, (2.42) with f^j replaced by $\hat{f}^j$ for $j = 0, \ldots, N$ yields

$$|\nabla u|^2_{2;x^0,\varrho} \le \hat{C}\left[\left(\frac{\varrho^{\mu_0}}{r^{\mu_0}} + r^2\right) |\nabla u|^2_{2;x^0,r} + r^\mu \varkappa_\mu(F; u)\right] \tag{2.44}$$

for $0 < \varrho \le r$, since

$$r^2 |\hat{f}^0|^2_{2;x^0,r} \le C(r^\mu |f^0|^2_{2,(\mu-2)^+;\omega_1} + r^{\mu+2} |u|^2_{2,\mu;\omega_1} + r^2 |\nabla u|^2_{2;x^0,r}),$$
$$\sum_{i=1}^N |\hat{f}^i|^2_{2;x^0,r} \le C r^\mu\left(\sum_{i=1}^N |f^i|^2_{2,\mu;\omega_1} + |u|^2_{2,\mu;\omega_1}\right).$$

For $1 < s < \infty$ let $H(s) \equiv s^{-\mu_0/2} \wedge d$. By (2.44) the function $\varphi(\varrho) \equiv |\nabla u|^2_{2;x^0,\varrho}$ satisfies

$$\varphi(\varrho) \le \hat{C}\left[2 \frac{\varrho^{\mu_0}}{r^{\mu_0}} \varphi(r) + \varrho^\mu \varkappa_\mu(F; u) s^\mu\right]$$

whenever $0 < r \le H(s)$ and $1 < r/\varrho \le s$. As in the proof of Theorem 1.17, we do not divide by ϱ^μ at this point, because ϱ is still restricted to vary away from 0. Instead, we apply Lemma 1.18 with $K = 2\hat{C}$, $\Phi(s) = \hat{C}\varkappa_\mu(F; u)s^\mu$ and $\varepsilon = \mu_0 - \mu$, so that

$$|\nabla u|^2_{2;x^0,\varrho} \le C\left[\frac{\varrho^\mu}{r^\mu} |\nabla u|^2_{2;x^0,r} + \varrho^\mu \varkappa_\mu(F; u)\right]$$

whenever $0 < \varrho \le r \le H(K^{1/\varepsilon})$, and finally

$$\varrho^{-\mu} |\nabla u|^2_{2;\omega[x^0,\varrho]} \le C\{[H(K^{1/\varepsilon})]^{-\mu} |\nabla u|^2_{2;\Omega} + \varkappa_\mu(F; u)\}.$$

We have proven that whenever $\omega \subset\subset \Omega$, all first derivatives of $u|_\omega$ belong to $L^{2,\mu}(\omega)$ with $|\nabla u|^2_{2,\mu;\omega} \le C\varkappa_\mu(F; u)$. Thus, if $\omega_1 \subset\subset \Omega$ with $\partial\omega_1$ of class C^1, Theorem 1.40 yields $u|_{\omega_1} \in L^{2,\mu+2}(\omega_1)$, $|u|^2_{2,\mu+2;\omega_1} \le C\varkappa_\mu(F; u)$. This shows that all the above considerations can be repeated with μ replaced by any $\mu' \le \mu + 2$, $\mu' < \mu_0$, and so on with a bootstrap argument.

Moreover, $u \in C^{0,(\mu-N+2)/2}(\Omega)$ if $\mu > N-2$.

Summing up, Lemma 2.15 leads to the following theorem.

THEOREM 2.16. *Let u solve (2.43) with $f^0 \in L^{2,(\mu-2)^+}(\Omega)$ and $f^1, \ldots, f^N \in L^{2,\mu}(\Omega)$, $0 < \mu < \mu_0$, μ_0 being defined as in Lemma 2.15. Whenever $\omega \subset\subset \Omega$, all first derivatives of $u|_\omega$ belong to $L^{2,\mu}(\omega)$ with norm estimate*

$$| \nabla u |_{2,\mu;\omega} \leq C\Big(| f^0 |_{2,(\mu-2)^+;\Omega} + \sum_{i=1}^{N} | f^i |_{2,\mu;\Omega} + | u |_{H^1(\Omega)} \Big),$$

the constant C (independent of u, F) depending on the coefficients of $a(u, v)$ only through the bound imposed on their $L^\infty(\Omega)$ norms, as well as through α. In particular, if $\mu > N-2$ then $u \in C^{0,\delta}(\Omega)$ with $\delta = (\mu - N + 2)/2$.

In its turn the proof of Lemma 2.15 will rely on the decomposition of u into a sum $w + z$, where w satisfies the homogeneous equation. We have the following lemma.

LEMMA 2.17. *There exists a constant C such that for any $\varrho \in \,]0, r]$*

$$| \nabla w |_{2;\varrho}^2 \leq C \frac{\varrho^{\mu_0}}{r^{\mu_0}} | \nabla w |_{2;r}^2$$

whenever w satisfies (2.36) with $\Omega = B_r$, μ_0 being defined as in Lemma 2.15. C is independent of r; it depends on the a^{ij}'s only through the bound imposed on their $L^\infty(B_r)$ norms as well as through α.

PROOF. Since $w - \int_{B_r} w\, dx$ solves the same equation as w, we may suppose $\int_{B_r} w\, dx = 0$. Then Lemma 1.35 yields

$$| w |_{2;r}^2 \leq C r^2 | \nabla w |_{2;r}^2,$$

hence

$$| w(x) - w(0) |^2 \leq C r^{2-N-2\delta_0} | x |^{2\delta_0} | \nabla w |_{2;r}^2$$

for $x \in \bar{B}_{r/2}$, by Theorem 2.14. Let $0 < \varrho \leq r/4$ and set $v = g^2[w - w(0)]$ with $g \in C^1(\bar{B}_r)$, supp $g \subset B_{2\varrho}$, $0 \leq g \leq 1$, $g = 1$ on $\bar{B}_\varrho$, $| \nabla g(x) | \leq 2\varrho^{-1}$. Then the equation yields

$$0 = \int_{B_r} a^{ij} w_{x_i} \{ g^2 w_{x_j} + 2 g g_{x_j} [w - w(0)] \}\, dx \geq \frac{\alpha}{2} \int_{B_{2\varrho}} g^2 | \nabla w |^2\, dx$$
$$- C \max_{\bar{B}_{2\varrho}} | w - w(0) |^2 \int_{B_{2\varrho}} | \nabla g |^2\, dx$$

by standard arguments. Therefore,

$$\int_{B_\varrho} |\nabla w|^2\,dx \le C\varrho^{N-2} \max_{\bar{B}_{2\varrho}} |w - w(0)|^2.$$

The conclusion follows easily: notice that whenever $r/4 < \varrho \le r$,

$$|\nabla w|^2_{2;\varrho} \le 4^{N-2+2\delta_0} \frac{\varrho^{N-2+2\delta_0}}{r^{N-2+2\delta_0}} |\nabla w|^2_{2;r}\,. \qquad \square$$

At this point we can proceed to the proof of Lemma 2.15.

Proof of Lemma 2.15. Solve

$$z \in H_0{}^1(B_r),$$
$$\int_{B_r} a^{ij} z_{x_i} v_{x_j}\,dx = \int_{B_r} (f^0 v + f^i v_{x_i})\,dx \qquad \text{for } v \in H_0{}^1(B_r)$$

with the help of Theorem 2.1 and of the corollary of Theorem 1.43. Then

$$\begin{aligned} |\nabla z|^2_{2;r} &\le \alpha^{-2} |F|^2_{H^{-1}(B_r)} \\ &\le C\Big(r^2 |f^0|^2_{2;r} + \sum_{i=1}^{N} |f^i|^2_{2;r}\Big), \end{aligned}$$

since Poincaré's inequality in $H_0{}^1(B_r)$ yields

$$\begin{aligned} |f^0 v|_{1;r} &\le |f^0|_{2;r} |v|_{2;r} \\ &\le C |f^0|_{2;r} r |\nabla v|_{2;r} \end{aligned}$$

when $v \in H_0{}^1(B_r)$. The function $w = u - z$ satisfies (2.36) with $\Omega = B_r$ so that

$$\begin{aligned} |\nabla u|^2_{2;\varrho} &\le 2(|\nabla w|^2_{2;\varrho} + |\nabla z|^2_{2;\varrho}) \\ &\le C \frac{\varrho^{\mu_0}}{r^{\mu_0}} |\nabla w|^2_{2;r} + 2 |\nabla z|^2_{2;r} \\ &\le C\Big[\frac{\varrho^{\mu_0}}{r^{\mu_0}} |\nabla u|^2_{2;r} + \Big(1 + \frac{\varrho^{\mu_0}}{r^{\mu_0}}\Big) |\nabla z|^2_{2;r}\Big] \\ &\le C\Big(\frac{\varrho^{\mu_0}}{r^{\mu_0}} |\nabla u|^2_{2;r} + r^2 |f^0|^2_{2;r} + \sum_{i=1}^{N} |f^i|^2_{2;r}\Big) \end{aligned}$$

for $0 < \varrho \le r$ by Lemma 2.17. $\square$

REMARK. Theorem 2.16 can be compared with the following important theorem by N. G. Meyers [109]:

There exists $p > 2$ such that, if u satisfies (2.42) with $f^0 \in L^q(\Omega)$, $q = pN/(N+p)$ and $f^1, \ldots, f^N \in L^p(\Omega)$, then $u_{x_1}|_\omega, \ldots, u_{x_N}|_\omega \in L^p(\omega)$, with the corresponding norm estimate, whenever $\omega \subset\subset \Omega$.

Notice that, when $N = 2$, Meyers' theorem implies the Hölder continuity of u in Ω thanks to Sobolev's inequalities (Theorem 1.41). In the bidimensional case Hölder continuity can, however, also be proven by techniques analogous to those of the present section: see J. Kadlec and J. Nečas [84]. (The one-dimensional case is obvious: see the preliminary considerations of Section 2.3).

Until now the results of this section have concerned only interior regularity of solutions (and of their derivatives). Hölder continuity up to $\partial\Omega$ of solutions of (2.6) for $V = H_0^1(\Omega \cup \Gamma)$ can be proven under rather mild regularity assumptions about $\partial\Omega \setminus \Gamma$ and Γ (as well as F): see G. Stampacchia [143]. We prefer instead to show that, if the assumptions about $\partial\Omega \setminus \Gamma$ and Γ are strong enough, global regularity can easily be deduced from previous interior results through an extension technique.

Beginning with the case $\Omega = B^+$, we investigate solutions of either equation

$$\begin{gathered} u \in H_0^1(B^+ \cup S^+), \\ a(u, v) = \langle F, v\rangle \equiv \int_{B^+} (f^0 v + f^i v_{x_i})\, dx \qquad \text{for } v \in H_0^1(B^+), \end{gathered} \tag{2.45}$$

or

$$\begin{gathered} u \in H^1(B^+), \\ a(u, v) = \langle F, v\rangle \equiv \int_{B^+} (f^0 v + f^i v_{x_i})\, dx \qquad \text{for } v \in H_0^1(B^+ \cup S^0). \end{gathered} \tag{2.46}$$

LEMMA 2.18. *Let u solve either (2.45) or (2.46) with $f^0 \in L^{2,(\mu-2)^+}(B^+)$ and $f^1, \ldots, f^N \in L^{2,\mu}(B^+)$, where μ is defined as in Theorem 2.16. Whenever $0 < R < 1$, all first derivatives of $u|_{B_R^+}$ belong to $L^{2,\mu}(B_R^+)$ with norm estimate*

$$|\nabla u|_{2,\mu;B_R^+} \leq C\Big(|f^0|_{2,(\mu-2)^+;B^+} + \sum_{i=1}^N |f^i|_{2,\mu;B^+} + |u|_{H^1(B^+)}\Big)$$

where C has the same dependence on the coefficients as in Theorem 2.16.

PROOF. For functions $w = w(x)$, $x \in B$, we define

$$\bar{w}(x', x_N) \equiv w(x', -x_N),$$

and denote by $D_i w$ the derivative w_{x_i}; notice that

$$D_i \bar{w} = \overline{D_i w} \qquad \text{for } i = 1, \ldots, N-1, \qquad D_N \bar{w} = -\overline{D_N w}.$$

We can suppose that the coefficients of $a(u, v)$ are bounded functions defined throughout B according to the rules

$$a^{iN} = -\overline{a^{iN}} \quad \text{and} \quad a^{Ni} = -\overline{a^{Ni}} \qquad \text{for } i = 1, \ldots, N-1,$$

$$d^N = -\overline{d^N}, \qquad b^N = -\overline{b^N},$$

$$a^{ij} = \overline{a^{ij}}, \qquad d^j = \overline{d^j}, \qquad b^i = \overline{b^i} \qquad \text{for all remaining values of } i, j,$$

$$c = \bar{c}.$$

Notice that for any $\xi \in \mathbb{R}^N$,

$$\overline{a^{ij}}\xi_i\xi_j = a^{ij}\eta_i\eta_j \geq \alpha \mid \xi \mid^2$$

a.e. in B^+, with $\eta_i = -\xi_i$ for $i = 1, \ldots, N-1$ and $\eta_N = \xi_N$.

Passing to the free term F, we consider $f^0, f^1, \ldots, f^N$ as functions of the corresponding Morrey spaces over B, defined by the rules

$$f^j = -\overline{f^j} \qquad \text{for } j = 0, 1, \ldots, N-1, \qquad f^N = \overline{f^N}$$

in the case (2.45), and

$$f^j = \overline{f^j} \qquad \text{for } j = 0, 1, \ldots, N-1, \qquad f^N = -\overline{f^N}$$

in the case (2.46) (see Lemma 1.16 and the remark after Theorem 1.17).

Finally, we consider u as a function of $H^1(B)$ defined by the rule

$$u = -\bar{u}$$

in the case (2.45) (see Lemma 1.44), and

$$u = \bar{u}$$

in the case (2.46) (see Lemma 1.29).

Now let $v \in C_o^1(B)$. We have

$$\int_{B^+} \overline{[(a^{ij}D_iu + d^ju)D_jv + (b^iD_iu + cu)v]}\,dx$$
$$= -\int_{B^+}\left(\sum_{i=1}^{N-1}\overline{a^{iN}}D_i\bar{u} + \overline{d^N}\bar{u}\right)D_N\bar{v}\,dx$$
$$-\int_{B^+}\left(\sum_{j=1}^{N-1}\overline{a^{Nj}}D_N\bar{u}D_j\bar{v} + \overline{b^N}D_N\bar{u}\bar{v}\right)dx$$
$$+\int_{B^+}\left[\left(\sum{}'\,\overline{a^{ij}}D_i\bar{u} + \sum_{j=1}^{N-1}\overline{d^j}\bar{u}\right)D_j\bar{v} + \left(\sum_{i=1}^{N-1}\overline{b^i}D_i\bar{u} + \bar{c}\bar{u}\right)\bar{v}\right]dx,$$

where $\sum'$ denotes summation over all remaining indices i, j. But then the quantity

$$\int_{B} [(a^{ij}D_iu + d^ju)D_jv + (b^iD_iu + cu)v]\,dx$$
$$= \int_{B^+} [(a^{ij}D_iu + d^ju)D_jv + (b^iD_iu + cu)v]\,dx$$
$$+ \int_{B^+} \overline{[(a^{ij}D_iu + d^ju)D_jv + (b^iD_iu + cu)v]}\,dx$$

equals

$$\int_{B^+} [(a^{ij}D_iu + d^ju)D_j(v - \bar{v}) + (b^iD_iu + cu)(v - \bar{v})]\,dx$$
$$= \int_{B^+} [f^0(v - \bar{v}) + f^iD_i(v - \bar{v})]\,dx$$

in the case (2.45) [notice that $(v - \bar{v})|_{B^+} \in H_0^1(B^+)$] and

$$\int_{B^+} [(a^{ij}D_iu + d^ju)D_j(v + \bar{v}) + (b^iD_iu + cu)(v + \bar{v})]\,dx$$
$$= \int_{B^+} [f^0(v + \bar{v}) + f^iD_i(v + \bar{v})]\,dx$$

in the case (2.46) [notice that $(v + \bar{v})|_{B^+} \in H_0^1(B^+ \cup S^0)$].

At this point we need only utilize the identities

$$\int_{B^+} [f^0(v - \bar{v}) + f^iD_i(v - \bar{v})]\,dx = \int_B (f^0v + f^iD_iv)\,dx$$

[in the case (2.45)] and

$$\int_{B^+} [f^0(v + \bar{v}) + f^iD_i(v + \bar{v})]\,dx = \int_B (f^0v + f^iD_iv)\,dx$$

[in the case (2.46)] to verify that u satisfies (2.42) with Ω replaced by B. The conclusion is now an immediate consequence of Theorem 2.16. ∎

At this point the following global regularity result can be proven.

THEOREM 2.19. *Let $\partial\Omega$ be of class C^1, the open portion $\Gamma \subseteq \partial\Omega$ being also closed. Let u solve* (2.6) *with $V = H_0^1(\Omega \cup \Gamma)$ and $\langle F, v\rangle = \int_\Omega (f^0 v + f^i v_{x_i})\,dx$ for $v \in V$, where $f^0, f^1, \ldots, f^N$ satisfy the same assumptions as in Theorem* 2.16. *Then all first derivatives of u belong to $L^{2,\mu}(\Omega)$ with the corresponding norm estimate; in particular, $u \in C^{0,\delta}(\bar{\Omega})$ with $\delta = (\mu - N + 2)/2$ if $\mu > N - 2$.*

PROOF. Let $x^0 \in \partial\Omega$ and let U be a bounded domain of R^N, $U \ni x^0$, such that $U \cap \partial\Omega$ is a portion of either $\partial\Omega\setminus\Gamma$ or Γ which is straightened by a C^1 diffeomorphism Λ: $\bar{U} \to \bar{B}$, $\Lambda(x^0) = 0$. Then $u' \equiv (u \circ \Lambda^{-1})|_{B^+}$ belongs to $H^1(B^+)$, with $|u'|_{H^1(B^+)} \leq C|u|_{H^1(U\cap\Omega)}$ (see Lemma 1.28). As a matter of fact, $u' \in H_0^1(B^+ \cup S^+)$ if $U \cap \partial\Omega \subset \partial\Omega\setminus\Gamma$, since u' is the limit in $H^1(B^+)$ of a sequence $\{u_n'\} \subset C_c^1(B^+ \cup S^+)$, $u_n' \equiv (u_n \circ \Lambda^{-1})|_{\overline{B^+}}$ with $\{u_n\} \subset C_c^1(\Omega \cup \Gamma)$, $u_n \to u$ in $H^1(\Omega)$. Moreover, the function $v' \equiv (v \circ \Lambda^{-1})|_{\overline{B^+}}$ belongs to $C_c^1(B^+ \cup S^0)$ if $v \in C^1(\bar{\Omega})$ with $\operatorname{supp} v \subset U \cap \bar{\Omega}$, and even to $C_c^1(B^+)$ if $\operatorname{supp} v \subset U \cap \Omega$. Vice versa, any function $v' \in C_c^1(B^+ \cup S^0)$, or even $v' \in C_c^1(B^+)$, can be obtained by inverting the above procedure. Thus a density argument and a change of variables in the equation yield

$$
\begin{aligned}
a'(u', v') &\equiv \int_{B^+} [(a'^{hk} u'_{y_h} + d'^k u') v'_{y_k} + (b'^h u'_{y_h} + c' u') v']\,dy \\
&= \langle F', v'\rangle \equiv \int_{B^+} (f'^0 v' + f'^h v'_{y_h})\,dy \qquad (2.47)
\end{aligned}
$$

for $v' \in H_0^1(B^+)$ (if $U \cap \partial\Omega \subset \partial\Omega\setminus\Gamma$) or $v' \in H_0^1(B^+ \cup S^0)$ (if $U \cap \partial\Omega \subset \Gamma$). In (2.47),

$$
\begin{aligned}
a'^{hk}(y) &\equiv a^{ij}[x(y)] y_{hx_i}[x(y)] y_{kx_j}[x(y)] J(y), \\
d'^k(y) &\equiv d^j[x(y)] y_{kx_j}[x(y)] J(y), \\
b'^h(y) &\equiv b^i[x(y)] y_{hx_i}[x(y)] J(y), \\
c'(y) &\equiv c[x(y)] J(y), \\
f'^0(y) &\equiv f^0[x(y)] J(y), \\
f'^h(y) &\equiv f^i[x(y)] y_{hx_i}[x(y)] J(y),
\end{aligned}
$$

where $y = y(x) \equiv \Lambda(x)$, $x = x(y) \equiv \Lambda^{-1}(y)$, and $J(y)$ denotes the absolute value of the Jacobian determinant of Λ^{-1} at y.

The coefficients of $a'(u', v')$ belong to $L^\infty(B^+)$ with norms bounded by some constant times the sum of the $L^\infty(\Omega)$ norms of the coefficients of $a(u, v)$. Moreover,

$$a'^{hk}\xi_h\xi_k = Ja^{ij}y_{hx_i}y_{kx_j}\xi_h\xi_k \geq (\min_{\overline{B^+}} J)\alpha \sum_{i=1}^N \left(\sum_{h=1}^N y_{hx_i}\xi_h\right)^2 \geq \alpha' \mid \xi \mid^2,$$

$\alpha' > 0$. As for $f'^0, f'^1, \dots, f'^N$, we can bound their respective Morrey norms with some constant times the sum of the norms of $f^0, f^1, \dots, f^N$ (see Lemma 1.15).

To (2.47) we apply Lemma 2.18. We thus arrive at the membership of $u'_{y_1}|_{B_R^+}, \dots, u'_{y_N}|_{B_R^+}$ in $L^{2,\mu}(B_R^+)$, $0 < R < 1$, with norm estimate. For $s = 1, \dots, N$ we set

$$z_s'(y', y_N) \equiv \begin{cases} u'_{y_s}(y', y_N) & \text{if } (y', y_N) \in B_R^+, \\ u'_{y_s}(y', -y_N) & \text{if } (y', -y_N) \in B_R^+. \end{cases}$$

Each function z_s' is in $L^{2,\mu}(B_R)$ (see Lemma 1.16). Therefore, each function $z_s(x) \equiv (z_s' \circ \Lambda)(x)$, $x \in B_r(x^0) \subset \Lambda^{-1}(B_R)$, belongs to $L^{2,\mu}(B_r(x^0))$ by Lemma 1.15, and the restrictions to $\Omega \cap B_r(x^0)$ of $u_{x_1}, \dots, u_{x_N}$ belong to $L^{2,\mu}(\Omega \cap B_r(x^0))$ by the chain rule.

We now cover $\partial\Omega$ with open spheres $B^1, \dots, B^m$ such as $B_r(x^0)$ in the preceding. Let $\{g_j\}_{j=0,1,\dots,m}$ be a partition of unity relative to the open covering $\{\omega_j\}_{j=0,1,\dots,m}$ of $\bar{\Omega}$, where $\omega_i = B^i$ for $i = 1, \dots, m$, $\Omega \supset\supset \omega_0 \supset \Omega \setminus \bigcup_{i=1}^m B^i$ (see the corollary of Lemma 1.4). Thus,

$$u = \sum_{j=0}^m g_j u,$$

and all first derivatives of $g_j u$ belong to $L^{2,\mu}(\Omega)$ by the above considerations for $j = 1, \dots, m$, by Theorem 2.16 for $j = 0$. □

2.5. H^k Regularity by the Method of Difference Quotients

2.5.1. Regularity in the Interior

The following lemma throws light on the results of the present section.

LEMMA 2.20. *Assume $a^{ij} \in C^{0,1}(\bar{\Omega})$ and let u satisfy*

$$u \in H^1(\Omega), \qquad \operatorname{supp} u \subset \Omega,$$
$$\int_\Omega a^{ij}u_{x_i}v_{x_j}\,dx = \langle F, v\rangle \equiv \int_\Omega (f^0 v + f^i v_{x_i})\,dx \qquad \textit{for } v \in H_0^1(\Omega) \tag{2.48}$$

with $f^0 \in L^2(\Omega)$, $f^1, \ldots, f^N \in H^1(\Omega)$. *Then* u *belongs to* $H^2(\Omega)$ *with norm estimate*

$$| u |_{H^2(\Omega)} \leq C\Big(| f^0 |_{2;\Omega} + \sum_{i=1}^{N} | f^i |_{H^1(\Omega)} + | u |_{H^1(\Omega)}\Big),$$

where the constant C *(independent of* u, F*) depends on the* a^{ij}*'s through the bound imposed on their* $C^{0,1}(\bar{\Omega})$ *norms, as well as through* α.

PROOF. As in Section 1.5.2 we utilize the notations

$$\tau_h w(x) \equiv \tau_h{}^s w(x) \equiv w(x + he^s), \qquad \delta_h w \equiv \delta_h{}^s w \equiv (\tau_h w - w)/h$$

for $h \in \mathbb{R}\setminus\{0\}$, e^s being the sth unit coordinate vector for an arbitrarily fixed value of s.

Let $\operatorname{supp} u \subset \omega \subset\subset \Omega$ and set

$$d \equiv \operatorname{dist}(\omega, \partial\Omega), \qquad \omega^h \equiv \{x \in \Omega \mid \operatorname{dist}(x, \omega) < | h |\},$$

$\omega' \equiv \omega^{d/2}$. For $0 < | h | < d/4$ we insert the admissible function $v = -\delta_{-h}\delta_h u$ in (2.48) and obtain

$$\int_{\omega^h} \delta_h(a^{ij}u_{x_i})\delta_h u_{x_j}\, dx = -\int_{\omega^h} [f^0\delta_{-h}\delta_h u - (\delta_h f^i)\delta_h u_{x_i}]\, dx.$$

From the bounds

$$| \delta_{-h}\delta_h u |_{2;\omega^h} \leq | \nabla\delta_h u |_{2;\omega'},$$
$$| \delta_h f^i |_{2;\omega^h} \leq | \nabla f^i |_{2;\omega'}$$

(see Lemma 1.21) and from the identity

$$\delta_h(a^{ij}u_{x_i}) = (\tau_h a^{ij})\delta_h u_{x_i} + (\delta_h a^{ij})u_{x_i}$$

we deduce

$$\begin{aligned}
\alpha \int_{\omega^h} | \nabla\delta_h u |^2\, dx &\leq \int_{\omega^h} (\tau_h a^{ij})(\delta_h u_{x_i})\delta_h u_{x_j}\, dx \\
&= -\int_{\omega^h} \{f^0\delta_{-h}\delta_h u + [(\delta_h a^{ji})u_{x_j} - \delta_h f^i]\delta_h u_{x_i}\}\, dx \\
&\leq C\Big(| f^0 |_{2;\Omega} + \sum_{i,j=1}^{N} | a^{ij} |_{C^{0,1}(\bar{\Omega})} | \nabla u |_{2;\Omega} \\
&\qquad + \sum_{i=1}^{N} | f^1 |_{H^1(\Omega)}\Big) | \nabla\delta_h u |_{2;\omega'}.
\end{aligned}$$

Since

$$\int_{\omega^h} |\nabla\delta_h u|^2\,dx = \int_{\omega'} |\nabla\delta_h u|^2\,dx,$$

we have bounded $|\nabla\delta_h u|_{2;\omega'}$ uniformly with respect to h for $0 < |h| < d/4$. By Lemma 1.21 u_{x_s} belongs to $H^1(\omega')$ with norm estimate, so that the conclusion follows immediately. □

Lemma 2.20 is utilized to prove the following more general result.

LEMMA 2.21. *Let k be a nonnegative integer. Suppose that the coefficients of the bilinear form* (2.11) *satisfy*

$$a^{ij}, d^j \in C^{k,1}(\bar{\Omega}), \qquad b^i, c \in H^{k,\infty}(\Omega),$$

and let u solve (2.43) *with $f^0 \in H^k(\Omega)$, $f^1, \ldots, f^N \in H^{k+1}(\Omega)$. Whenever $\omega \subset\subset \Omega$, $u|_\omega$ belongs to $H^{k+2}(\omega)$ with norm estimate*

$$|u|_{H^{k+2}(\omega)} \le C\Big(|f^0|_{H^k(\Omega)} + \sum_{i=1}^N |f^i|_{H^{k+1}(\Omega)} + |u|_{H^1(\Omega)}\Big).$$

The constant C (independent of u, F) depends on the coefficients of the bilinear form only through the bound imposed on their respective norms, as well as through α.

PROOF. For $h = 0, 1, \ldots$ we set $d_h \equiv \operatorname{dist}(\omega, \partial\Omega)/2^h$, $\omega_h \equiv \{x \in \mathbb{R}^N \mid \operatorname{dist}(x, \omega) < d_h\}$. We also set

$$\varkappa_k(F; u) \equiv |f^0|^2_{H^k(\Omega)} + \sum_{i=1}^N |f^i|^2_{H^{k+1}(\Omega)} + |u|^2_{H^1(\Omega)}$$

and proceed by induction.

Step 1: *The case* $k = 0$. To begin with, we get rid of lower order coefficients by writing (2.43) as

$$\int_\Omega a^{ij}u_{x_i}v_{x_j}\,dx = \int_\Omega (\hat{f}^0 v + \hat{f}^i v_{x_i})\,dx$$
$$\equiv \int_\Omega [(f^0 - b^i u_{x_i} - cu)v + (f^i - d^i u)v_{x_i}]\,dx; \qquad (2.49)$$

notice that

$$|\hat{f}^0|^2_{2;\Omega} + \sum_{i=1}^N |\hat{f}^i|^2_{H^1(\Omega)} \le C\varkappa_0(F; u).$$

Next, we apply the cut-off technique. Let $g \in C_c^\infty(\Omega)$, $g = 1$ on $\bar{\omega}$. Whenever $v \in H_0^1(\Omega)$, the function gu satisfies

$$\begin{aligned}\int_\Omega a^{ij}(gu)_{x_i} v_{x_j}\,dx &= \int_\Omega [a^{ij}u_{x_i}(gv)_{x_j} - a^{ij}u_{x_i}g_{x_j}v + a^{ij}ug_{x_i}v_{x_j}]\,dx\\ &= \int_\Omega [\hat{f}^0 gv + \hat{f}^i(gv)_{x_i} - a^{ij}u_{x_i}g_{x_j}v + a^{ij}ug_{x_i}v_{x_j}]\,dx\\ &= \int_\Omega [(\hat{f}^0 g - a^{ij}u_{x_i}g_{x_j} + \hat{f}^i g_{x_i})v + (\hat{f}^i g + a^{ji}ug_{x_j})v_{x_i}]\,dx.\end{aligned}$$

Since $\operatorname{supp}(gu) \subset \Omega$ and

$$|\hat{f}^0 g - a^{ij}u_{x_i}g_{x_j} + \hat{f}^i g_{x_i}|_{2;\Omega}^2 + \sum_{i=1}^N |\hat{f}^i g + a^{ji}ug_{x_j}|_{H^1(\Omega)}^2 \leq C\varkappa_0(F; u),$$

the conclusion for $k = 0$ follows from the previous lemma with u replaced by gu.

Step 2: *The case* $k \in \mathbb{N}$. Suppose that the sought-for result holds with k replaced by $k - 1$, k being some natural number. Then $u \in H^{k+1}(\omega_1)$ and, when restricted to ω_1, the functions $\hat{f}^0, \hat{f}^1, \ldots, \hat{f}^N$ from (2.49) satisfy

$$|\hat{f}^0|_{H^k(\omega_1)}^2 + \sum_{i=1}^N |\hat{f}^i|_{H^{k+1}(\omega_1)}^2 \leq C\varkappa_k(F; u).$$

Now let $v \in C_c^\infty(\omega_1)$. From (2.49), written with v replaced by v_{x_s}, we deduce

$$\begin{aligned}\int_{\omega_1} a^{ij}u_{x_s x_i}v_{x_j}\,dx &= -\int_{\omega_1} (a^{ij}u_{x_i}v_{x_s x_j} + a^{ij}_{x_s}u_{x_i}v_{x_j})\,dx\\ &= -\int_{\omega_1} (\hat{f}^0 v_{x_s} + \hat{f}^i v_{x_s x_i} + a^{ij}_{x_s}u_{x_i}v_{x_j})\,dx\\ &= \int_{\omega_1} [\hat{f}^0_{x_s}v + (\hat{f}^j_{x_s} - a^{ij}_{x_s}u_{x_i})v_{x_j}]\,dx\end{aligned}$$

for $s = 1, \ldots, N$. By density, the first and the last term above are equal for any $v \in H_0^1(\omega_1)$. The conclusion follows from the inductive assumption concerning the value $k - 1$, with Ω replaced by ω_1 and u by u_{x_s}; notice that

$$|\hat{f}^0_{x_s}|_{H^{k-1}(\omega_1)}^2 + \sum_{i=1}^N |\hat{f}^i_{x_s} - a^{ji}_{x_s}u_{x_j}|_{H^k(\omega_1)}^2 \leq C\varkappa_k(F; u). \qquad \square$$

Thanks to Theorem 1.41, Lemma 2.21 is immediately seen to admit the following corollary.

COROLLARY. *Suppose that all coefficients of the bilinear form* (2.11) *belong to* $C^\infty(\bar{\Omega})$, *and let u satisfy* (2.43) *with* $f^0, f^1, \ldots, f^N \in C^\infty(\bar{\Omega})$. *Then* $u \in C^\infty(\Omega)$.

REMARK. Whatever the nonnegative integer k in the assumptions of Lemma 2.21, any solution of (2.43) verifies the equation

$$Lu = f^0 - f^i_{x_i} \qquad \text{in } \Omega$$

almost everywhere.

2.5.2. Boundary and Global Regularity

We now want to extend the results of Section 2.5.1 up to the boundary of Ω. We begin with the case $\Omega = B^+$.

LEMMA 2.22. *Assume* $a^{ij} \in C^{0,1}(\overline{B^+})$ *and let u satisfy either*

$$\begin{gathered} u \in H_0^1(B^+), \qquad \operatorname{supp} u \subset B^+ \cup S^0, \\ \int_{B^+} a^{ij} u_{x_i} v_{x_j}\, dx = \langle F, v \rangle \equiv \int_{B^+} (f^0 v + f^i v_{x_i})\, dx \qquad \textit{for } v \in H_0^1(B^+), \end{gathered} \tag{2.50}$$

or

$$\begin{gathered} u \in H^1(B^+), \qquad \operatorname{supp} u \subset B^+ \cup S^0, \\ \int_{B^+} a^{ij} u_{x_i} v_{x_j}\, dx = \langle F, v \rangle \equiv \int_{B^+} (f^0 v + f^i v_{x_i})\, dx \qquad \textit{for } v \in H_0^1(B^+ \cup S^0) \end{gathered} \tag{2.51}$$

with $f^0 \in L^2(B^+), f^1, \ldots, f^N \in H^1(B^+)$. *Then* $u \in H^2(B^+)$ *with norm estimate*

$$|u|_{H^2(B^+)} \le C\Big(|f^0|_{2;+} + \sum_{i=1}^N |f^i|_{H^1(B^+)} + |u|_{H^1(B^+)}\Big);$$

the constant C (independent of u, F) depends on the a^{ij}*'s through the bound imposed on their* $C^{0,1}(\overline{B^+})$ *norms, as well as through* α.

PROOF. Let $h \in \mathbb{R}\setminus\{0\}$ with $|h| < \operatorname{dist}(\operatorname{supp} u, S^+)$. For $s = 1, \ldots, N-1$, $\delta_h \equiv \delta_h{}^s$, the functions $\delta_h u$ and $\delta_{-h}\delta_h u$ belong to $H_0^1(B^+)$ in the case (2.49), to $H_0^1(B^+ \cup S^0)$ in the case (2.51). We can therefore proceed as in the proof of Lemma 2.20 and demonstrate that $u_{x_1}, \ldots, u_{x_{N-1}} \in H^1(B^+)$ with norm estimates.

We now write the distributional identity

$$-(a^{ij}u_{x_i})_{x_j} = f^0 - f^i_{x_i} \qquad \text{in } B^+$$

as

$$(a^{NN}u_{x_N})_{x_N} = -\textstyle\sum'(a^{ij}u_{x_i})_{x_j} - f^0 + f^i_{x_i} \qquad \text{in } B^+,$$

where $\sum'$ denotes summation from 1 to N over all pairs of indices $(i, j) \neq (N, N)$. By the preceding considerations,

$$(a^{NN}u_{x_N})_{x_N} \in L^2(B^+).$$

Since the Lipschitzian function a^{NN} is $\geq \alpha > 0$ on $\overline{B^+}$, we have

$$\begin{aligned}\int_{B^+} u_{x_N}w_{x_N}\,dx &= \int_{B^+} u_{x_N}\Big(a^{NN}\frac{w}{a^{NN}}\Big)_{x_N}\,dx \\ &= \int_{B^+}\Big[u_{x_N}a^{NN}_{x_N}\frac{w}{a^{NN}} + u_{x_N}a^{NN}\Big(\frac{w}{a^{NN}}\Big)_{x_N}\Big]\,dx \\ &= \int_{B^+}[u_{x_N}a^{NN}_{x_N}(a^{NN})^{-1} - (a^{NN}u_{x_N})_{x_N}(a^{NN})^{-1}]w\,dx \\ &\qquad\qquad \text{for } w \in C_c^1(B^+),\end{aligned}$$

hence $u_{x_Nx_N} \in L^2(B^+)$ (with norm estimate). ☐

Remark 1. Inspection shows that the conclusion of the above lemma remains valid if (2.51) is weakened into the requirement that

$$u \in H^1(B^+), \qquad \operatorname{supp} u \subset B^+ \cup S^0,$$

$$\int_{B^+} a^{ij}u_{x_i}(\delta_{-h}\delta_h u)_{x_j}\,dx \geq \langle F, \delta_{-h}\delta_h u\rangle$$

for $h \in \mathbb{R}\setminus\{0\}$ with $|h| < \operatorname{dist}(\operatorname{supp} u, S^+)$, $\delta_h = \delta_h{}^s$ for $s = 1, \ldots, N-1$, and

$$\int_{B^+} a^{ij}u_{x_i}v_{x_j}\,dx = \langle F, v\rangle \qquad \text{for } v \in H_0{}^1(B^+).$$

This fact will be utilized later on (proof of Theorem 4.39).

Remark 2. In the case (2.51) of Lemma 2.22 we can construct a domain $\omega \subset B^+$ in such a way that $\operatorname{supp} u \cap B^+ \subset \omega$, $\bar{\omega} \subset B^+ \cup S^0$, and $\partial\omega$ is of class C^1. By the divergence theorem (see Theorem 1.53) the func-

er 2

) satisfies

$$\ldots_{x_j}\, dx = \int_\omega fv\, dx + \int_{\partial\omega} \zeta v|_\Gamma\, d\sigma \qquad \text{for } v \in H^1(\omega),$$

where $f \equiv (f^0 - f^i_{x_i})|_\omega \in L^2(\omega)$ and $\zeta \equiv f^i|_{\partial\omega}\mu^i \in H^{1/2}(\partial\omega)$, $(\mu^1, \ldots, \mu^N)$ denoting the outward unit normal to $\partial\omega$. From Lemmas 2.5 and 2.6 we can therefore deduce that

$$a^{ij}u_{x_i}\mu^j - f^i\mu^i \qquad \text{on } \partial\omega,$$

hence that

$$a^{iN}u_{x_i} = f^N \qquad \text{a.e. } [N-1] \text{ on } S^0.$$

As a counterpart to Lemma 2.21 we have the following lemma.

LEMMA 2.23. *Let k be a nonnegative integer. Suppose that the coefficients of the bilinear form* (2.11) (*for $\Omega = B^+$*) *satisfy*

$$a^{ij},\ d^j \in C^{k,1}(\overline{B^+}), \qquad b^i,\ c \in H^{k,\infty}(B^+),$$

and let u satisfy either (2.45) *or* (2.46) *with $f^0 \in H^k(B^+)$ and $f^1, \ldots, f^N \in H^{k+1}(B^+)$. Whenever $0 < R < 1$, $u|_{B_R^+}$ belongs to $H^{k+2}(B_R^+)$ with norm estimate*

$$| u |_{H^{k+2}(B_R^+)} \leq C\left(| f^0 |_{H^k(B^+)} + \sum_{i=1}^{N} | f^i |_{H^{k+1}(B^+)} + | u |_{H^1(B^+)}\right);$$

the constant C (independent of u, F) depends on the coefficients of $a(u, v)$ through the bound imposed on their respective norms, as well as through α.

PROOF. We proceed by induction on k.

Step 1: *The case $k = 0$.* The function u satisfies

$$\int_{B^+} a^{ij}u_{x_i}v_{x_j}\, dx = \int_{B^+} (\hat{f}^0 v + \hat{f}^i v_{x_i})\, dx$$

for $v \in H_0^1(B^+)$ in the case (2.45), for $v \in H_0^1(B^+ \cup S^0)$ in the case (2.46), with $\hat{f}^0 \equiv f^0 - b^i u_{x_i} - cu \in L^2(B^+)$ and $\hat{f}^i \equiv f^i - d^i u \in H^1(B^+)$ for $i = 1, \ldots, N$. Let $g \in C_c^\infty(B)$, $g = 1$ on $\bar{B}_R$: Lemma 2.22 can be applied to the function gu, with f^0 replaced by $\hat{f}^0 g - a^{ij}u_{x_i}g_{x_j} + \hat{f}^i g_{x_i}$, f^i by $\hat{f}^i g + a^{ji}ug_{x_j}$ for $i = 1, \ldots, N$ (see Step 1 of the proof of Lemma 2.21). The conclusion follows in the case at hand.

Step 2: *The case* $k \in \mathbb{N}$. Suppose that the lemma is valid with k replaced by $k-1$. The functions $u_{x_1}|_{B_r^+}, \ldots, u_{x_N}|_{B_r^+}$ then belong to $H^k(B_r^+)$, $r = (R+1)/2$.

Let $s = 1, \ldots, N-1$: then we also have $u_{x_s}|_{B_r^+} \in H_0^1(B_r^+ \cup S_r^+)$ if $u \in H_0^1(B^+ \cup S^+)$ [since $(\delta_h^s u)|_{B_r^+} \in H_0^1(B_r^+ \cup S_r^+)$ for $0 < |h| < (1-R)/2$]. As in Step 2 of the proof of Lemma 2.21 it is clear that u_{x_s} satisfies

$$\int_{B_r^+} a^{ij} u_{x_s x_i} v_{x_j}\, dx = \int_{B_r^+} [\hat{f}^0_{x_s} v + (\hat{f}^i_{x_s} - a^{ji}_{x_s} u_{x_j}) v_{x_i}]\, dx$$

for $v \in H_0^1(B_r^+)$ in the case (2.45), for $v \in H_0^1(B_r^+ \cup S_r^0)$ in the case (2.46). [Notice that $\int_{B_r^+} w_{x_s}\, dx = 0$ if $w \in H^1(B_r^+)$ with supp $w \subset B_r^+ \cup S_r^0$.] The inductive assumption concerning the value $k-1$, with B^+ replaced by B_r^+ and u by u_{x_s}, yields $u_{x_s}|_{B_R^+} \in H^{k+1}(B_R^+)$ (with norm estimate). At this point we need only utilize the distributional identity

$$(a^{NN} u_{x_N})_{x_N} = -\Sigma'(a^{ij} u_{x_i})_{x_j} - \hat{f}^0 + \hat{f}^i_{x_i} \qquad \text{in } B_R^+$$

to arrive at $u_{x_N x_N}|_{B_R^+} \in H^k(B_R^+)$ (with norm estimate). ☐

Corollary. *Take the coefficients of* $a(u, v)$ *in* $C^\infty(\overline{B^+})$ *and let* u *satisfy either* (2.45) *or* (2.46) *with* $f^0, f^1, \ldots, f^N \in C^\infty(\overline{B^+})$. *Then* $u \in C^\infty(\overline{B_R^+})$ *for any* $R \in]0, 1[$.

Remark. In the case (2.46) of Lemma 2.23 it is easy to verify that $a^{iN} u_{x_i} + d^N u = f^N$ a.e. $[N-1]$ on S^0 by using Remark 2 after Lemma 2.22.

For what concerns global regularity we have the following theorem.

Theorem 2.24. *Let* k *be a nonnegative integer. Suppose that* $\partial\Omega$ *is of class* $C^{k+1,1}$, *that its open portion* Γ *is also closed, and that the coefficients of the bilinear form* (2.11) *satisfy*

$$a^{ij},\ d^j \in C^{k,1}(\bar{\Omega}),\ b^i,\ c \in H^{k,\infty}(\Omega).$$

Then any solution u *of* (2.6) *with* $V = H_0^1(\Omega \cup \Gamma)$ *and* $\langle F, v\rangle = \int_\Omega (f^0 v + f^i v_{x_i})\, dx$ *for* $v \in V$, *where* $f^0 \in H^k(\Omega)$ *and* $f^1, \ldots, f^N \in H^{k+1}(\Omega)$, *belongs to* $H^{k+2}(\Omega)$ *with norm estimate*

$$|u|_{H^{k+2}(\Omega)} \leq C\Big(|f^0|_{H^k(\Omega)} + \sum_{i=1}^{N} |f^i|_{H^{k+1}(\Omega)} + |u|_{H^1(\Omega)}\Big).$$

The constant C is independent of u, F; it depends on the coefficients of a(u, v) only through the bound imposed on their respective norms and through α.

This theorem can be proven by the same technique adopted in the proof of Theorem 2.19, with some simplifications. Note that, if $\Lambda: \bar{U} \to \bar{B}$ is a $C^{k+1,1}$ diffeomorphism that straightens a portion $U \cap \partial\Omega$ of $\partial\Omega$, the data in equation (2.47) satisfy

$$a'^{hk},\ d'^{k} \in C^{k,1}(\overline{B^+}),\ b'^{h},\ c' \in H^{k,\infty}(B^+),$$
$$f'^{0} \in H^{k}(B^+),\ f'^{1}, \ldots, f'^{N} \in H^{k+1}(B^+),$$

and the membership in $H^{k+1}(B_R{}^+)$ of $u'_{y_1}|_{B_R{}^+}, \ldots, u'_{y_N}|_{B_R{}^+}$ implies membership in $H^{k+1}(\omega)$ of $u_{x_1}|_{\omega}, \ldots, u_{x_N}|_{\omega}$ [with $\omega \equiv \Lambda^{-1}(B_R{}^+)$, and $u' \equiv (u \circ \Lambda^{-1})|_{B^+}$], by Lemma 1.28.

COROLLARY. *Let the assumptions of Theorem* 2.24 *be satisfied for all values of k. Then* $u \in C^\infty(\bar{\Omega})$.

REMARK. By Lemmas 2.5 and 2.6, the function u from Theorem 2.24 satisfies

$$(a^{ij}u_{x_i} + d^{j}u)\nu^{j} = f^{i}\nu^{i} \qquad \text{a.e. } [N-1] \text{ on } \Gamma.$$

2.6. Interior Regularity for Nonlinear Equations

In the sequel we shall make use of the following terminology: $g(x, \zeta)$ is a *Carathéodory function* of $x \in \Omega$ and $\zeta \in \mathbb{R}^M$ (M being a natural number) if

- $g(\cdot, \zeta)\colon \Omega \to \mathbb{R}$ is measurable for any $\zeta \in \mathbb{R}^M$,
- $g(x, \cdot)\colon \mathbb{R}^M \to \mathbb{R}$ is continuous for a.a. $x \in \Omega$.

We now take $f^0, f^1, \ldots, f^N$ in $L^\infty(\Omega)$ and rewrite (2.43) as

$$u \in H^1(\Omega), \qquad -\frac{\partial}{\partial x_i} A^{i}(u, \nabla u) + A^{0}(u, \nabla u) = 0 \qquad \text{in } \Omega. \qquad (2.52)$$

Here, for $j = 0, 1, \ldots, N$, $A^{j}(\eta, \xi)$ is the function $x \mapsto a^{j}(x, \eta(x), \xi(x))$ if $\eta, \xi_1, \ldots, \xi_N$ denote measurable functions on Ω, $\xi \equiv (\xi_1, \ldots, \xi_N)$, with

$$\begin{aligned} a^{i}(x, \eta, \xi) &\equiv a^{ji}(x)\xi_j + d^{i}(x)\eta - f^{i}(x) \qquad \text{for } i = 1, \ldots, N,\\ a^{0}(x, \eta, \xi) &\equiv b^{j}(x)\xi_j + c(x)\eta - f^{0}(x). \end{aligned} \qquad (2.53)$$

The a^j's are Carathéodory functions of $x \in \Omega$ and $(\eta, \xi) \in \mathbb{R}^{1+N}$; moreover, there exist two constants C_1, C_2 such that

$$| a^j(x, \eta, \xi) | \leq C_1(| \eta | + | \xi | + 1), \qquad j = 0, 1, \ldots, N, \tag{2.54}$$

and

$$\begin{gathered} a^i(x, \eta, \xi)\xi_i \geq (\alpha/2) | \xi |^2 - C_2(\eta^2 + 1) \\ \text{for a.a. } x \in \Omega \qquad \text{and} \qquad \text{any } (\eta, \xi) \in \mathbb{R}^{1+N}. \end{gathered} \tag{2.55}$$

If the requirement that the a^j's be defined by (2.53) is dropped, (2.52) becomes a nonlinear equation; it still makes sense in $H^{-1}(\Omega)$ if the a^j's are Carathéodory functions of $x \in \Omega$ and $(\eta, \xi) \in \mathbb{R}^{1+N}$ satisfying (2.54). We shall investigate the solvability of (b.v.p.'s associated with) nonlinear equations such as (2.52) in Sections 4.3 and 4.9. In the present section we instead provide some interior regularity results for solutions u, assuming their existence.

We take $N \geq 3$.

2.6.1. Local Boundedness

The next result can be viewed as a nonlinear counterpart to Lemma 2.12.

LEMMA 2.25. *For* $j = 0, 1, \ldots, N$ *let* $a^j(x, \eta, \xi)$ *be a Carathéodory function of* $x \in \Omega$ *and* $(\eta, \xi) \in \mathbb{R}^{1+N}$ *satisfying* (2.54), (2.55). *Let* u *satisfy* (2.52) *and set*

$$u^{(k)} \equiv (-k) \vee u \wedge k \qquad \textit{for } 0 < k < \infty.$$

Assume $x^0 \in \Omega$, $B_{2r}(x^0) \subseteq \Omega$ $(0 < r \leq 1)$. *Then a bound*

$$\left[\int_{B_\varrho(x^0)} (1 + u^2 | u^{(k)} |^{p\lambda+2(\lambda-1)})\, dx\right]^{1/\lambda} \leq C \frac{(1 + p)^2}{(R - \varrho)^2} \int_{B_R(x^0)} (1 + u^2 | u^{(k)} |^p)\, dx, \tag{2.56}$$

where $\lambda \equiv N/(N - 2)$, *holds whenever* $0 \leq p < \infty$ *and* $0 < \varrho < R \leq 2r$, *the constant* C *being independent of* u, x^0, r, *and* k.

PROOF. We take $x^0 = 0$ and fix $g \in C^1(\bar{\Omega})$ with $\operatorname{supp} g \subset B_R$, $0 \leq g \leq 1$, $g = 1$ on $\overline{B_\varrho}$, $| \nabla g | \leq 2(R - \varrho)^{-1}$. Then $v \equiv g^2 | u^{(k)} |^p u$ belongs to

$H_0^1(\Omega)$ with

$$v_{x_i} = g^2 | u^{(k)} |^p (p u_{x_i}^{(k)} + u_{x_i}) + 2 g g_{x_i} | u^{(k)} |^p u,$$

and (2.52) yields

$$\int_\Omega A^i g^2 | u^{(k)} |^p (p u_{x_i}^{(k)} + u_{x_i})\, dx = -2 \int_\Omega A^i g g_{x_i} | u^{(k)} |^p u\, dx + \int_\Omega A^0 g^2 | u^{(k)} |^p u\, dx,$$

where we have written A^j for $A^j(u, \nabla u)$, $j = 0, 1, \ldots, N$. Since $A^i(u, \nabla u) u_{x_i}^{(k)} = A^i(u, \nabla u^{(k)}) u_{x_i}^{(k)}$ by Theorem 1.56, (2.54) and (2.55) yield

$$\begin{aligned} \frac{\alpha}{2} \int_\Omega g^2 | u^{(k)} |^p (p | \nabla u^{(k)} |^2 + | \nabla u |^2)\, dx &\le C_2(1 + p) \int_\Omega g^2 | u^{(k)} |^p (u^2 + 1)\, dx \\ &\quad + 2NC_1 \int_\Omega (| u | + | \nabla u | + 1) g | u^{(k)} |^p | u | | \nabla g |\, dx \\ &\quad + C_1 \int_\Omega (| u | + | \nabla u | + 1) g^2 | u^{(k)} |^p | u |\, dx. \end{aligned}$$

We majorize the quantities

$$2NC_1 \int_\Omega | \nabla u | g | u^{(k)} |^p | u | | \nabla g |\, dx$$

and

$$C_1 \int_\Omega | \nabla u | g^2 | u^{(k)} |^p | u |\, dx$$

with

$$\frac{\alpha}{8} \int_\Omega g^2 | u^{(k)} |^p | \nabla u |^2\, dx + C \int_\Omega | u^{(k)} |^p u^2 | \nabla g |^2\, dx$$

and

$$\frac{\alpha}{8} \int_\Omega g^2 | u^{(k)} |^p | \nabla u |^2\, dx + C \int_\Omega g^2 | u^{(k)} |^p u^2\, dx,$$

respectively. Set $\hat{u} \equiv | u^{(k)} |^{p/2} | u |$, so that

$$| u^{(k)} |^p \le | u^{(k)} |^{p+2} + 1 \le \hat{u}^2 + 1$$

and

$$| u^{(k)} |^p | u | \le | u^{(k)} |^p (u^2 + 1) \le 2\hat{u}^2 + 1:$$

since

$$C_2(1+p)\int_\Omega g^2\,|\,u^{(k)}\,|^p(u^2+1)\,dx$$
$$+\,2NC_1\int_\Omega(|\,u\,|+1)g\,|\,u^{(k)}\,|^p\,|\,u\,|\,|\,\nabla g\,|\,dx$$
$$+\,C_1\int_\Omega(|\,u\,|+1)g^2\,|\,u^{(k)}\,|^p\,|\,u\,|\,dx$$
$$\le C_2(1+p)\int_\Omega g^2(2\hat{u}^2+1)\,dx$$
$$+\,NC_1\int_\Omega(g^2+|\,\nabla g\,|^2)(3\hat{u}^2+1)\,dx+C_1\int_\Omega g^2(3\hat{u}^2+1)\,dx,$$

we arrive at the inequality

$$\frac{\alpha}{2}\int_\Omega g^2\,|\,u^{(k)}\,|^p(p\,|\,\nabla u^{(k)}\,|^2+\frac{1}{2}\,|\,\nabla u\,|^2)\,dx$$
$$\le C(1+p)\int_\Omega(g^2+|\,\nabla g\,|^2)(\hat{u}^2+1)\,dx.$$

But

$$|\,\nabla\hat{u}\,|^2\le\left(\frac{p}{2}+4\right)|\,u^{(k)}\,|^p\left(p\,|\,\nabla u^{(k)}\,|^2+\frac{1}{2}\,|\,\nabla u\,|^2\right),$$

so that

$$\int_\Omega|\,\nabla(g\hat{u})\,|^2\,dx\le C(1+p)^2\int_\Omega(g^2+|\,\nabla g\,|^2)(\hat{u}^2+1)\,dx. \tag{2.57}$$

From (2.57) we deduce that

$$|\,g\hat{u}\,|^2_{2\lambda;R}\le C(1+p)^2\int_{B_R}(g^2+|\,\nabla g\,|^2)(\hat{u}^2+1)\,dx$$

(see Theorem 1.33 and the corollary of Theorem 1.43). But then

$$\left(\int_{B_\varrho}\hat{u}^{2\lambda}\,dx\right)^{1/\lambda}\le C\frac{(1+p)^2}{(R-\varrho)^2}\int_{B_R}(\hat{u}^2+1)\,dx,$$

and (2.56) (with $x^0=0$) follows from the inequality

$$u^2\,|\,u^{(k)}\,|^{p\lambda+2(\lambda-1)}\le\hat{u}^{2\lambda}$$

since $\varrho^N\le[R^N/(R-\varrho)^2]^\lambda$. □

We now insert $R = r_m \equiv r(1 + 1/2^m)$, $\varrho = r_{m+1}$ and $p = p_m \equiv 2(\lambda^m - 1)$ in (2.56), thus obtaining

$$A_{m+1} \leq A_m \left(C \frac{2\lambda^m - 1}{r 2^{-m-1}} \right)^{2/\lambda^m},$$

hence

$$A_{m+1} \leq A_0 \prod_{i=0}^{m} \left(C \frac{2\lambda^i - 1}{r 2^{-i-1}} \right)^{2/\lambda^i} \tag{2.58}$$

where

$$A_m \equiv \left[\iint_{B_{r_m}(x^0)} (1 + u^2 \mid u^{(k)} \mid^{p_m})\, dx \right]^{1/\lambda^m}$$

$(m = 0, 1, 2, \ldots)$. The logarithm of the right-hand side of (2.58) is bounded by

$$\ln A_0 + 2 \sum_{i=0}^{\infty} \frac{i+1}{\lambda^i} \ln [(C 2 \lambda^i)^{1/(i+1)} 2] - 2 \ln r \sum_{i=0}^{\infty} \frac{1}{\lambda^i}$$

$$= \ln A_0 + C - N \ln r.$$

We thus arrive at

$$\left(\int_{B_r(x^0)} \mid u^{(k)} \mid^{2\lambda^{m+1}} dx \right)^{1/\lambda^{m+1}} \leq A_{m+1}$$
$$\leq C r^{-N} \int_{B_{2r}(x^0)} (1 + u^2)\, dx,$$

and finally at

$$\operatorname*{ess\,sup}_{B_r(x^0)} u^2 \leq C r^{-N} \int_{B_{2r}(x^0)} (1 + u^2)\, dx$$

after letting $m \to \infty$, $k \to \infty$. (Compare with the proof of Lemma 2.11.) By a straightforward compactness argument we can therefore conclude with the following theorem.

THEOREM 2.26. *Same assumptions about the functions* $a^j(x, \eta, \xi)$ *as in Lemma* 2.25. *Whenever* $\omega \subset\subset \Omega$, *the restriction to* ω *of any function* u *satisfying* (2.52) *belongs to* $L^\infty(\omega)$ *with*

$$\mid u \mid_{\infty;\omega} \leq C \left[\iint_{\Omega} (1 + u^2)\, dx \right]^{1/2}.$$

2.6.2. H^2 Regularity

For what concerns interior differentiability of solutions to (2.52) we have the following theorem.

THEOREM 2.27. *For* $i = 1, \ldots, N$ *let the functions* $a^i(x, \eta, \xi)$ *belong to* $C^1(\bar{\Omega} \times \mathbb{R} \times \mathbb{R}^N)$ *and satisfy*

$$\begin{gathered} | a^i(x, \eta, \xi) | \leq C(| \eta | + | \xi | + 1), \\ | a^i_{x_1}(x, \eta, \xi) |, \ldots, | a^i_{x_N}(x, \eta, \xi) | \leq C_1(| \eta |)(| \xi | + 1), \\ | a_\eta{}^i(x, \eta, \xi) |, | a^i_{\xi_1}(x, \eta, \xi) |, \ldots, | a^i_{\xi_N}(x, \eta, \xi) | \leq C_1(| \eta |) \\ \text{for } (x, \eta, \xi) \in \bar{\Omega} \times \mathbb{R} \times \mathbb{R}^N \end{gathered} \tag{2.59}$$

as well as

$$\begin{gathered} a^i_{\xi_j}(x, \eta, \xi)\tilde{\xi}_i\tilde{\xi}_j \geq \alpha | \tilde{\xi} |^2 \\ \text{for } (x, \eta, \xi) \in \bar{\Omega} \times \mathbb{R} \times \mathbb{R}^N, \qquad \tilde{\xi} \in \mathbb{R}^N \qquad (\alpha > 0), \end{gathered} \tag{2.60}$$

and let $a^0(x, \eta, \xi)$ *be a Carathéodory function of* $x \in \Omega$ *and* $(\eta, \xi) \in \mathbb{R}^{1+N}$ *satisfying*

$$\begin{gathered} | a^0(x, \eta, \xi) | \leq C(| \eta | + | \xi | + 1) \\ \text{for a.a. } x \in \Omega \qquad \text{and} \qquad \text{any } (\eta, \xi) \in \mathbb{R}^{1+N}; \end{gathered} \tag{2.61}$$

in (2.59) $C_1(s)$ *is an increasing function of* $s \in [0, \infty[$. *Whenever* $\omega \subset\subset \Omega$, *the restriction to* ω *of any solution* u *to* (2.52) *belongs to* $H^2(\omega)$ *with*

$$| u |_{H^2(\omega)} \leq C\left[\int_\Omega (1 + | \nabla u |^2)\, dx\right]^{1/2} \tag{2.62}$$

PROOF. It is easy to see [by writing $a^i(x, \eta, \xi)$ as $\int_0^1 a^i_{\xi_j}(x, \eta, t\xi)\xi_j\, dt + a^i(x, \eta, 0)$] that (2.60) implies (2.55). By Theorem 2.26, $u|_{\Omega'}$ belongs to $L^\infty(\Omega')$ whenever $\Omega' \subset\subset \Omega$, so that we can without loss of generality prove the present theorem under the additional assumption $u \in L^\infty(\Omega)$.

Nonlinearity forces us to introduce difference quotients at the same time as multiplication by a cutoff function. (Compare with Lemmas 2.20 and 2.21.) We utilize the notations

$$\tau_h w(x) \equiv \tau_h{}^s w(x) \equiv w(x + he^s), \qquad \delta_h w \equiv \delta_h{}^s w \equiv (\tau_h w - w)/h$$

for $h \in \mathbb{R}\setminus\{0\}$, e^s denoting the sth coordinate vector ($s = 1, \ldots, N$). Let $g \in C^\infty(\bar{\Omega})$, supp $g \subset \Omega' \subset\subset \Omega$, $0 \leq g \leq 1$, $g = 1$ on $\bar{\omega} \subset \Omega'$ and take $v =$

$-\delta_{-h}(g^2\delta_h u)$ with $0 < |h| < \frac{1}{2}\operatorname{dist}(\Omega', \partial\Omega)$. We have

$$\int_\Omega A^i(u, \nabla u) v_{x_i}\, dx = \int_{\Omega'} \delta_h A^i(u, \nabla u)(g^2 \delta_h u_{x_i} + 2 g g_{x_i} \delta_h u)\, dx$$

and

$$\begin{aligned} &\delta_h A^i(u, \nabla u)(x) \\ &\quad = \frac{1}{h}\int_0^1 \frac{d}{dt} a^i(x + the^s,\ u(x) + th\, \delta_h u(x),\ \nabla u(x) + th\, \nabla\delta_h u(x))\, dt \\ &\quad = \int_0^1 [a^i_{x_s} + a^i_\eta\, \delta_h u(x) + a^i_{\xi_j} \delta_h u_{x_j}(x)]\, dt, \end{aligned}$$

the argument of all partial derivatives of a^i in the last integral being

$$(x + the^s,\ u(x) + th\, \delta_h u(x),\ \nabla u(x) + th\, \nabla\delta_h u(x)).$$

Therefore, by utilizing (2.59) [with $C_1(|\eta|)$ replaced by $C_1(|u|_{\infty;\Omega})$] together with (2.60) and (2.61), we obtain

$$\begin{aligned} \int_\Omega A^i(u, \nabla u) v_{x_i}\, dx \geq \int_{\Omega'} \{ & g^2[-C(1 + |\delta_h u| + |\nabla u| \\ & + |h|\,|\nabla\delta_h u|)\,|\nabla\delta_h u| + \alpha\,|\nabla\delta_h u|^2] \\ & - C(1 + |\delta_h u| + |\nabla u| + |h|\,|\nabla\delta_h u| \\ & + |\nabla\delta_h u|) 2g\,|\nabla g|\,|\delta_h u|\}\, dx. \end{aligned}$$

For $|h|$ small enough the right-hand side of the above inequality is minorized by a quantity

$$\frac{\alpha}{2}\int_{\Omega'} g^2\,|\nabla\delta_h u|^2\, dx - C\int_\Omega (1 + |\nabla u|^2)\, dx$$

(see Theorem 1.21). On the other hand,

$$\begin{aligned} &\left| \int_\Omega A^0(u, \nabla u) v\, dx \right| \\ &\quad \leq C\int_{\Omega'} (|\nabla u| + 1)(|\tau_{-h} g|\,|\delta_{-h}(g\delta_h u)| + |\delta_{-h} g|\,|g\delta_h u|)\, dx \\ &\quad \leq \frac{\alpha}{8}\int_{\Omega'} |\nabla(g\delta_h u)|^2\, dx + C\int_{\Omega'} (|\nabla u| + 1)^2\, dx. \end{aligned}$$

From (2.52) we therefore deduce the estimate

$$\frac{\alpha}{4}\int_{\Omega'} g^2 |\nabla\delta_h u|^2 \, dx \le C\int_{\Omega} (1 + |\nabla u|^2) \, dx,$$

which implies $u|_\omega \in H^2(\omega)$ together with (2.62). □

2.6.3. $H^{1,\infty}$ and $C^{k,\delta}$ Regularity

LEMMA 2.28. *Same assumptions as in Theorem 2.27. Whenever $\omega \subset\subset \Omega$, the restrictions to ω of all first derivatives of any function u satisfying (2.52) belong to $L^\infty(\omega)$, with*

$$|\nabla u|_{\infty;\omega} \le C\left[\int_{\Omega} (1 + |\nabla u|^2) \, dx\right]^{1/2}.$$

PROOF. Thanks to Theorems 2.26 and 2.27 we can, without loss of generality, restrict ourselves to solutions of (2.52) which belong to $L^\infty(\Omega)$ as well as to $H^2(\Omega)$.

For $i, j = 1, \ldots, N$ we set

$$A^{ij}(x) \equiv a^j_{\xi_i}(x, u(x), \nabla u(x)).$$

Next we fix $s = 1, \ldots, N$ and put

$$\begin{aligned} B^i_{(s)}(x) \equiv {} & -a^i_{x_s}(x, u(x), \nabla u(x)) \\ & -a_\eta{}^i(x, u(x), \nabla u(x))u_{x_s}(x) + a^0(x, u(x), \nabla u(x))\delta^{is} \end{aligned}$$

with $\delta^{is} = 0$ for $i \ne s$, $= 1$ for $i = s$; note that $|B^i_{(s)}| \le C(1 + |\nabla u|)$. If $v \in C_c^\infty(\Omega)$, (2.52) yields

$$\begin{aligned} 0 &= \int_{\Omega} [A^i(u, \nabla u)v_{x_s x_i} + A^0(u, \nabla u)v_{x_s}] \, dx \\ &= \int_{\Omega} (-A^{ij}u_{x_s x_i}v_{x_j} + B^i_{(s)}v_{x_i}) \, ds, \end{aligned}$$

so that $w \equiv u_{x_s}$ satisfies the equation

$$w \in H^1(\Omega), \qquad -(A^{ij}w_{x_i})_{x_j} = B^i_{(s)x_i} \quad \text{in } \Omega \tag{2.63}$$

as an identity in $H^{-1}(\Omega)$.

The function $|\nabla u|^2$ belongs to $H^{1,1}(\Omega)$; by Theorem 1.56, the same is true of $z_{k\varepsilon} \equiv \varepsilon \vee |\nabla u|^2 \wedge k$, $0 \le \varepsilon < k < \infty$, with

$$z_{k\varepsilon x_j} = \begin{cases} 0 & \text{for } |\nabla u|^2 \le \varepsilon \quad \text{or} \quad |\nabla u|^2 \ge k, \\ 2u_{x_i}u_{x_i x_j} & \text{for } \varepsilon < |\nabla u|^2 < k, \end{cases} \tag{2.64}$$

so that $z_{k\varepsilon} \in H^1(\Omega) \cap L^\infty(\Omega)$. Take $\varepsilon > 0$ and let $p \ge 0$. Assuming, without loss of generality, that $\Omega \supseteq B_{2r}$ for some $r \in]0, 1]$, we denote by g the same cutoff function as in the proof of Lemma 2.25, and set

$$v \equiv v_{(s)} \equiv g^2 z_{k\varepsilon}^p w.$$

Thus, $v \in H_0^{1,1}(\Omega)$ with

$$v_{x_j} = g^2(z_{k\varepsilon}^p w_{x_j} + p z_{k\varepsilon}^{p-1} z_{k\varepsilon x_j} w) + 2g g_{x_j} z_{k\varepsilon}^p w,$$

and finally $v \in H_0{}^1(\Omega)$ because of (2.64).

Because of assumptions (2.59)–(2.61), the equation yields

$$\begin{aligned} \int_\Omega & g^2(\alpha z_{k\varepsilon}^p |\nabla w|^2 + p z_{k\varepsilon}^{p-1} A^{ij} w_{x_i} w z_{k\varepsilon x_j})\, dx - C \int_\Omega g\, |\nabla g|\, z_{k\varepsilon}^p\, |w|\, |\nabla w|\, dx \\ &\le \int_\Omega A^{ij} w_{x_i} v_{x_j}\, dx = \int_\Omega B_{(s)}^i v_{x_i}\, dx \\ &\le C \int_\Omega (|\nabla u| + 1)[g^2(z_{k\varepsilon}^p |\nabla w| + p z_{k\varepsilon}^{p-1} |\nabla z_{k\varepsilon}|\, |w|) \\ &\quad + 2g\, |\nabla g|\, z_{k\varepsilon}^p\, |w|]\, dx \end{aligned}$$

by the previous definitions of A^{ij} and $B_{(s)}^i$.

We now write u_{x_s} instead of w and sum over s from 1 to N. Since $u_{x_s x_i} u_{x_s} z_{k\varepsilon x_i} = \frac{1}{2} z_{k\varepsilon x_i} z_{k\varepsilon x_j}$ [see (2.64)] and $|\nabla z_{k\varepsilon}|\, |\nabla u| = |\nabla z_{k\varepsilon}|\, z_{k\varepsilon}^{1/2}$, we obtain

$$\begin{aligned} \alpha \int_\Omega & g^2\left(z_{k\varepsilon}^p \sum_{s=1}^N |\nabla u_{x_s}|^2 + \frac{p}{2} z_{k\varepsilon}^{p-1} |\nabla z_{k\varepsilon}|^2\right) dx \\ &\le \frac{\alpha}{4} \int_\Omega g^2 z_{k\varepsilon}^p \sum_{s=1}^N |\nabla u_{x_s}|^2\, dx + C \int_\Omega |\nabla g|^2 z_{k\varepsilon}^p |\nabla u|^2\, dx \\ &\quad + \frac{\alpha}{4} \int_\Omega g^2 z_{k\varepsilon}^p \sum_{s=1}^N |\nabla u_{x_s}|^2\, dx + C \int_\Omega (|\nabla u| + 1)^2 g^2 z_{k\varepsilon}^p\, dx \\ &\quad + \frac{\alpha}{4} p \int_\Omega g^2 z_{k\varepsilon}^{p-1} |\nabla z_{k\varepsilon}|^2\, dx + Cp \int_\Omega (|\nabla u| + 1)^2 g^2 z_{k\varepsilon}^p\, dx \\ &\quad + C \int_\Omega (g^2 + |\nabla g|^2) z_{k\varepsilon}^p (|\nabla u| + 1)^2\, dx, \end{aligned}$$

hence

$$\frac{\alpha}{2}\int_\Omega g^2\Big(z_{k\varepsilon}^p \sum_{s=1}^N |\nabla u_{x_s}|^2 + \frac{p}{2} z_{k\varepsilon}^{p-1} |\nabla z_{k\varepsilon}|^2\Big)\,dx$$
$$\leq C(1+p)\int_\Omega (g^2 + |\nabla g|^2)(1+|\nabla u|^2) z_{k\varepsilon}^p\,dx. \tag{2.65}$$

The function $\hat{z}_\varepsilon \equiv z_{k\varepsilon}^{p/2} |\nabla u|$ satisfies

$$\hat{z}_{\varepsilon x_s} = \frac{p}{2} z_{k\varepsilon}^{p/2-1} z_{k\varepsilon x_s} |\nabla u| + z_{k\varepsilon}^{p/2} u_{x_i} u_{x_i x_s} / |\nabla u|$$

(see Problem 1.24), hence

$$|\nabla \hat{z}_\varepsilon|^2 \leq 2\Big(\frac{p^2}{4} z_{k\varepsilon}^{p-1} |\nabla z_{k\varepsilon}|^2 + C z_{k\varepsilon}^p \sum_{s=1}^N |\nabla u_{x_s}|^2\Big)$$
$$\leq (p+C)\Big(\frac{p}{2} z_{k\varepsilon}^{p-1} |\nabla z_{k\varepsilon}|^2 + z_{k\varepsilon}^p \sum_{s=1}^N |\nabla u_{x_s}|^2\Big).$$

From (2.65) we therefore deduce the inequality

$$\int_\Omega |\nabla(g\hat{z}_\varepsilon)|^2\,dx \leq C(1+p)^2 \int_\Omega (g^2 + |\nabla g|^2)(z_{k\varepsilon}^p + \hat{z}_\varepsilon^2)\,dx.$$

Passing to a suitable subsequence of indices ε we utilize a weak convergence argument and ascertain that the above remains valid with $\varepsilon = 0$, $\hat{z}_0$ being of course $z_{k0}^{p/2} |\nabla u|$. But since

$$z_{k0}^p \leq z_{k0}^{p+1} + 1 \leq \hat{z}_0^2 + 1,$$

we have obtained (2.57) with $\hat{u}$ replaced by $\hat{z}_0$. The conclusion of the lemma can now be reached by proceeding as in Section 2.6.1. ☐

At this point we can easily show how the Hölder continuity results for linear equations play a pivotal role in the nonlinear theory. Indeed, consider a solution $u \in H^2(\Omega) \cap H^{1,\infty}(\Omega)$ of (2.52). The functions A^{ij} and $B^i_{(s)}$ appearing in (2.63) are in $L^\infty(\Omega)$, and the restrictions to any $\omega \subset\subset \Omega$ of the function $w \equiv u_{x_s}$ belongs to $C^{0,\delta}(\bar{\omega})$, for some $\delta \in]0, 1[$, by Theorem 2.16. In the general case $u \in H^1(\Omega)$ we need only apply Theorems 2.26 and 2.27 as well as Lemma 2.28, then replace Ω by any Ω' with $\omega \subset\subset \Omega' \subset\subset \Omega$. This demonstrates the following theorem.

THEOREM 2.29. *Under the same assumptions as in Theorem 2.27, every solution of* (2.52) *belongs to* $C^{1,\delta}(\Omega)$ *for some* $\delta \in]0, 1[$.

Thanks to the above result, regularity of derivatives of order > 1 can be deduced from the linear theory of the next chapter: if, for instance, the functions $a^j(x, \eta, \xi)$ are in $C^{1,\gamma}(\bar{\Omega} \times \mathbb{R} \times \mathbb{R}^N)$ for $j = 1, \ldots, N$ and in $C^{0,\gamma}(\bar{\Omega} \times \mathbb{R} \times \mathbb{R}^N)$ for $j = 0$, γ being any given number in $]0, 1[$, then any solution of (2.52) belongs to $C^{2,\gamma}(\Omega)$. [See Theorem 3.4(iii).]

Problems

2.1. The Lax–Milgram theorem can be generalized as follows. Let U, V be Hilbert spaces and let $u, v \mapsto a(u, v)$ be a functional on $U \times V$, linear in each variable, with $|a(u, v)| \leq M |u|_U |v|_V$ for $u \in U$, $v \in V$ $(M > 0)$, $\sup_{v \in V, |v|_V \leq 1} |a(u, v)| \geq \alpha_0 |u|_V$ for $u \in U$ $(\alpha_0 > 0)$, $\sup_{u \in U} a(u, v) > 0$ for $v \in V$, $v \neq 0$. Then for any choice of $F \in V'$ there exists a unique solution of $u \in U$, $a(u, v) = \langle F, v \rangle$ for $v \in V$, and $|u|_U \leq \alpha_0^{-1} |F|_{V'}$. See I. Babuška [7].

2.2. Let $N = 1$, $\Omega =]0, 1[$. Functions $u \in H^1(\Omega)$ satisfying $u'(0) = 0$, $u(1) = 0$, $-u'' = f'$ with $f \in H^1(\Omega)$, $|f|_{p;\Omega} \leq 1$ for a given p (finite) do not admit a common bound $|u|_{\infty;\Omega} \leq C$. [Note that, for any choice of $k \in \mathbb{R}$ and $\varepsilon > 0$, we can find f with $f(0) = k$ and $|f|_{p;\Omega} < \varepsilon$.] Compare with Theorem 2.7.

2.3. A bounded bilinear form $a(u, v)$ on $H_0^1(\Omega)$ is defined by (2.11) under the following assumptions:

- the d^j's and the b^i's belong to $L^N(\Omega)$ if $N > 2$, to $L^{2+\varepsilon}(\Omega)$ for some $\varepsilon > 0$ if $N = 2$, to $L^2(\Omega)$ if $N = 1$;
- c belongs to $L^{N/2}(\Omega)$ if $N > 2$, to $L^{1+\varepsilon}(\Omega)$ for some $\varepsilon > 0$ if $N = 2$, to $L^1(\Omega)$ if $N = 1$,

in addition to $a^{ij} \in L^\infty(\Omega)$. If, moreover, the uniform ellipticity condition is supposed to hold, then $a(u, v)$ as above is coercive on $H_0^1(\Omega)$ relative to $L^2(\Omega)$. [Note that given $\varepsilon > 0$, any function $h \in L^p(\Omega)$, $1 < p < \infty$, in particular any lower-order coefficient of $a(u, v)$, can be written as $h_1 + h_2$ with $|h_1|_{p;\Omega} < \varepsilon$ and $|h_2|_{\infty;\Omega} \leq k$ provided the positive real number $k = k(\varepsilon)$ is large enough.]

2.4. Throughout this and the next five problems $a(u, v)$ denotes the bilinear form (2.11) with coefficients in $L^\infty(\Omega)$.
The requirement that Ω be connected plays no role in the proof of Theorem 2.4 if $\Gamma = \varnothing$. Why?

2.5. If the injection $V \subsetneq L^2(\Omega)$ is compact and the weak maximum principle holds for A, Theorem 2.10 remains valid for nonnegative functions satisfying (2.23) instead of (2.6).

2.6. Let $u \in H^1(\Omega)$ satisfy $a(u, v) = \langle F, v \rangle$ for $v \in H_0^1(\Omega)$, with $F \in H^{-1}(\Omega)$. Whenever $\omega \subset\subset \Omega$, there exists a constant C independent of u and F such that

$$| u |_{H^1(\omega)} \leq C(| F |_{H^{-1}(\Omega)} + | u |_{2;\Omega}).$$

The same estimate, except for $| u |_{\infty;\Omega}$ instead of $| u |_{2;\Omega}$ on the right-hand side, remains valid if $u \in H^1(\Omega) \cap L^\infty(\Omega)$ satisfies $a(u, v) \leq \langle F, v \rangle$ for $v \in H_0^1(\Omega)$, $v \geq 0$. (Utilize the inequality $u + | u |_{\infty;\Omega} \geq 0$.)

2.7. Let $\partial\Omega$ be of class C^1, and let the assumptions of Theorem 2.16, $\mu > N - 2$, be satisfied together with $L1 \geq 0$, $u|_{\partial\Omega} = \eta \in C^{0,1}(\partial\Omega)$. Then $u \in C^{0,\delta}(\bar{\Omega})$ with

$$| u |_{C^{0,\delta}(\bar{\Omega})} \leq C\left(| f^0 |_{2,(\mu-2)^+;\Omega} + \sum_{i=1}^{N} | f^i |_{2,\mu;\Omega} + | \eta |_{C^{0,1}(\partial\Omega)}\right),$$

C being independent of $u, f^0, f^1, \ldots, f^N$; moreover,

$$| u |_{\infty;\Omega} \leq | \eta |_{\infty;\partial\Omega}$$

if in addition $f^0 = f^1 = \cdots = f^N = 0$.

2.8. Let $\partial\Omega$ be of class C^1 and $L1 \geq 0$. Whenever $\eta \in C^0(\partial\Omega)$ there exists $u \in H^1_{\text{loc}}(\Omega) \cap C^0(\bar{\Omega})$ satisfying $Lu = 0$ in Ω, $u|_{\partial\Omega} = \eta$; such a function u is unique, since, whenever $\varepsilon > 0$, any element of $C^0(\bar{\Omega})$ vanishing on $\partial\Omega$ is $\leq \varepsilon$ on $\partial\omega$, $\omega \subset\subset \Omega$, provided dist$(\omega, \partial\Omega)$ is small enough.

2.9. The solution of the b.v.p. considered in the preceding problem belongs to $C^{0,\gamma}(\bar{\Omega})$ for some $\gamma \in]0, 1[$ provided $\eta \in C^{0,\gamma_1}(\partial\Omega)$ for some $\gamma_1 \in]0, 1[$. To see this, consider a controlled C^{0,γ_1} extension w of η to $\mathbb{R}^N$ and introduce regularizations $w_n \equiv \varrho_n * w$. Letting $u_n \in H^1(\Omega)$ denote the solution of the b.v.p.

$$Lu_n = 0 \qquad \text{in } \Omega, \qquad u_n|_{\partial\Omega} = \eta_n \equiv w_n|_{\partial\Omega}$$

utilize the bound $| u_n |_{C^{0,\delta}(\bar{\Omega})} \leq C | w_n |_{C^1(\bar{\Omega})}$ to arrive at $| u_n |_{C^{0,\delta}(\bar{\Omega})} \leq Cn^{1-\gamma_1} | \eta |_{C^{0,\gamma_1}(\partial\Omega)}$, and the bound $| u_n - u |_{\infty;\Omega} \leq | w_n - w |_{\infty;\Omega}$ to arrive at $| u_n - u |_{\infty;\Omega} \leq Cn^{-\gamma_1} \times | \eta |_{C^{0,\gamma_1}(\partial\Omega)}$ (see Problem 2.7). The conclusion follows from the inequality

$$| u(x) - u(y) | \leq C | \eta |_{C^{0,\gamma_1}(\partial\Omega)}(n^{-\gamma_1} + n^{1-\gamma_1} | x - y |^{\delta})$$

for $x, y \in \bar{\Omega}$, $0 < | x - y | \leq 1$, after choosing n between $| x - y |^{-\delta}$ and $| x - y |^{-\delta} + 1$.

3

$H^{k,p}$ and $C^{k,\delta}$ Theory

The contents of the present chapter can be tersely illustrated by considering the mixed elliptic b.v.p.

$$-(a^{ij}u_{x_i} + d^j u)_{x_j} + b_i u_{x_i} + cu = f \qquad \text{in } \Omega,$$

$$u = 0 \qquad \text{on } \partial\Omega \setminus \Gamma, \qquad (a^{ij}u_{x_i} + d^j u)|_\Gamma \nu^j = 0 \qquad \text{on } \Gamma.$$

By Theorem 2.24 the membership of f in $L^2(\Omega)$ guarantees that a variational solution u to such a problem belongs to $H^2(\Omega)$ provided $\partial\Omega \setminus \Gamma$, Γ and the coefficients of the operator satisfy some suitable regularity assumptions. In this chapter we extend this result in the following directions: if $f \in L^p(\Omega)$ with $2 < p < \infty$ then u belongs to $H^{2,p}(\Omega)$, if $f \in C^{0,\delta}(\bar{\Omega})$ with $0 < \delta < 1$ then u belongs to $C^{2,\delta}(\bar{\Omega})$. In the same vein as in Section 2.4.3 we follow the approach of S. Campanato. Here the three main stages of this approach are as follows:

- Estimates on spheres (Section 3.1) and on hemispheres (Section 3.4); the latter estimates are considerably more difficult than the former ones.
- Application of Lemma 1.18 to the preceding estimates. This leads to $L^{2,\mu}$ regularity of derivatives in the interior (Section 3.2) and by using essentially the same technique, near the boundary (Section 3.5).
- Utilization of an interpolation theorem by J. Marcinkiewicz (proven in the appendix to this chapter, Section 3.8) which leads to L^p regularity in the interior (Section 3.3) as well as, by the same method, near the boundary (Section 3.5).

Global regularity is at this point easily obtained (Section 3.5).

When the previous b.v.p. is replaced by

$$-a^{ij}u_{x_ix_j} + a^iu_{x_i} + au = f \qquad \text{in } \Omega,$$

$$u = 0 \qquad \text{on } \partial\Omega\diagdown\Gamma, \qquad \beta^iu_{x_i}|_\Gamma + \beta u|_\Gamma = 0 \qquad \text{on } \Gamma,$$

where the a^{ij}'s are less than Lipschitzian, although at least continuous on $\bar{\Omega}$, and the vector field $(\beta^1, \ldots, \beta^N)$ is never tangent to Γ, the variational results developed until now cannot be directly applied. If, however, the a^{ij}'s are "frozen" at a point $x^0 \in \bar{\Omega}$, the new problem can be transformed into a variational one near x^0. The estimates of the preceding sections provide local bounds on existing solutions u, from which global bounds can be deduced (Section 3.6). Then uniqueness can be derived from maximum principles, and existence in $H^{2,p}(\Omega)$ or $C^{2,\delta}(\bar{\Omega})$ obtained by an approximation procedure (Section 3.7).

3.1. Estimates on Spheres

Throughout this chapter we shall assume the N^2 functions a^{ij} at least continuous on the closure $\bar{\Omega}$ of the bounded domain $\Omega \subset \mathbb{R}^N$, with

$$a^{ij}\xi_i\xi_j \geq \alpha \mid \xi \mid^2 \qquad \text{on } \bar{\Omega} \qquad \text{for } \xi \in \mathbb{R}^N \qquad (\alpha > 0).$$

By the Tietze extension theorem we can view the a^{ij}'s as the restrictions to $\bar{\Omega}$ of functions $\tilde{a}^{ij} \in C^0(\bar{\Omega}')$, where $\Omega' \supset\supset \Omega$ is another bounded domain: we denote by τ a common modulus of uniform continuity of the $\tilde{a}^{ij}$'s on $\bar{\Omega}'$, hence of the a^{ij}'s on $\bar{\Omega}$. Note that τ is also a modulus of uniform continuity for restrictions to $\bar{\Omega}$ of regularizations $\varrho_n * \tilde{a}^{ij}$.

As in Chapter 1, we shall denote by $(h)_\omega$ the average $(1/|\omega|)\int_\omega h(x)\,dx$ over a nonvoid bounded domain $\omega \subset \mathbb{R}^N$ of a function $h \in L^p(\omega)$ (or $h \in [L^p(\omega)]^N$), so that $(h)_\omega$ minimizes the real function $\int_\Omega |h(x) - \lambda|^2\,dx$ of $\lambda \in \mathbb{R}$ (or $\lambda \in \mathbb{R}^N$), and set $(h)_{x^0,\varrho} \equiv (h)_{\Omega[x^0,\varrho]}$, $(h)_\varrho \equiv (h)_{0,\varrho}$.

The estimates of the present section concern the case when $\Omega = B_r$.

3.1.1. Homogeneous Equations with Constant Coefficients

Beginning with homogeneous equations with constant coefficients $a^{ij}(x) \equiv a_0^{ij}$, we have the following lemma.

LEMMA 3.1. *There exists a constant C, depending on the a_0^{ij}'s through the bound imposed on their absolute values as well as through α, such that* *for any r and any* $\varrho \in]0, r]$

$$| \nabla w |_{2;\varrho}^2 \le C \frac{\varrho^N}{r^N} | \nabla w |_{2;r}^2 \tag{3.1}$$

and

$$| \nabla w - (\nabla w)_\varrho |_{2;\varrho}^2 \le C \frac{\varrho^{N+2}}{r^{N+2}} | \nabla w - (\nabla w)_r |_{2;r}^2 \tag{3.2}$$

whenever w satisfies

$$\begin{gathered} w \in H^1(B_r), \\ \int_{B_r} a_0^{ij} w_{x_i} v_{x_j}\, dx = 0 \qquad \text{for } v \in H_0^1(B_r). \end{gathered} \tag{3.3}$$

(Compare with Lemma 2.17.)

PROOF. We proceed in three steps.

Step 1: *Preliminary reductions.* It suffices to prove (3.1) and (3.2) for functions w satisfying

$$\begin{gathered} w \in C^\infty(\bar{B}_r), \\ a_0^{ij} w_{x_i x_j}(x) = 0 \qquad \text{for } x \in B_r. \end{gathered} \tag{3.4}$$

Indeed, by the corollary of Lemma 2.21 any solution of (3.3) satisfies also (3.4) provided r is replaced by εr, $0 < \varepsilon < 1$. On the other hand, once the inequalities

$$| \nabla w |_{2;\varepsilon\varrho}^2 \le C \frac{(\varepsilon\varrho)^N}{(\varepsilon r)^N} | \nabla w |_{2;\varepsilon r}^2$$

and

$$| \nabla w - (\nabla w)_{\varepsilon\varrho} |_{2;\varepsilon\varrho}^2 \le C \frac{(\varepsilon\varrho)^{N+2}}{(\varepsilon r)^{N+2}} | \nabla w - (\nabla w)_{\varepsilon r} |_{2;\varepsilon r}^2$$

have been ascertained, (3.1) and (3.2) follow from a passage to the limit as $\varepsilon \to 1^-$.

Next, let $r/2 < \varrho \le r$: then,

$$| \nabla w |_{2;\varrho}^2 \le 2^N \frac{\varrho^N}{r^N} | \nabla w |_{2;r}^2$$

and

$$| \nabla w - (\nabla w)_\varrho |_{2;\varrho}^2 \le 2^{N+2} \frac{\varrho^{N+2}}{r^{N+2}} | \nabla w - (\nabla w)_r |_{2;r}^2,$$

so that we can restrict our considerations to the range $0 < \varrho \le r/2$.

Step 2: *Proof of* (3.1). For $k = 2, 3, \ldots$ Lemma 2.21 provides H^k bounds on solutions to (3.3) [in particular on solutions to (3.4)], which for our present purposes it is convenient to write as follows:

$$| w |_{H^k(B_{r/2})} \leq C(k, r) \, | w |_{H^1(B_{3r/4})} . \tag{3.5}$$

If we fix a value of k sufficiently large with respect to N, $H^k(B_{r/2})$ is continuously imbedded into $C^0(\bar{B}_{r/2})$ (Theorem 1.41) and therefore

$$| w |_{\infty;r/2} \leq C(r) \, | w |_{H^1(B_{3r/4})} .$$

The right-hand side of the latter inequality can in turn be bounded by a quantity $C(r) \, | w |_{2;r}$. We can rapidly see this as follows: Fix a cutoff function $g \in C_c^\infty(B_r)$ with $0 \leq g \leq 1$ in B_r, $g = 1$ on $\bar{B}_{3r/4}$; then

$$\begin{aligned} 0 &= \int_{B_r} a_0{}^{ij} w_{x_i} (g^2 w)_{x_j} \, dx = \int_{B_r} g^2 a_0{}^{ij} w_{x_i} w_{x_j} \, dx + 2 \int_{B_r} a_0{}^{ij} w_{x_i} w g g_{x_j} \, dx \\ &\geq \alpha \int_{B_r} g^2 \, | \nabla w |^2 \, dx - C(r) \int_{B_r} g \, | \nabla w | \, | w | \, dx \\ &\geq \frac{\alpha}{2} \int_{B_r} g^2 \, | \nabla w |^2 \, dx - C(r) \, | w |_{2;r}^2 \end{aligned}$$

by standard arguments, hence the claimed bound. (See also Lemma 2.11.)

Summing up, w satisfies

$$| w |_{\infty;r/2} \leq C(r) \, | w |_{2;r} .$$

But then w satisfies also

$$| w |_{2;\varrho}^2 \leq C \varrho^N \, | w |_{\infty;\varrho}^2 \leq C(r) \varrho^N \, | w |_{2;r}^2$$

whenever $0 < \varrho \leq r/2$. In order to evaluate the dependence on r of the last constant above, we pass to new variables $y = x/r$ and define $w'(y) \equiv w(ry)$ for $y \in \bar{B}$. Thus (3.4) is equivalent to

$$w' \in C^\infty(\bar{B}),$$

$$a_0{}^{ij} w'_{y_i y_j}(y) = 0 \qquad \text{for } y \in B.$$

From the previous considerations it follows that, whenever $0 < \varrho/r \leq 1/2$,

$$| w' |_{2;\varrho/r}^2 \leq C(1) \frac{\varrho^N}{r^N} \, | w' |_{2;1}^2 ,$$

and also

$$| w |_{2;\varrho}^2 \le C(1) \frac{\varrho^N}{r^N} | w |_{2;r}^2$$

after the inverse change of variables $y \mapsto x = ry$.

If w satisfies (3.4), so does any of its derivatives. In particular, the estimate just obtained becomes

$$| w_{x_i} |_{2;\varrho}^2 \le C(1) \frac{\varrho^N}{r^N} | w_{x_i} |_{2;r}^2 \qquad \text{for } i = 1, \ldots, N$$

$(0 < \varrho \le r/2)$, so that (3.1) holds.

Step 3: Proof of (3.2). This time we utilize (3.5) to obtain a bound

$$| \nabla w |_{\infty;r/2} \le C(r) \, | w |_{H^1(B_{3r/4})}$$

via the continuous imbedding $H^k(B_{r/2}) \subsetneq C^1(\bar{B}_{r/2})$ for a sufficiently large fixed value of k. Thus we also have

$$| \nabla w |_{\infty;r/2} \le C(r) \, | w |_{2;r}.$$

Let $0 < \varrho \le r/2$: from the Lipschitz inequality

$$| w(x) - w(0) |^2 \le C\varrho^2 \, | \nabla w |_{\infty;r/2}^2,$$

valid for $x \in \bar{B}_\varrho$, we deduce

$$| w - (w)_\varrho |_{2;\varrho}^2 \le | w - w(0) |_{2;\varrho}^2 \le C\varrho^{N+2} \, | \nabla w |_{\infty;r/2} \le C(r)\varrho^{N+2} \, | w |_{2;r}^2.$$

A passage to new variables $y = x/r$ shows $C(r) = C(1)/r^{N+2}$, so that

$$| w - (w)_\varrho |_{2;\varrho}^2 \le C(1) \frac{\varrho^{N+2}}{r^{N+2}} | w |_{2;r}^2.$$

Any function $w_{x_i} - (w_{x_i})_r$, $i = 1, \ldots, N$, satisfies (3.4) whenever w does, and therefore

$$\begin{aligned} | w_{x_i} - (w_{x_i})_\varrho |_{2;\varrho}^2 &= | w_{x_i} - (w_{x_i})_r - \big(w_{x_i} - (w_{x_i})_r\big)_\varrho |_{2;\varrho}^2 \\ &\le C(1) \frac{\varrho^{N+2}}{r^{N+2}} | w_{x_i} - (w_{x_i})_r |_{2;r}^2 \qquad \text{for } i = 1, \ldots, N \end{aligned}$$

$(0 < \varrho \le r/2)$, so that (3.2) holds. □

3.1.2. Nonhomogeneous Equations with Variable Coefficients

LEMMA 3.2. *There exists a constant C independent of r, which depends on the a^{ij}'s through the bound imposed on $|a^{ij}(0)|$ as well as through α, such that for any $\varrho \in]0, r]$*

$$|\nabla u|^2_{2;\varrho} \le C\left\{\left[\frac{\varrho^N}{r^N} + \tau^2(r)\right] |\nabla u|^2_{2;r} + r^2 |f^0|^2_{2;r} + \sum_{i=1}^N |f^i|^2_{2;r}\right\} \tag{3.6}$$

and

$$|\nabla u - (\nabla u)_\varrho|^2_{2;\varrho} \le C\left[\frac{\varrho^{N+2}}{r^{N+2}} |\nabla u - (\nabla u)_r|^2_{2;r} + \tau^2(r) |\nabla u|^2_{2;r} + r^2 |f^0|^2_{2;r} + \sum_{i=1}^N |f^i - (f^i)_r|^2_{2;r}\right] \tag{3.7}$$

whenever u satisfies

$$u \in H^1(B_r),$$

$$\int_{B_r} a^{ij} u_{x_i} v_{x_j}\, dx = \langle F, v\rangle \equiv \int_{B_r} (f^0 v + f^i v_{x_i})\, dx \quad \textit{for } v \in H_0^1(B_r)$$

with $f^0, \ldots, f^N \in L^2(B_r)$.

PROOF. As in the proof of Lemma 2.15 we shall decompose u into a sum $w + z$, where w satisfies a homogeneous equation. In order to apply Theorem 3.1 to w, however, we need a preliminary passage from variable coefficients to constant ones.

We shall proceed in three steps.

Step 1: *A preliminary reduction.* We shall prove (3.6) and (3.7) in the special case when the a^{ij}'s are constant on $\bar{B}_r$: $a^{ij}(x) \equiv a_0^{ij}$. In the general case of variable coefficients we need only take into account that u can be viewed as a solution to the variational equation

$$\int_{B_r} a_0^{ij} u_{x_i} v_{x_j}\, dx = \int_{B_r} (f^0 v + \hat{f}^i v_{x_i})\, dx$$
$$\equiv \int_{B_r} \{f^0 v + [f^i + (a_0^{ji} - a^{ji}) u_{x_j}] v_{x_i}\}\, dx \qquad \text{for } v \in H_0^1(B_r),$$

$a_0^{ij} \equiv a^{ij}(0)$, and that

$$|\hat{f}^i|_{2;r} \le |f^i|_{2;r} + C\tau(r) |\nabla u|_{2;r},$$
$$|\hat{f}^i - (\hat{f}^i)_r|_{2;r} \le |f^i - (f^i)_r|_{2;r} + C\tau(r) |\nabla u|_{2;r}.$$

Step 2: *Proof of* (3.6). By Theorem 2.1 and the corollary of Theorem 1.43 the Dirichlet problem

$$z \in H_0^1(B_r),$$

$$\int_{B_r} a_0^{ij} z_{x_i} v_{x_j}\, dx = \int_{B_r} (f^0 v + f^i v_{x_i})\, dx \qquad \text{for } v \in H_0^1(B_r)$$

is uniquely solvable, and the solution z satisfies

$$|\nabla z|_{2;r}^2 \le C\left(r^2 |f^0|_{2;r}^2 + \sum_{i=1}^N |f^i|_{2;r}^2\right);$$

the above bound follows from the Poincaré inequality in $H_0^1(B_r)$, which yields

$$|f^0 v|_{1;r} \le |f^0|_{2;r} |v|_{2;r} \le C |f^0|_{2;r}\, r\, |\nabla v|_{2;r}$$

when $v \in H_0^1(B_r)$, so that

$$|F|_{H^{-1}(B_r)}^2 \le C\left(r^2 |f^0|_{2;r}^2 + \sum_{i=1}^N |f^i|_{2;r}^2\right).$$

The function $w \equiv u - z$ satisfies (3.3), so that (3.1) leads to

$$\begin{aligned}
|\nabla u|_{2;\varrho}^2 &\le 2(|\nabla w|_{2;\varrho}^2 + |\nabla z|_{2;\varrho}^2) \le C \frac{\varrho^N}{r^N} |\nabla w|_{2;r}^2 + 2 |\nabla z|_{2;r}^2 \\
&\le C\left[\frac{\varrho^N}{r^N} |\nabla u|_{2;r}^2 + \left(1 + \frac{\varrho^N}{r^N}\right) |\nabla z|_{2;r}^2\right] \\
&\le C\left(\frac{\varrho^N}{r^N} |\nabla u|_{2;r}^2 + r^2 |f^0|_{2;r}^2 + \sum_{i=1}^N |f^i|_{2;r}^2\right)
\end{aligned}$$

for $0 < \varrho \le r$, which amounts to (3.6) when $\tau(r) = 0$.

Step 3: *Proof of* (3.7). The unique solution to the variational Dirichlet problem

$$z \in H_0^1(B_r),$$

$$\int_{B_r} a_0^{ij} z_{x_i} v_{x_j}\, dx = \int_{B_r} \{f^0 v + [f^i - (f^i)_r] v_{x_i}\}\, dx \qquad \text{for } v \in H_0^1(B_r)$$

satisfies

$$|\nabla z|_{2;r}^2 \le C\left(r^2 |f^0|_{2;r}^2 + \sum_{i=1}^N |f^i - (f^i)_r|_{2;r}^2\right).$$

Since $\int_{B_r} v_{x_i}\,dx = 0$ $(i = 1, \ldots, N)$ whenever $v \in H_0^1(B_r)$, u satisfies

$$\int_{B_r} a_0^{ij} u_{x_i} v_{x_j}\,dx = \int_{B_r} \{f^0 v + [f^i - (f^i)_r] v_{x_i}\}\,dx \qquad \text{for } v \in H_0^1(B_r),$$

and therefore $w \equiv u - z$ solves (3.3). From (3.2) it follows that

$$\begin{aligned}
|\nabla u - (\nabla u)_\varrho|_{2;\varrho}^2 &\le |\nabla u - (\nabla w)_\varrho|_{2;\varrho}^2 \le 2(|\nabla w - (\nabla w)_\varrho|_{2;\varrho}^2 + |\nabla z|_{2;\varrho}^2) \\
&\le C\frac{\varrho^{N+2}}{r^{N+2}} |\nabla w - (\nabla w)_r|_{2;r}^2 + 2|\nabla z|_{2;r}^2 \\
&\le C\frac{\varrho^{N+2}}{r^{N+2}} |\nabla w - (\nabla u)_r|_{2;r}^2 + 2|\nabla z|_{2;r}^2 \\
&\le C\left[\frac{\varrho^{N+2}}{r^{N+2}} |\nabla u - (\nabla u)_r|_{2;r}^2 + \left(1 + \frac{\varrho^{N+2}}{r^{N+2}}\right)|\nabla z|_{2;r}^2\right] \\
&\le C\left(\frac{\varrho^{N+2}}{r^{N+2}} |\nabla u - (\nabla u)_r|_{2;r}^2 + r^2 |f^0|_{2;r}^2 + \sum_{i=1}^N |f^i - (f^i)_r|_{2;r}^2\right).
\end{aligned}$$

This proves (3.7) in the constant coefficient case. □

3.2. Interior $L^{2,\mu}$ Regularity of Derivatives

In this section we set out sufficient conditions in order that the first and possibly the second derivatives of variational solutions in a bounded domain Ω belong to $L^{2,\mu}(\omega)$ when restricted to $\omega \subset\subset \Omega$. The importance of $L^{2,\mu}$ regularity when $N < \mu < N + 2$ is self-evident, thanks to the isomorphism $L^{2,\mu}(\omega) \sim C^{0,(\mu-N)/2}(\bar{\omega})$ for ω of class (A) (Theorem 1.17). The $L^{2,N}$ regularity will play a fundamental role in the proofs of the L^p regularity results of Section 3.3. As for $L^{2,\mu}$ regularity when $0 < \mu < N$, we utilize it as a tool to arrive at the range $[N, N + 2[$ by a sort of bootstrap argument.

3.2.1. Regularity of First Derivatives

We want to prove a result analogous to Theorem 2.16. Since the proof is rather lengthy, we begin with the equation

$$\begin{gathered}
u \in H^1(\Omega), \\
\int_\Omega a^{ij} u_{x_i} v_{x_j}\,dx = \langle F, v\rangle \equiv \int_\Omega (f^0 v + f^i v_{x_i})\,dx \qquad \text{for } v \in H_0^1(\Omega),
\end{gathered} \tag{3.8}$$

involving no lower-order coefficients.

LEMMA 3.3. *Let u solve* (3.8) *with $f^0 \in L^{2,(\mu-2)^+}(\Omega)$ and $f^1, \ldots, f^N \in L^{2,\mu}(\Omega)$ for some $\mu \in]0, N+2[$.*

(i) *Let $\mu < N$. Whenever $\omega \subset\subset \Omega$, all first derivatives of $u|_\omega$ belong to $L^{2,\mu}(\omega)$ with norm estimate*

$$| \nabla u |_{2,\mu;\omega} \leq C\Big(| f^0 |_{2,(\mu-2)^+;\Omega} + \sum_{i=1}^N | f^i |_{2,\mu;\Omega} + | \nabla u |_{2;\Omega} \Big).$$

The constant C (independent of $u, f^0, \ldots, f^N$) depends on the a^{ij}'s through the bound imposed on their $L^\infty(\Omega)$ norms, as well as through α and τ.

(ii) *Let $\mu = N$. If $a^{ij} \in C^{0,\delta}(\Omega)$ for some $\delta \in]0, 1[$, the same conclusion as in* (i) *applies, except that now C depends on the a^{ij}'s through the bound imposed on their $C^{0,\delta}(\bar\Omega)$ norms, as well as through α.*

(iii) *Let $\mu > N$. If $a^{ij} \in C^{0,\delta}(\bar\Omega)$ with $\delta = (\mu - N)/2$, the same conclusion applies as in* (ii).

PROOF. Letting $0 < d \leq \text{dist}(\omega, \partial\Omega)$, $d \leq 1$, we set $d_h \equiv d/2^h$ and denote by ω_h the d_h-neighborhood of ω, that is, $\omega_h \equiv \{x \in \mathbb{R}^N \mid \text{dist}(x, \omega) < d_h\}$. We also set

$$\varkappa_\mu(F) \equiv | f^0 |^2_{2,(\mu-2)^+;\Omega} + \sum_{i=1}^N | f^i |^2_{2,\mu;\Omega}, \quad \varkappa_\mu(F; u) \equiv \varkappa_\mu(F) + | \nabla u |^2_{2;\Omega};$$

notice that whenever $x^0 \in \bar\omega$ and $0 < r \leq d_1$,

$$\varkappa_\mu(F) \geq r^{-\mu}\Big(r^2 | f^0 |^2_{2;x^0,r} + \sum_{i=1}^N | f^i |^2_{2;x^0,r} \Big) \qquad \text{if } 0 < \mu < N,$$

$$\varkappa_\mu(F) \geq r^{-\mu}\Big(r^2 | f^0 |^2_{2;x^0,r} + \sum_{i=1}^N | f^i - (f^i)_{x^0,r} |^2_{2;x^0,r} \Big) \qquad \text{if } N \leq \mu < N+2$$

(see Theorem 1.17).

Step 1: *Proof of* (i). After an inessential translation of the origin, Lemma 3.2 can be applied to any sphere $B_r(x^0)$ with $x^0 \in \bar\omega$, $0 < r \leq d_1$. If $0 < \mu < N$, (3.6) yields

$$| \nabla u |^2_{2;x^0,\varrho} \leq \hat{C}\Big\{ \Big[\frac{\varrho^N}{r^N} + \tau^2(r) \Big] | \nabla u |^2_{2;x^0,r} + r^\mu \varkappa_\mu(F) \Big\},$$

$0 < \varrho \leq r$. To any $s \in]1, \infty[$ we can associate a positive number $H(s) \leq d_1$ by the criterion $0 < r \leq H(s) \Rightarrow \tau^2(r) \leq s^{-N}$. The function $\phi(\varrho) \equiv$

$|\nabla u|^2_{2;x^0,\varrho}$ therefore satisfies

$$\phi(\varrho) \leq \hat{C}\Big[2\,\frac{\varrho^N}{r^N}\,\phi(r) + \varrho^\mu \varkappa_\mu(F) s^\mu\Big]$$

whenever $0 < r \leq H(s)$ and $1 < r/\varrho \leq s$. We can apply Lemma 1.18 with $K = 2\hat{C}$, $\Phi(s) = \hat{C}\varkappa_\mu(F)s^\mu$: taking $\varepsilon = N - \mu$ we obtain for $x^0 \in \bar{\omega}$ and $0 < \varrho \leq r \leq H(K^{1/\varepsilon})$,

$$|\nabla u|^2_{2;x^0,\varrho} \leq C\Big[\frac{\varrho^\mu}{r^\mu}\,|\nabla u|^2_{2;x^0,r} + \varrho^\mu \varkappa_\mu(F)\Big]$$

and also

$$\varrho^{-\mu}\,|\nabla u|^2_{2;\omega[x^0,\varrho]} \leq C\{[H(K^{1/\varepsilon})]^{-\mu}\,|\nabla u|^2_{2;\Omega} + \varkappa_\mu(F)\}.$$

We have thus obtained the desired bound on

$$\sup_{x^0\in\bar{\omega},\,0<\varrho\leq H(K^{1/\varepsilon})} \varrho^{-\mu}\,|\nabla u|^2_{2;\omega[x^0,\varrho]},$$

hence on $|\nabla u|^2_{L^{2,\mu}(\omega)}$.

Step 2: *An intermediate inequality for* $\mu \geq N$. If $\mu \geq N$ we can utilize Step 1 with ω replaced by ω_1 and μ replaced by $N - \eta$, where η can be arbitrarily fixed in $]0, N[$. Letting the a^{ij}'s belong to $C^{0,\delta}(\bar{\Omega})$, set $\eta = \delta$. Thus all first derivatives of u belong to $L^{2,N-\delta}(\omega_1)$ with norm estimate. Let $x^0 \in \bar{\omega}$, $0 < r \leq d_1$. Lemma 3.2 can again be applied to $B_r(x^0)$; this time we utilize (3.7) and obtain

$$\begin{aligned}
&|\nabla u - (\nabla u)_{x^0,\varrho}|^2_{2;x^0,\varrho}\\
&\quad\leq C\Big[\frac{\varrho^{N+2}}{r^{N+2}}\,|\nabla u - (\nabla u)_{x^0,r}|^2_{2;x^0,r} + r^{2\delta}\,|\nabla u|^2_{2;x^0,r} + r^\mu \varkappa_\mu(F)\Big]\\
&\quad\leq C\Big[\frac{\varrho^{N+2}}{r^{N+2}}\,|\nabla u - (\nabla u)_{x^0,r}|^2_{2;x^0,r} + r^{N+\delta}\varkappa_{N-\delta}(F;u) + r^\mu \varkappa_\mu(F)\Big]\\
&\quad\leq \hat{C}\Big[\frac{\varrho^{N+2}}{r^{N+2}}\,|\nabla u - (\nabla u)_{x^0,r}|^2_{2;x^0,r} + r^{(N+\delta)\wedge\mu}\varkappa_\mu(F;u)\Big], \qquad (3.9)
\end{aligned}$$

$0 < \varrho \leq r$, since $r \leq 1$ and $\varkappa_{N-\delta}(F) \leq C\varkappa_\mu(F)$.

Step 3: *Proof of* (ii). Let $\phi(\varrho) \equiv |\nabla u - (\nabla u)_{x^0,\varrho}|^2_{2;x^0,\varrho}$. When $\mu = N$ (3.9) yields

$$\phi(\varrho) \leq \hat{C}\Big[\frac{\varrho^{N+2}}{r^{N+2}}\,\phi(r) + \varrho^N \varkappa_N(F;u) s^N\Big]$$

provided $0 < \varrho \le r \le d_1$ and $1 < r/\varrho \le s$. We can apply Lemma 1.18, this time with $K = \hat{C}$, $\Phi(s) = \hat{C}\varkappa_N(F; u)s^N$. Choosing any $\varepsilon \in]0, 2[$ we obtain

$$| \nabla u - (\nabla u)_{x^0,\varrho} |^2_{2;x^0,\varrho} \le C\left[\frac{\varrho^{N+2-\varepsilon}}{r^{N+2-\varepsilon}} | \nabla u - (\nabla u)_{x^0,r} |^2_{2;x^0,r} + \varrho^N \varkappa_N(F; u)\right]$$

whenever $0 < \varrho \le r \le d_1$, hence

$$\begin{aligned} \varrho^{-N} | \nabla u - (\nabla u)_{\omega[x^0,\varrho]} |^2_{2;\omega[x^0,\varrho]} &\le C[d_1^{-N-2} | \nabla u |^2_{2;x^0,d_1} + \varkappa_N(F; u)] \\ &\le C\varkappa_N(F; u) \end{aligned}$$

whenever $0 < \varrho \le d_1$, and the conclusion follows.

Step 4: *Proof of* (iii). If $x^0 \in \overline{\omega_1}$ and $0 < r \le d_2$, (3.9) is still valid, so that for $\mu = N + 2\delta$ the function $\phi(\varrho) \equiv | \nabla u - (\nabla u)_{x^0,\varrho} |^2_{2;x^0,\varrho}$ satisfies

$$\phi(\varrho) \le \hat{C}\left[\frac{\varrho^{N+2}}{r^{N+2}} \phi(r) + \varrho^{N+\delta}\varkappa_\mu(F; u)s^{N+\delta}\right]$$

whenever $0 < \varrho \le r$ and $1 < r/\varrho \le s$. We again apply Lemma 1.18, this time with $K = \hat{C}$, $\Phi(s) = \hat{C}\varkappa_\mu(F; u)s^{N+\delta}$, and obtain for $0 < \varrho \le r \le d_2$ (after letting $\varepsilon = 2 - 2\delta$)

$$| \nabla u - (\nabla u)_{x^0,\varrho} |^2_{2;x^0,\varrho} \le C\left[\frac{\varrho^{N+2\delta}}{r^{N+2\delta}} | \nabla u - (\nabla u)_{x^0,r} |^2_{2;x^0,r} + \varrho^{N+\delta}\varkappa_\mu(F; u)\right].$$

From this inequality it is now easy to deduce that $\nabla u|_{\omega_1} \in [L^{2,N+\delta}(\omega_1)]^N$ with norm estimate

$$| \nabla u |^2_{2,N+\delta;\omega_1} \le C\varkappa_\mu(F; u).$$

To reach the sought-for conclusion in its full strength we utilize the isomorphism $L^{2,N+\delta}(\omega_1) \sim C^{0,\delta/2}(\bar{\omega}_1)$ [it is not restrictive to assume ω, and therefore also ω_1, of class (A)...]. The above inequality therefore yields

$$| \nabla u |^2_{\infty;\omega_1} \le C\varkappa_\mu(F; u).$$

Thus,

$$r^{2\delta} | \nabla u |^2_{2;x^0,r} \le Cr^{N+2\delta} | \nabla u |^2_{\infty;\omega_1} \le Cr^\mu\varkappa_\mu(F; u)$$

whenever $x^0 \in \bar{\omega}$ and $0 < r \le d_1$. By utilizing (3.7) we can reinforce (3.9) as

$$| \nabla u - (\nabla u)_{x^0,\varrho} |^2_{2;x^0,\varrho} \le C\left[\frac{\varrho^{N+2}}{r^{N+2}} | \nabla u - (\nabla u)_{x^0,r} |^2_{2;x^0,r} + r^\mu\varkappa_\mu(F; u)\right]$$

for x^0, r as above and $0 < \varrho \leq r$. At this point the same procedure leading to $L^{2,N+\delta}(\omega_1)$ regularity with norm estimate can be repeated to provide $L^{2,\mu}(\omega)$ regularity with norm estimate. □

We can now move from (3.8) to the complete equation

$$u \in H^1(\Omega),$$
$$a(u, v) = \langle F, v \rangle \equiv \int_\Omega (f^0 v + f^i v_{x_i})\, dx \qquad \text{for } v \in H_0^1(\Omega), \tag{3.10}$$

where

$$a(u, v) \equiv \int_\Omega [(a^{ij} u_{x_i} + d^j u) v_{x_j} + (b^i u_{x_i} + cu) v]\, dx; \tag{3.11}$$

the lower-order coefficients of the bilinear form (3.11) are assumed at least essentially bounded on Ω.

THEOREM 3.4. *Let u solve* (3.10) *with $f^0 \in L^{2,(\mu-2)^+}(\Omega)$, $f^1, \ldots, f^N \in L^{2,\mu}(\Omega)$ for some $\mu \in]0, N+2[$.*

(i) *Let $\mu < N$. Whenever $\omega \subset\subset \Omega$, $u|_\omega$ and all its first derivatives belong to $L^{2,\mu}(\omega)$ with norm estimate*

$$|u|_{2,\mu;\omega} + |\nabla u|_{2,\mu;\omega} \leq C\Big(|f^0|_{2,(\mu-2)^+;\Omega} + \sum_{i=1}^N |f^i|_{2,\mu;\Omega} + |u|_{H^1(\Omega)}\Big).$$

The constant C (independent of $u, f^0, \ldots, f^N$) depends on the coefficients of $a(u, v)$ through the bound imposed on their $L^\infty(\Omega)$ norms, as well as through α and τ.

(ii) *Let $\mu = N$. If $a^{ij} \in C^{0,\delta}(\bar{\Omega})$ for some $\delta \in]0, 1[$, the same conclusion as in* (i) *applies, except that now C depends on the coefficients through the bound imposed on $|a^{ij}|_{C^{0,\delta}(\bar{\Omega})}$ and $|d^j, b^i, c|_{\infty;\Omega}$, as well as through α.*

(iii) *Let $\mu > N$. If $a^{ij}, d^j \in C^{0,\delta}(\bar{\Omega})$ with $\delta = (\mu - N)/2$, the same conclusion as in* (i) *holds, except that now C depends on the coefficients through the bound on $|a^{ij}, d^j|_{C^{0,\delta}(\bar{\Omega})}$ and $|b^i, c|_{\infty;\Omega}$, as well as α.*

PROOF. We set

$$\varkappa_\mu(F; u) \equiv |f^0|^2_{2,(\mu-2)^+;\Omega} + \sum_{i=1}^N |f^i|^2_{2,\mu;\Omega} + |u|^2_{H^1(\Omega)}$$

and write $d_h \equiv d/2^h$ $[0 < d \leq \operatorname{dist}(\omega, \partial\Omega)]$, $\omega_h \equiv d_h$-neighborhood of ω;

we assume $\partial\omega$ smooth. Moreover, we rewrite (3.10) as

$$\int_\Omega a^{ij}u_{x_i}v_{x_j}\,dx = \int_\Omega (\hat{f}^0 v + \hat{f}^i v_{x_i})\,dx$$
$$= \int_\Omega [(f^0 - b^i u_{x_i} - cu)v + (f^i - d^i u)v_{x_i}]\,dx.$$

This will enable us to utilize a bootstrap argument based on Lemma 3.3.

Step 1: *Proof of* (i). Let $\mu < N$. We know (see Theorem 1.38) that whenever $w \in H^1(\Omega)$ with $w_{x_1}, \ldots, w_{x_N} \in L^{2,\lambda}(\Omega)$, $0 \le \lambda < N$, $w|_{\omega_1}$ belongs to $L^{2,\lambda+2}(\omega_1)$, with the corresponding norm estimate. Thus, $u|_{\omega_1}$ belongs to $L^{2,2}(\omega_1)$, and therefore $\hat{f}^1|_{\omega_1}, \ldots, \hat{f}^N|_{\omega_1}$ to $L^{2,\mu_1}(\omega_1)$, $\mu_1 \equiv \mu \wedge 2$, with

$$\sum_{i=1}^N |\hat{f}^i|^2_{2,\mu_1;\omega_1} \le C\varkappa_\mu(F; u).$$

Lemma 3.3(i) applies with Ω replaced by ω_1. If $\mu_1 = \mu$ we are done; if $\mu_1 < \mu$, we replace ω by ω_2, μ by μ_1, and so on for a convenient finite number of times.

Step 2: *Proof of* (ii) *and* (iii). Let $N \le \mu < N + 2$. We can utilize (i) with ω replaced by ω_1: for any $\mu' < N$, $u_{x_1}|_{\omega_1}, \ldots, u_{x_N}|_{\omega_1} \in L^{2,\mu'}(\omega_1)$, and $u|_{\omega_2} \in L^{2,\mu'+2}(\omega_2)$, $(d^i u)|_{\omega_2} \in L^\infty(\omega_2)$ in the case (ii), $(d^i u)|_{\omega_2} \in C^{0,\delta}(\bar{\omega}_2)$ in the case (iii), hence $\hat{f}^0|_{\omega_2} \in L^{2,(\mu-2)^+}(\omega_2)$ and $\hat{f}^1|_{\omega_2}, \ldots, \hat{f}^N|_{\omega_2} \in L^{2,\mu}(\omega_2)$ in either case, with

$$|\hat{f}^0|^2_{2,(\mu-2)^+;\omega_2} + \sum_{i=1}^N |\hat{f}^i|^2_{2,\mu;\omega_2} \le C\varkappa_\mu(F; u).$$

Lemma 3.3(ii), (iii) then applies with Ω replaced by ω_2. Therefore $u|_\omega$, $u_{x_1}|_\omega, \ldots, u_{x_N}|_\omega$ belong to $L^{2,\mu}(\omega)$ with $|u|^2_{2,\mu;\omega} + |\nabla u|^2_{2,\mu;\omega}$ bounded by $C\varkappa_\mu(F; u)$. □

3.2.2. Regularity of Second Derivatives

THEOREM 3.5. *Let u solve* (3.10) *with $f^0 \in L^{2,\mu}(\Omega)$ and $f^i, f^i_{x_j} \in L^{2,\mu}(\Omega)$ for $i, j = 1, \ldots, N$, where $0 < \mu < N + 2$.*

(i) *Let $\mu \le N$. If $a^{ij}, d^j \in C^{0,1}(\bar{\Omega})$, then, whenever $\omega \subset\subset \Omega$, all derivatives $u_{x_i x_j}|_\omega$ belong to $L^{2,\mu}(\omega)$ with norm estimate*

$$\sum_{i,j=1}^N |u_{x_i x_j}|_{2,\mu;\omega} \le C\Big[|f^0|_{2,\mu;\Omega} + \sum_{i=1}^N (|f^i|_{2,\mu;\Omega} + |\nabla f^i|_{2,\mu;\Omega}) + |u|_{H^1(\Omega)}\Big].$$

The constant C (independent of u, F) depends on the coefficients through the bound imposed on $|a^{ij}, d^j|_{C^{0,1}(\bar{\Omega})}$ *and* $|b^i, c|_{\infty;\Omega}$ *as well as through* α.

(ii) *Let* $\mu > N$. *If* $a^{ij}, d^j \in C^{1,\delta}(\bar{\Omega})$, $b^i, c \in C^{0,\delta}(\bar{\Omega})$ *with* $\delta = (\mu - N)/2$, *the same conclusion as in* (i) *is valid, except that now the constant C depends on the coefficients through the bound imposed on* $|a^{ij}, d^j|_{C^{1,\delta}(\bar{\Omega})}$ *and* $|b^i, c|_{C^{0,\delta}(\bar{\Omega})}$ *as well as through* α.

PROOF. The idea of the proof is to differentiate the equation for u and obtain the equations for $u_{x_1}, \ldots, u_{x_N}$, as in Step 2 of the proof of Lemma 2.21. Here are the details: For all values of μ, by Theorem 3.4 $u|_{\omega_2}, u_{x_1}|_{\omega_2}, \ldots, u_{x_N}|_{\omega_2} \in L^{2,\mu}(\omega_2)$ with norm estimate, whereas by Lemma 2.21 $u|_{\omega_2} \in H^2(\omega_2)$ with

$$|u|^2_{H^2(\omega_2)} \leq C\left(|f^0|^2_{2;\Omega} + \sum_{i=1}^N |f^i|^2_{H^1(\Omega)} + |u|^2_{H^1(\Omega)}\right) \leq C\varkappa_\mu'(F; u);$$

here ω_h is again the d_h-neighborhood of ω,

$$\varkappa_\mu'(F; u) \equiv |f^0|^2_{2,\mu;\Omega} + \sum_{i=1}^N (|f^i|^2_{2,\mu;\Omega} + |\nabla f^i|^2_{2,\mu;\Omega}) + |u|^2_{H^1(\Omega)}.$$

When $v \in C_c^\infty(\omega_2)$ we can rewrite the identity

$$a(u, v_{x_s}) = \langle F, v_{x_s}\rangle$$

as

$$\begin{aligned}\int_{\omega_2} (a^{ij}u_{x_s x_i} + d^j u_{x_s})v_{x_j}\,dx &= \int_{\omega_2} \left(\sum_{i\neq s} \hat{f}^i v_{x_i} + \hat{f}^s v_{x_s}\right) dx \\ &= \int_{\omega_2} \Big[\sum_{i\neq s} (f^i_{x_s} - d^i_{x_s}u - a^{ji}_{x_s}u_{x_j})v_{x_i} \\ &\quad + (f^s_{x_s} - d^s_{x_s}u - a^{js}_{x_s}u_{x_j} + b^i u_{x_i} + cu - f^0)v_{x_s}\Big]\,dx,\end{aligned}$$

$s = 1, \ldots, N$ (no summation over s). Both when $\mu < N$ and when $\mu > N$ the assumptions about the coefficients of $a(u, v)$ yield $\hat{f}^i|_{\omega_2} \in L^{2,\mu}(\omega_2)$ with

$$|\hat{f}^i|_{2,\mu;\omega_2} \leq C\varkappa_\mu'(F; u), \qquad i = 1, \ldots, N.$$

We can therefore apply Theorem 3.4 again, this time with ω_2 instead of Ω and $u_{x_s}|_{\omega_2}$ instead of u. Consequently $u_{x_i x_s}|_\omega \in L^{2,\mu}(\omega)$ for $i = 1, \ldots, N$,

and

$$| \nabla u_{x_s} |_{2,\mu;\omega} \leq C\varkappa_\mu'(F, u).$$

When $\mu = N$ we utilize the previous result with ω replaced by ω_1, μ by any $\mu' < N$: thus, $u|_{\omega_2}, u_{x_1}|_{\omega_2}, \ldots, u_{x_N}|_{\omega_2} \in L^{2,\mu'+2}(\omega_2)$ and $(d^i_{x_s}u + a^{ji}_{x_s}u_{x_j})|_{\omega_2} \in L^\infty(\omega_2)$ for $i = 1, \ldots, N$. The conclusion follows by the same argument as above. □

Theorem 3.5(ii) shows that if the coefficients of the bilinear form (3.11) are regular enough and $f^0, f_i, f^i_{x_j} \in C^{0,\delta}(\bar{\Omega})$ for some $\delta \in]0, 1[$, then all second derivatives of any solution u to (3.10) belong to $C^{0,\delta}(\Omega)$. It is natural to wonder whether an analogous result holds if $C^{0,\delta}$ is replaced by C^0. The following example shows that such is not the case (unless, of course, $N = 1$).

Example. Let $N = 2$, $\Omega = B_{1/2}$, $a(u, v) = \int_{B_{1/2}} u_{x_i}v_{x_i}\,dx$. The function

$$u(x) \equiv (x_1^2 - x_2^2)(-\ln | x |)^{1/2} \qquad \text{for } x \in \bar{B}_{1/2}\setminus\{0\},$$

$u(0) \equiv 0$ [notice that $u(x_1, x_2) = -u(x_2, x_1)$] belongs to $C^1(\bar{B}_{1/2})$, and $u|_{\bar{B}_{1/2}\setminus B_\varepsilon}$ to $C^\infty(\bar{B}_{1/2}\setminus B_\varepsilon)$ for any $\varepsilon \in]0, 1/2[$. Since

$$u_{x_1x_1}(x) = 2(-\ln|x|)^{1/2} + \frac{x_1^2(x_1^2 - x_2^2)}{|x|^4(-\ln|x|)^{1/2}} - \frac{2x_1^2}{|x|^2(-\ln|x|)^{1/2}} - \frac{x_1^2 - x_2^2}{2|x|^2(-\ln|x|)^{1/2}} - \frac{x_1^2(x_1^2 - x_2^2)}{4|x|^4(-\ln|x|)^{3/2}}$$

and therefore

$$-\Delta u(x) = f^0(x) \equiv \frac{x_1^2 - x_2^2}{2|x|^2}\left[\frac{4}{(-\ln|x|)^{1/2}} + \frac{1}{2(-\ln|x|)^{3/2}}\right]$$

for $x \in B_{1/2}\setminus\{0\}$, an application of the divergence theorem over $B_{1/2}\setminus\bar{B}_\varepsilon$ $(0 < \varepsilon < 1/2)$ followed by a passage to the limit as $\varepsilon \to 0^+$ shows that

$$\int_{B_{1/2}} u_{x_i}v_{x_i}\,dx = \int_{B_{1/2}} f^0 v\,dx \qquad \text{for } v \in H_0^1(B_{1/2}).$$

The function f^0 (set 0 at the origin) is in $C^0(\bar{B}_{1/2})$, although $f^0 \notin C^{0,\delta}(\bar{B}_{1/2})$ whenever $0 < \delta < 1$. Without any need of further direct inspection, the membership in $L^{2,N}(B_{1/2})$ of all second derivatives of u follows from Theorem 3.5(i). However, $u_{x_1x_1}$ is not even bounded near the origin.

3.3. Interior L^p Regularity of Derivatives

We now proceed to exhibit interior L^p regularity for first (and second) derivatives of solutions to (3.10). This we do by a technique of interpolation between L^2 and $L^{2,N}$ which requires a bit of the theory of *weak Lebesgue spaces.*

A measurable real function h on a bounded domain $\omega \subset \mathbb{R}^N$ is said to belong to $L^p(\omega)$-weak, $1 \leq p < \infty$, if a constant C can be associated to it in such a way that

$$\text{meas}_N\{x \in \omega \,\big|\, |h(x)| > s\} \leq (C/s)^p \qquad \text{for } 0 < s < \infty;$$

the infimum of all such constants C is denoted by $]h[_{p;\omega}$. A moment's thought shows that $L^p(\omega)$-weak is a linear space, and that

$$L^p(\omega) \subset L^p(\omega)\text{-weak} \qquad \text{with }]h[_{p;\omega} \leq |h|_{p;\omega}.$$

The mapping $h \to]h[_{p;\omega}$ does not, however, define a norm on $L_p(\omega)$-weak, since the triangle inequality need not be satisfied: for instance, if $N = 1$, $\omega =]0, 1[$ and $p = 1$, we have $]h_i[_{1;\omega} = 1/4$ both for $h_1(x) \equiv x$ and $h_2(x) \equiv 1 - x$, whereas $]h_1 + h_2[_{1;\omega} = 1$. Besides, $L^p(\omega)$ is a proper subset of $L^p(\omega)$-weak, as the simple example $h(x) \equiv 1/x$ shows for N, ω, and p as above. On the other hand, it is easy to ascertain that $L^{p+\varepsilon}(\omega)$-weak $\subset L^p(\omega)$ whenever $\varepsilon > 0$ (see Problem 3.1).

For the sake of notational homogeneity we also write $L^\infty(\omega)$-weak $\equiv L^\infty(\omega)$, $]\cdot[_{\infty;\omega} \equiv |\cdot|_{\infty;\omega}$.

Let $\omega' \subset \mathbb{R}^N$ be another bounded domain and $\mathscr{E}: L^p(\omega) \to L^p(\omega')$-weak be a *subadditive mapping*, that is,

$$]\mathscr{E}(f_1 + f_2)[_{p;\omega'} \leq]\mathscr{E}(f_1)[_{p;\omega'} +]\mathscr{E}(f_2)[_{p;\omega'}$$

for $f_1, f_2 \in L^p(\omega)$. We say that $\mathscr{E}$ is of the *weak type p* (from ω into ω') when a constant C can be associated to it so that

$$]\mathscr{E}(f)[_{p;\omega'} \leq C\,|f|_{p;\omega} \qquad \text{for } f \in L^p(\omega).$$

Of course, if in particular $\mathscr{E}: L^p(\omega) \to L^p(\omega')$ with $|\mathscr{E}(f)|_{p;\omega'} \leq C\,|f|_{p;\omega}$ —in which case $\mathscr{E}$ is said to be of the *strong type p*—, then $\mathscr{E}$ is also of the weak type p.

The proof of the next lemma makes crucial use of the notions just introduced. It also utilizes two fundamental results—one due to J. Marcinkiewicz, the other to F. John and L. Nirenberg—which are given in the Appendix to the present chapter.

LEMMA 3.6. *Let Q be a bounded open cube of $\mathbb{R}^N$. Denote by T: $L^2(\Omega) \to L^2(Q)$ a linear mapping such that T: $L^\infty(\Omega) \to L^{2,N}(Q)$, with*

$$| Tf |_{2;Q} \leq K_2 | f |_{2;\Omega} \qquad \text{for } f \in L^2(\Omega),$$

$$| Tf |_{2,N;Q} \leq K_\infty | f |_{\infty;\Omega} \qquad \text{for } f \in L^\infty(\Omega).$$

Then for $2 < p < \infty$ T maps $L^p(\Omega)$ into $L^p(Q)$, and there exists a constant C, depending on T only through K_2 and K_∞, such that

$$| Tf |_{p;Q} \leq C | f |_{p;\Omega} \qquad \text{for } f \in L^p(\Omega).$$

PROOF. Let us first remark that a constant C exists such that

$$\frac{1}{| Q' |} \int_{Q'} | h - (h)_{Q'} | \, dx \leq C | h |_{2,N;Q} \qquad \text{for } h \in L^{2,N}(Q) \tag{3.12}$$

whenever Q' is an open subcube of Q. In fact, $Q' \subset B_{\sqrt{N}\varrho}(x^0)$ if x^0 is the center of Q' and 2ϱ the length of its edges. Thus,

$$\begin{aligned}
\frac{1}{| Q' |^2} \left(\int_{Q'} | h - (h)_{Q'} | \, dx \right)^2 &\leq \frac{1}{| Q' |} \int_{Q'} | h - (h)_{Q'} |^2 \, dx \\
&\leq 2^{-N} \varrho^{-N} \int_{Q[x^0, \sqrt{N}\varrho]} | h - (h)_{Q(x^0, \sqrt{N}\varrho)} |^2 \, dx \\
&\leq C | h |^2_{2,N;Q} .
\end{aligned}$$

Let now Δ: $\bar{Q} = \bigcup_k \bar{Q}_k$ denote a countable decomposition of $\bar{Q}$, the Q_k's being mutually disjoint open cubes with edges parallel to those of Q. The subadditive mapping $\mathscr{C}_\Delta$: $L^2(\Omega) \to L^2(Q)$ defined by

$$\mathscr{C}_\Delta(f) \equiv \sum_k \chi_{Q_k} \frac{1}{| Q_k |} \int_{Q_k} | Tf - (Tf)_{Q_k} | \, dx \qquad \text{with } \chi_{Q_k} \equiv \begin{cases} 1 & \text{on } Q_k \\ 0 & \text{elsewhere,} \end{cases}$$

is also a mapping from $L^\infty(\Omega)$ into $L^\infty(Q)$, and

$$| \mathscr{C}_\Delta(f) |_{2;Q} \leq K_2 | f |_{2;\Omega} \qquad \text{for } f \in L^2(\Omega),$$

$$| \mathscr{C}_\Delta(f) |_{\infty;Q} \leq C K_\infty | f |_{\infty;\Omega} \qquad \text{for } f \in L^\infty(\Omega)$$

whatever the decomposition Δ [see (3.12)]. $\mathscr{C}_\Delta$ is therefore of both strong types 2 and ∞, so that Theorem 3.30 of the Appendix applies: $\mathscr{T}_\Delta$ maps $L^r(\Omega)$ into $L^r(\Omega)$ for any $r \in]2, \infty[$, with norm estimate

$$|\mathscr{T}_\Delta(f)|_{r;Q} \leq C(r)K_2^{2/r}K_\infty^{1-2/r}|f|_{r;\Omega} \qquad \text{for } f \in L^r(\Omega)$$

independent of Δ. Now let

$$[M_r(h)]^r \equiv \sup_\Delta \sum_k |Q_k|^{1-r}\left(\int_{Q_k} |h - (h)_{Q_k}|\, dx\right)^r$$

and fix $f \in L^r(\Omega)$. Since

$$|\mathscr{T}_\Delta(f)|_{r;Q}^r = \sum_k |Q_k|^{1-r}\left(\int_{Q_k} |Tf - (Tf)_{Q_k}|\, dx\right)^r,$$

we have

$$M_r(Tf) \leq C(r)K_2^{2/r}K_\infty^{1-2/r}|f|_{r;\Omega},$$

so that Lemma 3.31 of the Appendix applies. Thus the function $Tf - (Tf)_Q$ is in $L^r(Q)$-weak, and

$$]Tf - (Tf)_Q[_{r;Q} \leq C(r)K_2^{2/r}K_\infty^{1-2/r}|f|_{r;\Omega}.$$

This means that the linear mapping $\Phi\colon f \to Tf - (Tf)_Q$, besides being bounded from $L^2(\Omega)$ into $L^2(Q)$, is also of the weak type r. We can again apply Theorem 3.30 and conclude from the above that for any $p \in [2, \infty[$ Φ is bounded from $L^p(\Omega)$ into $L^p(Q)$ with norm estimate

$$|\Phi f|_{p;Q} = |Tf - (Tf)_Q|_{p;Q} \leq C(p, K_2, K_\infty)|f|_{p;Q} \qquad \text{for } f \in L^p(\Omega).$$

This completes the proof, since

$$\begin{aligned} |Tf|_{p;Q} &\leq |Tf - (Tf)_Q|_{p;Q} + |(Tf)_Q|_{p;Q} \\ &\leq |Tf - (Tf)_Q|_{p;Q} + |Q|^{1/p-1/2}|Tf|_{2;Q}, \end{aligned}$$

and

$$|Tf|_{2;Q} \leq K_2|f|_{2;\Omega} \leq |\Omega|^{1/2-1/p}|f|_{p;\Omega}.$$

□

At this point the desired regularity results can be demonstrated.

THEOREM 3.7. *Let u solve* (3.10) *with* $f^0, \ldots, f^N \in L^p(\Omega)$, $2 < p < \infty$, *and let* $a^{ij}, d^j \in C^{0,\delta}(\bar{\Omega})$ *for some* $\delta \in]0, 1[$. *Whenever* $\omega \subset\subset \Omega$, $u|_\omega$ *belongs to* $H^{1,p}(\omega)$ *with norm estimate*

$$|u|_{H^{1,p}(\omega)} \leq C\left(\sum_{j=0}^N |f^j|_{p;\Omega} + |u|_{H^1(\Omega)}\right).$$

The constant C (independent of u and F) depends on the coefficients through the bound imposed on $|\, a^{ij}, d^j \,|_{C^{0,\delta}(\bar{\Omega})}$ *and* $|\, b^i, c \,|_{\infty;\Omega}$ *as well as through* α.

Proof. We first assume that the bilinear form (3.11) is coercive on $H^1(\Omega)$ and solve

$$z^0 \in H_0^1(\Omega),$$

$$a(z^0, v) = \int_\Omega f^0 v \, dx \qquad \text{for } v \in H_0^1(\Omega),$$

as well as

$$z^i \in H_0^1(\Omega),$$

$$a(z^i, v) = \int_\Omega f^i v_{x_i} \, dx \qquad \text{for } v \in H_0^1(\Omega),$$

$i = 1, \ldots, N$ (no summation over i inside the integral sign). Notice that by coerciveness,

$$|\, z^j \,|_{H^1(\Omega)} \leq C \,|\, f^j \,|_{2;\Omega} \qquad \text{for } j = 0, 1, \ldots, N.$$

If $Q \subset\subset \Omega$ is a cube, we denote by $T_h^j \colon L^2(\Omega) \to L^2(Q)$ the bounded linear mapping $f^j \mapsto z^j_{x_h}|_Q$, $h = 1, \ldots, N$. By Theorem 3.4(ii) T_h^j is also continuous from $L^\infty(\Omega) \subsetneq L^{2,N}(\Omega)$ into $L^{2,N}(Q)$ and finally from $L^p(\Omega)$ into $L^p(Q)$ as a consequence of Lemma 3.6. Let $z \equiv \sum_{j=0}^N z^j$. The function $w \equiv u - z \in H^1(\Omega)$ satisfies $a(w, v) = 0$ for $v \in H_0^1(\Omega)$, so that Theorem 3.4(iii) yields $w|_{\bar{Q}}, w_{x_1}|_{\bar{Q}}, \ldots, w_{x_N}|_{\bar{Q}} \in C^{0,\delta}(\bar{Q})$, hence $w|_Q \in H^{1,p}(Q)$ with the corresponding norm estimate. This proves the theorem in the coercive case, since ω can be covered by a finite number of cubes such as Q.

In the noncoercive case we rewrite (3.10) as

$$a(u, v) + \lambda \int_\Omega uv \, dx = \langle F + \lambda u, v \rangle \qquad \text{for } v \in H_0^1(\Omega),$$

where λ is so large that the bilinear form on the left-hand side is coercive on $H_0^1(\Omega)$. With the usual notations $d_h \equiv d/2^h$ $[0 < d \leq \operatorname{dist}(\omega, \partial\Omega)]$ and $\omega_h \equiv d_h$-neighborhood of ω, we proceed by the following bootstrap argument. (For brevity's sake we take the case $N \geq 3$.) Utilizing the continuous imbedding $H_0^1(\Omega) \subsetneq L^{2^*}(\Omega)$ and the result just proven in the coercive case, this time with ω replaced by ω_1 and p by $p_1 \equiv p \wedge 2^*$, we obtain $u|_{\omega_1} \in H^{1,p_1}(\omega_1)$ with norm estimate. If $p_1 = p$ we are done; if not, we utilize the same procedure as above and obtain $u|_{\omega_2} \in H^{1,p_2}(\omega_2)$, $p_2 \equiv p \wedge p_1{}^*$, etc., thus arriving at the conclusion in a finite number of steps. ☐

The requirement that a^{ij}, d^j be Hölder continuous on $\bar{\Omega}$, although essential in the previous proof, is stronger than necessary for the validity of the above result (see C. B. Morrey, Jr. [118]). The next example, however, shows that discontinuities in the a^{ij}'s cannot be allowed if we want the range of validity to be the entire half-line $2 < p < \infty$.

EXAMPLE. Let $N = 2$, $\Omega = B$, and set $a_{11}(x) = 1 - (1 - \lambda^2)x_2^2 \mid x \mid^{-2}$, $a_{12}(x) = a_{21}(x) = (1 - \lambda^2)x_1x_2 \mid x \mid^{-2}$, $a_{22}(x) = 1 - (1 - \lambda^2)x_1^2 \mid x \mid^{-2}$ for $x \neq 0$, where $0 < \lambda < 1$. Thus, the a^{ij}'s belong to $L^\infty(B)$ but do not admit continuous extensions to $\bar{B}$.

Let $u(x) \equiv x_1 \mid x \mid^{\lambda-1}$, so that $u_{x_1}(x) = \mid x \mid^{\lambda-1} + (\lambda - 1)x_1^2 \mid x \mid^{\lambda-3}$, $u_{x_2}(x) = (\lambda - 1)x_1x_2 \mid x \mid^{\lambda-3}$. Then $u|_{\bar{B}\setminus B_\varepsilon}$ belongs to $C^\infty(\bar{B}\setminus B_\varepsilon)$ whenever $0 < \varepsilon < 1$; moreover, $u \in H^{1,p}(B)$ for $p < 2/(1 - \lambda)$, but $\mid \nabla u \mid \notin L^{2/(1-\lambda)}(B)$. However, since in $B\setminus\{0\}$ u satisfies

$$(a^{ij}u_{x_i})_{x_j} = [\lambda \mid x \mid^{\lambda-1} + (\lambda^2 - \lambda)x_2^2 \mid x \mid^{\lambda-3}]_{x_1} + [(\lambda - \lambda^2)x_1x_2 \mid x \mid^{\lambda-3}]_{x_2} = 0,$$

an application of the Green formula over $B\setminus\overline{B_\varepsilon}$ $(0 < \varepsilon < 1)$ followed by a passage to the limit as $\varepsilon \to 0^+$ shows that

$$\int_B a^{ij}u_{x_i}v_{x_j}\,dx = 0 \qquad \text{for } v \in H_0^1(B).$$

The passage to second derivatives is almost immediate:

THEOREM 3.8. *Let u satisfy* (3.10) *with $f^0 \in L^p(\Omega), f^1, \ldots, f^N \in H^{1,p}(\Omega)$, $2 \leq p < \infty$, and let a^{ij}, $d^j \in C^{0,1}(\bar{\Omega})$. Whenever $\omega \subset\subset \Omega$, $u|_\omega \in H^{2,p}(\omega)$ with norm estimate*

$$\mid u \mid_{H^{2,p}(\omega)} \leq C\Big(\mid f^0 \mid_{p;\Omega} + \sum_{i=1}^N \mid f^i \mid_{H^{1,p}(\Omega)} + \mid u \mid_{H^1(\Omega)}\Big).$$

The constant C (independent of u and F) depends on the coefficients through the bound imposed on $\mid a^{ij}, d^j \mid_{C^{0,1}(\bar{\Omega})}$ and $\mid b^i, c \mid_{\infty;\Omega}$ as well as through α.

PROOF. The case $p = 2$ is Lemma 2.21 for $k = 0$. If $p > 2$ we need only repeat the proof of Theorem 3.5(i), replacing $L^{2,\mu}$ $(0 < \mu < N)$ with L^p and utilizing Theorem 3.7 instead of Theorem 3.4. ∎

Notice that the example following Theorem 3.5 can also be utilized to show that the range of p in the above result cannot be extended to cover $p = \infty$.

3.4. Estimates on Hemispheres

Throughout this section and the first two subsections of the next one we take some hemisphere of $\mathbb{R}^N$ and investigate the regularity of functions that satisfy a variational equation in its interior together with a Dirichlet or a Neumann condition on the flat portion of its boundary. In such a setting we shall provide the counterparts of the interior regularity results proven in the previous three sections.

We shall utilize the notations

$$H_0^{1;+}(B_\varrho^+) \equiv H_0^1(B_\varrho^+ \cup S_\varrho^+), \qquad H_0^{1;0}(B_\varrho^+) \equiv H_0^1(B_\varrho^+ \cup S_\varrho^0).$$

Let $\Omega = B_r^+$.

3.4.1. Homogeneous Equations with Constant Coefficients

Take $a^{ij}(x) \equiv a_0^{ij}$.

Lemma 3.9. *There exists a constant C, depending on the a_0^{ij}'s through the bound imposed on their absolute values as well as through α, such that for any r and any $\varrho \in\,]0, r]$,*

$$|\nabla w|_{2;\varrho,+}^2 \leq C\frac{\varrho^N}{r^N}|\nabla w|_{2;r,+}^2 \tag{3.13}$$

and

$$|\nabla w - (\nabla w)_\varrho|_{2;\varrho,+}^2 \leq C\frac{\varrho^{N+2}}{r^{N+2}}|\nabla w - (\nabla w)_r|_{2;r,+}^2 \tag{3.14}$$

whenever w satisfies either

$$w \in H_0^{1;+}(B_r^+),$$

$$\int_{B_r^+} a_0^{ij} w_{x_i} v_{x_j}\, dx = 0 \qquad \textit{for } v \in H_0^1(B_r^+),$$

or

$$w \in H^1(B_r^{\,\prime}),$$

$$\int_{B_r^+} a_0^{ij} w_{x_i} v_{x_j}\, dx = 0 \qquad \textit{for } v \in H_0^{1;0}(B_r^+).$$

Proof. As in Step 1 of the proof of Lemma 3.1 it can be checked (this time through the corollary of Lemma 2.23) that it suffices to prove the lemma in the case of functions w satisfying either

$$\begin{aligned} &w \in C^\infty(\overline{B_r^+}), \qquad w = 0 \qquad \text{on } S_r^0, \\ &a_0^{ij} w_{x_i x_j}(x) = 0 \qquad \text{for } x \in B_r^+ \end{aligned} \tag{3.15}$$

or

$$\begin{aligned} &w \in C^\infty(\overline{B_r^+}), \qquad a_0^{iN} w_{x_i} = 0 \qquad \text{on } S_r^0, \\ &a_0^{ij} w_{x_i x_j}(x) = 0 \qquad \text{for } x \in B_r^+. \end{aligned} \tag{3.16}$$

In addition, it suffices to prove (3.13) and (3.14) for $0 < \varrho \leq r/2$.

After these preliminary observations we proceed in four steps.

Step 1: *Proof of* (3.13) *in the case* (3.15). We write the H^k bounds provided by Lemma 2.23 as follows:

$$| w |_{H^k(B^+_{r/2})} \leq C(k, r) \, | w |_{H^1(B^+_{3r/4})}. \tag{3.17}$$

Next we choose k so large that $H^k(B^+_{r/2}) \subsetneq C^1(\overline{B^+_{r/2}})$ and therefore

$$| \nabla w |_{\infty; r/2, +} \leq C(r) \, | w |_{H^1(B^+_{3r/4})}.$$

With obvious changes (such as integration over B_r^+ instead of B_r ...) we can proceed as in Step 2 of the proof of Lemma 3.1 to show that the right-hand side of the above inequality is bounded by $C(r) \, | w |_{2;r,+}$. Thus, if $0 < \varrho \leq r/2$, we have

$$| \nabla w |^2_{2;\varrho,+} \leq C\varrho^N \, | \nabla w |^2_{\infty;r/2,+} \leq C(r)\varrho^N \, | w |^2_{2;r,+};$$

for the sake of future reference (see Step 4 below) we emphasize that the inequality

$$| \nabla w |^2_{2;\varrho,+} \leq C(r)\varrho^N \, | w |^2_{2;3r/4,+} \tag{3.18}$$

can be proven analogously. Finally, we estimate $| w |^2_{2;r,+}$ by $Cr^2 \, | \nabla w |^2_{2;r,+}$ (thanks to the corollary of Theorem 1.45) and arrive at

$$| \nabla w |^2_{2;\varrho,+} \leq C(r)\varrho^N \, | \nabla w |^2_{2;r,+}.$$

To evaluate the dependence on r of $C(r)$ we pass to new variables $y = x/r$ and show that $C(r) = C(1)/r^N$.

Step 2: *Proof of* (3.13) *in the case* (3.16). For $s = 1, \ldots, N-1$ the derivative w_{x_s} of a solution to (3.16) is a solution as well, so that (3.17) becomes

$$| w_{x_s} |_{H^k(B^+_{r/2})} \leq C(k, r) \, | w_{x_s} |_{H^1(B^+_{3r/4})}.$$

For what concerns w_{x_N} we first notice that any of its derivatives of order h, except $\partial^h w_{x_N}/\partial x_N{}^h$, is a derivative of the same order of some w_{x_s}. Next wc utilize (3.16) to express $w_{x_N x_N}$ by means of

$$w_{x_N x_N}(x) = -(a_0{}^{NN})^{-1} \sum_{(i,j)\neq(N,N)} a_0{}^{ij} w_{x_i x_j}(x), \qquad x \in B_r{}^+;$$

this shows that any pure derivative $\partial^h w_{x_N}/\partial x_N{}^h$, $h = 1, 2, \ldots$, is a linear combination of derivatives of order h of $w_{x_1}, \ldots, w_{x_{N-1}}$. Summing up,

$$\sum_{i=1}^N \| w_{x_i} \|_{H^k(B^+_{r/2})} \leq C(k, r) \sum_{i=1}^N \| w_{x_i} \|_{H^1(B^+_{3r/4})},$$

hence

$$\| \nabla w \|_{\infty;r/2,+} \leq C(r) \sum_{i=1}^N \| w_{x_i} \|_{H^1(B^+_{3r/4})}, \tag{3.19}$$

if k is chosen so large that $H^k(B^+_{r/2}) \subsetneq C^0(\overline{B^+_{r/2}})$.

In order to find a convenient estimate of $\sum_{i=1}^N \| w_{x_i} \|_{H^1(B^+_{3r/4})}$ we take a cutoff function $g \in C^\infty(B_r)$ with $0 \leq g \leq 1$ in B_r, $\operatorname{supp} g \subset B_{7r/8}$, $g = 1$ on $\bar{B}_{3r/4}$ and obtain

$$\begin{aligned} 0 &= \int_{B_r{}^+} a_0{}^{ij} w_{x_i} [g^2(w - \lambda)]_{x_j}\, dx \\ &= \int_{B_r{}^+} g^2 a_0{}^{ij} w_{x_i} w_{x_j}\, dx + 2 \int_{B_r{}^+} a_0{}^{ij} w_{x_i} (w - \lambda) g g_{x_j}\, dx \\ &\geq \frac{\alpha}{2} \int_{B_r{}^+} g^2 \| \nabla w \|^2\, dx - C(r) \| w - \lambda \|^2_{2;7r/8,+}, \end{aligned}$$

hence

$$\| \nabla w \|_{2;3r/4,+} \leq C(r) \| w - \lambda \|_{2;7r/8,+} \leq C(r) \| w - \lambda \|_{2;r,+}$$

for any $\lambda \in \mathbb{R}$; in particular,

$$\sum_{i=1}^N \| w_{x_i} \|_{2;3r/4,+} \leq C(r) \| w - (w)_r \|_{2;r,+}.$$

Notice that the inequality

$$\sum_{i=1}^N \| w_{x_i} \|_{2;7r/8,+} \leq C(r) \| w - (w)_r \|_{2;r,+}$$

can be proven analogously. By the same token, for $s = 1, \ldots, N-1$ we have

$$| \nabla w_{x_s} |_{2;3r/4,+} \leq C(r) \, | w_{x_s} - \lambda_s |_{2;7r/8,+},$$

hence

$$\sum_{i=1}^{N} | \nabla w_{x_i} |_{2;3r/4,+} \leq C(r) \sum_{i=1}^{N} | w_{x_i} - \lambda_i |_{2;7r/8,+} \tag{3.20}$$

for $\lambda_1, \ldots, \lambda_N \in \mathbb{R}$, after expressing $w_{x_N x_N}$ as a linear combination of first derivatives of $w_{x_1}, \ldots, w_{x_{N-1}}$; in particular,

$$\sum_{i=1}^{N} | \nabla w_{x_i} |_{2;3r/4,+} \leq C(r) \sum_{i=1}^{N} | w_{x_i} |_{2;7r/8,+} \leq C(r) \, | w - (w)_r |_{2;r,+}.$$

Summing up,

$$\sum_{i=1}^{N} | w_{x_i} |_{H^1(B^+_{3r/4})} \leq C(r) \, | w - (w)_r |_{2;r,+}. \tag{3.21}$$

The two inequalities (3.19) and (3.21), combined with the Poincaré inequality in $H^1(B_r{}^+)$, yield

$$\begin{aligned} | \nabla w |^2_{2;\varrho,+} &\leq C\varrho^N \, | \nabla w |^2_{\infty;r/2,+} \leq C(r)\varrho^N \, | w - (w)_r |^2_{2;r,+} \\ &\leq C(r)\varrho^N \, | \nabla w |^2_{2;r,+} \end{aligned}$$

for $0 < \varrho \leq r/2$. The conclusion follows from the change of coordinates $y = x/r$, which shows that $C(r) = C(1)/r^N$.

Step 3: *Proof of* (3.14) *in the case* (3.15). If $w(x)$ satisfies (3.15) so does $w'(x) \equiv w(x) - \lambda x_N$, $\lambda \in \mathbb{R}$. Therefore

$$\sum_{i=1}^{N} | \nabla w'_{x_i} |_{\infty;r/2,+} \leq C(r) \, | w' |_{H^k(B^+_{r/2})} \leq C(r) \, | w' |_{H^1(B^+_{3r/4})} \leq C(r) \, | w' |_{2;r,+}$$

(see Step 1) once k has been fixed so large that $H^k(B^+_{r/2}) \subsetneq C^2(\overline{B^+_{r/2}})$. Using the Lipschitz inequality

$$\begin{aligned} | \nabla w(x) - \nabla w(0) |^2 &= | \nabla w'(x) - \nabla w'(0) |^2 \\ &\leq C\varrho^2 \sum_{i=1}^{N} | \nabla w'_{x_i} |^2_{\infty;r/2,+}, \qquad x \in \overline{B_\varrho{}^+} \end{aligned}$$

we obtain

$$\begin{aligned} | \nabla w - (\nabla w)_\varrho |^2_{2;\varrho,+} &\leq | \nabla w - \nabla w(0) |^2_{2;\varrho,+} \\ &\leq C\varrho^{N+2} \sum_{i=1}^{N} | \nabla w'_{x_i} |^2_{\infty;r/2,+} \leq C(r)\varrho^{N+2} \, | w' |^2_{2;r,+} \end{aligned}$$

for $0 < \varrho \le r/2$. Since (see the remark after the corollary of Theorem 1.45) Poincaré's inequality in $H_0^{1;+}(B_R^+)$ yields

$$| w' |^2_{2;r,+} \le r^2 | w_{x_N} - \lambda |^2_{2;r,+},$$

the conclusion follows by taking $\lambda = (w_{x_N})_r$ and computing the dependence on r of the final constant factor of $| w_{x_N} - (w_{x_N})_r |_{2;r;+}$ through the usual change of coordinates $y = x/r$.

Step 4: *Proof of* (3.14) *in the case* (3.16). Any derivative w_{x_s}, $s = 1, \ldots, N-1$, of a solution to (3.16) is a solution as well. Therefore [see (3.19), (3.21)]

$$| \nabla w_{x_s} |_{\infty;r/2,+} \le C(r) | w_{x_s} - (w_{x_s})_r |_{2;r,+},$$

and

$$\begin{aligned} \sum_{s=1}^{N-1} | w_{x_s} - (w_{x_s})_\varrho |^2_{2;\varrho,+} &\le \sum_{s=1}^{N-1} | w_{x_s} - w_{x_s}(0) |^2_{2;\varrho,+} \\ &\le C\varrho^{N+2} \sum_{s=1}^{N-1} | \nabla w_{x_s} |^2_{\infty;r/2,+} \\ &\le C(r)\varrho^{N+2} \sum_{s=1}^{N-1} | w_{x_s} - (w_{x_s})_r |^2_{2;r,+} \end{aligned} \tag{3.22}$$

for $0 < \varrho \le r/2$. Let $\hat{w} \equiv a_0^{iN} w_{x_i}$. Since $\hat{w}$ satisfies (3.15), from Step 1 [see (3.18)] we deduce

$$\begin{aligned} | \hat{w} - (\hat{w})_\varrho |^2_{2;\varrho,+} &\le C\varrho^2 | \nabla \hat{w} |^2_{2;\varrho,+} \le C(r)\varrho^{N+2} | \hat{w} |^2_{2;3r/4,+} \\ &\le C(r)\varrho^{N+2} | \nabla \hat{w} |^2_{2;3r/4,+} \end{aligned}$$

where use has also been made of the Poincaré inequality both in $H^1(B_\varrho^+)$ and in $H_0^{1;+}(B^+_{3r/4})$. From identity $\hat{w} - (\hat{w})_\varrho = a_0^{iN}[w_{x_i} - (w_{x_i})_\varrho]$ we can deduce (after dividing by $a_0^{NN} > 0$)

$$\begin{aligned} | w_{x_N} - (w_{x_N})_\varrho |^2_{2;\varrho,+} &\le C\left(\sum_{s=1}^{N-1} | w_{x_s} - (w_{x_s})_\varrho |^2_{2;\varrho,+} + | \hat{w} - (\hat{w})_\varrho |^2_{2;\varrho,+} \right) \\ &\le C(r)\varrho^{N+2}\left(\sum_{s=1}^{N-1} | w_{x_s} - (w_{x_s})_r |^2_{2;r,+} + | \nabla \hat{w} |^2_{2;3r/4,+} \right). \end{aligned} \tag{3.23}$$

Finally, from (3.20) we obtain

$$\begin{aligned} | \nabla \hat{w} |_{2;3r/4,+} &\le C \sum_{i=1}^{N} | \nabla w_{x_i} |_{2;3r/4,+} \le C(r) | \nabla w - (\nabla w)_{7r/8} |_{2;7r/8,+} \\ &\le C(r) | \nabla w - (\nabla w)_r |_{2;r,+}, \end{aligned}$$

so that the conclusion follows from (3.22) and (3.23) after the usual evaluation of the constant $C(r)$. □

3.4.2. Nonhomogeneous Equations with Variable Coefficients

LEMMA 3.10. *There exists a constant C independent of r, which depends on the a^{ij}'s through the bound imposed on $|a^{ij}(0)|$ as well as through α, such that for any $\varrho \in]0, r]$,*

$$|\nabla u|^2_{2;\varrho,+} \leq C\left\{\left[\frac{\varrho^N}{r^N} + \tau^2(r)\right] |\nabla u|^2_{2;r,+} + r^2 |f^0|^2_{2;r,+} + \sum_{i=1}^N |f^i|^2_{2;r,+}\right\} \tag{3.24}$$

and

$$|\nabla u - (\nabla u)_\varrho|^2_{2;\varrho,+} \leq C\left[\frac{\varrho^{N+2}}{r^{N+2}} |\nabla u - (\nabla u)_r|^2_{2;r,+} + \tau^2(r) |\nabla u|^2_{2;r,+} + r^2 |f^0|^2_{2;r,+} + \sum_{i=1}^N |f^i - (f^i)_r|^2_{2;r,+}\right] \tag{3.25}$$

whenever u satisfies either

$$\begin{gathered} u \in H_0^{1;+}(B_r^+), \\ \int_{B_r^+} a^{ij} u_{x_i} v_{x_j}\, dx = \langle F, v\rangle \equiv \int_{B_r^+} (f^0 v + f^i v_{x_i})\, dx \qquad \text{for } v \in H_0^1(B_r^+) \end{gathered} \tag{3.26}$$

or

$$\begin{gathered} u \in H^1(B_r^+), \\ \int_{B_r^+} a^{ij} u_{x_i} v_{x_j}\, dx = \langle F, v\rangle \equiv \int_{B_r^+} (f^0 v + f^i v_{x_i})\, dx \qquad \text{for } v \in H_0^{1;0}(B_r^+) \end{gathered} \tag{3.27}$$

with $f^0, f^1, \ldots, f^N \in L^2(B_r^+)$.

PROOF. As in Step 1 of the proof of Lemma 3.2, it can be easily checked that it suffices to prove (3.24) and (3.25) under the additional hypothesis that $a^{ij}(x) \equiv a_0^{ij}$ for $x \in \overline{B_r^+}$, which we shall assume valid throughout.

Step 1: *Proof of* (3.24) *in both cases* (3.26) *and* (3.27). The Dirichlet problem

$$\begin{gathered} z \in H_0^1(B_r^+), \\ \int_{B_r^+} a_0^{ij} z_{x_i} v_{x_j}\, dx = \langle F, v\rangle \qquad \text{for } v \in H_0^1(B_r^+) \end{gathered}$$

is uniquely solvable, and

$$|\nabla z|_{2;r,+}^2 \leq C\left(r^2 |f^0|_{2;r,+}^2 + \sum_{i=1}^{N} |f^i|_{2;r,+}^2\right).$$

Here use has been made of the Poincaré inequality in $H_0^1(B_r^+)$, exactly as in Step 2 of the proof of Lemma 3.2. Analogously, the mixed b.v.p.

$$z \in H_0^{1;0}(B_r^+),$$

$$\int_{B_r^+} a_0^{ij} z_{x_i} v_{x_j}\, dx = \langle F, v\rangle \qquad \text{for } v \in H_0^{1;0}(B_r^+)$$

is uniquely solvable, and

$$|\nabla z|_{2;r,+}^2 \leq C\left(r^2 |f^0|_{2;r,+}^2 + \sum_{i=1}^{N} |f^i|_{2;r,+}^2\right).$$

In this case the Poincaré inequality in $H_0^{1;0}(B_r^+)$ has to be utilized both to ensure the coerciveness of $\int_{B_r^+} a_0^{ij} z_{x_i} v_{x_j}\, dx$ on $H_0^{1;0}(B_r^+)$ and to provide the bound on $|F|_{[H_0^{1;0}(B_r^+)]'}$.

At this point we introduce the function $w \equiv u - z$, to which Lemma 3.9 applies. Thanks to (3.13), we can arrive at (3.24) by proceeding as in Step 2 of the proof of Lemma 3.2.

Step 2: *Proof of* (3.25) *in the case* (3.26). Since $\int_{B_r^+} v_{x_i}\, dx = 0$ ($i = 1, \ldots, N$) whenever $v \in H_0^1(B_r^+)$, u verifies

$$\int_{B_r^+} a_0^{ij} u_{x_i} v_{x_j}\, dx = \int_{B_r^+} \{f^0 v + [f^i - (f^i)_r] v_{x_i}\}\, dx \qquad \text{for } v \in H_0^1(B_r^+).$$

Therefore, if z is the unique solution to the Dirichlet problem

$$z \in H_0^1(B_r^+),$$

$$\int_{B_r^+} a_0^{ij} z_{x_i} v_{x_j}\, dx = \int_{B_r^+} \{f^0 v + [f^i - (f^i)_r] v_{x_i}\}\, dx \qquad \text{for } v \in H_0^1(B_r^+),$$

Lemma 3.9 applies to $w \equiv u - z$. Hence (3.14) is valid, and the conclusion follows as in Step 3 of the proof of Lemma 3.2.

Step 3: *Proof of* (3.25) *in the case* (3.27). Let $u'(x) \equiv u(x) - (f^N)_r \times x_N / a_0^{NN}$. Since $\int_{B_r^+} v_{x_s}\, dx = 0$ for $s = 1, \ldots, N-1$ whereas $\int_{B_r^+} v_{x_N}\, dx$

$= -\int_{S_r^0} v|_{S_r^0}\, dx'$ if $v \in H_0^{1;0}(B_r^+)$, u' satisfies

$$\int_{B_r^+} a_0^{ij} u'_{x_i} v_{x_j}\, dx = \int_{B_r^+} a_0^{ij} u_{x_i} v_{x_j}\, dx - \int_{B_r^+} \frac{a_0^{Nj}}{a_0^{NN}} (f^N)_r v_{x_j}\, dx$$

$$= \langle F, v\rangle + \int_{S_r^0} (f^N)_r v|_{S_r^0}\, dx'$$

$$= \int_{B_r^+} \{f^0 v + [f^i - (f^i)_r] v_{x_i}\}\, dx \qquad \text{for } v \in H_0^{1;0}(B_r^+).$$

Moreover, $\nabla u - (\nabla u)_\varrho = \nabla u' - (\nabla u')_\varrho$, which shows that for the purpose of proving (3.25) it is irrelevant to replace u by u' [since $\tau(r) = 0$].

Now let z solve

$$z \in H_0^{1;0}(B_r^+),$$

$$\int_{B_r^+} a_0^{ij} z_{x_i} v_{x_j}\, dx = \int_{B_r^+} \{f^0 v + [f^i - (f^i)_r] v_{x_i}\}\, dx \qquad \text{for } v \in H_0^{1;0}(B_r^+).$$

Then the inequality

$$|\nabla z|_{2;r,+}^2 \le C\left(r^2 |f^0|_{2;r,+}^2 + \sum_{i=1}^N |f^i - (f^i)_r|_{2;r,+}^2\right)$$

is satisfied, and $w \equiv u' - z$ satisfies (3.14) by Lemma 3.9. The conclusion follows by standard arguments. □

3.5. Boundary and Global Regularity of Derivatives

3.5.1. $L^{2,\mu}$ Regularity near the Boundary

Let $\Omega = B^+$. Beginning with the b.v.p.'s

$$u \in H_0^{1;+}(B^+),$$
$$\int_{B^+} a^{ij} u_{x_i} v_{x_j}\, dx = \langle F, v\rangle \equiv \int_{B^+} (f^0 v + f^i v_{x_i})\, dx \qquad \text{for } v \in H_0^1(B^+) \tag{3.28}$$

and

$$u \in H^1(B^+),$$
$$\int_{B^+} a^{ij} u_{x_i} v_{x_j}\, dx = \langle F, v\rangle \equiv \int_{B^+} (f^0 v + f^i v_{x_i})\, dx \qquad \text{for } v \in H_0^{1;0}(B^+) \tag{3.29}$$

we have the following lemma.

LEMMA 3.11. *Let u solve* (3.28) *or* (3.29) *with $f^0 \in L^{2,(\mu-2)^+}(B^+)$, f^1, $\ldots, f^N \in L^{2,\mu}(B^+)$ for some $\mu \in]0, N+2[$.*

(i) *Let $\mu < N$. Whenever $0 < R < 1$, all first derivatives of $u|_{B_R^+}$ belong to $L^{2,\mu}(B_R^+)$ with norm estimate*

$$| \nabla u |_{2,\mu;B_R^+} \leq C\Big(| f^0 |_{2,(\mu-2)^+;B^+} + \sum_{i=1}^{N} | f^i |_{2,\mu;B^+} + | \nabla u |_{2;+}\Big).$$

The constant C (independent of $u, f^0, \ldots, f^N$) depends on the a^{ij}'s through the bound imposed on their $L^\infty(B^+)$ norms, as well as through α and τ.

(ii) *Let $\mu = N$. If $a^{ij} \in C^{0,\delta}(\overline{B^+})$ for some $\delta \in]0, 1[$, the conclusion of* (i) *remains valid, except that now C depends on the a^{ij}'s through the bound imposed on their $C^{0,\delta}(\overline{B^+})$ norms, as well as through α.*

(iii) *Let $\mu > N$. If $a^{ij} \in C^{0,\delta}(\overline{B^+})$ with $\delta = (\mu - N)/2$, the conclusion of* (ii) *remains valid.*

PROOF. We do not need to distinguish between the two cases (3.28) and (3.29). After setting $R_h \equiv (1 - R)/2^h$ and

$$\varkappa_\mu(F) \equiv | f^0 |^2_{2,(\mu-2)^+;B^+} + \sum_{i=1}^{N} | f^i |^2_{2,\mu;B^+}, \qquad \varkappa_\mu(F; u) \equiv \varkappa_\mu(F) + | \nabla u |^2_{2;+},$$

we shall proceed in three steps.

Step 1: A preliminary reduction for any μ. As in the proof of Theorem 1.39 we shall now show that the crucial estimates over intersections $B_R^+ \cap B_\sigma(x)$ can be reduced to estimates over hemispheres $B_\varrho^+(x^0)$. If $x \in \overline{B_R^+}$ with $x_N > R_2$ and if $0 < \sigma \leq R_3$, then $B_R^+[x, \sigma] = B_R^+ \cap B_\sigma(x) \subseteq B_\sigma(x) \subset \omega \equiv \{y \in B^+_{R+R_3} | \, y_N > R_3\}$. Since $\omega \subset\subset B^+$, Lemma 3.3 applies: for $i = 1, \ldots, N$ $u_{x_i}|_\omega \in L^{2,\mu}(\omega)$ and

$$\begin{aligned}\sigma^{-\mu} | \nabla u - (\nabla u)_{B_R^+[x,\sigma]} |^2_{2;B_R^+[x,\sigma]} &\leq \sigma^{-\mu} | \nabla u - (\nabla u)_{x,\sigma} |^2_{2;x,\sigma} \\ &\leq | \nabla u |^2_{2,\mu;\omega} \leq \varkappa_\mu(F; u).\end{aligned}$$

This means that there remains to bound $\sigma^{-\mu} | \nabla u - (\nabla u)_{B_R^+[x,\sigma]} |^2_{2;B_R^+[x,\sigma]}$ only when $x \in \overline{B_R^+}$ with $x_N \leq R_2$, $0 < \sigma \leq R_3$. But then, $B_R^+[x, \sigma] \subset B_\varrho^+(x^0) \subset B^+$, where x^0 is the projection of x over S_R^0 and $\varrho \equiv 4\sigma$; therefore,

$$\sigma^{-\mu} | \nabla u - (\nabla u)_{B_R^+[x,\sigma]} |^2_{2;B_R^+[x,\sigma]} \leq 4^\mu \varrho^{-\mu} | \nabla u - (\nabla u)_{x^0,\varrho} |^2_{2;x^0,\varrho,+}.$$

Summing up, we need to prove that

$$\varrho^{-\mu} \mid \nabla u - (\nabla u)_{x^0,\varrho} \mid^2_{2;x^0,\varrho,+} \le C\varkappa_\mu(F; u) \tag{3.30}$$

for $x^0 \in S_R{}^0$ and $0 < \varrho \le R_1$.

Step 2: *Proof of* (3.30) *for* $\mu < N$. Let $0 < r \le R_1$. Since the translation of the origin in x^0 is irrelevant for what concerns the estimate (3.24), the latter holds for our function u in the sense that

$$\begin{aligned} \mid \nabla u \mid^2_{2;x^0,\varrho,+} &\le C\left\{\left[\frac{\varrho^N}{r^N} + \tau^2(r)\right] \mid \nabla u \mid^2_{2;x^0,r,+} + r^2 \mid f^0 \mid^2_{2;x^0,r,+} + \sum_{i=1}^N \mid f^i \mid^2_{2;x^0,r,+}\right\} \\ &\le C\left\{\left[\frac{\varrho^N}{r^N} + \tau^2(r)\right] \mid \nabla u \mid^2_{2;x^0,r,+} + r^\mu \varkappa_\mu(F)\right\} \end{aligned}$$

for $0 < \varrho \le r$. We can at this point proceed exactly as in Step 1 of the proof of Lemma 3.3, and prove the existence of $\bar r \in \,]0, R_1[$ such that $\varrho^{-\mu} \mid \nabla u \mid^2_{2;x^0,\varrho,+} \le \varkappa_\mu(F; u)$ whenever $x^0 \in S_R{}^0$ and $0 < \varrho \le \bar r$. This is sufficient for the proof of (3.30) when $0 < \mu < N$.

Step 3: *Proof of* (3.30) *for* $\mu \ge N$. As a consequence of (i) with R replaced by $R + R_1$, for $i = 1, \ldots, N$ $u_{x_i}|_{B^+_{R+R_1}} \in L^{2,N-\delta}(B^+_{R+R_1})$ with the corresponding norm estimate. From (3.25) we deduce that

$$\begin{aligned} &\mid \nabla u - (\nabla u)_{x^0,\varrho} \mid^2_{2;x^0,\varrho,+} \\ &\quad\le C\left[\frac{\varrho^{N+2}}{r^{N+2}} \mid \nabla u - (\nabla u)_{x^0,r} \mid^2_{2;x^0,r,+} + r^{2\delta} \mid \nabla u \mid^2_{2;x^0,r,+} + r^\mu \varkappa_\mu(F)\right] \\ &\quad\le C\left[\frac{\varrho^{N+2}}{r^{N+2}} \mid \nabla u - (\nabla u)_{x^0,r} \mid^2_{2;x^0,r,+} + r^{N+\delta} \varkappa_{N-\delta}(F; u) + r^\mu \varkappa_\mu(F)\right] \\ &\quad\le C\left[\frac{\varrho^{N+2}}{r^{N+2}} \mid \nabla u - (\nabla u)_{x^0,r} \mid^2_{2;x^0,r,+} + r^{(N+\delta)\wedge\mu} \varkappa_\mu(F; u)\right] \end{aligned}$$

for $x^0 \in S_R{}^0$ and $0 < \varrho \le r \le R_1$.

If $\mu = N$ we can proceed as in Step 3 of the proof of Lemma 3.3 to show (3.30).

If $\mu = N + 2\delta$, $\delta \in \,]0, 1[$, the regularity $u_{x_i}|_{B^+_{R+R_1}} \in L^{2,N+\delta}(B^+_{R+R_1})$ is first ascertained by a procedure analogous to that of the case $\mu = N$. Then, the isomorphism $L^{2,N+\delta}(B^+_{R+R_1}) \sim C^{0,\delta/2}(\overline{B^+_{R+R_1}})$ is utilized to obtain

$$\mid \nabla u - (\nabla u)_{x^0,\varrho} \mid^2_{2;x^0,\varrho,+} \le C\left[\frac{\varrho^{N+2}}{r^{N+2}} \mid \nabla u - (\nabla u)_{x^0,r} \mid^2_{2;x^0,r,+} + r^\mu \varkappa_\mu(F; u)\right]$$

for $x^0 \in S_R{}^0$ and $0 < \varrho < r \leq R_1$. From this inequality (3.30) can be deduced again (see Step 4 of the proof of Lemma 3.3). □

Lemma 3.11 can be extended to b.v.p.'s involving the complete form (3.11) (where $\Omega = B^+$), namely,

$$\begin{gathered} u \in H_0{}^{1;+}(B^+), \\ a(u, v) = \langle F, v\rangle \equiv \int_{B^+} (f^0 v + f^i v_{x_i})\, dx \qquad \text{for } v \in H_0{}^1(B^+) \end{gathered} \tag{3.31}$$

and

$$\begin{gathered} u \in H^1(B^+), \\ a(u, v) = \langle F, v\rangle \equiv \int_{B^+} (f^0 v + f^i v_{x_i})\, dx \qquad \text{for } v \in H_0{}^{1;0}(B^+). \end{gathered} \tag{3.32}$$

THEOREM 3.12. *Let u solve either* (3.31) *or* (3.32) *with* $f^0 \in L^{2,(\mu-2)^+}(B^+)$, $f^1, \ldots, f^N \in L^{2,\mu}(B^+)$ *for some* $\mu \in]0, N+2[$.

(i) *Let* $\mu < N$. *Whenever* $0 < R < 1$, $u|_{B_R^+}$ *and all its first derivatives belong to* $L^{2,\mu}(B_R{}^+)$ *with*

$$|u|_{2,\mu;B_R^+} + |\nabla u|_{2,\mu;B_R^+} \leq C\Big(|f^0|_{2,(\mu-2)^+;B^+} + \sum_{i=1}^N |f^i|_{2,\mu;B^+} + |u|_{H^1(B^+)}\Big).$$

C (independent of u, F) depends on the coefficients of the bilinear form through the bound imposed on their $L^\infty(B^+)$ *norms as well as through* α *and* τ.

(ii) *Let* $\mu = N$. *If* $a^{ij} \in C^{0,\delta}(\overline{B^+})$ *for some* $\delta \in]0, 1[$, *the same conclusion as in* (i) *is valid, except that now C depends on the coefficients through the bound imposed on* $|a^{ij}|_{C^{0,\delta}(\overline{B^+})}$ *and* $|d^j, b^i, c|_{\infty;+}$ *as well as through* α.

(iii) *Let* $\mu > N$. *If* $a^{ij}, d^j \in C^{0,\delta}(\overline{B^+})$ *with* $\delta = (\mu - N)/2$, *the conclusion of* (i) *remains valid with the obvious changes for what concerns C.*

(Compare with Lemma 2.18.)

The proof of this theorem is perfectly analogous to that of Theorem 3.4. The only real difference is that now the membership in $L^{2,\lambda+2}(B_R{}^+)$ of $w|_{B_R^+}$ if $w \in H^1(B_{R'}^+)$ with $w_{x_1}, \ldots, w_{x_N} \in L^{2,\lambda}(B_{R'}^+)$, $0 \leq \lambda < N$ and $R < R'$ (see Theorem 1.39) must be utilized. We leave the details to the reader.

Passing to second derivatives we have the following theorem.

THEOREM 3.13. *Let u solve either* (3.31) *or* (3.32) *with* $f^0 \in L^{2,\mu}(B^+)$ *and* $f^i, f^i_{x_j} \in L^{2,\mu}(B^+)$ *for* $i, j = 1, \ldots, N$.

(i) *Let* $0 < \mu \le N$. *If* $a^{ij}, d^j \in C^{0,1}(\overline{B^+})$, *then whenever* $0 < R < 1$, *all derivatives* $u_{x_ix_j}|_{B_R{}^+}$ *belong to* $L^{2,\mu}(B_R{}^+)$ *with*

$$\sum_{i,j=1}^{N} |u_{x_ix_j}|_{2,\mu;B_R{}^+} \le C\Big[|f^0|_{2,\mu;B^+} + \sum_{i=1}^{N}(|f^i|_{2,\mu;B^+} + |\nabla f^i|_{2,\mu;B^+}) + |u|_{H^1(B^+)}\Big].$$

The constant C *is independent of* u *and* F *but depends on the coefficients through the bound imposed on* $|a^{ij}, d^j|_{C^{0,1}(\overline{B^+})}$ *and* $|b^i, c|_{\infty;+}$ *as well as through* α.

(ii) *Let* $\mu > N$. *If* $a^{ij}, d^j \in C^{1,\delta}(\overline{B^+})$ *and* $b^i, c \in C^{0,\delta}(\overline{B^+})$ *with* $\delta = (\mu - N)/2$, *the same conclusion as in* (i) *is valid, except that now* C *depends on the coefficients through the bound imposed on* $|a^{ij}, d^j|_{C^{1,\delta}(\overline{B^+})}$ *and* $|b^i, c|_{C^{0,\delta}(\overline{B^+})}$ *as well as through* α.

Proof. We proceed simultaneously for all values of μ. Let $R_h \equiv (1 - R)/2^h$. By Lemma 2.23 $u|_{B^+_{R+R_2}} \in H^2(B^+_{R+R_2})$ with

$$|u|^2_{H^2(B^+_{R+R_2})} \le C\Big(|f^0|^2_{2;+} + \sum_{i=1}^{N}|f^i|^2_{H^1(B^+)} + |u|^2_{H^1(B^+)}\Big)$$
$$\le C\Big[|f^0|^2_{2,\mu;B^+} + \sum_{i=1}^{N}(|f^i|^2_{2,\mu;B^+} + |\nabla f^i|^2_{2,\mu;B^+}) + |u|^2_{H^1(B^+)}\Big]$$

whereas $u|_{B^+_{R+R_2}}, u_{x_1}|_{B^+_{R+R_2}}, \ldots, u_{x_N}|_{B^+_{R+R_2}} \in L^{2,\mu}(B^+_{R+R_2})$ by the previous theorem. Let $s = 1, \ldots, N-1$: the function $u_{x_s}|_{B^+_{R+R_2}}$ belongs to $H^1(B^+_{R+R_2})$, and even to $H_0{}^{1;+}(B^+_{R+R_2})$ in the case (3.31), and satisfies

$$\begin{aligned}\int_{B^+_{R+R_2}} (a^{ij}u_{x_sx_i} + d^j u_{x_s})v_{x_j}\,dx &= \int_{B^+_{R+R_2}}\Big(\sum_{i\ne s}\hat{f}^i v_{x_i} + \hat{f}^s v_{x_s}\Big)\,dx\\ &\equiv \int_{B^+_{R+R_2}}\Big[\sum_{i\ne s}(f^i_{x_s} - d^i_{x_s}u - a^{ji}_{x_s}u_{x_j})v_{x_i}\\ &\quad + (f^s_{x_s} - d^s_{x_s}u - a^{js}_{x_s}u_{x_j} + b^i u_{x_i}\\ &\quad + cu - f^0)v_{x_s}\Big]\,dx,\end{aligned} \tag{3.33}$$

in the case (3.31) whenever $v \in C_c^\infty(B^+_{R+R_2})$, in the case (3.32) whenever $v \in C_c^\infty(B^+_{R+R_2} \cup S^0_{R+R_2})$. Notice that $\int_{B^+_{R+R_2}} (hv_{x_j})_{x_s}\,dx = 0$ if $h \in H^1(B^+_{R+R_2})$ and $v \in C_c^\infty(B^+_{R+R_2} \cup S^0_{R+R_2})$. The conclusion about each derivative $u_{x_ix_s}$ follows as in the proof of Theorem 3.5. For what concerns $u_{x_Nx_N}$, the conclusion follows from the equation rewritten as

$$u_{x_N x_N} = (a^{NN})^{-1}\Big[-\sum_{(i,j)\neq(N,N)} (a^{ij}u_{x_i})_{x_j} - a^{NN}_{x_N} u_{x_N} - (d^j u)_{x_j} + b^i u_{x_i}$$

$$+ cu - f^0 + f^i_{x_i}\Big] \qquad \text{in } B_R^+. \tag{3.34}$$

□

3.5.2. L^p Regularity near the Boundary

We conclude the study of the regularity of solutions to either (3.31) or (3.32) with two theorems which correspond, respectively, to Theorem 3.7 and Theorem 3.8. Again we take $\Omega = B^+$.

THEOREM 3.14. *Let u solve either* (3.31) *or* (3.32) *with $f^0, \ldots, f^N \in L^p(B^+)$, $2 < p < \infty$, and let $a^{ij}, d^j \in C^{0,\delta}(\overline{B^+})$ for some $\delta \in\,]0, 1[$. Whenever $0 < R < 1$, $u|_{B_R^+}$ belongs to $H^{1,p}(B_R^+)$ with norm estimate*

$$|\, u\,|_{H^{1,p}(B_R^+)} \leq C\Big(\sum_{j=0}^{N} |\, f^j\,|_{p;+} + |\, u\,|_{H^1(B^+)}\Big).$$

The constant C is independent of u and F but depends on the coefficients of the bilinear form through the bound imposed on $|\, a^{ij}, d^j\,|_{C^{0,\delta}(\bar\Omega)}$ and $|\, b^i, c\,|_{\infty;+}$ as well as through α.

PROOF. Let u solve (3.31). We proceed as in the proof of Theorem 3.7. Namely, we first assume the bilinear form coercive on $H^1(B^+)$ and solve

$$z^0 \in H_0^1(B^+),$$

$$a(z^0, v) = \int_{B^+} f^0 v\, dx \qquad \text{for } v \in H_0^1(B^+),$$

$$z^i \in H_0^1(B^+),$$

$$a(z^i, v) = \int_{B^+} f^i v_{x_i}\, dx \qquad \text{for } v \in H_0^1(B^+),$$

$i = 1, \ldots, N$. Let Q be a cube with edges parallel to the coordinate axes, $Q \subset B^+_{R+R_1}$ $[R_1 \equiv (1 - R)/2]$. Each mapping $T_h{}^j : f^j \mapsto z^j_{x_h}|_Q$ $(j = 0, 1, \ldots, N$ and $h = 1, \ldots, N)$ goes from $L^2(B^+)$ into $L^2(Q)$ by the very definition of the z^j's, from $L^\infty(B^+) \subsetneq L^{2,N}(B^+)$ into $L^{2,N}(Q)$ by Theorem 3.12(ii) (with R replaced by $R + R_1$). Thus $T_h{}^j$: $L^p(B^+) \to L^p(Q)$ by Lemma 3.6. Let $z \equiv \sum_{j=0}^N z^j$: the function $w \equiv u - z \in H_0^{1;+}(B^+)$ satisfies $a(w, v) = 0$

for $v \in H_0^1(B^+)$, so that Theorem 3.12(iii) yields $w|_{\bar{Q}} \in C^{1,\delta}(\bar{Q}) \subsetneq H^{1,p}(Q)$ with the corresponding norm estimate. In the coercive case the conclusion follows after covering B_R^+ by a finite number of cubes Q as above. In the noncoercive case a bootstrap argument, based on the identity

$$a(u, v) + \lambda \int_{B^+} uv\, dx = \langle F + \lambda u, v\rangle \qquad \text{for } v \in H_0^1(B^+)$$

as well as on the result just proven for coercive bilinear forms, leads to the desired conclusion for u solution of (3.31).

The procedure for u solution of (3.32) is perfectly analogous. □

As an illustration of the sharpness of the above requirement about the range of p, consider the following example.

EXAMPLE. Let $N = 2$ and set $u(x) \equiv x_2(1 - \ln|x|)$. Since $u_{x_1} = -x_1x_2|x|^{-2}$ and $u_{x_2} = 1 - \ln|x| - x_2^2|x|^{-2}$, $u \in H_0^{1;+}(B^+)$, and as a matter of fact $u \in H^{1,p}(B^+)$ for any $p \in]2, \infty[$; however, $u|_{B_R^+}$ does not belong to $H^{1,\infty}(B_R^+)$. Let $v \in H_0^1(B^+)$. Since

$$\int_{B^+} (x_1x_2|x|^{-2}v_{x_1} + x_2^2|x|^{-2}v_{x_2})\, dx = -\int_{B^+} x_2|x|^{-2}v\, dx$$
$$= \int_{B^+} (-1 + \ln|x|)v_{x_2}\, dx,$$

u satisfies

$$\int_{B^+} u_{x_i}v_{x_i}\, dx = \int_{B^+} f^i v_{x_i}\, dx \equiv \int_{B^+} (-2x_1x_2|x|^{-2}v_{x_1} - 2x_2^2|x|^{-2}v_{x_2})\, dx,$$

and f^1, f^2 belong to $L^\infty(B^+)$.

THEOREM 3.15. *Let u satisfy either* (3.31) *or* (3.32) *with $f^0 \in L^p(B^+)$, $f^1, \ldots, f^N \in H^{1,p}(B^+)$, $2 \le p < \infty$, and let $a^{ij}, d^j \in C^{0,1}(\overline{B^+})$. Whenever $0 < R < 1$, $u|_{B_R^+}$ belongs to $H^{2,p}(B_R^+)$ with*

$$|u|_{H^{2,p}(B_R^+)} \le C\Big(|f^0|_{p;+} + \sum_{i=1}^{N} |f^i|_{H^{1,p}(B^+)} + |u|_{H^1(B^+)}\Big);$$

the constant C (independent of u and F) depends on the coefficients through the bound imposed on $|a^{ij}, d^j|_{C^{0,1}(\overline{B^+})}$ and $|b^i, c|_{\infty;+}$ as well as through α.

PROOF. The case $p = 2$ is Lemma 2.23 for $k = 0$. If $p > 2$ we repeat the proof of Theorem 3.13(i), with $L^{2,\mu}$ $(0 < \mu < N)$ replaced by L^p. □

3.5.3. Global Regularity

Setting $V = H_0{}^1(\Omega \cup \Gamma)$ and assuming Γ closed, we consider solutions to the mixed boundary valve problem

$$u \in V, \qquad\qquad (3.35)$$
$$a(u, v) = \langle F, v\rangle \equiv \int_\Omega (f^0 v + f^i v_{x_i})\, dx \qquad \text{for } v \in V,$$

with $a(u, v)$ given by (3.11). The following results can be proven by the same technique utilized in the proof of Theorem 2.19.

THEOREM 3.16. *Let u solve* (3.35).

(i) *Assume $\partial\Omega$ of class C^1 and $f^0 \in L^{2,(\mu-2)^+}(\Omega)$, $f^1, \ldots, f^N \in L^{2,\mu}(\Omega)$ with $0 < \mu < N$. Then u and all its first derivatives belong to $L^{2,\mu}(\Omega)$ with*

$$|\, u\,|_{2,\mu;\Omega} + |\,\nabla u\,|_{2,\mu;\Omega} \le C\Big(|\, f^0\,|_{2,(\mu-2)^+;\Omega} + \sum_{i=1}^N |\, f^i\,|_{2,\mu;\Omega} + |\, u\,|_{H^1(\Omega)}\Big).$$

(ii) *Assume $\partial\Omega$ of class $C^{1,\delta}$ and $a^{ij} \in C^{0,\delta}(\bar\Omega)$ for some $\delta \in\,]0, 1[$. Let $f^0 \in L^{2,(N-2)^+}(\Omega)$, $f^1, \ldots, f^N \in L^{2,N}(\Omega)$. Then the conclusion of* (i) *is valid for $\mu = N$.*

(iii) *Assume $\partial\Omega$ of class $C^{1,\delta}$, a^{ij}, $d^j \in C^{0,\delta}(\bar\Omega)$ for some $\delta \in\,]0, 1[$, $f^0 \in L^{2,(\mu-2)^+}(\Omega)$, $f^1, \ldots, f^N \in L^{2,\mu}(\Omega)$ for $\mu = N + 2\delta$. Then the conclusion of* (i) *is valid for the present value of μ.*

(iv) *Same assumptions about $\partial\Omega$, a^{ij}, d^j as in* (iii). *Let $f^0, \ldots, f^N \in L^p(\Omega)$ for some $p \in\,]2, \infty[$. Then $u \in H^{1,p}(\Omega)$ with*

$$|\, u\,|_{H^{1,p}(\Omega)} \le C\Big(\sum_{j=0}^N |\, f^j\,|_{p;\Omega} + |\, u\,|_{H^1(\Omega)}\Big).$$

In all estimates above the constants (independent of u, F) depend on the coefficients of the bilinear form through the bound imposed on their respective norms, as well as through α; in the estimate of (i) *it depends also on τ.*

THEOREM 3.17. *Let u solve* (3.35).

(i) *Assume $\partial\Omega$ of class $C^{1,1}$ and a^{ij}, $d^j \in C^{0,1}(\bar\Omega)$. Let $f^0 \in L^{2,\mu}(\Omega)$ and $f^i, f^i_{x_j} \in L^{2,\mu}(\Omega)$ for $i, j = 1, \ldots, N$ with $0 < \mu \le N$. Then all second derivatives of u belong to $L^{2,\mu}(\Omega)$ with*

$$\sum_{i,j=1}^N |\, u_{x_i x_j}\,|_{2,\mu;\Omega} \le C\Big[|\, f^0\,|_{2,\mu;\Omega} + \sum_{i=1}^N (|\, f^i\,|_{2,\mu;\Omega} + |\,\nabla f^i\,|_{2,\mu;\Omega}) + |\, u\,|_{H^1(\Omega)}\Big].$$

(ii) *Same assumptions about* $\partial\Omega$ *and* a^{ij}, d^j *as in* (i). *Let* $f^0 \in L^p(\Omega)$, $f^1, \ldots, f^N \in H^{1,p}(\Omega)$ *for some* $p \in [2, \infty[$. *Then* $u \in H^{2,p}(\Omega)$ *with*

$$| u |_{H^{2,p}(\Omega)} \leq C\Big(| f^0 |_{p;\Omega} + \sum_{i=1}^{N} | f^i |_{H^{1,p}(\Omega)} + | u |_{H^1(\Omega)}\Big).$$

(iii) *Assume* $\partial\Omega$ *of class* $C^{2,\delta}$ *and* a^{ij}, $d^j \in C^{1,\delta}(\bar{\Omega})$, b^i, $c \in C^{0,\delta}(\bar{\Omega})$ *for some* $\delta \in]0, 1[$. *Let* $f^0 \in L^{2,\mu}(\Omega)$ *and* f^i, $f^i_{x_j} \in L^{2,\mu}(\Omega)$ *for* $i, j = 1, \ldots, N$, $\mu = N + 2\delta$. *Then the conclusion of* (i) *is valid for the present value of* μ.

In all estimates above the constants (independent of u, F) depend on the coefficients of the bilinear form through the bound imposed on their respective norms, as well as through α.

Note that the case $p = 2$ in Theorem 3.17(ii) is Theorem 2.24 for $k = 0$.

3.6. *A priori* Estimates on Solutions to Nonvariational Boundary Value Problems

Consider the mixed b.v.p.

$$\begin{aligned} Lu &\equiv -a^{ij}u_{x_ix_j} + a^iu_{x_i} + au = f \qquad \text{in } \Omega, \\ u|_{\partial\Omega\setminus\Gamma} = 0, \qquad Bu &\equiv \beta^iu_{x_i}|_\Gamma + \beta u|_\Gamma = \zeta \qquad \text{on } \Gamma. \end{aligned} \tag{3.36}$$

In this section and the next Γ is closed and $\partial\Omega$ is of class $C^{1,1}$ for the $H^{2,p}$ theory, of class $C^{2,\delta}$ for the $C^{2,\delta}$ theory, $0 < \delta < 1$. We assume $a^{ij} = a^{ji}$ [an unrestrictive hypothesis: both a^{ij} and a^{ji} can be replaced, if necessary, by $(a^{ij} + a^{ji})/2$] and $\beta^i\nu^i \geq \varkappa$ on Γ, $\varkappa$ being some positive constant. L is said to be a *nonvariational (elliptic) operator* since its principal part cannot in general be put into divergence, or variational, form. Problem (3.36) is called a *nonvariational b.v.p.* The condition on $\partial\Omega\setminus\Gamma$ is, of course, the (homogeneous) Dirichlet condition; the one on Γ is called a (nonhomogeneous) *regular oblique derivative condition.*

3.6.1. The Case of Smooth Coefficients

If the regularity assumptions about the coefficients of L and B are suitably strong not only can the principal part of L be put into variational form, but indeed the whole problem (3.36) can be given a variational formulation to which previous results apply. This is illustrated by the next two lemmas.

LEMMA 3.18. (i) *If* $a^{ij} \in C^{0,1}(\bar{\Omega})$, $a^i, a \in L^\infty(\Omega)$, $\beta^i, \beta \in C^{0,1}(\Gamma)$, *there exists a constant* C, *depending on the coefficients of* L *and* B *through the bound imposed on their respective norms as well as through* α, *such that*

$$| u |_{H^{2,p}(\Omega)} \leq C(| Lu |_{p;\Omega} + | Bu |_{H^{1/p',p}(\Gamma)} + | u |_{H^{1,p}(\Omega)}) \qquad (3.37)$$

when $u \in H^{2,p}(\Omega)$, $2 \leq p < \infty$, *vanishes on* $\partial\Omega\setminus\Gamma$.

(ii) *If* $a^{ij} \in C^{1,\delta}(\bar{\Omega})$, $a^i, a \in C^{0,\delta}(\bar{\Omega})$, $\beta^i, \beta \in C^{1,\delta}(\Gamma)$, *a similar estimate*

$$| u |_{C^{2,\delta}(\bar{\Omega})} \leq C(| Lu |_{C^{0,\delta}(\bar{\Omega})} + | Bu |_{C^{1,\delta}(\Gamma)} + | u |_{C^{1,\delta}(\bar{\Omega})}) \qquad (3.38)$$

holds when $u \in C^{2,\delta}(\bar{\Omega})$ *vanishes on* $\partial\Omega\setminus\Gamma$.

PROOF. We set $f \equiv Lu$, $\zeta \equiv Bu$, and proceed in two steps.

Step 1: *The variational formulation of* (3.36). Consider the case (i) and set

$$\theta \equiv \frac{a^{ij}|_\Gamma \nu^i \nu^j}{\beta^k \nu^k}, \qquad \tau^k \equiv \theta\beta^k - a^{kj}|_\Gamma \nu^j.$$

Thus, $(\tau^1, \ldots, \tau^N)$ is a $C^{0,1}$ vector field on Γ satisfying $\nu^k\tau^k = 0$ identically. Let $\alpha^{ij} \in C^{0,1}(\bar{\Omega})$ be such that $\alpha^{ij}|_\Gamma = \nu^j\tau^i$; then

$$(\alpha^{ij}|_\Gamma - \alpha^{ji}|_\Gamma)\nu^j = \tau^i.$$

Finally, let $\hat{d}^j \in C^{0,1}(\bar{\Omega})$ be such that $\hat{d}^j|_\Gamma = \theta\beta\nu^j$. We define a bilinear form on $H^1(\Omega)$ by setting

$$\begin{aligned}\hat{a}(u, v) &\equiv \int_\Omega [(\hat{a}^{ij}u_{x_i} + \hat{d}^j u)v_{x_j} + (\hat{b}^i u_{x_i} + \hat{c}u)v]\, dx \\ &\equiv \int_\Omega \{[(a^{ij} + \alpha^{ij} - \alpha^{ji})u_{x_i} + \hat{d}^j u]v_{x_j} \\ &\quad + [(a^i + a^{ij}_{x_j} + \alpha^{ij}_{x_j} - \alpha^{ji}_{x_j} + \hat{d}^i)u_{x_i} + (a + \hat{d}^j_{x_j})u]v\}\, dx; \end{aligned} \qquad (3.39)$$

notice that $\hat{a}^{ij}\xi_i\xi_j = a^{ij}\xi_i\xi_j$. Inspection shows that whenever $u \in H^2(\Omega)$,

$$\hat{a}(u, v) = \int_\Omega (Lu)v\, dx \qquad \text{for } v \in H_0{}^1(\Omega)$$

and

$$(\hat{a}^{ij}u_{x_i} + \hat{d}^j u)|_\Gamma \nu^j = \theta Bu$$

[i.e., θBu is the conormal derivative of u with respect to the bilinear form (3.39)].

Notice also that

$$\int_\Omega (\hat{c}v + \hat{d}^j v_{x_j})\,dx = \int_\Omega [av + (\hat{d}^j v)_{x_j}]\,dx = \int_\Omega av\,dx + \int_\Gamma (\hat{d}^j v)|_\Gamma \nu^j\,d\sigma$$

$$= \int_\Omega av\,dx + \int_\Gamma \theta\beta v|_\Gamma\,d\sigma \qquad \text{for } v \in V. \tag{3.40}$$

Now write $\zeta = z|_\Gamma$, with $z \in H_0^{1,p}(\Omega \cup \Gamma)$.

If $\theta = t|_\Gamma$, $\nu^i = n^i|_\Gamma$ with $t, n^i \in C^{0,1}(\bar{\Omega})$, let $\hat{f}^i \equiv tzn^i$: then

$$|\hat{f}^i|_{H^{1,p}(\Omega)} \le C|z|_{H^{1,p}(\Omega)}.$$

Moreover,

$$\int_\Omega (\hat{f}^i_{x_i} v + \hat{f}^i v_{x_i})\,dx = \int_\Omega (\hat{f}^i v)_{x_i}\,dx = \int_\Gamma (\hat{f}^i v)|_\Gamma \nu^i\,d\sigma$$

$$= \int_\Gamma \theta\zeta v|_\Gamma\,d\sigma \qquad \text{for } v \in V.$$

Notice that, as a consequence, $\zeta \le 0$ implies $\int_\Omega (\hat{f}^i_{x_i} v + \hat{f}^i v_{x_i})\,dx \le 0$ if $v \in V$ is ≥ 0.

Lemma 2.6 can at this point be utilized to ascertain that $u \in H^2(\Omega)$ solves (3.36) if and only if

$$u \in V,$$

$$\hat{a}(u, v) = \int_\Omega (\hat{f}^0 v + \hat{f}^i v_{x_i})\,dx \equiv \int_\Omega [(f + \hat{f}^i_{x_i})v + \hat{f}^i v_{x_i}]\,dx \qquad \text{for } v \in V. \tag{3.41}$$

Analogous conclusions are easily obtained in the case (ii).

Step 2: Proof of (3.37) *and* (3.38). Let $u \in H^{2,p}(\Omega)$, $2 \le p < \infty$. To (3.41) we can apply the estimates of Theorem 3.17(ii):

$$|u|_{H^{2,p}(\Omega)} \le C\Big(|\hat{f}^0|_{p;\Omega} + \sum_{i=1}^N |\hat{f}^i|_{H^{1,p}(\Omega)} + |u|_{H^1(\Omega)}\Big)$$

$$\le C(|f|_{p;\Omega} + |z|_{H^{1,p}(\Omega)} + |u|_{H^{1,p}(\Omega)}).$$

By letting z vary in the equivalence class that defines ζ, we arrive at

$$|u|_{H^{2,p}(\Omega)} \le C(|f|_{p;\Omega} + |\zeta|_{H^{1/p',p}(\Gamma)} + |u|_{H^{1,p}(\Omega)}),$$

i.e., (3.37). The proof of (3.38) is perfectly analogous, since Theorem 3.17 (iii) provides the $C^{2,\delta}(\bar{\Omega})$ estimate on solutions to (3.36). ▯

LEMMA 3.19. *Suppose $a \geq 0$ in Ω and $\beta \geq 0$ on Γ, with in addition* $\operatorname{ess\,sup}_\Omega a + \max_\Gamma \beta > 0$ *if $\Gamma = \partial\Omega$.*

Under the assumptions of Lemma 3.18(i), (3.36) *admits a unique solution $u \in H^{2,p}(\Omega)$ if $f \in L^p(\Omega)$, $\zeta \in H^{1/p',p}(\Gamma)$, $2 \leq p < \infty$; under the assumptions of Lemma* 3.18(ii), (3.36) *admits a unique solution $u \in C^{2,\delta}(\bar{\Omega})$ if $f \in C^{0,\delta}(\bar{\Omega})$, $\zeta \in C^{1,\delta}(\Gamma)$.*

Additionally, $u \leq 0$ if $f \leq 0$ and $\zeta \leq 0$.

PROOF. Let $\hat{a}(u, v)$, $\hat{f}^0$, $\hat{f}^1, \ldots, \hat{f}^N$ be defined as in Step 1 of the previous proof. Thanks to (3.40), the present assumptions about a and β imply $\hat{a}(1, v) \geq 0$ whenever $v \in V$ is ≥ 0, and also $\hat{a}(1, v) \neq 0$ for some $v \in V$ if $\Gamma = \partial\Omega$, so that the assumptions of the corollary to Theorem 2.4 are satisfied. Therefore (3.41) admits a unique solution u. Let $f \leq 0$ and $\zeta \leq 0$: if $v \in V$ is ≥ 0, then $\int_\Omega (\hat{f}^0 v + \hat{f}^i v_{x_i})\, dx \leq 0$. Hence $u \leq 0$.

There remains to show that u has the required regularity [so that it satisfies (3.36)]. This can be done thanks to Theorem 3.17(ii) in the case (i) of Lemma 3.18, and to Theorem 3.17(iii) in the case (ii) of Lemma 3.18. ▯

3.6.2. The General Case

When the leading coefficients of L are less than Lipschitz continuous, there can be no hope of transforming (3.36) into a variational problem. Yet, Lemma 3.18 itself can be utilized to provide sufficient conditions in order that estimates such as (3.37) and (3.38) remain valid. Indeed we have the following lemma.

LEMMA 3.20. (i) *Let $a^{ij} \in C^0(\bar{\Omega})$, $a^i, a \in L^\infty(\Omega)$, $\beta^i, \beta \in C^{0,1}(\Gamma)$. There exists a constant C, depending on the coefficients of L and B through the bound imposed on their respective norms as well as through α and τ, such that* (3.37) *is satisfied whenever $u \in H^{2,p}(\Omega)$, $2 \leq p < \infty$, vanishes on $\partial\Omega \setminus \Gamma$.*

(ii) *Let $a^{ij}, a^i, a \in C^{0,\delta}(\bar{\Omega})$, $\beta^i, \beta \in C^{1,\delta}(\Gamma)$. There exists a constant C, depending on the coefficients of L and B through the bound imposed on their respective norms as well as through α, such that* (3.38) *is satisfied whenever $u \in C^{2,\delta}(\Omega)$ vanishes on $\partial\Omega \setminus \Gamma$.*

PROOF. We shall only prove (i), leaving the proof of (ii) to the reader (see Problem 3.9).

We fix $x^0 \in \Gamma$ and set $a_0^{ij} \equiv a^{ij}(x^0)$, $L_0 u \equiv -a_0^{ij} u_{x_i x_j} + a^i u_{x_i} + au$. Let $r > 0$ be so small that $B_{2r}(x^0) \cap \partial\Omega \subset \Gamma$. Denoting by $g_{x^0,r}$ a cutoff function from $C^\infty(\mathbb{R}^N)$ with $0 \leq g_{x^0,r} \leq 1$, $\operatorname{supp} g_{x^0,r} \subset B_{2r}(x^0)$ and $g_{x^0,r} = 1$ on $\bar{B}_r(x^0)$, let $u' \equiv u'_{x^0,r} \equiv g_{x^0,r}\, u$. We can apply Lemma 3.18(i) with L replaced by L_0 and u by u'. Thus,

$$\begin{aligned} |u'|_{H^{2,p}(\Omega)} &\leq C(|L_0 u'|_{p;\Omega} + |Bu'|_{H^{1/p',p}(\Gamma)} + |u'|_{H^{1,p}(\Omega)}) \\ &\leq C(|Lu'|_{p;\Omega} + |(L-L_0)u'|_{p;\Omega} + |Bu'|_{H^{1/p',p}(\Gamma)} + |u'|_{H^{1,p}(\Omega)}). \end{aligned}$$

Notice that C, while independent of r, depends on the a^{ij}'s through α and the bound on $|a^{ij}|_{\infty;\Omega}$.

We can majorize $|(L-L_0)u'|_{p;\Omega} = |(L-L_0)u'|_{p;\Omega \cap B_{2r}(x^0)}$ by $\tau(2r) \times \sum_{i,j=1}^N |u'_{x_i x_j}|_{p;\Omega}$. Thus,

$$|u'|_{H^{2,p}(\Omega)} \leq C[|Lu'|_{p;\Omega} + |Bu'|_{H^{1/p',p}(\Gamma)} + \tau(2r)\,|u'|_{H^{2,p}(\Omega)} + |u'|_{H^{1,p}(\Omega)}]$$

If now $r = r(x^0)$ is so small that in the above inequality $\tau(2r) < \varepsilon$ with $\varepsilon > 0$ suitably chosen we obtain

$$|u'|_{H^{2,p}(\Omega)} \leq C(|Lu'|_{p;\Omega} + |Bu'|_{H^{1/p',p}(\Gamma)} + |u'|_{H^{1,p}(\Omega)}).$$

Similar considerations show that if $x^0 \in \partial\Omega \setminus \Gamma$ (or $x^0 \in \Omega$), there exists a positive number $r = r(x^0)$ such that $B_{2r}(x^0) \cap \partial\Omega \subset \partial\Omega \setminus \Gamma$ (or $B_{2r}(x^0) \subset\subset \Omega$), and the function $u' \equiv u'_{x^0,r} \equiv g_{x^0,r}\, u$ satisfies

$$|u'|_{H^{2,p}(\Omega)} \leq C(|Lu'|_{p;\Omega} + |u'|_{H^{1,p}(\Omega)}).$$

Since $\bar{\Omega}$ is compact there exists a finite number m of pairs (x^{0h}, r^h) with $r^h \equiv r(x^{0h})$, chosen with the criterion illustrated above, such that $\bar{\Omega} \subset \bigcup_{h=1}^m B_{r^h}(x^{0h})$. By defining $u_h' \equiv u'_{x^{0h},r^h}$ we therefore have

$$\begin{aligned} |u|_{H^{2,p}(\Omega)} &\leq \sum_{h=1}^m |u_h'|_{H^{2,p}(\Omega)} \\ &< C \sum_{h=1}^m (|Lu_h'|_{p;\Omega} + |Bu_h'|_{H^{1/p',p}(\Gamma)} + |u_h'|_{H^{1,p}(\Omega)}). \end{aligned}$$

A straightforward computation shows that each term $|Lu_h'|_{p;\Omega}$ is bounded by $C(|Lu|_{p;\Omega} + |u|_{H^{1,p}(\Omega)})$, each term $|Bu_h'|_{H^{1,p'/p}(\Gamma)}$ by $C(|Bu|_{H^{1,p'/p}(\Gamma)} + |u|_{H^{1,p}(\Omega)})$ and each term $|u_h'|_{H^{1,p}(\Omega)}$ by $C|u|_{H^{1,p}(\Omega)}$, so that (3.37) obtains. □

The right-hand sides of both (3.37) and (3.38) depend not only on the norms of Lu and Bu but also on some norms of u itself. We can, however, get rid of such an inconvenient dependency through a suitable zero-order "penalization" of L, as the next lemma shows.

LEMMA 3.21. *There exists a positive constant $\bar{\lambda}$, dependent on the coefficients of L and B through the bound imposed on $|a^{ij}, a^i, a|_{\infty;\Omega}$ and on $|\beta^i, \beta|_{C^{0,1}(\Gamma)}$ as well as through α and τ, such that the following is true:*

(i) *Same assumptions as in Lemma* 3.20 (i). *For any $\lambda \geq \bar{\lambda}$ and any $p \in [2, \infty[$ there exists a constant C, dependent on the coefficients of L and B through the bound imposed on their respective norms as well as through α and τ, such that*

$$|u|_{H^{2,p}(\Omega)} \leq C(|Lu + \lambda u|_{p;\Omega} + |Bu|_{H^{1/p',p}(\Gamma)})$$

whenever $u \in H^{2,p}(\Omega)$ vanishes on $\partial\Omega \setminus \Gamma$.

(ii) *Same assumptions as in Lemma* 3.20(ii). *For any $\lambda \geq \bar{\lambda}$ there exists a positive constant C, dependent on the coefficients of L and B through the bound imposed on their respective norms as well as through α, such that*

$$|u|_{C^{2,\delta}(\bar{\Omega})} \leq C(|Lu + \lambda u|_{C^{0,\delta}(\bar{\Omega})} + |Bu|_{C^{1,\delta}(\Gamma)})$$

whenever $u \in C^{2,\delta}(\bar{\Omega})$ vanishes on $\partial\Omega \setminus \Gamma$.

PROOF. *Step* 1: *Proof of* (i) *for $p = 2$.* Let ${a_n}^{ij}$, $n \in N$, be the restrictions to $\bar{\Omega}$ of regularizations $\varrho_n * \tilde{a}^{ij}$ (see the introductory considerations of Section 3.1): thus, ${a_n}^{ij} \to a^{ij}$ in $C^0(\bar{\Omega})$. Let L_n be the operator obtained from L after replacing a a^{ij} by ${a_n}^{ij}$ and let $\hat{a}_n(u, v)$ be the bilinear form analogously obtained from $\hat{a}(u, v)$ [see (3.39); the function θ utilized in the definitions of the coefficients must be replaced by $\theta_n \equiv {a_n}^{ij}|_\Gamma \nu^i\nu^j/\beta^k\nu^k$]. We provide estimates on $|u|_{2;\Omega}$ and $|L_n u|_{2;\Omega}$ as follows.

Let $u \in H^2(\Omega)$, $u|_{\partial\Omega\setminus\Gamma} = 0$. By the Green formula,

$$\begin{aligned}\int_\Omega (L_n u)u\,dx &= \hat{a}_n(u, u) - \int_\Gamma \theta_n(Bu)u\,|_\Gamma\,d\sigma\\ &\geq \frac{\alpha}{2}|\nabla u|^2_{2;\Omega} - C(n)|u|^2_{2;\Omega} - \frac{\alpha}{2}|u|^2_{H^1(\Omega)} - C|Bu|^2_{H^{1/2}(\Gamma)},\end{aligned}$$

where we have minorized $\hat{a}_n(u, u)$ as in Section 2.2.1 and have majorized $|\int_\Gamma \theta_n(Bu)u|_\Gamma\,d\sigma|$ (independently of n) by

$$C|Bu|_{2;\Gamma}|u|_\Gamma|_{2;\Gamma} \leq \frac{\alpha}{2}|u|^2_{H^1(\Omega)} + C|Bu|^2_{H^{1/2}(\Gamma)}.$$

Thus,

$$\frac{1}{2} | L_n u + \lambda u |_{2;\Omega}^2 + \frac{1}{2} | u |_{2;\Omega}^2 \geq \int_\Omega (L_n u + \lambda u) u \, dx$$

$$\geq \left[\lambda - C(n) - \frac{\alpha}{2}\right] | u |_{2;\Omega}^2 - C | Bu |_{H^{1/2}(\Gamma)}^2,$$

[with $C(n)$ dependent on the a^{ij}'s only through α and the bound on $| a^{ij} |_{\infty;\Omega}$], so that there exists a positive constant $\lambda'(n)$ such that

$$| u |_{2;\Omega}^2 \leq | L_n u + \lambda u |_{2;\Omega}^2 + C | Bu |_{H^{1/2}(\Gamma)}^2 \qquad \text{for } \lambda \geq \lambda'(n).$$

Moreover,

$$| L_n u + \lambda u |_{2;\Omega}^2 = | L_n u |_{2;\Omega}^2 + 2\lambda \int_\Omega (L_n u) u \, dx + \lambda^2 | u |_{2;\Omega}^2$$

$$\geq | L_n u |_{2;\Omega}^2 + \lambda^2 | u |_{2;\Omega}^2 - 2\lambda\left[C(n) + \frac{\alpha}{2}\right] | u |_{2;\Omega}^2 - 2\lambda C | Bu |_{H^{1/2}(\Gamma)}^2,$$

so that there exists $\lambda''(n) > 0$ such that

$$| L_n u |_{2;\Omega}^2 \leq | L_n u + \lambda u |_{2;\Omega}^2 + 2\lambda C | Bu |_{H^{1/2}(\Gamma)}^2 \qquad \text{for } \lambda \geq \lambda''(n).$$

We can apply Lemma 3.20(i) with L replaced by L_n, the constant of the estimate being independent of n. Thanks to the interpolation inequality

$$| u |_{H^1(\Omega)} \leq \varepsilon \sum_{i,j=1}^{N} | u_{x_i x_j} |_{2;\Omega} + C(\varepsilon) | u |_{2;\Omega}^2$$

(see Lemma 1.37) we have

$$| u |_{H^2(\Omega)}^2 \leq C(| L_n u |_{2;\Omega}^2 + | Bu |_{H^{1/2}(\Gamma)}^2 + | u |_{2;\Omega}^2)$$

$$\leq C[| L_n u + \lambda u |_{2;\Omega}^2 + (2\lambda + 1) | Bu |_{H^{1/2}(\Gamma)}^2]$$

for $\lambda \geq \lambda(n) \equiv \lambda'(n) \vee \lambda''(n)$. At this point we fix an index $\bar{n}$ so large that

$$| a_{\bar{n}}^{ij} - a^{ij} |_{\infty;\Omega}^2 < \varepsilon, \text{ hence } | (L_{\bar{n}} - L) u |_{2;\Omega}^2 \leq C\varepsilon \sum_{i,j=1}^{N} | u_{x_i x_j} |_{2;\Omega}^2,$$

with ε suitably small. Note that $\bar{n}$ can be fixed so that it depends on the a^{ij}'s only through τ. Then the inequality

$$| L_{\bar{n}} u + \lambda u |_{2;\Omega}^2 \leq 2 | Lu + \lambda u |_{2;\Omega}^2 + 2 | (L_{\bar{n}} - L) u |_{2;\Omega}^2$$

yields

$$|u|^2_{H^2(\Omega)} \le C[|Lu + \lambda u|^2_{2;\Omega} + (2\lambda + 1)|Bu|^2_{H^{1/2}(\Gamma)}]$$

for $\lambda \ge \bar\lambda \equiv \lambda(\bar n)$, hence the desired conclusion in the case at hand.

Step 2: Completion of the proof. Let $\lambda \ge \bar\lambda$. By (3.37) we have

$$|u|_{H^{2,p}(\Omega)} \le C[|Lu + \lambda u|_{p;\Omega} + |Bu|_{H^{1/p',p}(\Gamma)} + (\lambda + 1)|u|_{H^{1,p}(\Omega)}].$$

For brevity's sake we restrict ourselves to $N \ge 3$. If $2 < p \le 2^*$, Theorem 1.33 yields $H^2(\Omega) \subsetneq H^{1,p}(\Omega)$, so that

$$\begin{aligned}|u|_{H^{2,p}(\Omega)} &\le C[|Lu + \lambda u|_{p;\Omega} + |Bu|_{H^{1/p',p}(\Gamma)} + (\lambda + 1)|u|_{H^2(\Omega)}]\\ &\le C(\lambda)(|Lu + \lambda u|_{p;\Omega} + |Bu|_{H^{1/p',p}(\Gamma)} + |Lu + \lambda u|_{2;\Omega} + |Bu|_{H^{1/2}(\Gamma)})\end{aligned}$$

by Step 1 of the present proof, and the inequality required by (i) holds. If instead $2^* < p \le 2^{**}$ we utilize the continuous imbedding $H^{2,2^*}(\Omega) \subsetneq H^{1,p}(\Omega)$ and the inequality just proven with p replaced by 2^*; proceeding thus, we conclude the proof of (i) in a finite number of steps for any choice of p. For the proof of (ii) it suffices to utilize (i) with p so large that $H^{2,p}(\Omega) \subsetneq C^{1,\delta}(\bar\Omega)$ and apply (3.38). □

3.7. Unique Solvability of Nonvariational Boundary Value Problems

3.7.1. Regularity of Solutions

Instead of (3.36) consider the nonvariational b.v.p.

$$\begin{gathered} Lu + \lambda u = f \quad \text{in } \Omega,\\ u|_{\partial\Omega\setminus\Gamma} = 0, \qquad Bu + \lambda u|_\Gamma = \zeta \quad \text{on } \Gamma. \end{gathered} \tag{3.42}$$

We have the following lemma.

LEMMA 3.22. *Let $\bar\lambda$ be the positive constant provided by Lemma 3.21, let $\lambda_0 \ge \bar\lambda$ be so large that $a + \lambda_0 \ge 0$ in Ω, $\beta + \lambda_0 > 0$ on Γ, and let $\lambda \ge \lambda_0$.*

(i) *If $a^{ij} \in C^0(\bar\Omega)$, $a^i, a \in L^\infty(\Omega)$, β^i, $\beta \in C^{0,1}(\Gamma)$ and $f \in L^p(\Omega)$, $\zeta \in H^{1/p',p}(\Gamma)$ with $2 \le p < \infty$, then (3.42) admits a unique solution $u \in H^{2,p}(\Omega)$.*

(ii) *If* $a^{ij}, a^i, a \in C^{0,\delta}(\bar{\Omega})$, $\beta^i, \beta \in C^{1,\delta}(\Gamma)$ *and* $f \in C^{0,\delta}(\bar{\Omega})$, $\zeta \in C^{1,\delta}(\Gamma)$, *then* (3.42) *admits a unique solution* $u \in C^{2,\delta}(\bar{\Omega})$.

In all cases $u \leq 0$ *in* Ω *if* $f \leq 0$ *in* Ω *and* $\zeta \leq 0$ *on* Γ.

PROOF. We shall give only the proof of (i), since the proof of (ii) is perfectly analogous.

Uniqueness follows immediately from the a priori estimate of Lemma 3.21(i) (with B replaced by $B + \lambda$). Passing to existence, we consider the same functions a_n^{ij} and operators L_n as in the proof of Lemma 3.21. Moreover, we fix $b^i, b \in C^{0,1}(\bar{\Omega})$ with $b^i|_\Gamma = \beta^i$, $b|_\Gamma = \beta$. By Lemma 3.19 each b.v.p.

$$L_n u_n + \lambda u_n = f \qquad \text{in } \Omega,$$

$$u_n|_{\partial\Omega \setminus \Gamma} = 0, \qquad Bu_n + \lambda u_n|_\Gamma = \zeta \qquad \text{on } \Gamma$$

is uniquely solvable in $H^{2,p}(\Omega)$, and $u_n \leq 0$ in Ω if $f \leq 0$ in Ω, $\zeta \leq 0$ on Γ. Moreover, Lemma 3.21(i) yields the existence of a uniform bound

$$| u_n |_{H^{2,p}(\Omega)} \leq C(| f |_{p;\Omega} + | \zeta |_{H^{1/p',p}(\Gamma)}).$$

By the reflexivity of $H^{2,p}(\Omega)$ we can extract from $\{u_n\}$ a subsequence, still denoted for simplicity's sake by the same symbol, such that $u_n \rightharpoonup u$ in $H^{2,p}(\Omega)$. Hence $u_n \to u$ in $H^{1,p}(\Omega)$ by Rellich's Theorem 1.34. Of course, $u \leq 0$ in Ω if $u_n \leq 0$ in Ω for all $n \in N$. Let $v \in C^0(\bar{\Omega})$. In the integral identities

$$\int_\Omega [-a_n^{ij} u_{nx_ix_j} + a^i u_{nx_i} + (a + \lambda)u_n]v \, dx = \int_\Omega fv \, dx$$

we can pass to the limit as $n \to \infty$ and obtain

$$\int_\Omega (Lu + \lambda u)v \, dx = \int_\Omega fv \, dx,$$

which shows that $Lu + \lambda u = f$ in Ω by the arbitrariness of v. On $\partial\Omega \setminus \Gamma$ u vanishes, as do all functions u_n. Finally, an easy application of the divergence theorem (whose details are left to the reader) yields

$$\int_\Gamma (Bu_n + \lambda u_n|_\Gamma)\eta \, d\sigma \to \int_\Gamma (Bu + \lambda u|_\Gamma)\eta \, d\sigma$$

whenever $\eta \in C^1(\Gamma)$, so that $Bu + \lambda u|_\Gamma = \zeta$ on Γ. □

We can now arrive at a regularity result for solutions of (3.36).

THEOREM 3.23. *Let* $u \in H^2(\Omega)$ *solve* (3.36).

(i) *If* $a^{ij} \in C^0(\bar{\Omega})$, $a^i, a \in L^\infty(\Omega)$, $\beta^i, \beta \in C^{0,1}(\Gamma)$ *and* $f \in L^p(\Omega)$, $\zeta \in H^{1/p',p}(\Gamma)$ *with* $p \in]2, \infty[$, *then* $u \in H^{2,p}(\Omega)$.

(ii) *If* $a^{ij}, a^i, a \in C^{0,\delta}(\bar{\Omega})$, $\beta^i, \beta \in C^{1,\delta}(\Gamma)$ *and* $f \in C^{0,\delta}(\bar{\Omega})$, $\zeta \in C^{1,\delta}(\bar{\Omega})$, *then* $u \in C^{2,\delta}(\bar{\Omega})$.

PROOF. Let us simply prove (i). We fix $\lambda \geq \lambda_0$ (see Lemma 3.22) and write (3.36) as

$$Lu + \lambda u = f + \lambda u \qquad \text{in } \Omega,$$

$$u|_{\partial\Omega \setminus \Gamma} = 0, \qquad Bu + \lambda u|_\Gamma = \zeta + \lambda u|_\Gamma \qquad \text{on } \Gamma.$$

Again, we take $N \geq 3$. Let $p_1 = p \wedge 2^*$. Since $H^2(\Omega) \subsetneq H^{1,2^*}(\Omega)$, we have $f + \lambda u \in L^{p_1}(\Omega)$ and $\zeta + \lambda u|_\Gamma \in H^{1/p_1',p_1}(\Gamma)$. Lemma 3.22 provides the existence of a unique solution $w \in H^{2,p_1}(\Omega)$ to the b.v.p.

$$Lw + \lambda w = f + \lambda u \qquad \text{in } \Omega,$$

$$w|_{\partial\Omega \setminus \Gamma} = 0, \qquad Bw + \lambda w|_\Gamma = \zeta + \lambda u|_\Gamma \qquad \text{on } \Gamma.$$

But then w is also the unique function from $H^2(\Omega)$ which solves the same problem; since u is already one such function, it must be $u = w$, hence $u \in H^{2,p_1}(\Omega)$. If $p_1 = p$ we are done. Otherwise, we repeat the above argument a suitable number of times to reach the desired conclusion. □

3.7.2. Maximum Principles

At this point we want to provide a suitable maximum principle for solutions to (3.36). The following observation casts light on the various stages through which we shall proceed: Let $a^i \in C^0(\Omega)$, $a = 0$, and let $u \in C^2(\Omega)$ satisfy $Lu < 0$ in Ω; then u cannot achieve a local maximum at a point $x^0 \in \Omega$. In order to show this all we have to do is recall that at an interior maximum point x^0 the gradient $\nabla u(x^0)$ vanishes, the Hessian matrix $[u_{x_i x_j}(x^0)]_{i,j=1,\ldots,N}$ is nonpositive, and therefore $0 > Lu(x^0) = -a^{ij}(x^0)u_{x_i x_j}(x^0)$, but also $a^{ij}(x^0)u_{x_i x_j}(x^0) \leq 0$ {by a well-known result of linear algebra which utilizes the fact that $[a^{ij}(x^0)]_{i,j=1,\ldots,N}$ is nonnegative}, hence a contradiction.

When u is only in some space $H^{2,p}(\Omega)$, so that the Hessian matrix of u at x^0 is not defined, a more sophisticated tool is needed, namely, the following *Bony maximum principle.*

LEMMA 3.24. *Let $u \in H^{2,p}(\Omega)$, $p > N$, achieve a local maximum at $x^0 \in \Omega$. Then,*

$$\operatorname{ess}\liminf_{x \to x^0} a^{ij}(x) u_{x_i x_j}(x) \leq 0.$$

PROOF. It suffices to proceed under the stronger assumption that $u(x^0)$ is a strict local maximum. This is certainly true if $u(x)$ is replaced by $u'(x) \equiv u(x) - |x - x^0|^4$; moreover,

$$\operatorname{ess}\liminf_{x \to x^0} a^{ij}(x) u_{x_i x_j}(x) = \operatorname{ess}\liminf_{x \to x^0} a^{ij}(x) u'_{x_i x_j}(x).$$

Let $U \subset\subset \Omega$ be an arbitrary open neighborhood of x^0 such that $u < u(x^0)$ in $\bar{U} \setminus \{x^0\}$, and let W be its (relatively closed) subset defined by

$$W \equiv \{y \in U \mid u(x) \leq u(y) + \nabla u(y) \cdot (x - y) \text{ for } x \in \bar{U}\};$$

notice that $H^{2,p}(\Omega) \subsetneq C^{1,1-N/p}(\bar{\Omega})$ by Theorem 1.41. Since $u(x^0)$ is a strict maximum we can fix $r > 0$ so small that

$$u(x) < u(x^0) + \eta \cdot (x - x^0) \qquad \text{for } x \in \partial U$$

whenever $\eta \in \mathbb{R}^N$ satisfies $|\eta| < r$. Thus for any η as above the maximum over $\bar{U}$ of the function $x \mapsto u(x) - \eta \cdot x$ can only be achieved in the interior of U: in other terms, there exists $y \in U$ such that

$$u(x) \leq u(y) + \eta \cdot (x - y) \qquad \text{for } x \in \bar{U}.$$

But this implies $\eta = \nabla u(y)$ and therefore $y \in W$.

Now let the mapping $\mathscr{F}: \bar{\Omega} \to \mathbb{R}^N$ be defined by $\mathscr{F}(y) \equiv \nabla u(y)$ so that by the above considerations $\mathscr{F}(W) \supseteq B_r$. We claim that the Lebesgue measure of W cannot be zero. Indeed, by Theorem 1.41 there exists a constant C such that, whenever $Q \subset \Omega$ is an open cube, each derivative $u_{x_i} \in C^{0,1-N/p}(\bar{Q})$ satisfies

$$|u_{x_i}(x') - u_{x_i}(x'')| \leq C(\operatorname{diam} Q)^{1-N/p} |u_{x_i}|_{H^{1,p}(Q)}$$

for $x', x'' \in \bar{Q}$ (see Problem 1.17). Thus,

$$|\mathscr{F}(\bar{Q})| \leq C(\operatorname{diam} Q)^{N(1-N/p)} |u|_{H^{2,p}(Q)}.$$

Assume $|W| = 0$. Then for any $\varepsilon > 0$ there exists a sequence $\{Q_n\}$ of pairwise disjoint cubes $Q_n \subset\subset \Omega$ such that $W \subset \bigcup_{n=1}^{\infty} \bar{Q}_n$, $\sum_{n=1}^{\infty} (\operatorname{diam} Q_n)^N < \varepsilon$.

But then, setting $a_n \equiv (\text{diam } Q_n)^N$, $b_n \equiv |\, u\,|^p_{H^{2,p}(Q_n)}$, we have

$$|\mathscr{F}(W)| \le \sum_{n=1}^{\infty} |\mathscr{F}(Q_n)| \le C \sum_{n=1}^{\infty} a_n^{1-N/p} b_n^{N/p} \le C\Big(\sum_{n=1}^{\infty} a_n\Big)^{1-N/p}\Big(\sum_{n=1}^{\infty} b_n\Big)^{N/p}$$
$$= C\Big[\sum_{n=1}^{\infty} (\text{diam } Q_n)^N\Big]^{1-N/p}\Big(\sum_{n=1}^{\infty} |\, u\,|^p_{H^{2,p}(Q_n)}\Big)^{N/p},$$

where use has been made of the Hölder inequality. This implies

$$|\mathscr{F}(W)| \le C\varepsilon^{1-N/p} |\, u\,|^N_{H^{2,p}(\Omega)}, \qquad \text{hence } |\mathscr{F}(W)| = 0,$$

which is absurd since $\mathscr{F}(W)$ contains a sphere.

By Theorem 1.20 $u_{x_i x_j}$ is a classical derivative at almost any point of Ω, hence in particular of W. Let ξ be fixed in the unit spherical surface S of $\mathbb{R}^N$: by the change of coordinates formula, at almost any point $y \in W$ each first derivative u_{x_i} admits a classical directional derivative with respect to ξ, and we can apply the MacLaurin formula to the function $\phi(t) \equiv u(y + t\xi)$, $t \in \mathbb{R}$ with $|\, t\,| < \text{dist}(y, \partial U)$. Thus,

$$u(x) = \phi(t) = \phi(0) + t\phi_t(0) + \frac{t^2}{2}[\phi_{tt}(0) + \sigma(t)]$$
$$= u(y) + \nabla u(y)\cdot(x-y) + \frac{t^2}{2}[u_{x_i x_j}(y)\xi_i\xi_j + \sigma(t)],$$
$$\sigma(t) \to 0 \qquad \text{as } t \to 0,$$

for $x = y + t\xi$. Since $u(x) \le u(y) + \nabla u(y)\cdot(x-y)$ for any x as above, it cannot be $u_{x_i x_j}(y)\xi_i\xi_j > 0$. Now let $\{\xi^{(n)}\}$ be a countable dense subset of S. From the above considerations it follows that at almost any point $y \in W$ we have $u_{x_i x_j}(y)\xi_i^{(n)}\xi_j^{(n)} \le 0$ for $n \in \mathbb{N}$, hence also $u_{x_i x_j}(y)\xi_i\xi_j \le 0$ whenever $\xi \in S$.

Summing up, we have proven that any open neighborhood U of x^0 has a subset of positive measure where the Hessian matrix of u is nonpositive and therefore $a^{ij}u_{x_i x_j} \le 0$, which concludes the proof. □

Next we have the *weak maximum principle* for the nonvariational operator L.

LEMMA 3.25. *Let $u \in H^{2,p}(\Omega)$, $p > N$, satisfy $Lu \le 0$ in Ω. Then,*

$$\max_{\bar{\Omega}} u = \max_{\partial\Omega} u$$

if $a = 0$,

$$\max_{\bar{\Omega}} u \le \max_{\partial\Omega} u^+$$

if $a \ge 0$ in Ω.

PROOF. *Step* 1: *The case* $a = 0$. Let us for just a minute strengthen our assumption about Lu into $Lu \leq \eta < 0$ in Ω (almost everywhere). If u achieved a local maximum at a point $x^0 \in \Omega$, the Bony maximum principle would imply ess lim $\sup_{x \to x^0} Lu(x) \geq 0$, hence a contradiction (see the observation preceding Lemma 3.24). In the general case $Lu \leq 0$ we fix $\gamma > 0$ so large that

$$\alpha\gamma^2 - | b^1 |_{\infty;\Omega}\gamma > 0;$$

then for any $\varepsilon > 0$ the function $u_\varepsilon'(x) \equiv u(x) + \varepsilon e^{\gamma x_1}$ satisfies

$$Lu_\varepsilon'(x) \leq \varepsilon[-a^{11}(x)\gamma^2 + b^1(x)\gamma]e^{\gamma x_1} \leq \varepsilon(-\alpha\gamma^2 + | b^1 |_{\infty;\Omega}\gamma)e^{\gamma x_1}$$
$$\leq \eta(\varepsilon) < 0$$

almost everywhere. Since u_ε' cannot achieve an interior maximum, the conclusion in the case at hand follows from the inequality

$$u(x) + \varepsilon e^{\gamma x_1} \leq \max_{y \in \partial\Omega} [u(y) + \varepsilon e^{\gamma y_1}] \qquad \text{for } x \in \bar{\Omega},$$

valid for any $\varepsilon > 0$. Notice that at no stage did we make use of any regularity assumption about $\partial\Omega$.

Step 2: *The case* $a \geq 0$ *in* Ω. If $u \leq 0$ in Ω there's nothing left to prove. Let $\Omega^+ \equiv \{x \in \Omega \mid u(x) > 0\} \neq \varnothing$. Since $L'u \equiv Lu - au \leq -au \leq 0$ in Ω^+ we can apply the conclusion of Step 1 with Ω replaced by Ω^+, L by L'. Thus,

$$\max_{\bar{\Omega}} u = \max_{\overline{\Omega^+}} u \leq \max_{\partial\Omega^+} u = \max_{\partial\Omega} u. \qquad \square$$

We can now prove the *Hopf boundary point lemma.*

LEMMA 3.26. *Let* $u \in H^{2,p}(\Omega)$, $p > N$, *satisfy* $Lu \leq 0$ *in* Ω. *If* u *achieves a strict local maximum at a point* $x^0 \in \partial\Omega$ *and if* $(\beta_0^1, \ldots, \beta_0^N) \in \mathbb{R}^N$ *with* $\beta_0^i\nu^i(x^0) > 0$, *then* $\beta_0^i u_{x_i}(x^0) > 0$ *provided either* $a = 0$, *or* $a \geq 0$ *in* Ω *and* $u(x^0) \geq 0$.

PROOF. Our regularity assumptions about the boundary of Ω imply the existence of a sphere $B_R(y) \subset \Omega$ which is tangent to $\partial\Omega$ at x^0. To see this it is enough to consider the case when x^0 is the origin of $\mathbb{R}^N$, $\Omega \cap B$ lies above $\partial\Omega \cap B$, and the latter is the graph of a $C^{1,1}$ function $x_N = \lambda(x')$ which vanishes together with all its first derivatives for $x' = 0$. The mean value theorem yields the existence of a constant C such that $| \lambda(x') |$

$\leq C\,|\,x'\,|^2$, hence the existence of a sufficiently small number $R \in\,]0, \tfrac{1}{2}[$ such that the distance of the point $y \equiv (0, \ldots, 0, R)$ from any point of $\partial\Omega \cap B$ is not less than R.

Now let $v(x) \equiv e^{-\gamma|x-y|^2} - e^{-\gamma R^2}$, where γ is a positive constant to be determined. Computation shows that

$$\begin{aligned} Lv(x) &= e^{-\gamma|x-y|^2}\{-4\gamma^2 a^{ij}(x)(x_i - y_i)(x_j - y_j) \\ &\qquad + 2\gamma[a^{ii}(x) - a^i(x)(x_i - y_i)]\} + a(x)v(x) \\ &\leq e^{-\gamma|x-y|^2}\{-4\alpha\gamma^2\,|\,x-y\,|^2 + 2\gamma[a^{ii}(x) + |\,a^i(x)\,|\,|\,x_i - y_i\,| + a(x)]\}. \end{aligned}$$

By the boundedness of the coefficients of L to any $\varrho \in\,]0, R[$ we can associate $\gamma > 0$ so that the last term of the above inequalities is ≤ 0 whenever $\varrho < |\,x - y\,| < R$. Since R can always be assumed so small that $u(x^0) > u(x)$ for $x \in B_R(y)$, on $S_\varrho(y)$ we have $u - u(x^0) < 0$ and therefore $u - u(x^0) + \varepsilon v \leq 0$ provided $\varepsilon > 0$ is also sufficiently small. The function $u - u(x^0) + \varepsilon v$ is ≤ 0 on $S_R(y)$ where $v = 0$. In addition, $L(u - u(x^0) + \varepsilon v) = Lu + \varepsilon Lv - au(x^0) \leq -au(x^0) \leq 0$ in $B_R(y)\setminus\overline{B_\varrho(y)}$ both if $a = 0$ and if $a \geq 0$ in Ω, $u(x^0) \geq 0$. We can apply Lemma 3.25 with Ω replaced by $B_R(y)\setminus\overline{B_\varrho(y)}$ to obtain $u - u(x^0) + \varepsilon v \leq 0$ throughout the annulus. Since x^0 is a maximum point for the function $u(x) - u(x^0) + \varepsilon v(x)$, elementary considerations show that

$$\beta_0{}^i u_{x_i}(x^0) \geq -\varepsilon\beta_0{}^i v_{x_i}(x^0) > 0. \qquad \square$$

The following result is the important *strong maximum principle.*

THEOREM 3.27. *Let* $u \in H^{2,p}(\Omega)$, $p > N$, *satisfy* $Lu \leq 0$ *in* Ω *and* $Bu \leq 0$ *on* Γ. *Unless* u *is a constant, and specifically a nonpositive one if* $\operatorname{ess\,sup}_\Omega a + \max_\Gamma \beta > 0$, *the maximum* M *of* u *on* $\bar\Omega$ *cannot be achieved on* $\Omega \cup \Gamma$ *if either* $a = 0$ *and* $\beta = 0$, *or* $a \geq 0$ *in* Ω, $\beta \geq 0$ *on* Γ, *and* $M \geq 0$.

PROOF. We need only rule out the possibility that u equals M at some point $x^0 \in \Omega \cup \Gamma$ without coinciding with M throughout $\bar\Omega$. To this end we assume the existence of a sphere $B_R(y) \subset \Omega$ and of a point $x^0 \in S_R(y) \cap (\Omega \cup \Gamma)$ such that $u(x) < M$ for $x \in B_R(y)$ and $u(x^0) = M$. Lemma 3.26 can be applied with Ω replaced by $B_R(y)$. Therefore $\beta_0{}^i u_{x_i}(x^0) > 0$ whenever $\beta_0{}^i(x_i{}^0 - y_i)/|\,x^0 - y\,| > 0$. But this is absurd because all first derivatives of u must vanish at the maximum point x^0 if the latter is interior to Ω, whereas $\beta_0{}^i u_{x_i}(x^0) \leq -\beta M \leq 0$ with $\beta_0{}^i \equiv \beta^i(x^0)$ if $x^0 \in \Gamma$. $\square$

on the data f and ζ, provided the coefficients of L and B satisfy a suitable set of assumptions. Among these the continuity of the a^{ij}'s plays an essential role, as the next example shows.

EXAMPLE. Let

$$a^{ij}(x) \equiv \delta_{ij} + \frac{N+\lambda-2}{1-\lambda}\,\frac{x_i x_j}{|x|^2} \qquad \text{for } |x| \neq 0.$$

If $\lambda < 1$ it is easy to ascertain that the matrix $[a^{ij}]_{i,j=1,\ldots,N}$ is uniformly elliptic in $B\setminus\{0\}$. Set $a^i = a = 0$, $\Omega = B$, and $\Gamma = \emptyset$. The corresponding problem (3.36) is not well-posed in $H^{2,p}(B)$ for $p < N/(2-\lambda)$, since the function $u(x) \equiv |x|^\lambda - 1$ belongs to that space and solves the homogeneous problem $Lu = 0$ in B, $u = 0$ on $S = \partial B$.

Notice that the first derivatives of u are not essentially bounded in B [compare with the regularity result provided by Theorem 3.23 in the case $a^{ij} \in C^0(\bar{\Omega})$].

3.8. The Marcinkiewicz Theorem and the John–Nirenberg Lemma

We shall now give two results that were utilized in the proof of Lemma 3.6.

The first one is known as the *Marcinkiewicz interpolation theorem*. We present it under the particular formulation adopted for our purposes.

THEOREM 3.30. *Let ω, ω' be bounded domains of $\mathbb{R}^N$, let $1 \leq q < r \leq \infty$, and let $\mathscr{T}$ be a subadditive mapping of both weak types q and r from ω into ω', with*

$$]\mathscr{T}(f)[_{q;\omega'} \leq C_q |f|_{q;\omega} \qquad \textit{for } f \in L^q(\omega),$$

$$]\mathscr{T}(f)[_{r;\omega'} \leq C_r |f|_{r;\omega} \qquad \textit{for } f \in L^r(\omega).$$

Then for any $p \in]q, r[$ $\mathscr{T}$ is of the strong type p, and

$$|\mathscr{T}(f)|_{p;\omega'} \leq C C_q^{1-\lambda} C_r^{\lambda} |f|_{p;\omega} \qquad \textit{for } f \in L^p(\omega), \tag{3.43}$$

where $C = C(p, q, r)$ and $\lambda \in]0, 1[$ is defined by $1/p \equiv (1-\lambda)/q + \lambda/r$ if $r < \infty$, $1/p \equiv (1-\lambda)/q$ if $r = \infty$.

PROOF. *Step* 1: *Preliminaries.* When $f: \omega \to \mathbb{R}$ is measurable we denote by $S(f, s)$ the set $\{x \in \omega \mid |f(x)| > s\}$ and call $\mu(s) \equiv |S(f, s)|$ the *distribution function* (over ω) of $|f|$. Notice that if $x \mapsto F(x, s)$ is the characteristic function of the set $S(f, s)$, Fubini's theorem yields

$$\int_\omega |f(x)|^p\,dx = \int_\omega dx \int_0^{|f(x)|} ps^{p-1}\,ds = p\int_\omega dx \int_0^\infty s^{p-1}F(x, s)\,ds$$
$$= p\int_0^\infty s^{p-1}\,ds \int_\omega F(x, s)\,dx = p\int_0^\infty s^{p-1}\mu(s)\,ds \tag{3.44}$$

whenever $f \in L^p(\omega)$ with $1 \leq p < \infty$.

At this point we fix f in $L^p(\omega)$ for $q < p < r \leq \infty$ and denote by h the function $\mathscr{T}(f)$, by $\nu(s)$ the distribution function (over ω') of $|h|$. For any choice of $\tau > 0$ we set $f_2(x) \equiv (-\tau) \vee f(x) \wedge \tau$, $f_1(x) \equiv f(x) - f_2(x)$: then $f_1 \in L^p(\omega) \subset L^q(\omega)$, $f_2 \in L^\infty(\omega) \subseteq L^r(\omega)$. We put $h_i \equiv \mathscr{T}(f_i)$ and denote by $\mu_i(s)$, $\nu_i(s)$ the respective distribution functions of $|f_i|$, $|h_i|$ for $i = 1, 2$. It is easy to verify that the subadditivity of $\mathscr{T}$ leads to

$$\nu(2s) \leq \nu_1(s) + \nu_2(s) \qquad \text{for } s > 0. \tag{3.45}$$

Since the integral $2^p p \int_0^\infty s^{p-1}\nu(2s)\,ds$ equals $p\int_0^\infty s^{p-1}\nu(s)\,ds = |h|_{p;\omega'}^p$ if finite, the conclusion of the theorem will follow from suitable estimates on the right-hand side of (3.45). By assumption

$$\nu_1(s) \leq C_q{}^q s^{-q} |f_1|_{q;\omega}^q.$$

Since $\mu_1(t) = \mu(t + \tau)$ for $t > 0$, we see that

$$\int_0^\infty s^{p-1}\nu_1(s)\,ds \leq C_q{}^q \int_0^\infty s^{p-q-1}\,ds \int_\omega |f_1|^q\,dx$$
$$= qC_q{}^q \int_0^\infty s^{p-q-1}\,ds \int_0^\infty t^{q-1}\mu_1(t)\,dt$$
$$= qC_q{}^q \int_0^\infty s^{p-q-1}\,ds \int_\tau^\infty (t-\tau)^{q-1}\mu(t)\,dt$$
$$\leq qC_q{}^q \int_0^\infty s^{p-q-1}\,ds \int_\tau^\infty t^{q-1}\mu(t)\,dt \equiv I_1.$$

For what concerns $\nu_2(s)$ we must distinguish between the two cases $r < \infty$ and $r = \infty$.

Step 2: *The case* $r < \infty$. Since r is finite we have

$$\nu_2(s) \le C_r^r s^{-r} \,|\, f_2 \,|_{r;\omega}^r,$$

and therefore

$$\begin{aligned}
\int_0^\infty s^{p-1}\nu_2(s)\,ds &\le C_r^r \int_0^\infty s^{p-r-1}\,ds \int_\omega |\, f_2 \,|^r\,dx \\
&= rC_r^r \int_0^\infty s^{p-r-1}\,ds \int_0^\infty t^{r-1}\mu_2(t)\,dt \\
&= rC_r^r \int_0^\infty s^{p-r-1}\,ds \int_0^\tau t^{r-1}\mu(t)\,dt \equiv I_2
\end{aligned}$$

because $\mu_2(t) = \mu(t)$ for $0 < t \le \tau$, $\mu_2(t) = 0$ for $t > \tau$.

Let $\tau = \tau(s) \equiv s/A$ with $A \equiv C_q^{q/(q-r)} C_r^{r/(r-q)}$. The quantities I_1 and I_2 are, respectively, computed as follows:

$$\begin{aligned}
I_1 &= qC_q^q \int_0^\infty t^{q-1}\mu(t)\,dt \int_0^{At} s^{p-q-1}\,ds \\
&= \frac{q}{p-q}\, C_q^q \int_0^\infty t^{q-1}\mu(t)[s^{p-q}]_0^{At}\,dt = \frac{q}{p-q}\, C_q^q A^{p-q} \int_0^\infty t^{p-1}\mu(t)\,dt \\
&= \frac{q}{p-q}\, C_q^{q+q(p-q)/(q-r)} C_r^{r(p-q)/(r-q)} \int_0^\infty t^{p-1}\mu(t)\,dt \\
&= \frac{q}{p-q}\, C_q^{p(1-\lambda)} C_r^{p\lambda} \int_0^\infty t^{p-1}\mu(t)\,dt,
\end{aligned}$$

$$\begin{aligned}
I_2 &= rC_r^r \int_0^\infty t^{r-1}\mu(t)\,dt \int_{At}^\infty s^{p-r-1}\,ds = \frac{r}{p-r}\, C_r^r \int_0^\infty t^{r-1}\mu(t)[s^{p-r}]_{At}^\infty\,dt \\
&= \frac{r}{r-p}\, C_r^r A^{p-r} \int_0^\infty t^{p-1}\mu(t)\,dt \\
&= \frac{r}{r-p}\, C_q^{q(p-r)/(q-r)} C_r^{r+r(p-r)/(r-q)} \int_0^\infty t^{p-1}\mu(t)\,dt \\
&= \frac{r}{r-p}\, C_q^{p(1-\lambda)} C_r^{p\lambda} \int_0^\infty t^{p-1}\mu(t)\,dt
\end{aligned}$$

[notice $p\lambda = (pr - qr)/(r - q)$, $p(1 - \lambda) = (qr - pq)/(r - q)$].

From (3.45) we deduce

$$
\begin{aligned}
| h |_{p;\omega'}^{p} &= 2^p p \int_0^\infty s^{p-1}\nu(2s)\, ds \\
&\leq 2^p p\left(\frac{q}{p-q} + \frac{r}{r-p}\right) C_q^{p(1-\lambda)} C_r^{p\lambda} \int_0^\infty t^{p-1}\mu(t)\, dt \\
&= 2^p p\left(\frac{1}{p-q} + \frac{1}{r-p}\right) C_q^{p(1-\lambda)} C_r^{p\lambda} p \int_0^\infty t^{p-1}\mu(t)\, dt \\
&= C C_q^{p(1-\lambda)} C_r^{p\lambda} | f |_{p;\omega}^{p},
\end{aligned}
$$

which amounts to (3.43) in the case at hand.

Step 3: *The case* $r = \infty$. Let $\tau = \tau(s) \equiv s/C_\infty$. Then $| h_2 |_{\infty;\omega} \leq C_\infty | f_2 |_{\infty;\omega} \leq s$ and therefore $\nu_2(s) = 0$. Since I_1 is evaluated as in Step 2 except for A replaced by C_∞, we can again arrive at the desired conclusion. □

We now turn to the fundamental result known as the *John–Nirenberg lemma*:

LEMMA 3.31. *Let Q be an open cube of $\mathbb{R}^N$ and let $h \in L^1(Q)$. Assume that for a fixed $r \in]1, \infty[$ the quantity*

$$\sum_k | Q_k |^{1-r}\left(\int_{Q_k} | h - (h)_{Q_k} |\, dx\right)^r$$

is uniformly bounded whatever the countable decomposition $\Delta : \bar{Q} = \bigcup_k \bar{Q}_k$, the Q_k's being mutually disjoint open cubes with edges parallel to those of Q. Let $M(h) \equiv M_r(h)$ be defined by

$$[M(h)]^r \equiv \sup_\Delta \sum_k | Q_k |^{1-r}\left(\int_{Q_k} | h - (h)_{Q_k} |\, dx\right)^r.$$

Then the function $h - (h)_Q$ belongs to $L^r(Q)$-weak, and there exists a constant $C = C(r)$ independent of h such that $]h - (h)_Q[_{r;Q} \leq CM(h)$.

SKETCH OF THE PROOF. Since $\big((h)_Q\big)_{Q_k} = (h)_Q$ the function $h - (h)_Q$ satisfies the same hypotheses as h, so that it can without loss of generality be assumed that $(h)_Q = 0$.

Introducing the distribution function $\nu(s) \equiv \text{meas}_N\{x \in Q \mid | h(x) | > s\}$ of $| h |$ we reformulate the thesis of the lemma by requiring that

$$\nu(s) \leq C[M(h)/s]^r \tag{3.46}$$

for $s > 0$. Since $\nu(s) \leq |Q|$ we can associate to any $d > 0$ a constant C such that (3.46) holds for $0 < s \leq d$. Consequently we only need to show the validity of (3.46) for all s larger than some suitable $d > 0$.

Assume now that a constant C has been found with the property

$$\nu(s) \leq C[M(h)/s]^{r(1-1/\varrho^{j+1})} \, |Q|^{1/\varrho^{j+1}} \tag{3.47}$$

whenever

$$2^{-N}s/[r(\varrho^j - 1) + 1] \geq M(h) \, |Q|^{-1/r}, \tag{3.48}$$

$j = 0, 1, 2, \ldots$, where $\varrho \equiv r/(r-1)$ is the conjugate exponent of r. Let $d \equiv 2^N M(h) \, |Q|^{-1/r}$ and take any $s \geq d$. If j is the largest integer such that (3.48) holds, for $j+1$ we have the opposite inequality and therefore

$$s \, |Q|^{1/r}/M(h) < 2^N[r(\varrho^{j+1} - 1) + 1] \leq 2^N r \varrho^{j+1}$$

[we consider only the nontrivial case $M(h) > 0$]. But then (3.47) yields (3.46) since

$$\begin{aligned} \nu(s) &\leq C[M(h)/s]^r[s \, |Q|^{1/r}/M(h)]^{r/\varrho^{j+1}} \leq C[M(h)/s]^r(2^N r\varrho^{j+1})^{r/\varrho^{j+1}} \\ &\leq C[M(h)/s]^r. \end{aligned}$$

We are thus left with the task of proving (3.48) $\Rightarrow$ (3.47). Notice that a constant C as required certainly exists if for any nonnegative integer j, (3.48) implies

$$\nu(s) \leq 2^{-N}\varrho^{1/\varrho + \cdots + j/\varrho^j}[2^N r(1 - \varrho^{-1-j})M(h)/s]^{r(1-1/\varrho^{j+1})}\left[\frac{1}{M(h)} \int_Q |h| \, dx\right]^{1/\varrho^j}.$$

This is obviously true when $j = 0$. The proof of the lemma depends on showing that the above implication holds for a natural number j provided it does so for $j - 1$. We omit this part of the proof not on the grounds that it is only computational (it is not), but because the computations involved are rather cumbersome. However, we mention that the inductive assumption concerning the value $j - 1$ is applied with Q replaced by K_n and h by $h - (h)_{K_n}$, where $\{K_n\}$ is a countable family of disjoint open cubes of Q with the following properties:

- $|h| \leq t \quad$ a.e. in $Q \setminus \bigcup_n K_n$,
- $|(h)_{K_n}| \leq 2^N t$,
- $\sum_n |K_n| \leq t^{-1} \int_Q |h| \, dx$

where $t = 2^{-N}s/[r(\varrho^j - 1) + 1]$, hence

$$t \geq |Q|^{-1} \int_Q |h|\, dx$$

by (3.48). The existence of such a family $\{K_n\}$ can be demonstrated as follows. Let Q be divided into 2^N equal subcubes and denote by K_{1m} those among them for which $(|h|)_{K_{1m}} \geq t$, hence $t\,|K_{1m}| \leq \int_{K_{1m}} |h|\, dx \leq 2^N t\,|K_{1m}|$ by the choice of t. Next, apply the same procedure to each remaining subcube of Q and denote by K_{2m} those, among all the subcubes of this second decomposition, for which $(|h|)_{K_{2m}} \geq t$ and therefore

$$t\,|K_{2m}| \leq \int_{K_{2m}} |h|\, dx \leq 2^N t\,|K_{2m}|.$$

By iteration a sequence of subcubes of Q, renamed $\{K_n\}$, is constructed with the property that $t\,|K_n| \leq \int_{K_n} |h|\, dx \leq 2^N t\,|K_n|$. Almost any point $x \in Q \setminus \bigcup_n K_n$ belongs to cubes K_i' with edge-length 2^{-i}, $i = 1, 2, \ldots$, such that $(|h|)_{K_i'} < t$. The first property required on the part of $\{K_n\}$ is therefore satisfied, and so obviously are the second and the third ones. Notice that for a.a. $x \in Q$ the inequalities $|h(x)| > s \geq 2^N t$ imply $x \in K_n$, hence $|h(x) - (h)_{K_n}| > s - 2^N t$, for some $n \in \mathbb{N}$. Consequently, $\nu(s) \leq \sum_n \text{meas}_N\{x \in K_n \mid |h(x) - (h_{K_n})| > s - 2^N t\}$. □

Problems

3.1. Use (3.44) to prove that $L^q(\omega)$-weak $\subset L^p(\omega)$ whenever ω is a bounded domain of $\mathbb{R}^N$ and $1 \leq p < q \leq \infty$.

3.2. This and the next six problems develop the $H^{k,p}$ theory for p in the range $]1, 2[$.

Let $\partial\Omega$ be of class $C^{1,\delta}$ (with Γ closed) and take $a^{ij}, d^j \in C^{0,\delta}(\bar{\Omega})$, for some $\delta \in\,]0, 1[$. If $a(u, v)$ (from 3.11) is coercive on $V = H_0^{\,1}(\Omega \cup \Gamma)$ and $f^1, \ldots, f^N \in L^p(\Omega)$ with $1 < p < 2$, there exists a unique solution to the variational b.v.p.

$$u \in H_0^{\,1,p}(\Omega \cup \Gamma), \qquad a(u, v) = \int_\Omega f^i v_{x_i}\, dx \qquad \text{for } v \in H_0^{\,1,p'}(\Omega \cup \Gamma).$$

To see this, begin with the proof of existence for $f^2 = \cdots = f^N = 0$. Let $f, g \in L^2(\Omega)$, and define bounded linear operators T_j, S^j: $L^2(\Omega) \to L^2(\Omega)$, $j = 0, 1, \ldots, N$, as follows:

- $T_0 f \equiv u$, $T_i f \equiv u_{x_i}$ for $i = 1, \ldots, N$, where

$$u \in V, \qquad a(u, v) = \int_\Omega f v_{x_1}\, dx \qquad \text{for } v \in V;$$

- $S^0 g \equiv z^0_{x_1}$, $S^i g \equiv z^i_{x_1}$ for $i = 1, \ldots, N$, where

$$z^0 \in V, \qquad a(v, z^0) = \int_\Omega g v\, dx \qquad \text{for } v \in V,$$

$$z^i \in V, \qquad a(v, z^i) = \int_\Omega g v_{x_i}\, dx \qquad \text{for } v \in V.$$

Then $\langle T_j f, g\rangle = \langle S^j g, f\rangle$. Each S^j is continuous from $L^{p'}(\Omega)$ into $L^{p'}(\Omega)$, and each T_j has a continuous extension $L^p(\Omega) \to L^p(\Omega)$. If now f^1 is the limit in $L^p(\Omega)$ of $\{f_n\} \subset L^2(\Omega)$, solve

$$u_n \in V, \qquad a(u_n, v) = \int_\Omega f_n v_{x_1}\, dx \qquad \text{for } v \in H_0^{1,p'}(\Omega \cup \Gamma)$$

and pass to the limit. As for uniqueness: if u is a solution of the b.v.p. for $f^1 = \cdots = f^N = 0$, take f in $L^{p'}(\Omega)$, solve

$$v \in H_0^{1,p'}(\Omega \cup \Gamma), \qquad a(w, v) = \int_\Omega f w\, dx \qquad \text{for } w \in V,$$

and replace w by u through a continuity argument: thus, $\int_\Omega f u\, dx = 0$.

3.3. Let $a^{ij}, d^j \in C^{0,1}(\bar\Omega)$. If u satisfies

$$u \in H^{1,p}(\Omega), \qquad a(u, v) = \langle F, v\rangle \equiv \int_\Omega (f^0 v + f^i v_{x_i})\, dx \qquad \text{for } v \in H_0^{1,p'}(\Omega)$$

with $f^0 \in L^p(\Omega)$, $f^1, \ldots, f^N \in H^{1,p}(\Omega)$, $1 < p < 2$, the conclusion of Theorem 3.8 remains valid (with the understanding, here as well as in Problems 3.4 and 3.5 below, that in the estimates the H^1 norm of u is replaced by the $H^{1,p}$ norm). To see this, it suffices to consider the case $d^j = b^i = c = 0$, $\operatorname{supp} u \subset \Omega$, $\operatorname{supp} f^j \subset \Omega$, so that the above equation holds for $v \in H^{1,p'}(\Omega)$; without loss of generality, $\partial\Omega$ can be assumed of class $C^{1,1}$. Fix any $s = 1, \ldots, N$ and solve the Dirichlet b.v.p.

$$w \in H_0^{1,p}(\Omega),$$

$$\int_\Omega a^{ij} w_{x_i} v_{x_j}\, dx = \int_\Omega [-f^0 v_{x_s} + (f^i_{x_s} - a^{ji}_{x_s} u_{x_j}) v_{x_i}]\, dx \qquad \text{for } v \in H_0^{1,p'}(\Omega).$$

Let $g \in C_c^\infty(\Omega)$: the solution $\hat v$ of the Dirichlet b.v.p.

$$\hat v \in H_0^1(\Omega), \qquad \int_\Omega a^{ij} z_{x_i} \hat v_{x_j}\, dx = \int_\Omega g z\, dx \qquad \text{for } z \in H_0^1(\Omega)$$

belongs to $H^{2,q}(\Omega)$ for any finite q. Since $\int_\Omega u_{x_i}(a^{ij}\hat v_{x_j})_{x_s}\, dx$ equals $-\int_\Omega a^{ij} \times w_{x_i} \hat v_{x_j}\, dx$ as well as $-\int_\Omega g u_{x_s}\, dx$ [after an approximation of u in $H^{1,p}(\Omega)$ with functions from $C_c^\infty(\Omega)$], the identity $u_{x_s} = w$ follows by the arbitrariness of g.

3.4. Let a^{ij}, d^j, F be as in Problem 3.3 except for Ω replaced by B^+. Let u solve either

$$u \in H_0^{1,p}(B^+ \cup S^+), \qquad a(u, v) = \langle F, v\rangle \qquad \text{for } v \in H_0^{1,p'}(B^+)$$

or

$$u \in H^{1,p}(B^+), \qquad a(u, v) = \langle F, v\rangle \qquad \text{for } v \in H_0^{1,p'}(B^+ \cup S^0)$$

Then the conclusion of Theorem 3.15 remains valid. (The same technique as in the preceding problem can be followed for $s = 1, \ldots, N - 1$.)

3.5. Suppose $\partial\Omega$ of class $C^{1,1}$ (with Γ closed) and a^{ij}, d^j, F as in Problem 3.3. If u satisfies

$$u \in H_0^{1,p}(\Omega \cup \Gamma), \qquad a(u, v) = \langle F, v\rangle \qquad \text{for } v \in H_0^{1,p'}(\Omega \cup \Gamma),$$

the conclusion of Theorem 3.17(ii) remains valid. Note that, as a consequence, both Lemmas 3.18(i) and 3.20(i) hold true if $1 < p < 2$.

3.6. If $\zeta \in H^{1/p',p}(\Gamma)$, $1 < p < 2$, there exists $u \in H^{2,p}(\Omega)$, $u = 0$ on $\partial\Omega\setminus\Gamma$, such that $Bu = \zeta$. Indeed, let $Lu \equiv -\Delta u + u$, and correspondingly construct $\hat{a}(u, v)$ as in (3.39): the solution of the b.v.p.

$$u \in H_0^{1,p}(\Omega \cup \Gamma), \qquad \hat{a}(u, v) = \int_\Omega (\tilde{f}^i v)_{x_i}\, dx \qquad \text{for } v \in H_0^{1,p'}(\Omega \cup \Gamma),$$

where the $\tilde{f}^i$'s are chosen as in Step 1 of the proof of Lemma 3.18, has the required properties.

3.7. For $1 < p < 2$ Lemma 3.21(i) can be proven under the additional assumption $Bu = 0$, as follows. Let $\hat{a}_n(u, v)$ be defined as in Step 1 of the proof of Lemma 3.21, and solve

$$w \in H_0^{1,p'}(\Omega \cup \Gamma),$$

$$\hat{a}_n(v, w) + \lambda \int_\Omega wv\, dx = \int_\Omega |u|^{p-2} uv\, dx \qquad \text{for } v \in H_0^{1,p}(\Omega \cup \Gamma),$$

λ large enough. Choose $v = |w|^{p'-2} w$ and utilize the inequality

$$\int_\Omega |\nabla w|\, |w|^{p'-1}\, dx \le \left(\int_\Omega |w|^{p'-2} |\nabla w|^2\, dx\right)^{1/2} \left(\int_\Omega |w|^{p'}\, dx\right)^{1/2}:$$

then there exists λ_n such that

$$|w|_{p';\Omega} \le (\lambda - \lambda_n)^{-1} |u|_{p;\Omega}^{p-1} \qquad \text{for } \lambda > \lambda_n.$$

On the other hand,

$$\int_\Omega (L_n u + \lambda u) w\, dx = \hat{a}_n(u, w) + \lambda \int_\Omega wu\, dx = \int_\Omega |u|^p\, dx.$$

At this point, utilize the inequalities

$$|Lu|_{p;\Omega} \le |Lu + \lambda u|_{p;\Omega} + \lambda |u|_{p;\Omega},$$

$$|u|_{H^{1,p}(\Omega)} \le C(\varepsilon) |u|_{p;\Omega} + \varepsilon \sum_{i,j=1}^N |u_{x_i x_j}|_{p;\Omega},$$

$$|u|_{p;\Omega} \le (\lambda - \lambda_n)^{-1} (|Lu + \lambda u|_{p;\Omega} + |(L - L_n)u|_{p;\Omega})$$

in (3.37), and choose a suitably large value of n.

3.8. Prove Theorems 3.28(i) and 3.29(i) for $1 < p < 2$.

3.9. Prove (ii) of Lemma 3.20 by proceeding as in the proof of (i); in order to majorize $|(L - L_0)u'|_{C^{0,\delta}(\bar{\Omega})}$ use the inequality

$$\max_{\overline{\Omega \cap B_{2r}(x^0)}} |u'_{x_ix_j}| \leq (4r)^\delta [u'_{x_ix_j}]_{\delta;\overline{\Omega \cap B_{2r}(x^0)}} + |\Omega|^{-1} |u'_{x_ix_j}|_{1;\Omega}$$

and estimate $|u'_{x_ix_j}|_{1;\Omega}$ through (i).

3.10. Use a cutoff technique and a bootstrap argument to demonstrate the following local counterpart of Theorem 3.23(i): if $\Omega'' \equiv \Omega' \cap \Omega$, where Ω' is an open subset of $\mathbb{R}^N$, and if $u \in H^2(\Omega'')$ with $Lu = f \in L^p(\Omega'')$, $u|_{\Omega' \cap (\partial\Omega \setminus \Gamma)} = 0$ and $(Bu)|_{\Omega' \cap \Gamma} = \zeta|_{\Omega' \cap \Gamma}$, where $\zeta \in H^{1/p',p}(\Gamma)$ $(2 < p < \infty)$, then $u \in H^{2,p}(\omega)$ whenever $\omega \subset \Omega''$ is open with $\operatorname{dist}(\omega, \Omega \setminus \Omega'') > 0$. An analogous statement can be given as a local counterpart to Theorem 3.23(ii).

3.11. Suppose that for some $k \in N$, $\partial\Omega$ is of class $C^{k+1,1}$, a^{ij}, a^i, $a \in C^{k-1,1}(\bar{\Omega})$, β^i, $\beta \in C^{k,1}(\Gamma)$, $f \in H^{k,p}(\Omega)$ and ζ is the trace on Γ of some function $z \in H^{k+1,p}(\Omega)$, $2 \leq p < \infty$. Then any solution $u \in H^2(\Omega)$ of (3.36) belongs to $H^{k+2,p}(\Omega)$. To see this, consider the case $k = 1$. Take difference quotients of u (in all directions near a point $x^0 \in \Omega$, and in all tangential directions, after straightening a suitable portion of the boundary, near a point $x^0 \in \partial\Omega$). Utilize a local counterpart of Lemma 3.20(i) to obtain L^p uniform bounds on the corresponding difference quotients of all second derivatives of u.

3.12. State and prove the regularity result in $C^{k,\delta}(\bar{\Omega})$ analogous to that of Problem 3.11 in $H^{k,p}(\Omega)$.

3.13. Let $u \in H^2_{\text{loc}}(\Omega) \cap C^0(\bar{\Omega})$ solve (3.36) with $a^{ij} \in C^0(\bar{\Omega})$, $f \in L^p(\Omega)$ $(N/2 < p < \infty)$, $\Gamma = \varnothing$. The equation $-a^{ij}v_{x_ix_j} + a^i v_{x_i} = f - au$ in Ω can have at most one solution $v \in H^2_{\text{loc}}(\Omega) \cap C^0(\bar{\Omega})$ vanishing on $\partial\Omega$, and u belongs to $H^{2,p}(\Omega)$.

3.14. Thanks to Sobolev inequalities, Lemma 3.20(i) (see Problem 3.5 as well) remains valid if β^i, β are taken in some space $H^{1/q',q}(\Gamma)$, for a suitable choice of $q > N/2$ depending on $p \in]1, \infty[$, instead of $C^{0,1}(\Gamma)$. This can be ascertained (after fixing $x^0 \in \Gamma$) by replacing Bu' with $B_0u' \equiv \tilde{\beta}_0{}^i u'_{x_i}|_\Gamma$, where $(\tilde{\beta}_0{}^1, \ldots, \tilde{\beta}_0{}^N)$ is the $C^{0,1}$ vector field on Γ defined by

$$\tilde{\beta}_0{}^i \equiv g_{x^0,R}\beta_0{}^i + (1 - g_{x^0,R})\nu^i, \qquad \beta_0{}^i \equiv \beta^i(x^0);$$

R is so small that $B_{2R}(x^0) \cap \partial\Omega \subset \Gamma$ and $\nu^i(x)\beta_0{}^i \geq \varkappa/2$ for $x \in B_{2R}(x^0) \cap \Gamma$. Note that $(B - B_0)u' = [(b^i - \beta_0{}^i)u'_{x_i} + bu']|_\Gamma$ if $b^i, b \in H^{1,q}(\Omega)$ are such that $b^i|_\Gamma = \beta^i$, $b|_\Gamma = \beta$ and r is small enough.

4

Variational Inequalities

The minimum problem we mentioned in the introduction to Chapter 2 can be generalized as follows:

$$\text{minimize } \mathscr{J}(v) \equiv \frac{1}{2}\int_\Omega (|\nabla v|^2 + v^2)\,dx - \int_\Omega fv\,dx$$

$$\text{over a convex subset } \mathbb{K} \text{ of } H_0{}^1(\Omega \cup \Gamma)$$

[with $f \in L^2(\Omega)$, Γ of class C^1]. If u is a solution to this problem, for any choice of v in $\mathbb{K}$ the function $\mathscr{J}(u + \lambda(v-u))$ of $\lambda \in [0, 1]$ must attain its minimum at $\lambda = 0$; hence, u must satisfy the condition

$$u \in \mathbb{K}, \qquad \frac{d}{d\lambda}\mathscr{J}(u + \lambda(v-u))|_{\lambda=0} \geq 0 \qquad \text{for } v \in \mathbb{K},$$

which amounts to

$$u \in \mathbb{K}, \qquad a(u, v-u) \geq \int_\Omega f(v-u)\,dx \qquad \text{for } v \in \mathbb{K} \tag{4.1}$$

[where $a(u, v)$ denotes the symmetric bilinear form $\int_\Omega (u_{x_i}v_{x_i} + uv)\,dx$]. Vice versa, a solution of (4.1) necessarily minimizes $\mathscr{J}(v)$ over $\mathbb{K}$ (see Lemma 4.1 below). These simple observations are sufficient to introduce the content of the present chapter.

In Section 4.1 we study the existence and uniqueness of solutions to a wide class of problems which includes (4.1) and involves bilinear forms, not necessarily symmetric, on a Hilbert space V. In Section 4.2 we generalize further and replace bilinear forms by mappings $\langle A(u), v\rangle$, with u, v varying in a Banach space V, and A (not necessarily linear) going from V

to V'; in particular, whenever the choice of $\mathbb{K} = V$ is admissible, we extend the corresponding theory of equations for linear operators A (see Section 2.1).

At this point we focus on $V = H_0^1(\Omega \cup \Gamma)$ or $V = H_0^{1,p}(\Omega \cup \Gamma)$.

In Section 4.3 we investigate the applicability of previous abstract results to more concrete examples of convex subsets of V, of bilinear forms and, especially, of nonlinear operators. We also show how to formulate certain types of problems like (4.1) as differential ones. In Section 4.4 we prove existence and uniqueness of solutions in some cases that are not covered by the general abstract theory.

Sections 4.5–4.8 are devoted to the study of conditions ensuring some regularity properties of solutions u, such as

$$u \in H^{2,p}(\Omega), \qquad u \in C^{0,\delta}(\bar{\Omega}), \qquad u \in C^{1,\delta}(\bar{\Omega}).$$

In Section 4.9 we tackle instead a class of nonlinear operators that do not enter the abstract theory of Section 4.2, and prove the existence of solutions to problems involving either a special type of proper convex subset $\mathbb{K}$ of V, or $\mathbb{K} = V$.

4.1. Minimization of Convex Functionals, and Variational Inequalities for Linear Operators

4.1.1. A Class of Minimum Problems

Consider the problem of finding a vector u such that

$$u \in \mathbb{K}, \qquad \mathscr{J}(u) \le \mathscr{J}(v) \qquad \text{for } v \in \mathbb{K}, \tag{4.2}$$

where $\mathbb{K}$ is a subset of a Hilbert space V and

$$\mathscr{J}(v) \equiv \frac{1}{2} | v |_V^2 - \langle F, v \rangle \qquad \text{for } v \in V$$

with F given in V'. To investigate the minimum problem (4.2) we single out the following properties of the quadratic functional $\mathscr{J}$:

- $\mathscr{J}$ is *convex*, that is,

$$\mathscr{J}(u + (1 - \lambda)v) \le \lambda \mathscr{J}(u) + (1 - \lambda)\mathscr{J}(v)$$
$$\text{for } u, v \in V, \qquad 0 \le \lambda \le 1;$$

and, more precisely,

- $\mathscr{J}$ is *strictly convex*, that is,

$$\mathscr{J}(u + (1-\lambda)v) < \lambda\mathscr{J}(u) + (1-\lambda)\mathscr{J}(v)$$
$$\text{for } u, v \in V \quad \text{with} \quad u \neq v, \qquad 0 < \lambda < 1$$

[note that $2(u, v)_V < |u|_V^2 + |v|_V^2$ if $u \neq v$];

- $\mathscr{J}$ is *coercive*, that is,

$$\mathscr{J}(v) \to \infty \qquad \text{as } |v|_V \to \infty;$$

- $\mathscr{J}$ is *weakly lower semicontinuous* in the sense that

$$\liminf_{n\to\infty} \mathscr{J}(v_n) \geq \mathscr{J}(v) \qquad \text{when } v_n \rightharpoonup v \qquad \text{in } V$$

(a consequence of the analogous property of the norm $|\cdot|_V$);

- $\mathscr{J}$ is *Gateaux differentiable* at any $u \in V$.

This means that there exists an element of V', denoted by $\mathscr{J}'(u)$ and called the *Gateaux derivative* of $\mathscr{J}$ at u, with the property

$$\langle \mathscr{J}'(u), v\rangle = \frac{d}{d\lambda}\mathscr{J}(u + \lambda v)|_{\lambda=0} \qquad \text{for } v \in V;$$

computation shows that

$$\langle \mathscr{J}'(u), v\rangle = (u, v)_V - \langle F, v\rangle \qquad \text{for } v \in V. \tag{4.3}$$

When $\mathbb{K}$ is convex the first and the last of the above properties of $\mathscr{J}$ lead to the following characterization of solutions to (4.2). Let u solve (4.2), and let $v \in \mathbb{K}$ be arbitrarily fixed. Then $u + \lambda(v - u) \in \mathbb{K}$ for $0 \leq \lambda \leq 1$; the real function $\lambda \mapsto \lambda^{-1}[\mathscr{J}(u + \lambda(v-u)) - \mathscr{J}(u)]$ of $\lambda \in]0, 1]$ is nonnegative and so too is its limit $\langle \mathscr{J}'(u), v - u\rangle$ as $\lambda \to 0^+$. This means that u satisfies

$$u \in \mathbb{K}, \ \langle \mathscr{J}'(u), v - u\rangle \geq 0 \qquad \text{for } v \in \mathbb{K}. \tag{4.4}$$

Vice versa, for $v \in \mathbb{K}$ and $0 < \lambda \leq 1$ the convexity of $\mathscr{J}$ implies

$$\lambda^{-1}[\mathscr{J}(u + \lambda(v-u)) - \mathscr{J}(u)] \leq \mathscr{J}(v) - \mathscr{J}(u),$$

so that another passage to the limit as $\lambda \to 0^+$ yields

$$\langle \mathscr{J}'(u), v - u\rangle \leq \mathscr{J}(v) - \mathscr{J}(u)$$

and consequently (4.2) if u solves (4.4). Summing up, we have proved the following lemma.

LEMMA 4.1. *Let* $\mathbb{K}$ *be a convex subset of* V. *Then* (4.2) *is equivalent to* (4.4).

Passing to existence and uniqueness we have the following lemma.

LEMMA 4.2. *If* $\mathbb{K}$ *is a nonvoid closed and convex subset of* V, (4.2) *admits a unique solution.*

PROOF. Let $\{u_n\}_n \subset \mathbb{K}$ be a minimizing sequence for $\mathscr{J}$ over $\mathbb{K}$, that is,

$$\mathscr{J}(u_n) \to \inf_{v \in \mathbb{K}} \mathscr{J}(v) \qquad \text{as } n \to \infty.$$

By the coerciveness of $\mathscr{J}$, $\{u_n\}$ is bounded and therefore contains a weakly convergent subsequence $\{u_{n_k}\}_k$ thanks to the reflexivity of V. Let u denote the weak limit of the u_{n_k}'s. Since $\mathbb{K}$ is closed and convex, $u \in \mathbb{K}$ (see Lemma 1.C). The weak lower semicontinuity of $\mathscr{J}$ implies

$$\mathscr{J}(u) \leq \liminf_{k \to \infty} \mathscr{J}(u_{n_k}) = \inf_{v \in \mathbb{K}} \mathscr{J}(v),$$

so that u solves (4.2).

As for uniqueness: if the minimum of $\mathscr{J}$ over $\mathbb{K}$ were also attained at another vector $w \in \mathbb{K}$, the strict convexity of $\mathscr{J}$ would imply

$$\mathscr{J}(\lambda u + (1 - \lambda)w) < \inf_{v \in \mathbb{K}} \mathscr{J}(v) \qquad \text{for } 0 < \lambda < 1,$$

hence a contradiction. □

REMARK. In the particular case $\mathbb{K} = V$ any vector $v = u \pm w$, $w \in V$, is admissible in (4.4). By the two lemmas above, therefore, to any given $F \in V'$ there corresponds a unique vector $u \in V$ such that $(u, w)_V = \langle F, w \rangle$ for $w \in V$. This amounts to a new proof of the Riesz representation theorem.

Because of (4.3) we can rewrite (4.4) as

$$u \in \mathbb{K}, \qquad (u, v - u)_V \geq \langle F, v - u \rangle \qquad \text{for } v \in \mathbb{K}, \tag{4.5}$$

or as

$$u \in \mathbb{K}, \qquad \langle \mathscr{T}^{-1}u - F, v - u \rangle \geq 0 \qquad \text{for } v \in \mathbb{K} \tag{4.6}$$

(where the operator $\mathcal{T}: V' \to V$ is the Riesz isomorphism), or as

$$u \in \mathbb{K}, \qquad (u - z, v - u)_V \geq 0 \qquad \text{for } v \in \mathbb{K} \tag{4.7}$$

with $z \equiv \mathcal{T}F$. When $\mathbb{K} \neq \emptyset$ is closed and convex, the solution u to (4.7), being the minimum point for $\mathcal{J}(v) = \frac{1}{2} | v |_V{}^2 - (z, v)_V$ over $\mathbb{K}$, is also the solution to the least distance problem (so familiar from calculus in the case $V = \mathbb{R}^N$)

$$u \in \mathbb{K}, \qquad | u - z |_V \leq | v - z |_V \qquad \text{for } v \in \mathbb{K}.$$

We call u the *projection* of z over $\mathbb{K}$ and write it as $P_{\mathbb{K}}(z)$. The mapping $P_{\mathbb{K}}: V \to \mathbb{K}$ so defined is not linear unless $\mathbb{K}$ is a linear subspace of V (a case that can be investigated as an easy exercise). $P_{\mathbb{K}}$ is, however, continuous, and even more than that: it is *nonexpansive*, in the sense of the next result.

LEMMA 4.3. *Let $\mathbb{K}$ be a nonvoid closed and convex subset of V. Then*

$$| P_{\mathbb{K}}(z_1) - P_{\mathbb{K}}(z_2) |_V \leq | z_1 - z_2 |_V \qquad \text{for } z_1, z_2 \in V.$$

PROOF. For $h = 1, 2$, (4.7) becomes

$$\big(P_{\mathbb{K}}(z_h) - z_h, v - P_{\mathbb{K}}(z_h)\big)_V \geq 0 \qquad \text{for } v \in \mathbb{K}.$$

Take $v = P_{\mathbb{K}}(z_2)$ when $h = 1$ and $v = P_{\mathbb{K}}(z_1)$ when $h = 2$: the sum of the two inequalities so obtained yields

$$\begin{aligned} | P_{\mathbb{K}}(z_1) - P_{\mathbb{K}}(z_2) |_V{}^2 &= \big(P_{\mathbb{K}}(z_1) - P_{\mathbb{K}}(z_2), P_{\mathbb{K}}(z_1) - P_{\mathbb{K}}(z_2)\big)_V \\ &\leq \big(z_1 - z_2, P_{\mathbb{K}}(z_1) - P_{\mathbb{K}}(z_2)\big)_V \\ &\leq | z_1 - z_2 |_V \, | P_{\mathbb{K}}(z_1) - P_{\mathbb{K}}(z_2) |_V, \end{aligned}$$

hence the desired result. □

4.1.2. Variational Inequalities

We now proceed to generalize (4.5), (4.6) as follows. On V we introduce a bilinear form $a(u, v)$ which we assume to satisfy the boundedness (i.e., continuity) requirement

$$| a(u, v) | \leq M \, | u |_V \, | v |_V \qquad \text{for } u, v \in V \qquad (M > 0);$$

a bounded linear operator $A: V \to V'$ is consequently defined by

$$\langle Au, v\rangle \equiv a(u, v) \qquad \text{for } u, v \in V. \tag{4.8}$$

We are interested in the *variational inequality* (henceforth, v.i.)

$$u \in \mathbb{K}, \qquad a(u, v-u) \geq \langle F, v-u\rangle \qquad \text{for } v \in \mathbb{K}, \tag{4.9}$$

which can also be written as

$$u \in \mathbb{K}, \qquad \langle Au - F, v-u\rangle \geq 0 \qquad \text{for } v \in \mathbb{K}, \tag{4.10}$$

where $\mathbb{K} \subseteq V$ and $F \in V'$ are given: (4.9) is the customary formulation of the v.i. for a bilinear form on a Hilbert space, whereas (4.10) is a formulation which can be extended to cover the case of a nonlinear operator on a Banach space (see Section 4.2).

When the coerciveness and symmetry requirements

$$a(u, u) \geq \alpha_0 \mid u \mid_V{}^2 \qquad \text{for } u \in V \qquad (\alpha_0 > 0)$$

and

$$a(u, v) = a(v, u) \qquad \text{for } u, v \in V$$

are satisfied in addition to boundedness, $a(u, v)$ is a scalar product on V equivalent to $(u, v)_V$ and A is the inverse of the corresponding Riesz isomorphism. Replace the definition (4.2) of $\mathscr{J}$ by

$$\mathscr{J}(v) \equiv \frac{1}{2}\, a(v, v) - \langle F, v\rangle \qquad \text{for } v \in V:$$

because of symmetry,

$$\langle \mathscr{J}'(u), v\rangle = a(u, v) - \langle F, v\rangle = \langle Au - F, v\rangle \qquad \text{for } v \in V,$$

$u \in V$, and (4.4) is nothing but the v.i. introduced above. The latter is therefore uniquely solvable whenever $a(u, v)$ is continuous, coercive, and symmetric and $\mathbb{K} \neq \emptyset$ is closed and convex, thanks to Lemmas 4.1 and 4.2.

When the symmetry assumption is dropped, (4.9) is no longer equivalent to a minimum problem such as (4.1). We can, however, directly provide the following existence and uniqueness result for v.i.'s.

THEOREM 4.4. *Let $a(u, v)$ be a continuous and coercive bilinear form on V and let $\mathbb{K} \neq \emptyset$ be a closed and convex subset of V. Then, for any choice of $F \in V'$,* (4.9) *admits a unique solution u and the mapping $F \mapsto u$*

so defined is (with an obvious extension of terminology from the special case $V = \mathbb{R}$) Lipschitzian from V' into V with Lipschitz constant α_0^{-1}.

PROOF. *Step* 1: *Existence and uniqueness.* By making use of the Riesz isomorphism $\mathscr{I}$ we rewrite (4.10) as

$$u \in \mathbb{K}, \qquad (\mathscr{I}(Au - F),\ v - u)_V \geq 0 \qquad \text{for } v \in \mathbb{K},$$

or equivalently as

$$u \in \mathbb{K}, \qquad (u - T_\varrho u,\ v - u)_V \geq 0 \qquad \text{for } v \in \mathbb{K}$$

with $T_\varrho u \equiv u - \varrho\mathscr{I}(Au - F)$, $\varrho > 0$. Thus u solves (4.10) if and only if it satisfies $u = P_{\mathbb{K}}(T_\varrho u)$ for some (and consequently for every) $\varrho > 0$.

Let us show the existence of some positive number ϱ such that the mapping $P_{\mathbb{K}} \circ T_\varrho$ is a contraction on V: this will prove the existence of a unique fixed point $u = P_{\mathbb{K}}(T_\varrho u)$, hence of a unique solution to (4.10).

For $u_1, u_2 \in V$ Lemma 4.3 yields

$$\begin{aligned} |P_{\mathbb{K}}(T_\varrho u_1) - P_{\mathbb{K}}(T_\varrho u_2)|_V^2 &\leq |T_\varrho u_1 - T_\varrho u_2|_V^2 \\ &= (u_1 - u_2 - \varrho\mathscr{I}A(u_1 - u_2),\ u_1 - u_2 \\ &\quad - \varrho\mathscr{I}A(u_1 - u_2))_V \\ &= |u_1 - u_2|_V^2 - 2\varrho\langle A(u_1 - u_2),\ u_1 - u_2\rangle \\ &\quad + \varrho^2 |A(u_1 - u_2)|_{V'}^2. \end{aligned}$$

By coerciveness

$$\varrho\langle A(u_1 - u_2),\ u_1 - u_2\rangle \geq \varrho\alpha_0 |u_1 - u_2|_V^2,$$

whereas by boundedness

$$|A(u_1 - u_2)|_{V'} \leq M |u_1 - u_2|_V.$$

Therefore,

$$|P_{\mathbb{K}}(T_\varrho u_1) - P_{\mathbb{K}}(T_\varrho u_2)|_V^2 \leq (1 - 2\varrho\alpha_0 + \varrho^2 M^2) |u_1 - u_2|_V^2.$$

This shows that $P_{\mathbb{K}} \circ T_\varrho$ is a contraction provided $1 - 2\varrho\alpha_0 + \varrho^2 M^2 < 1$, that is, $0 < \varrho < 2\alpha_0/M^2$.

Step 2: *Lipschitz dependence.* For $h = 1, 2$ fix $F_h \in V'$ and solve

$$u_h \in \mathbb{K}, \qquad a(u_h,\ v - u_h) \geq \langle F_h,\ v - u_h\rangle \qquad \text{for } v \in \mathbb{K}.$$

In the above v.i. take $v = u_2$ when $h = 1$ and $v = u_1$ when $h = 2$: by coerciveness, the sum of the two inequalities so obtained yields

$$\alpha_0 \, | u_1 - u_2 |_V^2 \le a(u_1 - u_2, u_1 - u_2) \le \langle F_1 - F_2, u_1 - u_2 \rangle \le | F_1 - F_2 |_{V'} \, | u_1 - u_2 |_V,$$

hence the sought-for Lipschitz inequality

$$| u_1 - u_2 |_V \le \alpha_0^{-1} \, | F_1 - F_2 |_{V'}$$

(compare with the proof of Lemma 4.3). □

Remark 1. When $\mathbb{K} = V$ any vector $v = u \pm w$, $w \in V$, is admissible in (4.9), which becomes

$$u \in V, \qquad a(u, w) = \langle F, w \rangle \qquad \text{for } w \in V.$$

Theorem 4.4 contains a proof of the unique solvability of the above equation, hence a new proof of the Lax–Milgram theorem.

Remark 2. Let $a(u, v)$ be bounded and coercive and let each element v of the nonvoid closed and convex set $K \subseteq V$ be the strong limit in V of a sequence $\{v_n\}$, each v_n belonging to a closed and convex set $\mathbb{K}_n \subseteq V$.

The reader may easily verify (by an argument that is also utilized for Theorem 4.5) that the solution u of (4.9) is the weak limit in V of the sequence $\{u_n\}$, each u_n being the solution of (4.9) with $\mathbb{K}$ replaced by $\mathbb{K}_n$.

This is the simplest example of convergence of solutions to v.i.'s under perturbations of the convex sets. For this aspect of the theory of v.i.'s we refer to U. Mosco [119].

For what concerns existence of solutions, the assumption of coerciveness can be weakened by requiring that $a(u, v)$ be nonnegative, i.e.,

$$a(u, u) \ge 0 \qquad \text{for } u \in V,$$

and satisfy a growth condition such as

$$\begin{aligned} &\text{there exist} \quad R \in \,]0, \infty[\quad \text{and} \quad v_0 \in \mathbb{K}, \quad | v_0 |_V < R, \\ &\text{such that} \\ &a(u, v_0 - u) < \langle F, v_0 - u \rangle \qquad \text{for } u \in \mathbb{K}, \quad | u |_V = R; \end{aligned} \tag{4.11}$$

as a matter of fact, even (4.11) can be dispensed with if $\mathbb{K}$ is bounded. This is illustrated by the next two results.

THEOREM 4.5. *Let $a(u, v)$ be a continuous nonnegative bilinear form on V and let $\mathbb{K} \neq \emptyset$ be a closed and convex bounded subset of V. Then for any choice of $F \in V'$* (4.9) *admits at least one solution.*

PROOF. For each $n \in N$ consider the v.i.

$$u_n \in \mathbb{K}, \qquad a(u_n, v - u_n) + \frac{1}{n}(u_n, v - u_n)_V \geq \langle F, v - u_n \rangle \qquad \text{for } v \in \mathbb{K}, \tag{4.12}$$

which enters the coercive case dealt with in Theorem 4.4 and is therefore uniquely solvable. By the boundedness of $\mathbb{K}$, a subsequence $\{u_{n_k}\}_k$ of $\{u_n\}_n \subseteq \mathbb{K}$ converges weakly in V to some vector u; $u \in \mathbb{K}$, because the convex set $\mathbb{K}$ is closed (see Lemma 1.C). By weak lower semicontinuity

$$a(u, u) \leq \liminf_{k\to\infty} \left[a(u_{n_k}, u_{n_k}) + \frac{1}{n_k} | u_{n_k} |_V^2 \right],$$

so that (4.12) yields

$$\begin{aligned} a(u, u) &\leq \liminf_{k\to\infty} \left[a(u_{n_k}, v) + \frac{1}{n_k} (u_{n_k}, v)_V - \langle F, v - u_{n_k} \rangle \right] \\ &= a(u, v) - \langle F, v - u \rangle \qquad \text{for } v \in \mathbb{K}, \end{aligned}$$

hence (4.9). □

THEOREM 4.6. *Let $a(u, v)$ be a continuous nonnegative bilinear form on V satisfying* (4.11) *where $\mathbb{K} \neq \emptyset$ is a closed and convex subset of V. Then for any choice of $F \in V'$* (4.9) *admits at least one solution.*

PROOF. Set

$$\mathbb{K}_R \equiv \{v \in \mathbb{K} \,\big|\, | v |_V \leq R\},$$

where R is the positive number appearing in (4.11). The previous theorem provides the existence of a solution u to the v.i.

$$u \in \mathbb{K}_R, \qquad a(u, v - u) \geq \langle F, v - u \rangle \qquad \text{for } v \in \mathbb{K}_R. \tag{4.13}$$

From (4.11) it follows that $| u |_V < R$. Let $w \in \mathbb{K}$ be arbitrarily fixed and correspondingly let $\lambda \in]0, 1[$ be so small that $v \equiv u + \lambda(w - u) \in \mathbb{K}$

satisfies $| v |_V < R$. With this choice of v, (4.13) yields $a(u, w - u) \geq \langle F, w - u \rangle$ so that u solves (4.9). □

Theorem 4.6 can in its turn be utilized to deal with the following situation. Let the norm $|\cdot|_V$ on V be equivalent to $[\cdot]_V + |\cdot|_H$, where $[\cdot]_V$ is a seminorm on V and $|\cdot|_H$ is a norm on another Hilbert space H, into which V is compactly injected.

The assumption

$$a(u, u) \geq \alpha_0 [u]_V^2 \qquad \text{for } u \in V \quad (\alpha_0 > 0),$$

referred to as *semicoerciveness*, is intermediate between coerciveness and nonnegativity (and implies coerciveness relative to H).

THEOREM 4.7. *Let $a(u, v)$ be a bounded semicoercive bilinear form and let $\mathbb{K} \ni 0$ be a closed and convex subset of V. Then* (4.9) *admits at least one solution for $F \in V'$ if either*

$$W \cap \mathbb{K} \text{ is bounded} \tag{4.14}$$

or

$$\begin{aligned} F = F_0 + F_1, \qquad &\text{with } |\langle F_0, v\rangle| \leq C[v]_V \qquad \text{for } v \in V \\ &\text{and } \langle F_1, w\rangle < 0 \qquad \text{for } w \in W \cap \mathbb{K}\setminus\{0\}, \end{aligned} \tag{4.15}$$

where $W \equiv \{w \in V \mid [w]_V = 0\}$.

PROOF. By Theorem 4.5 we need only consider the case when $\mathbb{K}$ is not bounded; by Theorem 4.6 we can limit ourselves to prove (4.11) with $v_0 = 0$. We shall proceed in two steps.

Step 1: *The case* (4.14). Suppose that (4.11) is not satisfied. Then to each $n \in N$ we can associate $u_n \in \mathbb{K}$ with

$$| u_n |_V = n, \qquad a(u_n, u_n) \leq \langle F, u_n \rangle.$$

By semicoerciveness,

$$\alpha_0 [u_n]_V^2 \leq \langle F, u_n \rangle. \tag{4.16}$$

Set $w_n \equiv u_n | u_n |_V^{-1}$. From the bounded sequence $\{w_n\}_n$ we can extract a subsequence $\{w_{n_k}\}_k$ which converges weakly in V, hence strongly in H, toward some vector w.

Fix any $\lambda > 0$: (4.16) yields

$$\alpha_0[\lambda w_{n_k}]_V^2 = \lambda^2 | u_{n_k} |_V^{-2} \alpha_0 [u_{n_k}]_V^2$$
$$\leq \lambda^2 | u_{n_k} |_V^{-1} \langle F, w_{n_k} \rangle = \lambda n_k^{-1} \langle F, \lambda w_{n_k} \rangle,$$

hence

$$[\lambda w_{n_k}]_V \to 0 \qquad \text{as } k \to \infty,$$

and finally $[\lambda w]_V = 0$ by the weak lower semicontinuity of the norm $|\cdot|_H + [\cdot]_V$ (see Problem 1.2), hence also of the seminorm $[\cdot]_V$. Therefore $\lambda w \in W$; moreover, $\lambda w_{n_k} \to \lambda w$ in V thanks to the equivalence of the norms $|\cdot|_V$ and $[\cdot]_V + |\cdot|_H$.

Since $| \lambda w_{n_k} |_V = \lambda$, w is $\neq 0$; moreover, since both 0 and u_{n_k} are in the convex set $\mathbb{K}$, λw_{n_k} belongs to $\mathbb{K}$ whenever $| u_{n_k} |_V = n_k > \lambda$, and consequently λw belongs to $\mathbb{K}$ for all $\lambda > 0$. In the case (4.14) this is a contradiction.

Step 2: *The case* (4.15). If (4.11) is not satisfied, consider the same sequences $\{u_{n_k}\}$ and $\{w_{n_k}\}$ as in Step 1, and take $\lambda = 1$. From (4.16) and our assumption about F_0 we deduce that

$$| u_{n_k} |_V \alpha_0 [w_{n_k}]_V^2 - C[w_{n_k}]_V \leq \langle F_1, w_{n_k} \rangle.$$

Letting $k \to \infty$ we see that

$$\langle F_1, w \rangle \geq 0,$$

since $[w_{n_k}]_V \to 0$. This contradicts our assumption about F_1 since $w \in W \cap \mathbb{K} \setminus \{0\}$. ∎

REMARK. Theorem 4.7 can be greatly improved: see G. Fichera [48], C. Baiocchi, F. Gastaldi and F. Tomarelli [9, 10].

4.2. Variational Inequalities for Nonlinear Operators

Many of the results of the previous section can be extended, sometimes with no substantial change (or even no change at all) in their proofs, to much more general settings. In the present section we shall show this, assuming at the outset that V is a reflexive Banach space.

4.2.1. Monotone and Pseudomonotone Operators

Beginning with minimum problems such as (4.1), we remark that (strict) convexity, coerciveness, weak lower semicontinuity, and Gateaux differentiability can be defined for nonlinear functionals $\mathscr{J}$ on V exactly as in the Hilbert case. Therefore the proofs of Lemmas 4.1 and 4.2 can be repeated word by word to yield the following more general results.

LEMMA 4.8. *Let $\mathbb{K}$ be a convex subset of V and let $\mathscr{J}$ be a Gateaux differentiable convex functional on V. Then* (4.2) *is equivalent to* (4.4).

LEMMA 4.9. *Let $\mathbb{K}$ be a nonvoid closed and convex subset of V and let $\mathscr{J}$ be a weakly lower semicontinuous, coercive and convex functional on V. Then* (4.2) *admits at least one solution; uniqueness holds if $\mathscr{J}$ is strictly convex.*

By the two lemmas above we can tackle the analog of (4.10), which we rewrite as

$$u \in \mathbb{K}, \qquad \langle A(u) - F, v - u\rangle \geq 0 \qquad \text{for } v \in \mathbb{K} \tag{4.17}$$

for a nonlinear operator $A: V \to V'$, whenever $u \in V$ and $A(u)$, hence also $A(u) - F$, is the Gateaux derivative at u of some convex functional on V. In order to pass from this setting to more general ones, we introduce the following definitions. We say that a nonlinear operator $A: V \to V'$ is

- *hemicontinuous* if each real function

  $$\lambda \mapsto \langle A((1-\lambda)u + \lambda v), v - u\rangle$$

  with $u, v \in V$, is continuous on $\mathbb{R}$;

- *monotone*, if

  $$\langle A(u) - A(v), u - v\rangle \geq 0 \qquad \text{for } u, v \in V;$$

- *strictly monotone*, if the requirement

  $$\langle A(u) - A(v), u - v\rangle = 0 \Rightarrow u = v$$

  is added to monotonicity.

Note that each of these three properties holds for $u \mapsto A(u) - F$, $F \in V'$, if and only if it does for $u \mapsto A(u)$. Note also that, when V is a Hilbert space, the linear operator $u \mapsto A(u) = Au$ associated to a continuous bilinear form $a(u, v)$ [see (4.8)] is automatically hemicontinuous; it is monotone if and only if $a(u, v)$ is nonnegative, and strictly monotone if $a(u, v)$ is coercive.

The next result casts light upon the above definitions.

LEMMA 4.10. *Let $\mathscr{J}$ be a Gateaux differentiable functional on V. If $\mathscr{J}$ is convex (strictly convex), then $\mathscr{J}'$ is both hemicontinuous and monotone (strictly monotone); if $\mathscr{J}'$ is monotone (strictly monotone), then $\mathscr{J}$ is convex (strictly convex).*

PROOF. Fix $u, v \in V$, $u \neq v$ and set $\varphi(\lambda) \equiv \mathscr{J}((1 - \lambda)u + \lambda v)$. Then $\varphi'(\lambda)$ exists and equals $\langle \mathscr{J}'((1 - \lambda)u + \lambda v), v - u\rangle$ for $\lambda \in \mathbb{R}$.

If $\mathscr{J}$ is convex on V, so is φ on $\mathbb{R}$. By well-known properties of convex functions on $\mathbb{R}$, φ' is continuous, and consequently $\mathscr{J}'$ is hemicontinuous. Moreover, φ' is nondecreasing: therefore

$$\varphi'(0) = \langle \mathscr{J}'(u), v - u\rangle \leq \langle \mathscr{J}'(v), v - u\rangle = \varphi'(1),$$

and $\mathscr{J}'$ is monotone.

Vice versa, if $\mathscr{J}'$ is monotone, φ' is nondecreasing, and consequently φ is convex. The conclusion about the convexity of $\mathscr{J}$ follows easily.

The proof of the "strict" case is perfectly analogous. □

Strict monotonicity immediately leads to a uniqueness result for (4.17), since the latter implies

$$\langle A(u_1) - A(u_2), u_1 - u_2\rangle \leq 0$$

whenever u_1 and u_2 are solutions. Therefore, we have the following lemma.

LEMMA 4.11. *If $A\colon V \to V'$ is strictly monotone,* (4.17) *can have at most one solution.*

Before passing to the existence of solutions we move a step further in generality. An operator $A\colon V \to V'$ is *pseudomonotone* if it is bounded (i.e., it maps bounded subsets of V into bounded subsets of V') and satisfies

$$\liminf_{n\to\infty} \langle A(u_n), u_n - v\rangle \geq \langle A(u), u - v\rangle \qquad \text{for } v \in V \tag{4.18}$$

whenever the sequence $\{u_n\}$ converges weakly in V toward u with

$$\limsup_{n\to\infty} \langle A(u_n), u_n - u\rangle \leq 0.$$

When V is finite-dimensional, A is pseudomonotone if it is continuous [and vice versa: see Lemma 4.14(ii) below].

LEMMA 4.12. *Let $A: V \to V'$ be bounded, hemicontinuous, and monotone. Then A is pseudomonotone.*

PROOF. Let $\{u_n\}$ be a sequence as required in the definition of pseudomonotonicity. Since A is monotone,

$$\langle A(u_n), u_n - u \rangle \geq \langle A(u), u_n - u \rangle;$$

letting $n \to \infty$, we see that the right-hand side of the above inequality tends to 0. Therefore,

$$\langle A(u_n), u_n - u \rangle \to 0 \qquad \text{as } n \to \infty. \tag{4.19}$$

Again by monotonicity, we have

$$\langle A(u_n) - A(w), u_n - w \rangle \geq 0$$

for $w = (1 - \lambda)u + \lambda v$, $0 < \lambda < 1$. Hence,

$$\begin{aligned} \lambda \langle A(u_n), u - v \rangle \geq & - \langle A(u_n), u_n - u \rangle \\ & + \langle A(w), u_n - u \rangle + \lambda \langle A(w), u - v \rangle. \end{aligned}$$

Let $n \to \infty$: by (4.19) the above inequality yields

$$\lambda \liminf_{n \to \infty} \langle A(u_n), u - v \rangle \geq \lambda \langle A(w), u - v \rangle,$$

and also

$$\begin{aligned} \liminf_{n \to \infty} \langle A(u_n), u_n - v \rangle &= \liminf_{n \to \infty} [\langle A(u_n), u_n - u \rangle + \langle A(u_n), u - v \rangle] \\ &\geq \langle A(w), u - v \rangle. \end{aligned}$$

We let λ (in the definition of w) tend to 0^+: (4.18) follows from hemicontinuity. □

We now introduce another class of nonlinear operators $A: V \to V'$. To wit, we say that A is a *Leray–Lions operator* if it is bounded and satisfies

$$A(u) = \mathscr{A}(u, u) \qquad \text{for } u \in V,$$

where $\mathscr{A}: V \times V \to V'$ has the following properties:

(i) whenever $u \in V$, the mapping $v \mapsto \mathscr{A}(u, v)$ is bounded and hemicontinuous from V to V', with

$$\langle \mathscr{A}(u, u) - \mathscr{A}(u, v), u - v \rangle \geq 0 \qquad \text{for } v \in V;$$

(ii) whenever $v \in V$, the mapping $u \mapsto \mathscr{A}(u, v)$ is bounded and hemicontinuous from V to V';

(iii) whenever $v \in V$, $\mathscr{A}(u_n, v)$ converges weakly to $\mathscr{A}(u, v)$ in V' if $\{u_n\} \subset V$ is such that $u_n \rightharpoonup u$ in V and

$$\langle \mathscr{A}(u_n, u_n) - \mathscr{A}(u_n, u), \ u_n - u \rangle \to 0;$$

(iv) whenever $v \in V$, $\langle \mathscr{A}(u_n, v), u_n \rangle$ converges to $\langle F, u \rangle$ if $\{u_n\} \subset V$ is such that $u_n \rightharpoonup u$ in V, $\mathscr{A}(u_n, v) \rightharpoonup F$ in V'.

LEMMA 4.13. *Every Leray–Lions operator $A: V \to V'$ is pseudomonotone.*

PROOF. Let $u_n \rightharpoonup u$ in V, with

$$\limsup_{n\to\infty} \langle A(u_n), \ u_n - u \rangle \leq 0.$$

Since $\{\mathscr{A}(u_n, u)\}_n$ is bounded in V', we can extract a subsequence $\{\mathscr{A}(u_{n_k}, u)\}_k$ which converges weakly in V' toward some functional F. Thus, $\langle \mathscr{A}(u_{n_k}, u), u_{n_k} \rangle \to \langle F, u \rangle$ by (iv), and also $\langle \mathscr{A}(u_{n_k}, u), \ u_{n_k} - u \rangle \to 0$.

Let

$$X_k \equiv \langle \mathscr{A}(u_{n_k}, u_{n_k}) - \mathscr{A}(u_{n_k}, u), \ u_{n_k} - u \rangle :$$

we have $X_k \geq 0$ [by (i)] as well as

$$\limsup_{k\to\infty} X_k \leq 0,$$

hence

$$X_k \to 0 \qquad \text{as } k \to \infty.$$

But then (iii) implies

$$\mathscr{A}(u_{n_k}, w) \rightharpoonup \mathscr{A}(u, w) \qquad \text{in } V' \tag{4.20}$$

whenever $w \in V$, and (iv) yields

$$\langle \mathscr{A}(u_{n_k}, w), u_{n_k} \rangle \to \langle \mathscr{A}(u, w), u \rangle,$$

hence

$$\langle \mathscr{A}(u_{n_k}, w), u_{n_k} - u \rangle \to 0. \tag{4.21}$$

Since $X_k \geq 0$ we arrive at

$$\liminf_{k \to \infty} \langle \mathscr{A}(u_{n_k}, u_{n_k}), u_{n_k} - u \rangle \geq 0$$

by taking $w = u$ in (4.21), hence

$$\langle \mathscr{A}(u_{n_k}, u_{n_k}), u_{n_k} - u \rangle \to 0. \tag{4.22}$$

We now arbitrarily fix $v \in V$ and take

$$w = u + \lambda(v - u), \qquad \lambda \in]0, 1[.$$

From the inequality

$$\langle \mathscr{A}(u_{n_k}, u_{n_k}) - \mathscr{A}(u_{n_k}, w), u_{n_k} - w \rangle \geq 0$$

we deduce that

$$\begin{aligned} \lambda \langle \mathscr{A}(u_{n_k}, u_{n_k}), u - v \rangle \geq &- \langle \mathscr{A}(u_{n_k}, u_{n_k}), u_{n_k} - u \rangle \\ &+ \langle \mathscr{A}(u_{n_k}, w), u_{n_k} - u \rangle + \lambda \langle \mathscr{A}(u_{n_k}, w), u - v \rangle, \end{aligned}$$

hence that

$$\begin{aligned} \lambda \liminf_{k \to \infty} \langle \mathscr{A}(u_{n_k}, u_{n_k}), u_{n_k} - v \rangle &= \lambda \liminf_{k \to \infty} [\langle \mathscr{A}(u_{n_k}, u_{n_k}), u_{n_k} - u \rangle \\ &\qquad + \langle \mathscr{A}(u_{n_k}, u_{n_k}), u - v \rangle] \\ &\geq \lambda \lim_{k \to \infty} \langle \mathscr{A}(u_{n_k}, w), u - v \rangle \\ &= \lambda \langle \mathscr{A}(u, u + \lambda(v - u)), u - v \rangle \end{aligned}$$

by (4.20), (4.21), and (4.22).

At this point we first divide by λ, then let $\lambda \to 0^+$: by hemicontinuity the result is that

$$\liminf_{k \to \infty} \langle A(u_{n_k}), u_{n_k} - v \rangle \geq \langle A(u), u - v \rangle,$$

and pseudomonotonicity easily follows. ∎

When V is a Hilbert space and $u \mapsto A(u) = Au$ is the linear operator associated with a bounded bilinear form, the weak (strong) convergence in V of a sequence $\{u_n\}$ toward a vector u implies the weak (strong) convergence of $\{Au_n\}$ toward Au. For pseudomonotone operators on reflexive Banach spaces we have instead the following lemma.

LEMMA 4.14. *Let $A: V \to V'$ be a pseudomonotone operator.*

(i) *If $u_n \rightharpoonup u$ in V and $A(u_n) \rightharpoonup F$ in V', with*

$$\limsup_{n\to\infty} \langle A(u_n), u_n\rangle \leq \langle F, u\rangle,$$

then $F = A(u)$.

(ii) *If $u_n \to u$ in V, then $A(u_n) \rightharpoonup A(u)$ in V'.*

PROOF. *Step* 1: *Proof of* (i). Since

$$\limsup_{n\to\infty} \langle A(u_n), u_n - u\rangle \leq \limsup_{n\to\infty} \langle A(u_n), u_n\rangle - \lim_{n\to\infty} \langle A(u_n), u\rangle \leq 0,$$

pseudomonotonicity yields (4.18). Therefore,

$$\begin{aligned}\langle A(u), u - v\rangle &\leq \limsup_{n\to\infty} \langle A(u_n), u_n - v\rangle \\ &\leq \langle F, u - v\rangle \qquad \text{for } v \in V,\end{aligned}$$

and finally $A(u) = F$ by taking $v = u \pm w$, $w \in V$.

Step 2: *Proof of* (ii). Since the image under A of the bounded sequence $\{u_n\}$ is bounded in the reflexive Banach space V', there exists a subsequence of indices such that $A(u_{n_k}) \rightharpoonup F$ in V' as $k \to \infty$. Since $\langle A(u_{n_k}), u_{n_k} - u\rangle \to 0$ as $k \to \infty$,

$$\begin{aligned}\langle F, u - v\rangle &= \liminf_{k\to\infty} \langle A(u_{n_k}), u_{n_k} - v\rangle \\ &\geq \langle A(u), u - v\rangle \qquad \text{for } v \in V,\end{aligned}$$

hence $F = A(u)$. By uniqueness, the whole sequence $\{Au_n\}$ converges weakly in V toward $A(u)$. □

Finally we provide the following criterion for the stability of the class of pesudomonotone operators under perturbations.

LEMMA 4.15. *Let $A_1, A_2: V \to V'$, with A_1 pseudomonotone and A_2 bounded, hemicontinuous, and monotone. Then $A \equiv A_1 + A_2$ is pseudomonotone.*

PROOF. Let $\{u_n\}$ converge weakly in V toward u, with

$$\limsup_{n\to\infty} \langle A(u_n), u_n - u\rangle \leq 0.$$

Then,

$$\begin{aligned}\langle A_1(u_n), u_n - u\rangle &= \langle A(u_n), u_n - u\rangle - \langle A_2(u_n), u_n - u\rangle \\ &\leq \langle A(u_n), u_n - u\rangle - \langle A_2(u), u_n - u\rangle\end{aligned}$$

(by the monotonicity of A_2), and

$$\limsup_{n\to\infty} \langle A_1(u_n), u_n - u\rangle \leq 0,$$

which implies

$$\liminf_{n\to\infty} \langle A_1(u_n), u_n - v\rangle \geq \langle A_1(u), u - v\rangle \qquad \text{for } v \in V \tag{4.23}$$

(by the pseudomonotonicity of A_1).

This implies that $\langle A_1(u_n), u_n - u\rangle \to 0$ as $n \to \infty$, hence

$$\begin{aligned}\limsup_{n\to\infty} \langle A_2(u_n), u_n - u\rangle = \limsup_{n\to\infty} [&\langle A(u_n), u_n - u\rangle \\ &- \langle A_1(u_n), u_n - u\rangle] \leq 0\end{aligned}$$

and

$$\liminf_{n\to\infty} \langle A_2(u_n), u_n - v\rangle \geq \langle A_2(u), u - v\rangle \qquad \text{for } v \in V \tag{4.24}$$

by Lemma 4.12. Summing (4.23) and (4.24) we obtain the inequality

$$\liminf_{n\to\infty} \langle A(u_n), u_n - v\rangle \geq \langle A(u), u - v\rangle \qquad \text{for } v \in V,$$

which completes the proof since A is obviously bounded. □

4.2.2. Existence and Approximation of Solutions

Returning to (4.17) we prove the following generalization of Theorem 4.5.

THEOREM 4.16. *Let A be a pseudomonotone operator $V \to V'$ and let $\mathbb{K} \neq \emptyset$ be a closed and convex bounded subset of V. Then for any choice of $F \in V'$* (4.17) *admits at least one solution.*

PROOF. We shall proceed in two steps.

Step 1: *The finite-dimensional case.* If V is a finite-dimensional space, it can be endowed with a scalar product $(\cdot, \cdot)_V$. Then (4.17) can be rewritten as

$$u \in \mathbb{K}, \qquad (\mathscr{T}(A(u) - F), v - u)_V \geq 0 \qquad \text{for } v \in \mathbb{K},$$

hence as (4.7) with $z \equiv u - \mathscr{T}(A(u) - F)$.

We need to prove that the mapping $\Pi_{\mathbb{K}}: u \mapsto P_{\mathbb{K}}(u - \mathscr{T}(A(u) - F))$ has a fixed point.

Since A is continuous from V into V' [by Lemma 4.14(ii): in the finite-dimensional case weak = strong], $\mathscr{T} \circ A$ is continuous from V into V, and so is $P_{\mathbb{K}}$ (by Lemma 4.3). Thus $\Pi_{\mathbb{K}}: \mathbb{K} \to \mathbb{K}$ is continuous, and the Brouwer fixed point Theorem 1.I yields the existence of $u = \Pi_{\mathbb{K}}(u)$.

Step 2: *The general case.* We proceed under the additional assumption that V is separable; if it is not, the proof requires a few minor modifications as in H. Brézis [18].

Every subspace of a separable metric space is separable. By the separability of V we can therefore construct a sequence $\{V_n\}$ of Banach subspaces of V, $\dim V_n \leq n$, and a sequence $\{\mathbb{K}_n\}$ of nonvoid closed and convex sets $\mathbb{K}_n \subseteq V_n$, with $\mathbb{K}_n \subseteq \mathbb{K}_{n+1}$, so that $\bigcup_{n=1}^{\infty} \mathbb{K}_n$ is dense in $\mathbb{K}$. For each $n \in \mathbb{N}$, Step 1 enables us to solve the v.i. (which we write with a slight abuse of notation)

$$u_n \in \mathbb{K}_n, \qquad \langle A(u_n) - F, v - u_n\rangle \geq 0 \qquad \text{for } v \in \mathbb{K}_n. \tag{4.25}$$

Since $\mathbb{K}$ is bounded, a suitable subsequence of the bounded sequence $\{u_n\}$, say the original sequence itself, converges weakly in V toward some vector u; since $\mathbb{K}$ is convex and closed, $u \in \mathbb{K}$. Let $\varepsilon > 0$ be arbitrarily fixed and let $\hat{n} \in \mathbb{N}$, $\hat{u} \in \mathbb{K}_{\hat{n}}$ be such that $| u - \hat{u} |_V < \varepsilon$. Then

$$\begin{aligned}\langle A(u_n), u_n - u\rangle &= \langle A(u_n), u_n - \hat{u}\rangle + \langle A(u_n), \hat{u} - u\rangle \\ &\leq \langle F, u_n - \hat{u}\rangle + \varepsilon \sup_{n \in \mathbb{N}} | A(u_n) |_{V'}\end{aligned}$$

for $n \geq \hat{n}$, hence

$$\begin{aligned}\limsup_{n \to \infty} \langle A(u_n), u_n - u\rangle &\leq \langle F, u - \hat{u}\rangle + \varepsilon \sup_{n \in \mathbb{N}} | A(u_n) |_{V'} \\ &\leq \varepsilon(| F |_{V'} + \sup_{n \in \mathbb{N}} | A(u_n) |_{V'}).\end{aligned}$$

By the arbitrariness of ε, $\limsup_{n\to\infty} \langle A(u_n), u_n - u\rangle \leq 0$, so that pseudomonotonicity yields

$$\liminf_{n\to\infty} \langle A(u_n), u_n - v\rangle \geq \langle A(u), u - v\rangle \qquad \text{for } v \in V.$$

By taking v in some $\mathbb{K}_{\hat{n}_0}$ we deduce from (4.25), written for $n \geq \hat{n}_0$, that

$$\liminf_{n\to\infty} \langle A(u_n), u_n - v\rangle \leq \langle F, u - v\rangle,$$

hence

$$\langle A(u), u - v\rangle \leq \langle F, u - v\rangle \qquad \text{for } v \in \bigcup_{n=1}^{\infty} \mathbb{K}_n.$$

By density u solves (4.17). □

If now the analog of the growth condition (4.11) is formulated for a nonlinear operator A as follows:

$$\begin{aligned} &\text{there exist} \quad R \in \,]0, \infty[\quad \text{and} \quad v_0 \in \mathbb{K}, \qquad |v_0|_V < R, \\ &\text{such that} \\ &\langle A(u) - F, v_0 - u\rangle < 0 \qquad \text{for } u \in \mathbb{K}, \qquad |u|_V = R, \end{aligned} \tag{4.26}$$

we can proceed as in the proof of Theorem 4.6, this time by making use of Theorem 4.16, and prove the following theorem.

THEOREM 4.17. *Let A be a pseudomonotone operator $V \to V'$ satisfying* (4.26), *where $\mathbb{K} \neq \varnothing$ is a closed and convex subset of V. Then for any choice of $F \in V'$* (4.17) *admits at least one solution.*

REMARK. Since $\mathbb{K} = V$ is admissible in Theorem 4.17, the latter contains an existence result for the equation

$$u \in V, \qquad A(u) = F.$$

Also Theorem 4.7 admits a generalization to the case of a nonlinear operator A. To see this we consider the following situation: V is compactly imbedded into a Banach space X, and $|\cdot|_V \sim [\cdot]_V + |\cdot|_X$, where $[\cdot]_V$ is a seminorm on V. Calling *semicoercive* a nonlinear operator $A: V \to V'$ such that

$$\langle A(u), u\rangle \geq \alpha_0 [u]_V{}^q \qquad \text{for } u \in V \quad (\alpha_0 > 0,\ q > 1),$$

we can proceed as in the proof of Theorem 4.7 with a few, obvious changes;

notice that now Theorem 4.17 must be utilized. Thus we have the following theorem.

THEOREM 4.18. *Let* $A: V \to V'$ *be a pseudomonotone semicoercive operator and let* $\mathbb{K} \ni 0$ *be a closed and convex subset of* V. *Then* (4.17) *admits at least one solution for* $F \in V'$ *if either*

$$W \cap \mathbb{K} \text{ is bounded}$$

or

$$F = F_0 + F_1, \qquad \text{with } |\langle F_0, v\rangle| \leq C[v]_V \qquad \text{for } v \in V$$
$$\text{and } \langle F_1, w\rangle < 0 \qquad \text{for } w \in W \cap \mathbb{K}\setminus\{0\},$$

where $W \equiv \{w \in V \mid [w]_V = 0\}$.

We conclude this section by showing how solutions to v.i.'s can be approximated by solutions to suitable equations (necessarily nonlinear, even when the v.i.'s are of the form (4.10), A linear).

First, we call a bounded operator $A: V \to V'$ *coercive* (relative to $\mathbb{K}$) if there exists $v_0 \in \mathbb{K}$ such that

$$|u|_V^{-1}\langle A(u), u - v_0\rangle \to \infty \qquad \text{as } |u|_V \to \infty.$$

This terminology reflects the fact that a linear operator $u \mapsto A(u) = Au$ associated with a bounded bilinear form on a Hilbert space V is coercive, in the above sense, if $a(u, v)$ is coercive in the usual sense. Note that (4.26) holds if $\mathbb{K}$ is unbounded and A coercive.

Next, we say that a bounded, hemicontinuous and monotone operator $\beta: V \to V'$ is a *penalty operator* associated with $\mathbb{K} \subseteq V$ if

$$\beta(u) = 0 \Leftrightarrow u \in \mathbb{K}.$$

THEOREM 4.19. *Let* $\mathbb{K} \neq \emptyset$ *be a closed and convex subset of* V, *let* A *be a pseudomonotone and coercive operator* $V \to V'$, *and let* β *be a penalty operator associated with* $\mathbb{K}$. *Then there exists a sequence* $\{u_n\}$, *where each* u_n *satisfies*

$$u_n \in V, \qquad A(u_n) + \frac{1}{\varepsilon(n)}\,\beta(u_n) = F$$

with $\varepsilon(n) \to 0^+$ *as* $n \to \infty$, *which converges weakly in* V *toward a solution to* (4.17).

PROOF. Let $\varepsilon > 0$ be arbitrarily fixed. The operator $u \mapsto A(u) + (1/\varepsilon)\beta(u)$ is pseudomonotone by Lemma 4.15. It is also coercive since the coerciveness of A implies

$$\begin{aligned} &|u|_V^{-1}\langle A(u) + \frac{1}{\varepsilon}\beta(u), u - v_0\rangle \\ &\quad = |u|_V^{-1}\langle A(u) + \frac{1}{\varepsilon}[\beta(u) - \beta(v_0)], u - v_0\rangle \\ &\quad \geq |u|_V^{-1}\langle A(u), u - v_0\rangle \to \infty \qquad \text{as } |u|_V \to \infty \end{aligned}$$

by the membership of v_0 in $\mathbb{K}$ and the monotonicity of β. Theorem 4.17 can therefore be applied with $\mathbb{K} = V$, and the equation

$$u_\varepsilon \in V, \qquad A(u_\varepsilon) + \frac{1}{\varepsilon}\beta(u_\varepsilon) = F$$

admits at least one solution u_ε. Moreover, the above inequality implies the existence of a constant C, independent of the choice of $\varepsilon > 0$, such that $|u_\varepsilon|_V \leq C$. Therefore $|A(u_\varepsilon)|_{V'}$ is also bounded independently of ε, and finally the equation implies

$$\beta(u_\varepsilon) = \varepsilon[F - A(u_\varepsilon)] \to 0 \qquad \text{in } V' \qquad \text{as } \varepsilon \to 0^+.$$

A sequence $\{\varepsilon(n)\}$ can be found, with the property that $\varepsilon(n) \to 0^+$ and $u_n \equiv u_{\varepsilon(n)} \rightharpoonup u$ in V as $n \to \infty$.

Let $v \in V$ be arbitrarily fixed; then the inequality

$$\langle \beta(u_n) - \beta(v), u_n - v\rangle \geq 0$$

yields $\langle \beta(v), u - v\rangle \leq 0$, hence $\langle \beta(u - \lambda w), w\rangle \leq 0$ with the choice of $v = u - \lambda w$ with $\lambda > 0$ and $w \in V$. By hemicontinuity we can let $\lambda \to 0^+$ and obtain $\langle \beta(u), w\rangle \leq 0$, hence $\beta(u) = 0$ by the arbitrariness of w. Therefore $u \in \mathbb{K}$. Next we fix $v \in \mathbb{K}$, so that $\beta(v) = 0$. From the equation we deduce

$$\langle A(u_n) - F, v - u_n\rangle = \frac{1}{\varepsilon(n)}\langle \beta(v) - \beta(u_n), v - u_n\rangle \geq 0,$$

hence

$$\limsup_{n\to\infty} \langle A(u_n), u_n - u\rangle \leq \limsup_{n\to\infty} \langle F, u_n - u\rangle = 0,$$

and finally

$$\liminf_{n\to\infty} \langle A(u_n), u_n - v\rangle \geq \langle A(u), u - v\rangle$$

by pseudomonotonicity. This suffices to show that u satisfies (4.17). ☐

REMARK. Theorem 4.19 provides a constructive approximation method if (4.17) is uniquely solvable. Notice that if this is the case, the solution u is the weak limit in V of $\{u_\varepsilon\}$ as $\varepsilon \to 0^+$.

4.3. Variational Inequalities in Sobolev Spaces

We now proceed to exhibit fundamental examples of convex sets and bilinear forms, or nonlinear operators, entering the theory of v.i.'s when the underlying space V is some closed subspace of $H^1(\Omega)$, or of $H^{1,p}(\Omega)$ with $1 < p < \infty$. We are also going to show that v.i.'s associated with unilateral or bilateral constraints (in the sense specified in Section 4.3.1 below) can be interpreted as obstacle problems.

4.3.1. Convex Sets

The abstract results of Section 4.1 can be applied to any closed and convex subset $\mathbb{K} \neq \emptyset$ of a closed linear subspace V of $H^1(\Omega)$.

The most important instance of a convex subset of V occurring in the theory of v.i.'s is

$$\mathbb{K} = \{v \in V \mid v \leq \psi \text{ in } \Omega\} \tag{4.27}$$

with ψ measurable. $\mathbb{K}$ is obviously closed (see Theorem 1.Q). Sufficient conditions in order that $\mathbb{K} \neq \emptyset$ can easily be given in the special case when $V = H_0^1(\Omega \cup \Gamma)$, with Γ of class C^1. If $\psi \in H^1(\Omega)$ with $\psi \geq 0$ on $\partial\Omega \setminus \Gamma$ in the sense of $H^1(\Omega)$, then $\mathbb{K} \ni \psi \wedge 0$. If instead $\psi \in C^0(\bar{\Omega})$ with $\psi > 0$ on $\partial\Omega \setminus \Gamma$, we can easily construct $\psi' \in C^1(\bar{\Omega})$ with $\psi' > 0$ on $\partial\Omega \setminus \Gamma$ and $\psi' \leq \psi$ (see Theorem 1.N and Lemma 1.7), so that $\psi' \wedge 0$ belongs to $H^1(\Omega)$ with $\operatorname{supp}(\psi' \wedge 0) \subset \Omega \cup \Gamma$ and finally $\psi' \wedge 0 \in \mathbb{K}$. Note that the above requirement that $\psi > 0$ on $\partial\Omega \setminus \Gamma$ cannot be weakened by replacing $>$ with $\geq$, as the following example shows.

EXAMPLE. Let $N = 1$, $\Omega =]0, 1[$, $\Gamma = \emptyset$, $\psi \in C^0(\bar{\Omega})$, $\psi(x) = -|x|^\delta$ with $0 < \delta < 1/2$ for $0 \leq x \leq 1/2$, $\psi(1) \geq 0$. Any function $v \in \mathbb{K}$ would simultaneously belong to $C^{0,1/2}([0, 1])$ (by Theorem 1.41) and satisfy $|v(x) - v(0)| = -v(x) \geq |x|^\delta$ for $x \in [0, 1]$, which is contradictory.

We introduce another important class of nonvoid closed and convex subsets of $V = H_0^1(\Omega \cup \Gamma)$ by setting

$$\mathbb{K} = \{v \in V \mid v \leq \hat{\psi} \text{ on } E \text{ in the sense of } H^1(\Omega)\}, \tag{4.28}$$

where $\bar{E} \subset \Omega \cup \Gamma$ and $\hat{\psi} \in C^0(E)$ with $\inf_E \hat{\psi} > -\infty$. If E equals Γ and the latter is also closed, (4.28) becomes

$$\mathbb{K} = \{v \in V \mid v \leq \hat{\psi} \text{ on } \Gamma \text{ in the sense of } H^1(\Omega)\}. \tag{4.29}$$

Note that the requirement $\hat{\psi} \in C^0(\Gamma)$ can then be replaced by $\hat{\psi} \in H^{1/2}(\Gamma)$, in which case $\mathbb{K}$ contains any function $\psi \in V$ with $\psi|_\Gamma = \hat{\psi}$.

The convex sets considered up to now are defined by *unilateral constraints* (the latter ones being placed above: unilateral constraints placed below can be dealt with through obvious changes). They are cones (i.e., they verify $v \in \mathbb{K} \Rightarrow \lambda v \in \mathbb{K}$ whenever $0 \leq \lambda < \infty$) if ψ, or $\hat{\psi}$, vanishes identically.

A closed convex set defined by *bilateral constraints* is the following:

$$\mathbb{K} = \{v \in V \mid \varphi \leq v \leq \psi \text{ in } \Omega\} \tag{4.30}$$

with φ and ψ measurable on Ω, $\varphi \leq \psi$. By considerations analogous to those developed about (4.27) it can be checked that when $V = H_0{}^1(\Omega \cup \Gamma)$, $\mathbb{K} \neq \emptyset$ if φ and ψ belong either to $H^1(\Omega)$ with $\varphi \leq 0 \leq \psi$ on $\partial\Omega \setminus \Gamma$ in the sense of $H^1(\Omega)$ [so that $\mathbb{K} \ni \varphi \vee (\psi \wedge 0)$], or to $C^0(\bar{\Omega})$ with $\varphi < \psi$ in $\Omega \cup \Gamma$, $\varphi|_{\partial\Omega \setminus \Gamma} < 0 < \psi|_{\partial\Omega \setminus \Gamma}$.

An important example of a nonvoid, closed, and convex subset of V which is not of the obstacle type is given by

$$\mathbb{K} = \{v \in V \,\big|\, |\nabla v| \leq 1 \text{ in } \Omega\} \tag{4.31}$$

(see the Notes to this chapter).

All the above considerations can be easily extended to the case when the Hilbert space $H^1(\Omega)$ is replaced by the reflexive Banach space $H^{1,p}(\Omega)$ for some $p \in]1, \infty[$.

4.3.2. Bilinear Forms and Nonlinear Operators

As in Section 2.2.1 we introduce a bounded bilinear form $a(u, v)$ on $H^1(\Omega)$, as well as bounded linear operators $A\colon H^1(\Omega) \to [H^1(\Omega)]'$ and $L\colon H^1(\Omega) \to H^{-1}(\Omega)$, by setting

$$\begin{aligned}\langle Au, v\rangle &\equiv a(u, v)\\ &\equiv \int_\Omega [(a^{ij}u_{x_i} + d^j u)v_{x_j} + (b^i u_{x_i} + cu)v]\,dx \qquad \text{for } u, v \in H^1(\Omega)\end{aligned} \tag{4.32}$$

and

$$\langle Lu, v\rangle \equiv a(u, v) \qquad \text{for } u \in H^1(\Omega),\ v \in H_0{}^1(\Omega), \tag{4.33}$$

under the assumptions

$$a^{ij}, d^j, b^i, c \in L^\infty(\Omega),$$

$$a^{ij}\xi_i\xi_j \geq \alpha \mid \xi \mid^2 \qquad \text{a.e. in } \Omega \qquad \text{for } \xi \in \mathbb{R}_N \qquad (\alpha > 0).$$

Again, we also view A as a bounded linear operator $H^1(\Omega) \to V'$ whenever V is a closed subspace of $H^1(\Omega)$, $V \supseteq H_0{}^1(\Omega)$.

The example of Section 2.2.1 provides us with a sufficient condition in order that $a(u, v)$ be coercive on V, so that the v.i. (4.9) can be investigated in the light of Theorem 4.4.

Other abstract results of Section 4.1 can be utilized to investigate (4.9) in some cases when coerciveness does not hold for the bilinear form (4.30). For instance, we illustrate Theorem 4.7 with the following example.

Example. Let $\partial\Omega$ be of class C^1, so that Rellich's theorem holds, and set $d^j = b^i = c = 0$. Then $a(u, v)$ is semicoercive on $V = H^1(\Omega)$, with $H = L^2(\Omega)$ and $[u]_V = \mid \nabla u \mid_{2;\Omega}$.

We must, however, mention that in applications of the theory of v.i.'s the most relevant semicoercive examples involve bilinear forms of types different from (4.32): see G. Fichera [48], C. Baiocchi, G. Gastaldi, and F. Tomarelli [9, 10].

Passing from $H^1(\Omega)$ to $H^{1,p}(\Omega)$ with p arbitrarily fixed in $]1, \infty[$, we denote by V a closed subspace of $H^{1,p}(\Omega)$, $V \supseteq H_0{}^{1,p}(\Omega)$. We define a nonlinear operator A: $H^{1,p}(\Omega) \to V'$ by setting

$$\langle A(u), v\rangle \equiv \int_\Omega [A^i(u, \nabla u)v_{x_i} + A^0(u, \nabla u)v]\, dx \qquad \text{for } u \in H^{1,p}(\Omega),\ v \in V, \tag{4.34}$$

where, for $j = 0, 1, \ldots, N$, $A^j(\eta, \xi)$ is the function $x \mapsto a^j(x, \eta(x), \xi(x))$ if $\eta, \xi_1, \ldots, \xi_N$ denote measurable functions on Ω, $\xi \equiv (\xi_1, \ldots, \xi_N)$; a^j is supposed to be a Carathéodory function of $x \in \Omega$ and $(\eta, \xi) \in \mathbb{R}^{1+N}$, with

$$\begin{gathered} \mid a^j(x, \eta, \xi) \mid \leq C(\mid \eta \mid^{p-1} + \mid \xi \mid^{p-1}) + h(x) \\ \text{for a.a. } x \in \Omega \qquad \text{and} \qquad \text{any } (\eta, \xi) \in \mathbb{R}^{1+N} \qquad [h \in L^{p'}(\Omega)]. \end{gathered} \tag{4.35}$$

We also define $L\colon H^{1,p}(\Omega) \to H^{-1,p'}(\Omega)$ by setting

$$\langle L(u), v\rangle \equiv \langle A(u), v\rangle \qquad \text{for } u \in H^{1,p}(\Omega),\ v \in H_0^{1,p}(\Omega),$$

that is

$$L(u) = -\frac{\partial}{\partial x_i} A^i(u, \nabla u) + A^0(u, \nabla u).$$

In the sequel we shall call $A\colon H^{1,p}(\Omega) \to V'$ bounded, or hemicontinuous, or monotone, if the restriction of A to V is such. For what concerns hemicontinuity (and boundedness) note that whenever $\eta, \xi_1, \ldots, \xi_N$ are functions from $L^p(\Omega)$, (4.35) implies $A^j(\eta, \xi) \in L^{p'}(\Omega)$ (and even

$$| A^j(\eta, \xi) |_{p';\Omega} \leq C \text{ if } | \eta, \xi_1, \ldots, \xi_N |_{p;\Omega} \leq C).$$

Thus, by a theorem of M. A. Krasnosel'skii, A^j is continuous from $[L^p(\Omega)]^{1+N}$ into $L^{p'}(\Omega)$. This circumstance, which implies hemicontinuity of A, can also be ascertained as a consequence of the following simple lemma.

LEMMA 4.20. *Let g be a Carathéodory function of $x \in \Omega$ and $\zeta \in \mathbb{R}^M$ such that*

$$| g(x, \zeta) | \leq C | \zeta |^r + h(x) \qquad \text{for a.a. } x \in \Omega \quad \text{and} \quad \text{any } \zeta \in \mathbb{R}^M, \qquad \text{with } h \in L^q(\Omega), \quad h \geq 0, \tag{4.36}$$

$1 \leq q < \infty$, $1 \leq r < \infty$. *Then the operator* $G\colon [L^p(\Omega)]^M \to L^q(\Omega)$, $p \equiv rq$, *defined by* $G(\zeta)\colon x \mapsto g(x, \zeta(x))$ *for* $\zeta \equiv (\zeta_1, \ldots, \zeta_M) \in [L^p(\Omega)]^M$ *is continuous.*

PROOF. For $j = 1, \ldots, M$ let $\zeta_{nj} \to \zeta_j$ in $L^p(\Omega)$ as $n \to \infty$, hence also

$$\zeta'_{kj} \equiv \zeta_{n_k j} \to \zeta_j \qquad \text{a.e. in } \Omega \qquad \text{as } k \to \infty,$$

$$| \zeta'_{kj} | \leq \zeta_j^* \in L^p(\Omega) \qquad \text{a.e. in } \Omega \qquad \text{for } h \in N$$

for a suitable subsequence of indices (see Theorem 1.Q). Because of (4.36) this implies that the sequence $\{| G(\zeta_k') - G(\zeta) |^q\}_k$, where $\zeta_k' \equiv (\zeta'_{k1}, \ldots, \zeta'_{kM})$, is dominated by $C[\sum_{j=1}^M (\zeta_j^*)^p + h^p] \in L^1(\Omega)$; by Lebesgue's theorem,

$$G(\zeta_k') \to G(\zeta) \qquad \text{in } L^q(\Omega) \qquad \text{as } k \to \infty.$$

Since the passage to a subsequence is at this point nugatory, we have proved that, as $n \to \infty$,

$$G(\zeta_n) \to G(\zeta) \qquad \text{in } L^q(\Omega)$$

whenever

$$\zeta_n \to \zeta \qquad \text{in } [L^p(\Omega)]^M. \qquad \square$$

If the requirement

$$\begin{aligned}&[a^0(\cdot, \eta, \xi) - a^0(\cdot, \eta', \xi')](\eta - \eta')\\ &\qquad + [a^i(\cdot, \eta, \xi) - a^i(\cdot, \eta', \xi')](\xi_i - \xi_i') \geq 0 \qquad (4.37)\\ &\text{a.e. in } \Omega \qquad \text{for } \eta, \eta' \in \mathbb{R} \qquad \text{and} \qquad \xi, \xi' \in \mathbb{R}^N\end{aligned}$$

is added to (4.35), A is rapidly seen to be monotone; as for strict monotonicity, it holds if a.e. in Ω the equality sign in (4.37) implies $\eta = \eta'$ and $\xi = \xi'$, or even if it only implies $\xi = \xi'$ provided Poincaré's inequality (as in Lemma 1.46) is valid in V. As a matter of fact, (4.37) implies more than monotonicity. To wit, let $u, v \in H^{1,p}(\Omega)$ with $(u - v)^+ \in V$: then, Theorem 1.56 implies

$$\begin{aligned}\langle A(u) - A(v), (u - v)^+\rangle = \int_{\Omega^+} &\{[A^i(u, \nabla u) - A^i(v, \nabla v)](u_{x_i} - v_{x_i})\\ &+ [A^0(u, \nabla u) - A^0(v, \nabla v)](u - v)\}\, dx\end{aligned}$$

with $\Omega^+ \equiv \{x \in \Omega \mid u(x) > v(x)\}$.

Consequently, $\langle A(u) - A(v), (u - v)^+\rangle$ is ≥ 0, and > 0 for $|\Omega^+| > 0$ if a.e. in Ω the equality sign in (4.37) implies $\eta = \eta'$, $\xi = \xi'$, or even if it only implies $\xi = \xi'$ and Poincaré's inequality holds in V. This property of the specific operator A defined by (4.34) underlies the following definition. A nonlinear operator $A: H^{1,p}(\Omega) \to V'$ is said to be

- *T-monotone*, if

$$\langle A(u) - A(v), (u - v)^+\rangle \geq 0 \qquad \text{for } u, v \in H^{1,p}(\Omega) \text{ with } (u - v)^+ \in V,$$

- *strictly T-monotone*, if the equality sign in the above inequality can only hold when $u \leq v$ in Ω;

T-monotonicity implies monotonicity, since

$$\begin{aligned}\langle A(u) - A(v), u - v\rangle = \langle A(u) - A(v), (u - v)^+\rangle&\\ + \langle A(v) - A(u), (v - u)^+\rangle& \qquad \text{for } u, v \in V.\end{aligned}$$

EXAMPLE. Let $\Phi(x, \eta, \xi)$ be continuously differentiable and convex with respect to $(\eta, \xi) \in \mathbb{R}^{1+N}$ for a.a. $x \in \Omega$, measurable with respect to $x \in \Omega$ for any $(\eta, \xi) \in \mathbb{R}^{1+N}$. By Lemma 4.10 in $\mathbb{R}^{1+N}$ the functions $a^0 = \Phi_\eta$, $a^1 = \Phi_{\xi_1} \dots, a^N = \Phi_{\xi_N}$ satisfy (4.37), with the strict inequality sign for $(\eta, \xi) \neq (\eta', \xi')$ if convexity is required to be strict. Under assumption (4.35), $A(u)$ from (4.37) is the Gateaux derivative of the convex functional

$$H^{1,p}(\Omega) \ni u \mapsto \int_\Omega \Phi(x, u(x), \nabla u(x))\, dx.$$

If (4.37) is weakened into

$$\begin{gathered}[a^i(\cdot, \eta, \xi) - a^i(\cdot, \eta, \xi')](\xi_i - \xi_i') \geq 0 \\ \text{a.e. in } \Omega \qquad \text{for } \eta \in \mathbb{R} \qquad \text{and} \qquad \xi, \xi' \in \mathbb{R}^N\end{gathered} \tag{4.38}$$

(which is the case when, in the above example, Φ is assumed to be convex with respect to ξ only), monotonicity can no longer be claimed. However, we have the following theorem.

THEOREM 4.21. *Let V be compactly imbedded into $L^p(\Omega)$ and let A be defined by* (4.34) *under assumption* (4.35). *Suppose that* (4.38) *holds, with the strict inequality sign for $\xi \neq \xi'$. Then A is a Leray–Lions operator, hence a pseudomonotone operator, when restricted to V.*

[For what concerns sufficient conditions in order that the injection $V \hookrightarrow L^p(\Omega)$ be compact, see Theorem 1.34 and the remark following Lemma 1.46.]

The proof of Theorem 4.21 is by no means straightforward. It requires the first part of the following technical result. (The second part will be utilized for Theorem 4.47.)

LEMMA 4.22. *Same assumptions as in Theorem* 4.21. If $\{u_n\} \subset V$ *is such that $u_n \rightharpoonup u$ in V and $\int_\Omega D_n\, dx \to 0$, where*

$$D_n \equiv [A^i(u_n, \nabla u_n) - A^i(u_n, \nabla u)](u_n - u)_{x_i}, \tag{4.39}$$

then $A^j(u_n, \nabla u_n) \rightharpoonup A^j(u, \nabla u)$ in $L^{p'}(\Omega)$ for $j = 0, 1, \dots, N$. If, moreover,

$$\begin{gathered}a^i(x, \eta, \xi)\xi_i \geq \alpha\, |\, \xi\, |^p - \lambda\, |\, \eta\, |^p - g(x) \\ \textit{for a.a. } x \in \Omega \qquad \textit{and} \qquad \textit{any } (\eta, \xi) \in \mathbb{R}^{1+N} \\ (\alpha > 0,\ \lambda \geq 0 \qquad \textit{and} \qquad g \in L^1(\Omega), \qquad g \geq 0),\end{gathered} \tag{4.40}$$

then $u_n \to u$ *in* V *and therefore*

$$A^j(u_n, \nabla u_n) \to A^j(u, \nabla u) \qquad in\ L^{p'}(\Omega) \qquad for\ j = 0, 1, \ldots, N.$$

[For $p = 2$ compare (4.35), (4.40) with (2.54), (2.55).]

PROOF. *Step 1: The general case.* By the strong convergence of $\{u_n\}_n$ in $L^p(\Omega)$ and $\{D_n\}_n$ in $L^1(\Omega)$ (D_n being ≥ 0), we can find a measurable subset Z of Ω, $|Z| = 0$, with the property that throughout $\Omega \setminus Z$ every function at hand is well-defined, and

$$u_k' \equiv u_{n_k} \to u, \qquad D_k' \equiv D_{n_k} \to 0 \qquad \text{as } k \to \infty$$

for a suitable subsequence of indices (see Theorem 1.Q). We fix $x \in \Omega \setminus Z$ and set

$$\eta \equiv u(x),\ \eta_k' \equiv u_k'(x),\ \xi \equiv \nabla u(x),\ \xi_{ki}' \equiv u_{kx_i}'(x),\ \xi_k' \equiv (\xi_{k1}', \ldots, \xi_{kN}').$$

We claim that $\{\xi_k'\}$ remains bounded. Suppose the contrary. Then for a subsequence of indices we have

$$|\xi_{k_h}' - \xi| > 1, \qquad (\xi_{k_h}' - \xi)/|\xi_{k_h}' - \xi| \to \xi^* \neq 0 \qquad \text{as } h \to \infty.$$

But then (4.38) yields

$$0 \leq \left[a^i(x, \eta_{k_h}', \xi_{k_h}') - a^i\left(x, \eta_{k_h}', \xi + \frac{\xi_{k_h}' - \xi}{|\xi_{k_h}' - \xi|}\right)\right](\xi_{k_h i}' - \xi_i),$$

hence also

$$\begin{aligned} 0 &\leq \left[a^i\left(x, \eta_{k_h}', \xi + \frac{\xi_{k_h}' - \xi}{|\xi_{k_h}' - \xi|}\right) - a^i(x, \eta_{k_h}', \xi)\right](\xi_{k_h i}' - \xi_i) \\ &= \left[a^i\left(x, \eta_{k_h}', \xi + \frac{\xi_{k_h}' - \xi}{|\xi_{k_h}' - \xi|}\right) - a^i(x, \eta_{k_h}', \xi_{k_h}')\right. \\ &\quad \left. + a^i(x, \eta_{k_h}', \xi_{k_h}') - a^i(x, \eta_{k_h}', \xi)\right](\xi_{k_h i}' - \xi_i) \leq D_{k_h}'(x). \end{aligned}$$

Letting $h \to \infty$ we obtain

$$[a^i(x, \eta, \xi + \xi^*) - a^i(x, \eta, \xi)]\xi_i^* = 0,$$

hence $\xi^* = 0$, a contradiction.

If now $\hat{\xi}$ is the limit of a converging subsequence of the bounded sequence $\{\xi'_k\}$, we have

$$0 = [a^i(x, \eta, \hat{\xi}) - a^i(x, \eta, \xi)](\hat{\xi}_i - \xi_i).$$

But then $\xi = \hat{\xi}$, so that $\xi_k' \to \xi$ and

$$a^j(x, \eta_k', \xi_k') \to a^j(x, \eta, \xi) \qquad \text{for } j = 0, 1, \ldots, N.$$

We have proved that, as $k \to \infty$,

$$A^j(u_k', \nabla u_k') \to A^j(u, \nabla u) \qquad \text{a.e. in } \Omega;$$

since the functions $A^j(u_k', \nabla u_k')$, $k \in N$, are uniformly bounded in $L^{p'}(\Omega)$ by (4.35), we arrive at

$$A^j(u_k', \nabla u_k') \rightharpoonup A^j(u, \nabla u) \qquad \text{in } L^{p'}(\Omega)$$

(see Problem 1.12), so that the weak convergence of the whole sequence $\{A^j(u_n, \nabla u_n)\}_n$ follows easily.

Step 2: *The case* (4.40). We return to the subsequence $\{u_k'\}$ of Step 1, which verifies $u_k' \to u$ in $L^p(\Omega)$ as well as

$$u_k' \to u,\ u'_{kx_1} \to u_{x_1}, \ldots, u'_{kx_N} \to u_{x_N} \qquad \text{a.e. in } \Omega.$$

The functions

$$\delta_k \equiv A^i(u_k', \nabla u_k')u'_{kx_i} + \lambda \mid u_k' \mid^p + g$$

are integrable and verify

$$\delta_k \to \delta \equiv A^i(u, \nabla u)u_{x_i} + \lambda \mid u \mid^p + g \qquad \text{a.e. in } \Omega;$$

note that, by (4.40)

$$\delta_k(x) \geq \alpha \mid \nabla u_k'(x) \mid^p. \tag{4.41}$$

By assumption, as $k \to \infty$ the quantity

$$\int_\Omega D_k'\,dx = \int_\Omega \delta_k\,dx - \lambda \int_\Omega \mid u_k' \mid^p dx - \int_\Omega g\,dx - \int_\Omega A^i(u_k', \nabla u_k')u_{x_i}\,dx$$
$$- \int_\Omega A^i(u_k', \nabla u)(u'_{kx_i} - u_{x_i})\,dx$$

tends to 0. But

$$\int_\Omega A^i(u_k', \nabla u_k')u_{x_i}\,dx \to \int_\Omega A^i(u, \nabla u)u_{x_i}\,dx$$

by Step 1, whereas

$$\int_\Omega A^i(u_k', \nabla u)(u'_{kx_i} - u_{x_i})\,dx \to 0$$

since $A^i(u_k', \nabla u) \to A^i(u, \nabla u)$ in $L^{p'}(\Omega)$ (see Lemma 4.20) and $u'_{kx_i} \rightharpoonup u_{x_i}$ in $L^p(\Omega)$. Thus,

$$\int_\Omega \delta_k\,dx \to \int_\Omega \delta\,dx.$$

We now set $\hat{\delta}_k \equiv \delta_k \wedge \delta = \delta - (\delta - \delta_k)^+$. Since $0 \leq \hat{\delta}_k \leq \delta$ and $\hat{\delta}_k \to \delta$ a.e. in Ω, the dominated convergence theorem yields

$$\hat{\delta}_k \to \delta \qquad \text{in } L^1(\Omega).$$

But then,

$$(\delta - \delta_k)^+ = \delta - \hat{\delta}_k \to 0 \qquad \text{in } L^1(\Omega),$$

and finally

$$\int_\Omega |\,\delta - \delta_k\,|\,dx = 2\int_\Omega (\delta - \delta_k)^+\,dx - \int_\Omega (\delta - \delta_k)\,dx \to 0.$$

Thus the sequence $\{\delta_k\}$ converges to δ in $L^1(\Omega)$ as well as a.e. in Ω, and has uniformly absolutely continuous integrals by Vitali's theorem; because of (4.41) this is also true of the sequence $\{|\,\nabla u_k'\,|^p\}$. Vitali's theorem can therefore be applied to the sequences

$$\{|\,u'_{kx_1} - u_{x_1}\,|^p\}, \ldots, \{|\,u'_{kx_N} - u_{x_N}\,|^p\},$$

so that for $i = 1, \ldots, N$

$$u'_{kx_i} \to u_{x_i} \qquad \text{in } L^p(\Omega) \qquad \text{as } k \to \infty.$$

The strong convergence of u_n to u in V [hence also of $A^i(u_n, \nabla u_n)$ to $A^i(u, \nabla u)$ in $L^{p'}(\Omega)$, by Lemma 4.20] follows immediately. ∎

Proof of Theorem 4.21. A is obviously bounded. For $u, v, w \in V$ we set

$$\langle \mathscr{A}'(u, v), w\rangle \equiv \int_\Omega A^i(u, \nabla v) w_{x_i}\, dx,$$

$$\langle \mathscr{A}''(u), w\rangle \equiv \int_\Omega A^0(u, \nabla u) w\, dx,$$

$$\mathscr{A}(u, v) \equiv \mathscr{A}'(u, v) + \mathscr{A}''(u),$$

so that $\mathscr{A}(u, u) = A(u)$, and proceed to verify requirements (i)–(iv) from the definition of Leray–Lions operators.

The boundedness properties required in (i) and (ii) are obviously satisfied; as for the hemicontinuity properties, they immediately follow from the stronger property ensured by Lemma 4.20. Finally, (4.38) implies that

$$\langle \mathscr{A}(u, u) - \mathscr{A}(u, v),\ u - v\rangle = \langle \mathscr{A}'(u, u) - \mathscr{A}'(u, v),\ u - v\rangle \geq 0.$$

Thus it remains to prove (iii) and (iv).

Let $\{u_n\} \subset V$ be such that $u_n \rightharpoonup u$ in V and $\int_\Omega D_n\, dx \to 0$, with D_n defined by (4.39). Then in particular $A^0(u_n, \nabla u_n) \rightharpoonup A^0(u, \nabla u)$ in $L^{p'}(\Omega)$ by Lemma 4.22, hence

$$\mathscr{A}''(u_n) \rightharpoonup \mathscr{A}''(u) \qquad \text{in } V'.$$

Since the convergence of $\mathscr{A}'(u_n, v)$ to $\mathscr{A}'(u, v)$ is ensured by Lemma 4.20 because $u_n \to u$ in $L^p(\Omega)$, (iii) follows.

Now let $\{u_n\} \subset V$ be such that $u_n \rightharpoonup u$ in V and $\mathscr{A}(u_n, v) \rightharpoonup F$ in V'. Then

$$\langle \mathscr{A}'(u_n, v), u_n\rangle \to \langle \mathscr{A}'(u, v), u\rangle$$

by Lemma 4.20. On the other hand, the inequality

$$|\langle \mathscr{A}''(u_n), u_n - u\rangle| \leq C\, |u_n - u|_{p;\Omega}$$

yields $\langle \mathscr{A}''(u_n), u_n - u\rangle \to 0$. Since

$$\langle \mathscr{A}''(u_n), u\rangle = \langle \mathscr{A}(u_n, v), u\rangle - \langle \mathscr{A}'(u_n, v), u\rangle \to \langle F, u\rangle - \langle \mathscr{A}'(u, v), u\rangle,$$

we have $\langle \mathscr{A}''(u_n), u_n\rangle \to \langle F, u\rangle - \langle \mathscr{A}'(u, v), u\rangle$; finally, $\langle \mathscr{A}(u_n, v), u_n\rangle \to \langle F, u\rangle$. ☐

The considerations developed in this section up till now provide us with a satisfactorily wide class of concrete cases entering the abstract setting of Theorem 4.17, except possibly for what concerns the growth condition (4.26). We therefore conclude this subsection with the following nonlinear extension of the example of Section 2.2.1.

EXAMPLE. Assume (4.35) for $j = 0, 1, \ldots, N$ and let (4.40) hold with $\lambda = 0$. Suppose that $a^0(x, \eta, \xi)\eta \geq c_0 \mid \eta \mid^p$ for a.a. $x \in \Omega$ and any $(\eta, \xi) \in \mathbb{R}^{1+N} (c_0 > 0)$.

Then, whatever the choice of V (and $\mathbb{K}$), A is coercive, hence *a fortiori* satisfies (4.26). The same conclusion remains valid even when it is only supposed that C from (4.35) for $j = 0$ and λ from (4.40) are $\leq \varepsilon$, with $\varepsilon > 0$ sufficiently small, provided Poincaré's inequality holds in V. We leave details to the reader.

4.3.3. Interpretation of Solutions

Fix $V = H_0^1(\Omega \cup \Gamma)$ with Γ of class C^1 and let $a(u, v)$ and A be defined by (4.32), L by (4.33), B as the conormal derivative operator [see (2.16)].

We first consider the v.i. (4.9) associated with the convex set (4.27) (supposed $\neq \varnothing$), that is, the *unilateral v.i.*

$$\begin{aligned} &u \in V, \qquad u \leq \psi \qquad \text{in } \Omega, \\ &a(u, v - u) \geq \langle F, v - u \rangle \qquad \text{for } v \in V, \qquad v \leq \psi \qquad \text{in } \Omega. \end{aligned} \tag{4.42}$$

If w is a nonnegative element of V the choice of $v = u - w$ is admissible in (4.42), so that $a(u, w) \leq \langle F, w \rangle$ and therefore $Au \leq F$ (in the sense of V') by the arbitrariness of w. Denote by f the restriction of F to $H_0^1(\Omega)$; then, $Lu \leq f$ in Ω [in the sense of $H^{-1}(\Omega)$].

Of course, a function u satisfying (4.42) vanishes on $\partial\Omega \setminus \Gamma$ [in the sense of $H^1(\Omega)$] by its mere membership in V.

In order to proceed further with the interpretation of (4.42) we tackle a special situation.

LEMMA 4.23. *Let Γ be closed and let $\psi \in H^1(\Omega)$,*

$$\begin{aligned} &\langle F, v \rangle = \int_\Omega fv \, dx + \langle \zeta, v|_\Gamma \rangle \qquad \textit{for } v \in V, \\ &\textit{with } f \in L^2(\Omega) \qquad \textit{and} \qquad \zeta \in [H^{1/2}(\Gamma)]'. \end{aligned} \tag{4.43}$$

A function $u \in H^1(\Omega)$ *with* $Lu \in L^2(\Omega)$ *solves* (4.42) *if and only if it satisfies*

$$\begin{gathered} u \le \psi, \qquad Lu \le f \ \ \text{and} \ \ (Lu - f)(u - \psi) = 0 \qquad \text{in } \Omega, \\ u = 0 \qquad \text{on } \partial\Omega \setminus \Gamma \text{ in the sense of } H^1(\Omega), \\ u|_\Gamma \le \psi|_\Gamma, \qquad Bu \le \zeta \qquad \text{on } \Gamma, \ \ \text{and} \ \ \langle Bu - \zeta, (u - \psi)|_\Gamma \rangle = 0. \end{gathered} \tag{4.44}$$

PROOF. *Step* 1: *The "if" part.* Let u satisfy (4.44). Then $u \in V$ and

$$\langle Au, v - u \rangle = a(u, v - u) = \int_\Omega Lu(v - u)\, dx + \langle Bu, (v - u)|_\Gamma \rangle$$

for $v \in V$ by the definition of the conormal derivative Bu. Take in particular $v \le \psi$: then $(v - u)^+|_\Gamma \le (\psi - u)|_\Gamma$, and $Bu \le \zeta$ on Γ {in the sense of $[H^{1/2}(\Gamma)]'$} implies

$$\langle Bu - \zeta, (v - u)^+|_\Gamma \rangle \le 0$$

as well as

$$\langle Bu - \zeta, (v - u)^+|_\Gamma \rangle \ge \langle Bu - \zeta, (\psi - u)|_\Gamma \rangle = 0.$$

On the other hand, $(v - u)^+$ can differ from 0 only where $\psi - u$ is >0 and therefore $Lu - f$ vanishes. Thus,

$$\begin{aligned} \langle Au - F, v - u \rangle &= \int_\Omega (Lu - f)(v - u)\, dx + \langle Bu - \zeta, (v - u)|_\Gamma \rangle \\ &= -\int_\Omega (Lu - f)(v - u)^- \, dx - \langle Bu - \zeta, (v - u)^-|_\Gamma \rangle \ge 0 \end{aligned}$$

because $(v - u)^-$ is a nonnegative element of V. This suffices to show that u solves (4.42). Note that, when $\Gamma = \varnothing$, the requirement $\psi \in H^1(\Omega)$ can be relinquished.

Step 2: *The "only if" part.* Under the assumptions of the implication we want to prove, the functions $u - \psi$ and $Lu - f$ are ≤ 0. We fix any measurable subset I of Ω having a positive distance from $\partial\Omega$ and denote by $\{\chi_n\}$ a sequence of functions from $C_c^\infty(\Omega)$ satisfying $0 \le \chi_n \le 1$, $\chi_n \to \chi_I$ (characteristic function of I) a.e. in Ω; note that χ_I is the limit of the sequence of its regularizations in every $L^p(\Omega)$, $p < \infty$, hence also (after passing to a subsequence, see Theorem 1.Q) a.e. in Ω. We can insert $v = u + \chi_n(\psi - u)$ in (4.42) and obtain

$$0 \le \langle Au - F, \chi_n(\psi - u) \rangle = \int_\Omega (Lu - f)\chi_n(\psi - u)\, dx,$$

hence also

$$0 \leq \int_\Omega (Lu - f)\chi_I(\psi - u)\, dx = \int_I (Lu - f)(\psi - u)\, dx$$

after passing to the limit as $n \to \infty$ with the help of the dominated convergence theorem. By the arbitrariness of I, $(Lu - f)(\psi - u) \geq 0$ and finally $(Lu - f)(\psi - u) = 0$ a.e. in Ω.

Next, let $\{\hat{\chi}_n\} \subset C_c^\infty(\Omega \cup \Gamma)$ satisfy $0 \leq \hat{\chi}_n \leq 1$ in $\bar{\Omega}$ and $\hat{\chi}_n = 1$ near Γ, $\hat{\chi}_n \to 0$ a.e. in Ω. Since $Au \leq F$ we have

$$0 \geq \langle Au - F, \hat{\chi}_n w\rangle = \int_\Omega (Lu - f)\hat{\chi}_n w\, dx + \langle Bu - \zeta, w|_\Gamma\rangle$$

whenever $w \in V$, $w \geq 0$. We can again pass to the limit under the integral sign and verify that $\int_\Omega (Lu - f)\hat{\chi}_n w\, dx \to 0$ as $n \to \infty$: hence $Bu \leq \zeta$ follows from $\langle Bu - \zeta, w|_\Gamma\rangle \leq 0$ by the arbitrariness of w, hence of $w|_\Gamma$. We now insert $v = u + \hat{\chi}_n(\psi - u)$ in (4.42) and obtain

$$\begin{aligned} 0 &\leq \langle Au - F, \hat{\chi}_n(\psi - u)\rangle \\ &= \int_\Omega (Lu - f)\hat{\chi}_n(\psi - u)\, dx + \langle Bu - \zeta, (\psi - u)|_\Gamma\rangle, \end{aligned}$$

hence

$$\langle Bu - \zeta, (\psi - u)|_\Gamma\rangle \geq 0$$

after a passage to the limit as $n \to \infty$, and finally

$$\langle Bu - \zeta, (\psi - u)|_\Gamma\rangle = 0$$

because $(\psi - u)|_\Gamma$ is a nonnegative element of $H^{1/2}(\Gamma)$. □

We call (4.44) an *obstacle problem*, more precisely a *variational unilateral problem*; we say that the condition on Γ is a *unilateral Neumann condition*, the one on $\partial\Omega \setminus \Gamma$ being of course the homogeneous Dirichlet condition.

We do not develop here the study of (4.9) in the case when $\mathbb{K}$ is given by (4.28) under general assumptions about the set E: the reader is referred to G. Stampacchia [141] for an illustration of connections to potential theory. Let us take up instead the special case (4.29), that is,

$$\begin{aligned} &u \in V, \quad u \leq \hat{\psi} \quad \text{on } \Gamma \text{ in the sense of } H^1(\Omega), \\ &\qquad a(u, v - u) \geq \langle F, v - u\rangle \qquad\qquad (4.45) \\ &\text{for } v \in V, \quad v \leq \hat{\psi} \quad \text{on } \Gamma \text{ in the sense of } H^1(\Omega), \end{aligned}$$

with Γ closed and either $\hat{\varphi} \in C^0(\Gamma)$ or $\hat{\varphi} \in H^{1/2}(\Gamma)$. Again, $Au \leq F$. Moreover, any function $v = u + w$, $w \in H_0^1(\Omega)$, is admissible in (4.45), which therefore implies $Lu = f$ with $f \equiv$ restriction of F to $H_0^1(\Omega)$. At this point the following result can be proven by proceeding as in the proof of Lemma 4.23.

LEMMA 4.24. *Let Γ be closed, let $\hat{\varphi} = \psi|_\Gamma$, $\psi \in H^1(\Omega)$, and assume* (4.43). *A function $u \in H^1(\Omega)$ solves* (4.45) *if and only if it satisfies*

$$\begin{gathered} Lu = f \quad \text{in } \Omega, \\ u = 0 \quad \text{on } \partial\Omega\setminus\Gamma \text{ in the sense of } H^1(\Omega), \\ u|_\Gamma \leq \psi|_\Gamma, \quad Bu \leq \zeta \quad \text{on } \Gamma, \text{ and } \langle Bu - \zeta, (u-\psi)|_\Gamma\rangle = 0. \end{gathered} \tag{4.46}$$

We now consider the v.i. (4.9) associated with the convex set (4.30), that is, the *bilateral v.i.*

$$\begin{gathered} u \in V, \quad \varphi \leq u \leq \psi \quad \text{in } \Omega, \\ a(u, v-u) \geq \langle F, v-u\rangle \quad \text{for } v \in V, \quad \varphi \leq v \leq \psi \quad \text{in } \Omega. \end{gathered} \tag{4.47}$$

LEMMA 4.25. *Let Γ be closed, let $\varphi, \psi \in H^1(\Omega)$, and assume* (4.43). *A solution u of* (4.47) *such that $Lu \in L^2(\Omega)$ is also a solution of*

$$\begin{gathered} \varphi \leq u \leq \psi, \quad (Lu - f)(u-\varphi) \leq 0 \\ \text{and } (Lu-f)(u-\psi) \leq 0 \quad \text{in } \Omega, \\ u = 0 \quad \text{on } \partial\Omega\setminus\Gamma \text{ in the sense of } H^1(\Omega), \\ \varphi|_\Gamma \leq u|_\Gamma \leq \psi|_\Gamma, \quad \langle Bu-\zeta, (u-\varphi)|_\Gamma\rangle \leq 0 \\ \text{and } \langle Bu - \zeta, (u-\psi)|_\Gamma\rangle \leq 0. \end{gathered} \tag{4.48}$$

For the proof, see Step 2 of the proof of Lemma 4.23: note that whenever χ is a function of $C_c^\infty(\Omega \cup \Gamma)$ that lies between 0 and 1, the functions $u + \chi(\varphi - u)$ and $u + \chi(\psi - u)$ are in V and lie between φ and ψ.

We say that the *obstacle problem* (4.48) is a *variational bilateral problem* with the homogeneous Dirichlet condition on $\partial\Omega\setminus\Gamma$ and a *bilateral Neumann condition* on Γ.

REMARK. As in Lemma 2.6, likewise in Lemmas 4.23–4.25, an important role is played by the circumstance that $Lu \in L^2(\Omega)$. It must, however, be noted that in Lemmas 2.6 and 4.24 such a circumstance is a straightforward consequence of the assumption about F, whereas in Lemmas 4.23 and 4.25 it must be assumed at the outset: that this assumption is not unnatural will be seen in Section 4.5 below.

Considerations analogous to preceding ones of this subsection can be made when $V = H_0^{1,p}(\Omega \cup \Gamma)$, $1 < p < \infty$, and (4.9) is replaced by (4.17) with $A(u)$ defined by (4.34), Lu being then replaced by $L(u)$.

For what in particular concerns the behavior of a solution u on Γ when the latter is closed, it suffices to introduce the functional $B(u) \in [H^{1/p',p}(\Gamma)]'$ defined by

$$\langle B(u), v|_\Gamma\rangle \equiv \langle A(u), v\rangle - \int_\Omega L(u)v\,dx \qquad \text{for } v \in V$$

if $L(u) \in L^{p'}(\Omega)$. We leave the details to the reader.

4.4. Existence and Uniqueness Results for a Class of Noncoercive Bilinear Forms

Throughout this and the next four sections we shall specialize in the study of (4.9) with $a(u, v)$ given by (4.32), the underlying space being $V = H_0^1(\Omega \cup \Gamma)$ with Γ of class C^1.

If $a(u, v)$ is of the most general noncoercive type [i.e., if it is merely coercive on V with respect to $L^2(\Omega)$], but V is compactly injected in $L^2(\Omega)$, and the operator $A: H^1(\Omega) \to V'$ verifies the weak maximum principle (see for instance Theorem 2.4), we know that the equation

$$\hat{u} \in V, \qquad a(\hat{u}, v) = \langle F, v\rangle \qquad \text{for } v \in V \tag{4.49}$$

($F \in V'$) admits a unique solution by dint of the Fredholm alternative (see Theorem 2.2). When dealing with v.i.'s instead of equations, we do not have any counterpart of Fredholm's theory at our disposal; yet, we can again arrive at existence and uniqueness results, at least for convex sets of either unilateral or bilateral type. This will be seen in the present section.

4.4.1. Unilateral Variational Inequalities

We say that $z \in H^1(\Omega)$ is a *subsolution* of (4.42) if $z \leq \psi$ in Ω, $z \leq 0$ on $\partial\Omega \setminus \Gamma$ in the sense of $H^1(\Omega)$, $Az \leq F$ [i.e., $a(z, w) \leq \langle F, w \rangle$ for $w \in V$, $w \geq 0$]; of course, a solution of (4.42) is a subsolution as well.

LEMMA 4.26. *Let $a(u, v)$ be coercive on V and let $F \in V'$ be given. If u solves* (4.42) *with ψ measurable in Ω, then u is maximal among all subsolutions.*

PROOF. Let z be a subsolution of (4.42). The function $v = z \vee u$ belongs to V and satisfies $v \leq \psi$ in Ω so that (4.42) yields

$$a\big(u, (z-u)^+\big) \geq \langle F, (z-u)^+ \rangle$$

by the identity $z \vee u - u = (z-u)^+$. But we also have

$$a\big(z, (z-u)^+\big) \leq \langle F, (z-u)^+ \rangle,$$

so that

$$0 \geq a\big(z-u, (z-u)^+\big) = a\big((z-u)^+, (z-u)^+\big).$$

By coerciveness, $| (z-u)^+ |_{H^1(\Omega)} = 0$ and therefore $z \leq u$ in Ω. ☐

The preceding lemma admits the following straightforward corollary.

COROLLARY. *Let $a(u, v)$ be coercive on V. For $h = 1, 2$ let F_h belong to V' and ψ_h be measurable in Ω, with $F_1 \leq F_2$ and $\psi_1 \leq \psi_2$. If $u = u_h$ solves* (4.42) *with $F = F_h$ and $\psi = \psi_h$, then $u_1 \leq u_2$.*

REMARK. Lemma 4.26 and its corollary have obvious extensions to the case when $H^1(\Omega)$ is replaced by $H^{1,p}(\Omega)$, $1 < p < \infty$, and (4.42) by its analog for a strictly T-monotone operator $A: H^{1,p}(\Omega) \to V'$.

We shall utilize Lemma 4.26 and its corollary for the following theorem, which generalizes Theorem 4.4 under the present choice of V, $\mathbb{K}$, and $a(u, v)$ (see also Theorem 2.3).

THEOREM 4.27. *Let Γ be such that V injects compactly into $L^2(\Omega)$. Let the weak maximum principle hold for $A: H^1(\Omega) \to V'$, let $F \in V'$ be given, and let the closed and convex set* (4.27) *be nonvoid. Then the v.i.* (4.42) *admits a unique solution, which in addition is maximal among all subsolutions.*

PROOF. We shall proceed in three steps.

Step 1: *Existence of subsolutions.* Fix $\lambda \geq 0$ so that

$$a_\lambda(u, v) \equiv a(u, v) + \lambda \int_\Omega uv\, dx \tag{4.50}$$

is coercive. By Theorem 4.4 there exists a unique solution to the v.i.

$$\hat{z} \in V, \qquad \hat{z} \leq \psi \qquad \text{in } \Omega,$$
$$a_\lambda(\hat{z}, v - \hat{z}) \geq 0 \qquad \text{for } v \in V, \qquad v \leq \psi \qquad \text{in } \Omega.$$

Since $A\hat{z} + \lambda\hat{z} \leq 0$, the weak maximum principle for $A + \lambda$ yields $\hat{z} \leq 0$ in Ω. Now let z solve the equation

$$z \in V, \qquad a(z, v) = a_\lambda(\hat{z}, v) \qquad \text{for } v \in V.$$

Since $Az = A\hat{z} + \lambda\hat{z} \leq A\hat{z}$, the weak maximum principle for A implies $z \leq \hat{z}$, hence $z \leq \psi$. This shows that at least when $F = 0$, (4.42) admits a subsolution z. When F is arbitrarily fixed in V', construct a subsolution z' to (4.42) with F replaced by 0 and ψ by $\psi - \hat{u}$ [see (4.49)]: the original v.i. then admits the subsolution $z \equiv z' + \hat{u}$.

Step 2: *Existence of solutions.* Define by recurrence: $u_0 = \hat{u}$,

$$\begin{gathered} u_n \in V, \qquad u_n \leq \psi \qquad \text{in } \Omega, \\ a_\lambda(u_n, v - u_n) \geq \langle F + \lambda u_{n-1}, v - u_n \rangle \\ \text{for } v \in V, \qquad v \leq \psi \qquad \text{in } \Omega, \end{gathered} \tag{4.51}$$

$n \in \mathbb{N}$, with $a_\lambda(u, v)$ as in (4.50). Lemma 4.26 and its corollary can be applied with $a(u, v)$ replaced by $a_\lambda(u, v)$ so that

$$u_{n-1} \geq u_n \geq z \qquad \text{in } \Omega \qquad \text{for } n \in \mathbb{N}$$

whenever z is a subsolution of (4.42): note that the weak maximum principle for A implies $z \leq u_0$, hence $Az + \lambda z \leq F + \lambda u_0$.

By fixing v in (4.51) we see that the sequence $\{u_n\}$, being bounded in $L^2(\Omega)$, is also bounded in V because the bilinear form (4.50) is coercive; thus $\{u_n\}$ converges weakly in V and strongly in $L^2(\Omega)$ toward some function u (no need to pass to a subsequence, thanks to monotonicity). Since

$$a_\lambda(u, u) \leq \liminf_{n\to\infty} a_\lambda(u_n, u_n),$$

we have

$$u \in V, \qquad u \leq \psi \qquad \text{in } \Omega,$$

$$a_\lambda(u, v-u) \geq \langle F + \lambda u, v-u \rangle \qquad \text{for } v \in V, \qquad v \leq \psi \qquad \text{in } \Omega$$

as well as

$$u \geq z \qquad \text{in } \Omega$$

whatever the subsolution z of (4.42).

Conclusion: (4.42) admits a solution that is maximal among all subsolutions.

Step 3: *Uniqueness.* We shall prove uniqueness by showing that any solution of (4.42) is maximal among subsolutions, more precisely that, whenever u_1 and u_2 are respectively a subsolution and a solution, the function $\hat{u} \equiv (u_1 - u_2)^+ \in V$ satisfies $A\hat{u} \leq 0$, hence $\hat{u} \leq 0$ by the weak maximum principle and finally $u_1 \leq u_2$. The compactness of $V \subsetneq L^2(\Omega)$ will play no role.

Suppose that the inequality $A\hat{u} \leq 0$ does not hold. Then there exists a function $w \in C_c^1(\Omega \cup \Gamma)$, $0 \leq w \leq 1$, such that $a(\hat{u}, w) > 0$. For $\varepsilon > 0$ consider the nonnegative function $w_\varepsilon \equiv \hat{u}w/(\hat{u} + \varepsilon)$: note that $w_\varepsilon \in V$ with $w_{\varepsilon x_j} = \hat{u}w_{x_j}/(\hat{u} + \varepsilon) + \varepsilon w \hat{u}_{x_j}/(\hat{u} + \varepsilon)^2$ (see Lemma 1.57). Since $\varepsilon w_\varepsilon \leq \hat{u}$, the function $v = u_2 + \varepsilon w_\varepsilon \leq u_2 + \hat{u} = u_1 \vee u_2$ can be inserted into (4.42) written for $u = u_2$. We thus obtain the inequality

$$a(u_2, w_\varepsilon) \geq \langle F, w_\varepsilon \rangle$$

which, together with the other inequality

$$a(u_1, w_\varepsilon) \leq \langle F, w_\varepsilon \rangle,$$

yields

$$\begin{aligned} 0 &\geq a(u_1 - u_2, w_\varepsilon) = a(\hat{u}, w_\varepsilon) \\ &= \int_\Omega \frac{\hat{u}}{\hat{u} + \varepsilon} [(a^{ij}\hat{u}_{x_i} + d^j\hat{u})w_{x_j} + (b^i\hat{u}_{x_i} + c\hat{u})w]\, dx \\ &\quad + \varepsilon \int_\Omega w\left[a^{ij} \frac{\hat{u}_{x_i}\hat{u}_{x_j}}{(\hat{u} + \varepsilon)^2} + d^j \frac{\hat{u}}{\hat{u} + \varepsilon} \frac{\hat{u}_{x_j}}{\hat{u} + \varepsilon}\right] dx \\ &\equiv I_1(\varepsilon) + \varepsilon I_2(\varepsilon) \end{aligned}$$

(since $w_\varepsilon = 0$ whenever $u_1 - u_2 < 0$). As $\varepsilon \to 0^+$, $I_1(\varepsilon)$ tends toward the positive quantity $a(\hat{u}, w)$ so that

$$\limsup_{\varepsilon\to 0^+} \varepsilon I_2(\varepsilon) < 0 \tag{4.52}$$

Let us prove that (4.52) is self-contradictory. Indeed, by uniform ellipticity the inequality $I_2(\varepsilon) \leq 0$ implies the following integral estimate on the function $G_\varepsilon(\hat{u}) \equiv w^{1/2} | \nabla \hat{u} |/(\hat{u} + \varepsilon)$:

$$\alpha \int_\Omega G_\varepsilon(\hat{u})^2 \, dx \leq \int_\Omega w a^{ij} \frac{\hat{u}_{x_i}\hat{u}_{x_j}}{(\hat{u}+\varepsilon)^2} \, dx$$

$$\leq - \int_\Omega w d^j \frac{\hat{u}}{\hat{u}+\varepsilon} \frac{\hat{u}_{x_j}}{\hat{u}+\varepsilon} \, dx \leq C \int_\Omega G_\varepsilon(\hat{u}) \, dx.$$

By the Cauchy–Schwarz inequality, therefore, $| G_\varepsilon(\hat{u}) |_{2;\Omega}$ can be bounded independently of $\varepsilon > 0$. But then the inequality

$$| I_2(\varepsilon) | \leq C\big(| G_\varepsilon(\hat{u}) |^2_{2;\Omega} + | G_\varepsilon(\hat{u}) |_{2;\Omega}\big)$$

implies $\varepsilon I_2(\varepsilon) \to 0$ as $\varepsilon \to 0^+$, which contradicts (4.52). This proves that the assumption $a(\hat{u}, w) > 0$ was absurd. □

The maximality property ascertained in Step 3 of the proof of Theorem 4.27 leads to the following corollary.

COROLLARY. *The conclusion of the corollary to Lemma 4.26 remains valid if the coerciveness assumption about $a(u, v)$ is weakened into the requirement that A satisfies the weak maximum principle.*

REMARK. All considerations developed until now can be repeated, with obvious changes, if (4.42) is replaced by (4.45).

Another consequence of the maximality property (more precisely of Lemma 4.26) is the following result, which we already utilized in Section 2.4.1.

LEMMA 4.28. *Let $\varphi = \bigvee_{h=1}^m \varphi^h$, $m \in \mathbb{N}$, where each φ^h belongs to $H^1(\Omega)$, and let there exist $\bigvee_{h=1}^m A\varphi^h \in V'$. Then*

$$A\varphi \leq \bigvee_{h=1}^{m} A\varphi^h \qquad \text{(in the sense of } V'\text{)}. \tag{4.53}$$

An analogous statement is valid if $\vee$ is replaced by $\wedge$, provided $\leq$ in (4.53) is replaced by $\geq$.

PROOF. Suppose first that $a(u, v)$ is coercive on V and consider (4.42) with $\psi = 0$, $F = \bigvee_{h=1}^{m} A\varphi^h - A\varphi$. Then the unique solution u is maximal among all subsolutions; in particular $u \geq \varphi^h - \varphi$ because $A(\varphi^h - \varphi) \leq \bigvee_{k=1}^{m} A\varphi^k - A\varphi$. Passing to the supremum over h we obtain $u \geq \bigvee_{h=1}^{m} \varphi^h - \varphi = 0$, hence $u = 0$ and $0 = Au \leq F$.

If $a(u, v)$ is not coercive, fix $\lambda > 0$ so that the bilinear form (4.50) is coercive. Then the family $\{A\varphi^h + \lambda\varphi^h\}_{h=1,\dots,m}$ is order bounded from above by $G \equiv \bigvee_{h=1}^{m} A\varphi^h + \lambda\varphi$, and there exists $\bigvee_{h=1}^{m} (A\varphi^h + \lambda\varphi^h)$ by Lemma 1.54. By the preceding part of this proof,

$$A\varphi + \lambda\varphi \leq \bigvee_{h=1}^{m} (A\varphi^h + \lambda\varphi^h) \leq G,$$

and (4.53) again holds. □

Under the same assumption about A as in Theorem 4.27 a result stronger that uniqueness holds:

THEOREM 4.29. *Let the weak maximum principle hold for* $A\colon H^1(\Omega) \to V'$. *For* $h = 1, 2$ *let* ψ_h *be measurable, with* $\psi_1 - \psi_2 \in L^\infty(\Omega)$. *If* $u = u_h$ *solves* (4.42) *with* $\psi = \psi_h$, *then* $u_1 - u_2$ *belongs to* $L^\infty(\Omega)$ *and verifies*

$$| u_1 - u_2 |_{\infty;\Omega} \leq C | \psi_1 - \psi_2 |_{\infty;\Omega}, \tag{4.54}$$

where $C \geq 1$ *depends only on* A, *and* $C = 1$ *if* $A1 \geq 0$.

PROOF. Solve the v.i.

$$z_0 \in V, \qquad z_0 \geq 0 \quad \text{in } \Omega,$$

$$a(z_0, v - z_0) \geq \langle -A1, v - z_0 \rangle \quad \text{for } v \in V, \quad v \geq 0 \quad \text{in } \Omega$$

with the help of Theorem 4.27. Note that $z_0 = 0$ when $A1 \geq 0$; even when the latter requirement is not fulfilled, z_0 still belongs to $L^\infty(\Omega)$ because $v = z_0 - (z_0 - k)^+$ is admissible in the above v.i., which therefore yields

$$a(z_0, (z_0 - k)^+) \leq \langle -A1, (z_0 - k)^+ \rangle$$

whenever $k \geq 0$: see Lemma 2.8 and the remark following it.

Let $\hat{z} \equiv z_0 + 1$, so that $\hat{z} \geq 1$, $\hat{z} = 1$ if $A1 \geq 0$, and $A\hat{z} \geq 0$. Next, let $k \equiv | \psi_1 - \psi_2 |_{\infty;\Omega}$, $\hat{u} \equiv (u_1 - u_2 - k\hat{z})^+$ and $w_\varepsilon \equiv \hat{u}w/(\hat{u} + \varepsilon)$, where $\varepsilon > 0$ and $w \in C_c^1(\Omega \cup \Gamma)$, $0 \leq w \leq 1$. Since $Au_1 \leq F$ we have

$$a(u_1 - k\hat{z}, w_\varepsilon) \leq \langle F, w_\varepsilon \rangle;$$

on the other hand, the function

$$v = u_2 + \varepsilon w_\varepsilon \leq u_2 + \hat{u} = (u_1 - k\hat{z}) \vee u_2$$
$$\leq (u_1 - k) \vee u_2 \leq (u_1 - \psi_1 + \psi_2) \vee u_2 \leq \psi_2$$

is admissible in (4.42) written for $\psi = \psi_2$, $u = u_2$, and therefore

$$a(u_2, w_\varepsilon) \geq \langle F, w_\varepsilon \rangle.$$

Summing up,

$$0 \geq a(u_1 - u_2 - k\hat{z}, w_\varepsilon) = a(\hat{u}, w_\varepsilon)$$

(because $w_\varepsilon = 0$ whenever $u_1 - u_2 - k\hat{z} < 0$). We can at this point proceed as in the proof of Theorem 4.27 and show that $A\hat{u} \leq 0$, hence $\hat{u} \leq 0$ by the weak maximum principle, and finally $u_1 - u_2 \leq k\hat{z} \leq k \mid \hat{z} \mid_{\infty;\Omega}$.

Since the roles of u_1 and u_2 can be interchanged, (4.54) holds with $C = \mid \hat{z} \mid_{\infty;\Omega}$. ∎

REMARK. Dependence of solutions on free terms could be tackled through an argument utilized for a special case in the proof of Theorem 5.5 below.

4.4.2. Bilateral Variational Inequalities

THEOREM 4.30. *Let Γ be such that V is compactly imbedded into $L^2(\Omega)$. Let the weak maximum principle hold for $A: H^1(\Omega) \to V'$, let φ and ψ be measurable functions in Ω, and let $F \in V'$. If the closed and convex set* (4.30) *is $\neq \emptyset$, the v.i.* (4.47) *admits a unique solution.*

PROOF. We shall proceed in two steps.

Step 1: Existence. To φ and ψ we associate a bounded, closed and convex subset $\mathscr{K} \neq \emptyset$ of $L^2(\Omega)$ as follows. If $\varphi, \psi \in L^2(\Omega)$, we set $\mathscr{K} \equiv \{v \in L^2(\Omega) \mid \varphi \leq v \leq \psi \text{ in } \Omega\}$. If $\psi \in L^2(\Omega)$ but $\varphi \notin L^2(\Omega)$, we utilize Theorem 4.27 to solve (4.42), call u' the solution, and set

$$\mathscr{K} \equiv \{v \in L^2(\Omega) \mid \varphi \vee u' \leq v \leq \psi \text{ in } \Omega\}.$$

The definition of $\mathscr{K}$ in each remaing case, that is, when either $\varphi \in L^2(\Omega)$, $\psi \notin L^2(\Omega)$ or $\varphi \notin L^2(\Omega)$, $\psi \notin L^2(\Omega)$, is at this point obvious.

Let the bilinear form (4.50) be coercive, and define a continuous mapping $S: L^2(\Omega) \to V$ by

$$\begin{gathered} S(u) \in V, \qquad \varphi \le S(u) \le \psi \qquad \text{in } \Omega, \\ a_\lambda\big(S(u), v - S(u)\big) \ge \langle F + \lambda u, v - S(u)\rangle \\ \text{for } v \in V, \qquad \varphi \le v \le \psi \qquad \text{in } \Omega, \end{gathered} \tag{4.55}$$

where Theorem 4.4 is taken into account.

We claim that S maps $\mathscr{K}$ into itself. To see this we can safely restrict our considerations to the case when $\varphi \notin L^2(\Omega)$, $\psi \in L^2(\Omega)$; let us prove that $S(u) \ge \varphi \vee u'$ for $u \in \mathscr{K}$. The function $v \equiv S(u) + [u' - S(u)]^+$ is admissible in (4.55), so that obvious passages lead to

$$\begin{aligned} 0 \ge \lambda \int_\Omega (u' - u)[u' - S(u)]^+\, dx &\ge a_\lambda\big(u' - S(u), [u' - S(u)]^+\big) \\ &= a_\lambda\big([u' - S(u)]^+, [u' - S(u)]^+\big) \end{aligned}$$

and finally to $[u' - S(u)]^+ = 0$ by coerciveness. This proves that $S(u) \ge u'$, hence $S(u) \in \mathscr{K}$.

Again by coerciveness, S maps bounded subsets of $L^2(\Omega)$ into bounded subsets of V, hence into relatively compact subsets of $L^2(\Omega)$. Thus, the Schauder theorem (see Theorem 1.J) yields the existence of a fixed point $u = S(u) \in \mathscr{K}$, hence of a solution to (4.47).

Step 2: *Uniqueness.* Let u_1 and u_2 solve (4.47) and set $\hat{u} \equiv (u_1 - u_2)^+$, $w_\varepsilon \equiv \hat{u}w/(\hat{u} + \varepsilon)$ with $\varepsilon > 0$ and $w \in C_c^1(\Omega \cup \Gamma)$, $0 \le w \le 1$. Then both functions $u_1 - \varepsilon w_\varepsilon \ge u_1 - \hat{u}$ and $u_2 + \varepsilon w_\varepsilon \le u_2 + \hat{u}$ lie between φ and ψ: by proceeding as in Step 3 of the proof of Theorem 4.27 it can be proved that $A\hat{u} \le 0$, hence that $u_1 \le u_2$ by the weak maximum principle. □

REMARK. For what concerns existence, the weak maximum principle plays a role only when $\varphi \notin L^2(\Omega)$ and/or $\psi \notin L^2(\Omega)$. As for uniqueness, note that the compactness of the imbedding $V \subsetneq L^2(\Omega)$ plays no role.

Just as the proof of uniqueness carries over from Theorem 4.27 to Theorem 4.30, so does the proof of the L^∞ estimate from Theorem 4.29 to the following theorem.

THEOREM 4.31. *Let the weak maximum principle hold for* $A: H^1(\Omega) \to V'$. *For* $h = 1, 2$ *let* φ_h, ψ_h *be measurable with* $\varphi_1 - \varphi_2, \psi_1 - \psi_2 \in L^\infty(\Omega)$.

If $u = u_h$ *solves* (4.47) *with* $\varphi = \varphi_h$, $\psi = \psi_h$, *then* $u_1 - u_2$ *belongs to* $L^\infty(\Omega)$ *and verifies*

$$| u_1 - u_2 |_{\infty;\Omega} \leq C \max(| \varphi_1 - \varphi_2 |_{\infty;\Omega}, | \psi_1 - \psi_2 |_{\infty;\Omega}),$$

where $C \geq 1$ *depends only on* A, $C = 1$ *if* $A1 \geq 0$.

4.5. Lewy–Stampacchia Inequalities and Applications to Regularity

We now turn to regularity of solutions to v.i.'s. Let us mention at the outset that in our study an essential role will be played not only by the regularity required on Γ, $\partial\Omega \setminus \Gamma$, F, and the coefficients of $a(u, v)$ (which was the case for equations: see Chapters 2 and 3), but also by specific features of the convex sets $\mathbb{K}$ under consideration. As a matter of fact, even when solution to the corresponding equation [i.e., to (4.9) with $\mathbb{K}$ replaced by V] would belong to $C^\infty(\bar{\Omega})$, a solution to the v.i. need not belong, say, to $H^2_{\text{loc}}(\Omega)$. This is illustrated by the following simple example.

Example. Let $N = 1$ and take $\Omega =]-1, 1[$, $V = H_0{}^1(\Omega)$, $a(u, v) = \int_{-1}^1 u'v'\,dx$, $\psi(x) = |x| - 1/2$, $F = 0$.

The function $u(x) = \frac{1}{2}(|x| - 1)$ satisfies (4.42) since $u \in V$, $u \leq \psi$, and

$$\begin{aligned} a(u, v - u) &= -\tfrac{1}{2}\int_{-1}^{0} (v' + 1/2)\,dx + \tfrac{1}{2}\int_{0}^{1} (v' - 1/2)\,dx \\ &= -\tfrac{1}{2}[v(0) + 1/2] - \tfrac{1}{2}[v(0) + 1|2] \geq 0 \end{aligned}$$

whenever $v \in V$ satisfies $v \leq \psi$ in Ω.

The regularity of u does not go beyond its being Lipschitz continuous, since u'' is not a function, but is instead the Dirac measure concentrated at 0 (see Problem 1.10).

In the light of the above, we shall tackle problems of regularity by separately considering various classes of convex sets, though all of obstacle type.

Beginning with convex sets defined by unilateral constraints we have the following theorem.

Theorem 4.32. *Let* $F \in V'$ *and* $\psi \in H^1(\Omega)$, $\psi \geq 0$ *on* $\partial\Omega \setminus \Gamma$ *in the sense of* $H^1(\Omega)$, *be such that there exists* $(A\psi) \wedge F \in V'$. *Then a solution u of* (4.42), *if existing, satisfies*

$$(A\psi) \wedge F \leq Au \leq F \qquad (\textit{in the sense of } V'). \tag{4.56}$$

More generally: if $\psi = \bigwedge_{h=1}^{m} \psi^h$ *with* $\psi^h \in H^1(\Omega)$, $\psi^h \geq 0$ *on* $\partial\Omega \setminus \Gamma$ *in the sense of* $H^1(\Omega)$, *and there exists* $\bigwedge_{h=1}^{m} (A\psi^h) \wedge F \in V'$, *then a solution of* (4.42), *if existing, satisfies*

$$\bigwedge_{h=1}^{m} (A\psi^h) \wedge F \leq Au \leq F \qquad (\textit{in the sense of } V'). \tag{4.57}$$

PROOF. Since u is already known to be a subsolution of (4.42), there remains to prove the left-hand-side inequalities of (4.56) and (4.57). Beginning with the former, we first assume that the bilinear form is coercive on V and solve

$$\begin{aligned} &u' \in V, \qquad u' \geq u \qquad \text{in } \Omega, \\ &a(u', v - u') \geq \langle (A\psi) \wedge F, v - u' \rangle \qquad \text{for } v \in V, \qquad v \geq u \qquad \text{in } \Omega \end{aligned} \tag{4.58}$$

in the light of Theorem 4.4. The function u' satisfies

$$Au' \geq (A\psi) \wedge F \qquad (\text{in the sense of } V'). \tag{4.59}$$

As a matter of fact, by (the obvious analog of) Lemma 4.26, u' minimizes the family of all functions $z \in H^1(\Omega)$ satisfying $z \geq 0$ on $\partial\Omega \setminus \Gamma$, $z \geq \psi$ in Ω, $Az \geq (A\psi) \wedge F$; in particular, $u' \leq \psi$. But then the choice of $v = u'$ is admissible in (4.42), which yields

$$a(u, u' - u) \geq \langle F, u' - u \rangle \geq \langle (A\psi) \wedge F, u' - u \rangle,$$

whereas (4.58) yields

$$a(u', u - u') \geq \langle (A\psi) \wedge F, u - u' \rangle$$

with the choice of $v = u$; thus,

$$a(u' - u, u' - u) \leq 0.$$

By coerciveness $u' = u$, so that (4.59) amounts to the left-hand side inequality of (4.56).

Let us now drop the assumption that $a(u, v)$ is coercive on V. Fix λ so large that the bilinear form (4.50) is coercive: since $G \equiv (A\psi) \wedge F + \lambda u$ is a lower bound for $A\psi + \lambda\psi$ and $F + \lambda u$, the previous part of the proof yields

$$Au + \lambda u \geq (A\psi + \lambda\psi) \wedge (F + \lambda u) \geq G$$

(see Lemma 1.54), hence again the desired conclusion.

For what concerns (4.57) it suffices to take into account that, by Lemma 4.28, $G \equiv \bigwedge_{h=1}^{m} (A\psi^h) \wedge F$ is a lower bound for $A\psi$ and F. □

The estimates from below provided by (4.56) and (4.57) are called the *(unilateral) Lewy–Stampacchia inequalities*. Their interpretation presents no difficulty when $\Gamma = \varnothing$, since A then coincides with the bounded linear functional L: $H^1(\Omega) \to H^{-1}(\Omega)$ defined in (4.33). Thus, say, (4.57) amounts to

$$\bigwedge_{h=1}^{m} (L\psi^h) \wedge f \leq Lu \leq f \qquad [\text{in the sense of } H^{-1}(\Omega)], \tag{4.60}$$

where we have written f instead of F for consistency with the general case. From (4.60) we can deduce regularity properties of the distribution Lu, in the sense that the latter belongs to some space $L^p(\Omega)$, $p > 1$, whenever f and $\bigwedge_{h=1}^{m} (L\psi^h) \wedge f$ do. The next example shows that the membership of Lu in $L^\infty(\Omega)$ is a regularity threshold for solutions of (4.42).

EXAMPLE. Take $N = 1$ and let $\Omega =]-1, 1[$, $a(u, v) = \int_{-1}^{1} u'v'\,dx$, $f = 0$, $\psi(x) = 4x^2 - 1$. We choose $\xi \in]-1, 0[$ so that the tangent line $y = \psi(\xi) + \psi'(\xi)(x - \xi)$ passes through the point $(-1, 0)$ of the (x, y) plane. Analogously, we choose $\eta \in]0, 1[$ so that the line $y = \psi(\eta) + \psi'(\eta)(x - \eta)$ passes through the point $(1, 0)$. Then the function

$$u(x) \equiv \begin{cases} \psi(\xi) + \psi'(\xi)(x - \xi) & \text{for } -1 \leq x \leq \xi, \\ \psi(x) & \text{for } \xi < x < \eta, \\ \psi(\eta) + \psi'(\eta)(x - \eta) & \text{for } \eta \leq x \leq 1 \end{cases}$$

is in $H_0{}^1(\Omega)$, lies below ψ and satisfies

$$\begin{aligned} a(u, v - u) &= \int_{-1}^{\xi} \psi'(\xi)[v'(x) - \psi'(\xi)]\,dx \\ &\quad + \int_{\xi}^{\eta} \psi'(x)[v'(x) - \psi'(x)]\,dx + \int_{\eta}^{1} \psi'(\eta)[v'(x) - \psi'(\eta)]\,dx \\ &= \psi'(\xi)[v(\xi) - \psi'(\xi)(\xi + 1)] \\ &\quad + \psi'(x)[v(x) - \psi(x)]\Big|_{\xi}^{\eta} - \int_{\xi}^{\eta} \psi''(x)[v(x) - \psi(x)]\,dx \\ &\quad + \psi'(\eta)[-v(\eta) - \psi'(\eta)(1 - \eta)] \end{aligned}$$

whenever $v \in H_0^1(\Omega)$. But $\psi'(\xi)(\xi+1) = \psi(\xi)$ and $\psi'(\eta)(1-\eta) = -\psi(\eta)$ so that

$$a(u, v-u) = -\int_\xi^\eta \psi''(x)[v(x) - \psi(x)]\,dx$$
$$= -8\int_\xi^\eta [v(x) - \psi(x)]\,dx,$$

and finally $a(u, v-u) \geq 0$ if $v \leq \psi$.

We have $Lu = -u'' \in L^\infty(\Omega)$, but $Lu \notin C^0(\Omega)$ although the free term and the obstacle are analytic functions.

When $\Gamma \neq \varnothing$ the interpretation of the Lewy–Stampacchia inequalities requires, so to speak, that they be decomposed into a part inside Ω and a part on Γ. In its generality this procedure requires the full machinery of order dual spaces as developed by B. Hanouzet and J. L. Joly [72]. We can, however, handle it rather simply under slightly restrictive assumptions, as the next result shows.

LEMMA 4.33. *Suppose that F is defined as in (4.43), that $\psi = \bigwedge_{h=1}^m \psi^h$, where $\psi^h \in H^1(\Omega)$, $\psi^h \geq 0$ on $\partial\Omega \setminus \Gamma$ in the sense of $H^1(\Omega)$, $L\psi^h \in L^2(\Omega)$, and that $B\psi^1, \ldots, B\psi^m, \zeta$ admit a greatest lower bound $\bigwedge_{h=1}^m (B\psi^h) \wedge \zeta \in [H^{1/2}(\Omega)]'$. If u solves (4.42), then Lu and Bu satisfy*

$$\bigwedge_{h=1}^m (L\psi^h) \wedge f \leq Lu \leq f \qquad \text{in } \Omega \tag{4.61}$$

and

$$\bigwedge_{h=1}^m (B\psi^h) \wedge \zeta \leq Bu \leq \zeta \qquad \text{in the sense of } [H^{1/2}(\Omega)]' \tag{4.62}$$

respectively.

PROOF. $A\psi^1, \ldots, A\psi^m$ and F admit a lower bound $G \in V'$ defined by

$$\langle G, v\rangle = \int_\Omega \left[\bigwedge_{h=1}^m (L\psi^h) \wedge f\right] v\,dx + \left\langle \bigwedge_{h=1}^m (B\psi^h) \wedge \zeta, v|_\Gamma \right\rangle \qquad \text{for } v \in V.$$

Theorem 4.32 therefore yields

$$G \leq Au \leq F \qquad \text{in the sense of } V',$$

which immediately implies (4.61). As a consequence, Lu belongs to $L^2(\Omega)$,

u admits a conormal derivative Bu on Γ, and (4.62) follows (see the proof of Lemma 4.23). □

Under the assumptions of Lemma 4.33 we can view u as a variational solution of a b.v.p.

$$Lu = f' \qquad \text{in } \Omega,$$

$$u = 0 \qquad \text{on } \partial\Omega\setminus\Gamma, \qquad Bu = \zeta' \qquad \text{on } \Gamma,$$

where $f' \in L^2(\Omega)$ and $\zeta' \in [H^{1/2}(\Gamma)]'$ are of course unknown. Local and even global regularity results for u can at this point be deduced from the regularity theory for b.v.p.'s developed in Chapters 2 and 3. Indeed, we have the following lemma.

LEMMA 4.34. *In addition to the hypotheses of Lemma* 4.33 *suppose that the coefficients* a^{ij}, d^j *of* $a(u, v)$ *are Lipschitz continuous on* $\bar{\Omega}$ *and that* f, $\bigwedge_{h=1}^m (L\psi^h) \wedge f \in L^p(\Omega)$, $2 \le p < \infty$. *Then a solution* u *of* (4.42) *is in* $H^{2,p}_{\text{loc}}(\Omega)$. *If in addition* Γ *is closed and* $\partial\Omega\setminus\Gamma$ *is of class* $C^{1,1}$, *the* $H^{2,p}$ *regularity of* u *extends up to* $\partial\Omega\setminus\Gamma$. *Finally, if it is also assumed that* Γ *is of class* $C^{1,1}$ *and that* $\zeta \in H^{1/p',p}(\Gamma)$ *with* $\zeta \le B\psi^h$ *in the sense of* $[H^{1/2}(\Gamma)]'$ *for* $h = 1, \ldots, m$, *then* $u \in H^{2,p}(\Omega)$.

PROOF. By (4.61) interior regularity follows from Theorem 3.8; the regularity up to $\partial\Omega\setminus\Gamma$ follows from Theorem 3.17(ii) via a cutoff argument. As for regularity up to Γ, notice that under our present assumptions, (4.62) implies $Bu = \zeta$, so that the conclusion follows from Theorem 3.17(ii). (See Step 1 of the proof of Lemma 3.18.) □

Passing from unilateral to bilateral v.i.'s we can easily arrive at *(bilateral) Lewy–Stampacchia inequalities* by observing that, in the proof of Theorem 4.32, the solution u' of (4.58) satisfies not only $u' \le \psi$ but also $u' \ge \varphi$ if $u \ge \varphi$. Thus we have the following theorem.

THEOREM 4.35. *Let* $F \in V'$ *and* $\varphi, \psi \in H^1(\Omega)$, $\varphi \le 0 \le \psi$ *on* $\partial\Omega\setminus\Gamma$ *in the sense of* $H^1(\Omega)$, *be such that there exist* $(A\varphi) \vee F$, $(A\psi) \wedge F \in V'$. *Then a solution* u *of* (4.47), *if existing, satisfies*

$$(A\psi) \wedge F \le Au \le (A\varphi) \vee F \qquad (\textit{in the sense of } V').$$

More generally: let $\varphi = \bigvee_{h=1}^m \varphi^h$ *with* $\varphi^h \in H^1(\Omega)$, $\varphi^h \le 0$ *on* $\partial\Omega\setminus\Gamma$ *in the sense of* $H^1(\Omega)$, $\psi = \bigwedge_{h=1}^m \psi^h$ *with* $\psi^h \in H^1(\Omega)$, $\psi^h \ge 0$ *on* $\partial\Omega\setminus\Gamma$ *in the sense of* $H^1(\Omega)$, *and let there exist*

$$\bigvee_{h=1}^{m} (A\varphi^h) \vee F, \; \bigwedge_{h=1}^{m} (A\psi^h) \wedge F \in V'.$$

Then a solution of (4.47), *if existing, satisfies*

$$\bigwedge_{h=1}^{m} (A\psi^h) \wedge F \leq Au \leq \bigvee_{h=1}^{m} (A\varphi^h) \vee F \qquad (\textit{in the sense of } V').$$

Consequently (see Lemma 4.33), we have the following lemma.

LEMMA 4.36. *Suppose that F is defined as in* (4.43), *that* $\varphi = \bigvee_{h=1}^{m} \varphi^h$, $\psi = \bigwedge_{h=1}^{m} \psi^h$ *where* $\varphi^h, \psi^h \in H^1(\Omega)$, $\varphi^h \leq 0 \leq \psi^h$ *on* $\partial\Omega \setminus \Gamma$ *in the sense of* $H^1(\Omega)$, $L\varphi^h$, $L\psi^h \in L^2(\Omega)$, *that* $B\varphi^1, \ldots, B\varphi^m$, ζ *admit a least upper bound* $\bigvee_{h=1}^{m} (B\varphi^h) \vee \zeta \in [H^{1/2}(\Gamma)]'$, *and that* $B\psi^1, \ldots, B\psi^m$, ζ *admit a greatest lower bound* $\bigwedge_{h=1}^{m} (B\psi^h) \wedge \zeta \in [H^{1/2}(\Gamma)]'$. *Let u solve* (4.47). *Then Lu and Bu satisfy*

$$\bigwedge_{h=1}^{m} (L\psi^h) \wedge f \leq Lu \leq \bigvee_{h=1}^{m} (L\varphi^h) \vee f \qquad \textit{in } \Omega$$

and

$$\bigwedge_{h=1}^{m} (B\psi^h) \wedge \zeta \leq Bu \leq \bigvee_{h=1}^{m} (B\varphi^h) \vee \zeta \qquad \textit{in the sense of } [H^{1/2}(\Gamma)]',$$

respectively.

Finally (see Lemma 4.34), we have the following lemma.

LEMMA 4.37. *In addition to the hypotheses of Lemma* 4.36 *suppose that the coefficients* a^{ij}, d^j *of* $a(u, v)$ *belong to* $C^{0,1}(\bar{\Omega})$ *and that* $\bigvee_{h=1}^{m} (L\varphi^h) \vee f$, $\bigwedge_{h=1}^{m} (L\psi^h) \wedge f \in L^p(\Omega)$, $2 \leq p < \infty$. *Then a solution u of* (4.47) *is in* $H^{2,p}_{\text{loc}}(\Omega)$. *If in addition* Γ *is closed and* $\partial\Omega \setminus \Gamma$ *is of class* $C^{1,1}$, *the* $H^{2,p}$ *regularity of u extends up to* $\partial\Omega \setminus \Gamma$. *Finally, if it is also assumed that* Γ *is of class* $C^{1,1}$ *and that* $\zeta \in H^{1/p',p}(\Gamma)$ *with* $B\varphi^h \leq \zeta \leq B\psi^h$ *in the sense of* $[H^{1/2}(\Gamma)]'$ *for* $h = 1, \ldots, m$, *then* $u \in H^{2,p}(\Omega)$.

REMARK 1. By the same techniques utilized in the proofs of Theorems 4.32 and 4.35, unilateral and bilateral Lewy–Stampacchia inequalities can be proven when V equals $H_0^{1,p}(\Omega \cup \Gamma)$, $1 < p < \infty$, and A denotes a bounded, hemicontinuous, strictly T-monotone (hence pseudomonotone) and coercive operator $H^{1,p}(\Omega) \to V'$: see Theorem 4.17, as well as Lemma 4.26 and the remark following it. Decompositions analogous to those of Lemmas 4.33 and 4.36 can also be proven by taking the final observation of Section 4.3 into account. Much more complex is instead the task of arriving at regularity results extending those of Lemmas 4.34 and 4.37 to the nonlinear case, e.g., we refer to G. Stampacchia [143].

REMARK 2. Both Lemmas 4.34 and 4.37 can be extended to the case of any $p \in]1, 2[$ such that $L^p(\Omega) \subsetneq V'$: see Problem 3.8.

4.6. Further $H^{2,p}$ Regularity

In Section 4.5 we entirely based our approach to $H^{2,p}$ regularity on Lewy–Stampacchia inequalities, thus automatically sidestepping two important questions that we are now going to discuss.

4.6.1. $H^{2,\infty}$ Regularity

In Lemma 4.34 we did not take into consideration the case $p = \infty$. The reason for this is that for $N > 1$ the membership of Lu in $L^\infty(\Omega)$ [which follows from (4.61) under suitable assumptions about Ω and f] does not suffice to guarantee that $u \in H^{2,\infty}_{\text{loc}}(\Omega)$, no matter how regular the coefficients of L are: see the example following Theorem 3.5. On the other hand, we already stressed that essential boundedness is a regularity threshold for Lu. Thus, the membership of u in $H^{2,\infty}(\Omega)$ does not follow from the regularity of Lu via the regularity theory for equations; it can instead be proven by directly exhibiting L^∞ bounds on second derivatives, as we shall do in the proof of the next result.

THEOREM 4.38. *Take $\partial\Omega$ of class $C^{3,1}$, $\Gamma = \emptyset$, and $a^{ij} \in C^{1,1}(\bar{\Omega})$, $d^j \in C^{1,\delta}(\bar{\Omega})$, b^i, $c, f \in C^{0,\delta}(\bar{\Omega})$ $(0 < \delta < 1)$. Let $\psi \in H^{2,\infty}(\Omega)$ with $\psi|_{\partial\Omega} \geq 0$. Then a solution of the unilateral v.i.* (4.42) *(with $F = f$) belongs to $H^{2,\infty}(\Omega)$.*

PROOF. *Step* 1: *Preliminary reductions.* Since

$$\sum_{i=1}^{N-1} \int_\Omega (a^{iN} u_{x_i} v_{x_N} + a^{Ni} u_{x_N} v_{x_i})\, dx$$
$$= \sum_{i=1}^{N-1} \int_\Omega [(a^{iN} + a^{Ni}) u_{x_i} v_{x_N} + (a^{Ni}_{x_N} u_{x_i} - a^{Ni}_{x_i} u_{x_N}) v]\, dx \quad \text{for } v \in H_0^1(\Omega),$$

we can without loss of generality assume

$$a^{Ni} = 0 \qquad \text{for } i = 1, \ldots, N-1. \tag{4.63}$$

Moreover, it is not restrictive to take $a(u, v) = \int_\Omega a^{ij} u_{x_i} v_{x_j}\, dx$, since in the general case f can be replaced by $f + (d^j u)_{x_j} - b^i u_{x_i} - cu$ which still belongs to $C^{0,\delta}(\bar{\Omega})$ by Lemma 4.34 and Theorem 1.41. The solution of (4.42) is then unique by the coerciveness of $a(u, v)$ on $H_0^1(\Omega)$.

Let G be a smooth function of $t \in \mathbb{R}$ such that $G' \in L^\infty(\mathbb{R})$ and

$$\begin{gathered} G(t) = 0 \qquad \text{if and only if } t \leq 0, \\ G'(t) \geq 0 \qquad \text{for all } t, \qquad G''(t) \geq 0 \qquad \text{for } t \leq 1. \end{gathered} \tag{4.64}$$

It is easy to check that the mapping $\beta: w \mapsto G(w - \psi)$ is a penalty operator associated with the convex set intervening in our v.i., so that the solution u is the weak limit in $H_0^1(\Omega)$ of the sequence $\{u_\varepsilon\}$ defined by

$$u_\varepsilon \in H_0^1(\Omega), \qquad Lu_\varepsilon + G_\varepsilon(u_\varepsilon - \psi) = f \qquad \text{in } \Omega,$$

with $G_\varepsilon \equiv \varepsilon^{-1} G$, $\varepsilon > 0$ (see Theorem 4.19 and the remark following it). A straightforward bootstrap argument based on Lemma 1.57 and Theorem 3.17 shows that $u_\varepsilon \in C^{2,\delta}(\bar{\Omega})$: we shall demonstrate the theorem by providing a bound, independent of ε, on $| u_\varepsilon |_{H^{2,\infty}(\Omega)}$. For the moment we claim that

$$| Lu_\varepsilon |_{\infty;\Omega} \leq C \tag{4.65}$$

and therefore [Theorem 3.17(ii)]

$$| u_\varepsilon |_{H^{2,p}(\Omega)} \leq C(p) \qquad \text{for all finite } p. \tag{4.66}$$

To prove (4.65) we fix any $q > 2$ and utilize the equation for u_ε to obtain

$$\begin{aligned} &\int_\Omega L(u_\varepsilon - \psi)[G_\varepsilon(u_\varepsilon - \psi)]^{q-1}\, dx + \int_\Omega [G_\varepsilon(u_\varepsilon - \psi)]^q\, dx \\ &\qquad = \int_\Omega (f - L\psi)[G_\varepsilon(u_\varepsilon - \psi)]^{q-1}\, dx. \end{aligned}$$

Since $(u_\varepsilon - \psi)|_{\partial\Omega} = -\psi|_{\partial\Omega} \leq 0$ and therefore $G_\varepsilon(u_\varepsilon - \psi)|_{\partial\Omega} = 0$, the function $G_\varepsilon(u_\varepsilon - \psi)$ belongs to $H_0^1(\Omega)$ and the first integral on the left-hand side above equals

$$(q-1)\int_\Omega a^{ij}(u_\varepsilon - \psi)_{x_i}(u_\varepsilon - \psi)_{x_j}[G_\varepsilon(u_\varepsilon - \psi)]^{q-2} G_\varepsilon'(u_\varepsilon - \psi)\, dx.$$

By uniform ellipticity and (4.64) we arrive at

$$\int_\Omega [G_\varepsilon(u_\varepsilon - \psi)]^q\, dx \leq \int_\Omega | f - L\psi | \, [G_\varepsilon(u_\varepsilon - \psi)]^{q-1}\, dx$$
$$\leq \left(\int_\Omega | f - L\psi |^q\, dx\right)^{1/q} \left\{\int_\Omega [G_\varepsilon(u_\varepsilon - \psi)]^q\, dx\right\}^{1-1/q},$$

that is,

$$| G_\varepsilon(u_\varepsilon - \psi) |_{q;\Omega} \leq | f - L\psi |_{q;\Omega},$$

and finally at

$$| Lu_\varepsilon - f |_{\infty;\Omega} = | G_\varepsilon(u_\varepsilon - \psi) |_{\infty;\Omega} \leq | f - L\psi |_{\infty;\Omega}$$

after letting $q \to \infty$. This yields (4.65). Note that, by Sobolev inequalities, (4.66) implies $u_\varepsilon \to u$ in $C^{1,\gamma}(\bar{\Omega})$ for all $\gamma \in [0, 1[$, in particular $u_\varepsilon \leq \psi + 1$ and therefore $G_\varepsilon''(u_\varepsilon - \psi) \geq 0$ if ε is small enough.

It will be convenient to have L replaced by $L_0 \equiv -a^{ij}\partial^2/\partial x_i \partial x_j = L + a^{ij}_{x_j}\partial/\partial x_i$. Thus u_ε solves

$$u_\varepsilon \in H_0^1(\Omega), \qquad L_0 u_\varepsilon + G_\varepsilon(u_\varepsilon - \psi) = f_\varepsilon \qquad \text{in } \Omega$$

with $f_\varepsilon \equiv f + a^{ij}_{x_j} u_{\varepsilon x_i}$; note that by (4.65) and (4.66),

$$| L_0 u_\varepsilon |_{\infty;\Omega} \leq C \tag{4.67}$$

and

$$| f_\varepsilon |_{C^{0,\delta}(\bar{\Omega})} \leq C$$

with constants independent of ε.

Step 2: *Interior bounds*. We fix $\omega \subset\subset \Omega$ and proceed to obtain a uniform bound on $| u_\varepsilon |_{H^{2,\infty}(\omega)}$. This we do in the special case of solutions to

$$L_0 u_\varepsilon + G_\varepsilon(u_\varepsilon - \psi) = K \qquad \text{in } \Omega, \tag{4.68}$$

K being some constant, under the assumption

$$L_0 \psi \geq 1 \qquad \text{in } \Omega. \tag{4.69}$$

[In the general case we need only replace u_ε by $u_\varepsilon + w_{\varepsilon K}$ and ψ by $\psi + w_{\varepsilon K}$, where

$$w_{\varepsilon K} \in H_0^1(\Omega), \qquad L_0 w_{\varepsilon K} = K - f_\varepsilon$$

with K large enough: $w_{\varepsilon K}$ is in $C^{2,\delta}(\bar{\Omega})$ with norm bounded independently of ε.] By a bootstrap argument the validity of (4.68) implies $u_\varepsilon \in H^{4,p}(\Omega)$ for every finite p (Problem 3.11).

Let r, s be arbitrarily fixed in the range from 1 to N and let $\lambda > 0$ be so small that

$$a^{ij}\xi_i\xi_j \pm \lambda\xi_r\xi_s \geq 0 \text{ in } \bar{\Omega} \text{ for } \xi \in \mathbb{R}^N, \quad \pm\lambda\psi_{x_rx_s} \geq -1 \text{ a.e. in } \Omega. \tag{4.70}$$

By taking partial derivatives in (4.68) we easily see that the function $\tilde{u} \equiv L_0u_\varepsilon + \lambda u_{\varepsilon x_rx_s}$ satisfies an equation of the form

$$L_0\tilde{u} + G_\varepsilon'\tilde{u} = G_\varepsilon'\tilde{\psi} + G_\varepsilon''\tilde{a}^{ij}(u_\varepsilon - \psi)_{x_i}(u_\varepsilon - \psi)_{x_j} + h^0 + h^i_{x_i} \text{ in } \Omega.$$

Here G_ε' stands for $G_\varepsilon'(u_\varepsilon - \psi)$ and G_ε'' for $G_\varepsilon''(u_\varepsilon - \psi)$, whereas $\tilde{\psi}$ is given by $L_0\psi + \lambda\psi_{x_rx_s}$, $\tilde{a}^{ij}$ by $a^{ij} - \lambda\delta^{ir}\delta^{js}$ ($= a^{rs} - \lambda$ if $i = r$ and $j = s$, $= a^{ij}$ otherwise); finally, the functions $h^0 \in C^0(\bar{\Omega})$, $h^i \in C^0(\bar{\Omega}) \cap H^1(\Omega)$ depend on $\tilde{u}$, hence on ε, with

$$| h^j |_{p;\Omega} \leq C(1 + | u_\varepsilon |_{H^{2,p}(\Omega)}) \qquad \text{for any } p \leq \infty, \qquad j = 0, 1, \ldots, N,$$

with C independent of ε. Since both G_ε' and G_ε'' are ≥ 0, so are also $G_\varepsilon'\tilde{\psi}$ and $G_\varepsilon''\tilde{a}^{ij}(u_\varepsilon - \psi)_{x_i}(u_\varepsilon - \psi)_{x_j}$ by (4.69), (4.70). Let $\tilde{U} \equiv g\tilde{u}$ with $g \in C_c^\infty(\Omega)$, $0 \leq g \leq 1$ in Ω, $g = 1$ on $\bar{\omega}$. Since

$$L_0\tilde{U} = gL_0\tilde{u} + \tilde{u}L_0g + 2\tilde{u}(a^{ij}g_{x_j})_{x_i} - 2(a^{ij}\tilde{u}g_{x_j})_{x_i},$$

$\tilde{U}$ satisfies a differential inequality of the form

$$L_0\tilde{U} + G_\varepsilon'\tilde{U} \geq H^0 + H^i_{x_i} \tag{4.71}$$

in Ω, where the H^j's have the same properties as the h^j's. This means that $\tilde{U}$ is a function from $H_0^1(\Omega)$ satisfying

$$\langle L_0\tilde{U} + G_\varepsilon'\tilde{U}, v\rangle = \int_\Omega [a^{ij}\tilde{U}_{x_i}v_{x_j} + (a^{ij}_{x_j}\tilde{U}_{x_i} + G_\varepsilon'\tilde{U})v]\,dx$$
$$\geq \int_\Omega (H^0v + H^iv_{x_i})\,dx \qquad \text{for } v \in H_0^1(\Omega), \qquad v \geq 0.$$

We have no control over $| G_\varepsilon' |_{\infty;\Omega}$ as ε varies, but since $G_\varepsilon' \geq 0$, we can avail ourselves of the remark after Lemma 2.8. Thus $\tilde{U}$, and by (4.67) $\lambda gu_{\varepsilon x_rx_s}$ as well, is bounded from below by a constant depending on ε only through the bound on the norms $| H^j |_{p;\Omega}$, hence on the norms $| u_\varepsilon |_{H^{2,p}(\Omega)}$, for p finite and large enough. Since the same conclusion holds with λ replaced by $-\lambda$, we have obtained a uniform bound on $| u_\varepsilon |_{H^{2,\infty}(\omega)}$.

Step 3: *Global bounds.* Let $\Lambda: \bar{U} \to \bar{B}$ be a $C^{3,1}$ diffeomorphism that straightens a portion $U \cap \partial\Omega$ of $\partial\Omega$, U being some bounded open subset of $\mathbb{R}^N$. In the new local coordinates $y = \Lambda(x)$ the function $\hat{u}_\varepsilon \equiv (u \circ \Lambda^{-1})|_{B^+}$ satisfies

$$\hat{u}_\varepsilon \in H_0^1(B^+ \cup S^+), \qquad \hat{L}_0\hat{u}_\varepsilon + G_\varepsilon(\hat{u}_\varepsilon - \hat{\psi}) = \hat{f}_\varepsilon \qquad \text{in } B^+:$$

here $\hat{L}_0 \equiv -\hat{a}^{hk}\partial^2/\partial y_h \partial y_k$ with $\hat{a}^{hk}(y) \equiv a^{ij}[\Lambda^{-1}(y)]y_{hx_i}[\Lambda^{-1}(y)]y_{kx_j}[\Lambda^{-1}(y)]$, $\hat{\psi} \equiv (\psi \circ \Lambda^{-1})|_{B^+}$, and $\hat{f}_\varepsilon$ is a function from $C^{0,\delta}(\overline{B^+})$ bounded in norm by a quantity $C(1 + |\hat{u}_\varepsilon|_{C^{1,\delta}(\overline{B^+})})$, hence [see (4.66)] by a constant independent of ε. For the purpose of providing a bound on $|u_\varepsilon|_{H^{2,\infty}(\Lambda^{-1}(B^+_{1/2}))}$, or equivalently on $|\hat{u}_\varepsilon|_{H^{2,\infty}(B^+_{1/2})}$, we may safely replace $\hat{u}_\varepsilon$ by $\hat{u}_\varepsilon + \hat{w}_{\varepsilon K}$ and $\hat{\psi}$ by $\hat{\psi} + \hat{w}_{\varepsilon K}$, where $\hat{w}_{\varepsilon K}$ solves

$$\hat{w}_{\varepsilon K} \in H_0^1(B^+), \qquad \hat{L}_0\hat{w}_{\varepsilon K} = K - \hat{f}_\varepsilon \qquad \text{in } B^+$$

and therefore belongs to $C^{2,\delta}(\overline{B_R{}^+})$, with

$$|\hat{w}_{\varepsilon K}|_{C^{2,\delta}(\overline{B_R{}^+})} \leq C(R) \text{ independent of } \varepsilon,$$

whatever $R \in]0, 1[$ [Theorem 3.13(ii)]. These considerations show that the bound on $H^{2,\infty}(B^+_{1/2})$ norms need only be proven in the special case when $U \cap \Omega = B^+$, $U \cap \partial\Omega = S^0$, and u_ε satisfies

$$u_\varepsilon \in H_0^1(B^+ \cup S^+), \qquad L_0u_\varepsilon + G_\varepsilon(u_\varepsilon - \psi) = K \qquad \text{in } B^+$$

with K so large that

$$L_0\psi \geq 1 \qquad \text{in } B^+;$$

for $R \in]0, 1[$ u_ε belongs to $H^{4,p}(B_R{}^+)$ whenever p is finite (Problem 3.11), and the above equation is satisfied at every point of $B^+ \cup S^0$.

We now take $g \in C_c^\infty(B)$ with $0 \leq g \leq 1$ in B, $g = 1$ on $\bar{B}_{1/2}$ and arrive again at (4.71), this time in B^+, for the function $\tilde{U}$ defined correspondingly.

When r and s are both fixed in the range from 1 to $N - 1$, minor changes in the techniques of Step 2 yield a uniform bound on $|u_{\varepsilon x_r x_s}|_{\infty;1/2;+}$. Indeed, $\tilde{U}$ vanishes near S^+ and is $\geq -|L_0u_\varepsilon|_{\infty;+}$ on S^0 because $u_{\varepsilon x_r x_s}|_{S^0} = 0$: since $\tilde{U} + |L_0u_\varepsilon|_{\infty;+}$ satisfies the same inequality (4.71) as $\tilde{U}$, the weak maximum principle (Theorem 2.4) implies $\tilde{U} + |L_0u_\varepsilon|_{\infty;+} \geq z$, where

$$z \in H_0^1(B^+), \quad \langle L_0z + G_\varepsilon'z, v\rangle = \int_{B^+} (H^0v + H^iv_{x_i})\,dx \quad \text{for } v \in H_0^1(B^+).$$

Lemma 2.8 applies to z, so that we arrive at a uniform bound from below for $\tilde{U}(x)$, hence for $\pm\lambda g(x)u_{\varepsilon x_r x_s}(x)$, $x \in \overline{B^+}$.

Things become considerably more difficult if one of the two indices r and s, say s, equals N. This is the case we are going to take up now.

Since $(u_\varepsilon - \psi)|_{S^0} \leq 0$ and $G_\varepsilon'(t) = 0$ for $t \leq 0$, we have

$$(L_0 u_\varepsilon)_{x_i} = -G_\varepsilon'(u_\varepsilon - \psi)_{x_i} = 0 \qquad \text{on } S^0$$

for $i = 1, \ldots, N$. This implies, first of all, that the conormal derivative

$$BL_0 u_\varepsilon = -[a^{iN}(L_0 u_\varepsilon)_{x_i}]|_{S^0}$$

vanishes identically. Moreover, since $u_{\varepsilon x_i x_j}|_{S^0} = 0$ for $i, j = 1, \ldots, N-1$, the identities

$$\begin{aligned} a^{iN}u_{\varepsilon x_r x_N x_i} &= (a^{iN}u_{\varepsilon x_i x_N})_{x_r} - a^{iN}_{x_r}u_{\varepsilon x_i x_N} \\ &= -(L_0 u_\varepsilon)_{x_r} - \Big(\sum_{i,j=1}^{N-1} a^{ij}u_{\varepsilon x_i x_j}\Big)_{x_r} - a^{iN}_{x_r}u_{\varepsilon x_i x_N} \end{aligned}$$

[see (4.63)] show that

$$Bu_{\varepsilon x_r x_N} = -(a^{iN}u_{\varepsilon x_r x_N x_i})|_{S^0} = a^{iN}_{x_r}u_{\varepsilon x_i x_N}|_{S^0}.$$

Summing up, we have

$$|\, B\tilde{U}\,|_{\infty;S^0} \leq C\,|\, u_\varepsilon\,|_{H^{2,\infty}(B^+)}.$$

$\tilde{U}$ is a function from $H_0^1(B^+ \cup S^0)$ satisfying (4.71) in B^+, hence

$$\begin{aligned} &\int_{B^+} [a^{ij}\tilde{U}_{x_i}v_{x_j} + (a^{ij}_{x_j}\tilde{U}_{x_i} + G_\varepsilon'\tilde{U})v]\,dx \\ &\qquad \geq \int_{B^+} (H^0 v - H^i v_{x_i})\,dx - \int_{S^0} (H^N|_{S^0} + B\tilde{U})v|_{S^0}\,dx' \\ &\qquad\qquad \text{for } v \in H_0^1(B^+ \cup S^0), \qquad v \geq 0. \end{aligned}$$

In the above inequality the right-hand side is minorized by

$$\int_{B^+} (H^0 v - H^i v_{x_i})\,dx - M_\varepsilon \int_{S^0} v|_{S^0}\,dx' = \int_{B^+} (H^0 v - H^i v_{x_i} + M_\varepsilon v_{x_N})\,dx,$$

where M_ε is a quantity $C(1 + |\, u_\varepsilon\,|_{H^{2,\infty}(B^+)})$.

Without loss of generality we now assume $|\,\tilde{U}\,|_{2;+} = 1$. Denoting by $B^+(k)$ the set $\{x \in B^+ \mid \tilde{U}(x) < -k\}$ we recall, from the proof of Lemma

2.8 (see Lemma 2.9) that $| B^+(k_1 + \hat{k}) | = 0$, i.e., $\tilde{U} \geq -k_1 - \hat{k}$, provided k_1 is suitably large, say $k_1 \geq C_0$, and $\hat{k}$ is chosen accordingly. We can always suppose $C_0 \geq 1$, $C_0 \geq 2 | L_0 u_\varepsilon |_{\infty;+}$ for every ε. Take $\lambda > 0$. Then in $B^+(k_1)$ we have

$$g u_{\varepsilon x_r x_N} \leq (-k_1 - g L_0 u_\varepsilon)/\lambda \leq -k_1/2\lambda,$$

so that

$$\begin{aligned} | B^+(k_1) | &\leq \text{meas}\{x \in B^+ \big| \, | g u_{\varepsilon x_r x_N}(x) | \geq k_1/2\lambda\} \\ &\leq (2\lambda)^p k_1^{-p} \int_{B^+} | u_{\varepsilon x_r x_N} |^p \, dx \end{aligned}$$

for any finite p. As for $\hat{k}$, in the present situation it is bounded from above by a quantity

$$C(k_1 + | u_\varepsilon |_{H^{2,\infty}(B^+)}) \, | B^+(k_1) |^\eta,$$

where C and η are some positive constants independent of ε. Therefore,

$$\tilde{U} \geq -k_1 - C_1(1 + | u_\varepsilon |_{H^{2,\infty}(B^+)}) k_1^{1-\eta p} \qquad \text{(where } C_1 \text{ depends on } p).$$

At this point we take

$$k_1 = [(C_0^2 + C_1)(1 + | u_\varepsilon |_{H^{2,\infty}(B^+)})]^{1/2}$$

and arrive at

$$\begin{aligned} \tilde{U} \geq &-[(C_0^2 + C_1)(1 + | u_\varepsilon |_{H^{2,\infty}(B^+)})]^{1/2} \\ &-[(C_0^2 + C_1)(1 + | u_\varepsilon |_{H^{2,\infty}(B^+)})]^{1+(1-\eta p)/2}, \end{aligned}$$

hence

$$\tilde{U} \geq -C(1 + | u_\varepsilon |_{H^{2,\infty}(B^+)})^{1/2}$$

if p is chosen large enough, and finally

$$\pm \lambda g(x) u_{\varepsilon x_r x_N}(x) \geq -C(1 + | u_\varepsilon |_{H^{2,\infty}(B^+)})^{1/2}, \qquad x \in \overline{B^+}, \tag{4.72}$$

for $r = 1, \ldots, N-1$. But since

$$a^{NN} u_{\varepsilon x_N x_N} = -\sum_{r=1}^{N-1} a^{rN} u_{\varepsilon x_r x_N} - K \qquad \text{on } S^0,$$

the same technique utilized for the case $r, s < N$ shows that (4.72) holds for $r = N$ as well. Summing up, we have proved

$$| u_\varepsilon |_{H^{2,\infty}(B^+_{1/2})} \leq C(1 + | u_\varepsilon |_{H^{2,\infty}(B^+)})^{1/2}. \tag{4.73}$$

Now let $\Omega = \bigcup_{j=0}^{m} \omega_j$, where $\omega_0 \subset\subset \Omega$ and $\omega_i = U_i \cap \Omega$ with $\bar{U}_i = \Lambda_i^{-1}(\bar{B}_{1/2})$, Λ_i being a diffeomorphism of class $C^{3,1}$, for $i = 1, \ldots, m$. From Step 2 and (4.73) it is easy to deduce

$$| u_\varepsilon |_{H^{2,\infty}(\Omega)} \leq C(1 + | u_\varepsilon |_{H^{2,\infty}(\Omega)})^{1/2},$$

and this yields the desired bound on $| u_\varepsilon |_{H^{2,\infty}(\Omega)}$. □

4.6.2. H^2 Regularity up to Γ under General Conditions

In Lemma 4.34 we imposed a very strong condition on $B\psi^1, \ldots, B\psi^m$ and ζ in order to extend the $H^{2,p}$ regularity of u up to Γ. We cannot do any better than this as long as we confine ourselves to the use of (4.62), which can at most yield $Bu \in L^\infty(\Gamma)$ if the inequalities $B\psi^h \geq \zeta$ are not assumed: (4.62) is therefore insufficient to guarantee the H^2 regularity of u near Γ no matter how regular Lu is. The following question now naturally arises: can a general criterion be given for the $H^{2,p}$ regularity near Γ of a solution u to (4.42), aside from (4.62) degenerating into $Bu = \zeta \in H^{1/p',p}(\Gamma)$?

The next example shows that the answer is negative when p is "too large."

EXAMPLE. Let $N = 2$ and set

$$u(x_1, x_2) \equiv -\chi(| z |^2) \operatorname{Re} z^{3/2} = -\chi(\varrho^2)\varrho^{3/2} \cos(3\theta/2)$$

with $z = x_1 + ix_2 = \varrho \exp(i\theta)$ $(i^2 = -1)$, χ being a smooth function on $[0, \infty[$ such that $0 \leq \chi \leq 1$, $\chi(r) = 1$ for $0 \leq r \leq 1/4$, $\chi(r) = 0$ for $r \geq 1$. In $\mathbb{R}^2 \setminus \{0\}$ the function $\operatorname{Re} z^{3/2}$ is harmonic; moreover, when $x_2 = 0$ the Cauchy–Riemann equations yield

$$\frac{\partial}{\partial x_2} \operatorname{Re} z^{3/2} = -\frac{\partial}{\partial x_1} \operatorname{Im} z^{3/2} = \begin{cases} 0 & \text{for } x_1 > 0, \\ -\frac{3}{2} | x_1 |^{1/2} & \text{for } x_1 < 0. \end{cases}$$

Let Ω be such that $\Omega \cap B = B^+$, with $\partial\Omega$ regular: then u belongs to $C^1(\bar{\Omega}) \cap H^2(\Omega)$, and by the above considerations the function $f \equiv -\Delta u$ is in $C^1(\bar{\Omega})$. Moreover, u satisfies

$$u|_{\partial\Omega} \leq 0, \qquad \left.\frac{\partial u}{\partial \nu}\right|_{\partial\Omega} \leq 0, \qquad u|_{\partial\Omega} \left.\frac{\partial u}{\partial \nu}\right|_{\partial\Omega} = 0.$$

[Note that for $-1 < x_1 < 1$ and $x_2 = 0$,

$$\left.\frac{\partial u}{\partial \nu}\right|_{\partial\Omega} = -\frac{\partial u}{\partial x_2} = \chi(|z|^2)\frac{\partial}{\partial x_2}\operatorname{Re} z^{3/2}.\Big]$$

Now let $\psi \in C^{2,\delta}(\bar{\Omega})$, $0 < \delta < 1$, solve the b.v.p.

$$-\Delta\psi = f \text{ in } \Omega, \qquad \psi = 0 \qquad \text{on } \partial\Omega$$

[see Theorem 3.17(iii)]: by the weak maximum principle (see Theorem 2.4), $\psi \geq u$ on $\bar{\Omega}$. This shows that u satisfies (4.44) with $\Gamma = \partial\Omega$, $L = -\Delta$ (and therefore $B = \partial/\partial\nu$), $\zeta = 0$. However, u does not belong to $H^{2,4}(\Omega)$ —nor to $C^{1,\delta}(\bar{\Omega})$ if $\delta > 1/2$, for the matter.

In the light of the above, we are left with the task of investigating $H^{2,p}$ regularity near Γ only when p is close to 2. We take $p = 2$ and begin by studying (4.45) instead of (4.42). Of course, we need only consider the case $V = H^1(\Omega)$.

THEOREM 4.39. *Let $\partial\Omega = \Gamma$ be of class $C^{1,1}$. Assume that a^{ij}, d^j are Lipschitz continuous on $\bar{\Omega}$, that F is given by* (4.43) *with $\zeta \in H^{1/2}(\partial\Omega)$, and that $\hat{\psi} = \psi|_{\partial\Omega}$ with $\psi \in H^2(\Omega)$. Then any solution u of* (4.45) *belongs to $H^2(\Omega)$; moreover,*

$$|u|_{H^2(\Omega)} \leq C(|f|_{2;\Omega} + |\zeta|_{H^{1/2}(\partial\Omega)} + |\psi|_{H^2(\Omega)} + |u|_{H^1(\Omega)}), \qquad (4.74)$$

where C (independent of u, F, ψ) depends on the coefficients of $a(u, v)$ through the constant of uniform ellipticity α and the bound imposed on

$$|a^{ij}, d^j|_{C^{0,1}(\bar{\Omega})}, \qquad |b^i, c|_{\infty;\Omega}.$$

PROOF. For $i = 1, \ldots, N$ let $n^i \in C^{0,1}(\bar{\Omega})$ be such that $n^i|_{\partial\Omega} = \nu^i$. If $z \in H^1(\Omega)$ is such that $z|_{\partial\Omega} = \zeta$ and $|z|_{H^1(\Omega)} = |\zeta|_{H^{1/2}(\partial\Omega)}$, the functions $f^i \equiv n^i z$ satisfy

$$\int_\Omega (f^i_{x_i} v + f^i v_{x_i})\,dx = \int_{\partial\Omega} \zeta v|_{\partial\Omega}\,d\sigma \qquad \text{for } v \in V$$

and therefore

$$\langle F, v\rangle = \int_\Omega (f^0 v + f^i v_{x_i})\,dx \qquad \text{for } v \in V$$

with $f^0 \equiv f + f^i_{x_i}$, as well as

$$|f^0|_{2;\Omega} + \sum_{i=1}^{N} |f^i|_{H^1(\Omega)} \leq C(|f|_{2;\Omega} + |\zeta|_{H^{1/2}(\partial\Omega)}).$$

Set

$$\varkappa(F; u) \equiv |f^0|^2_{2;\Omega} + \sum_{i=1}^{N} |f^i|^2_{H^1(\Omega)} + |u|^2_{H^1(\Omega)}.$$

Since $u - \psi$ satisfies (4.46) with f replaced by $f - L\psi$, $\psi|_{\partial\Omega}$ by 0 and ζ by $\zeta - B\psi$, we can without loss of generality assume $\hat{\psi} = 0$.

Moreover, since $f^0 - b^i u_{x_i} - cu \in L^2(\Omega)$ and $f^i - d^i u \in H^1(\Omega)$ with norms bounded by $C\varkappa(F; u)^{1/2}$, we can also assume that all coefficients of $a(u, v)$ except the leading ones vanish identically.

Let us consider the special case when $\Omega \cap B = B^+$, $\partial\Omega \cap B = S^0$. It is clear that u, when restricted to B^+, satisfies

$$u \in H^1(B^+), \qquad u \leq 0 \qquad \text{on } S^0,$$

$$\int_{B^+} a^{ij} u_{x_i} (v - u)_{x_j}\, dx \geq \langle F, v - u \rangle \tag{4.75}$$

$$\text{for } v \in H^1(B^+), \qquad \operatorname{supp}(v - u) \subset B^+ \cup S^0, \qquad v \leq 0 \qquad \text{on } S^0.$$

Fix $s = 1, \ldots, N - 1$ and write

$$\delta_h w(x) \equiv \delta_h^s w(x) \equiv h^{-1}[w(x + he^s) - w(x)]$$

for $h \in \mathbb{R}\setminus\{0\}$, e^s being the sth unit coordinate vector. Let $g \in C^\infty(\mathbb{R}^N)$ with $\operatorname{supp} g \subset B$, $0 \leq g \leq 1$, $g = 1$ on $\overline{B_R}$ with $0 < R < 1$. For $0 < |h| < \operatorname{dist}(\operatorname{supp} g, S^+)$ the function

$$v_h(x) \equiv u(x) + \frac{\varepsilon g(x)}{h^2} [u(x + he^s)g(x + he^s) + u(x - he^s)g(x - he^s) - 2u(x)g(x)]$$

satisfies

$$v_h \leq u[1 - 2\varepsilon g^2/h^2] \leq 0 \qquad \text{on } S^0$$

provided $\varepsilon \leq h^2/2$. We can therefore insert $v = v_h$ in (4.75) and obtain

$$\int_{B^+} a^{ij} u_{x_i} [g\delta_{-h}\delta_h(gu)]_{x_j}\, dx \geq \langle F, g\delta_{-h}\delta_h(gu) \rangle.$$

On the other hand,

$$\int_{B^+} a^{ij} u_{x_i} (gv)_{x_j}\, dx = \langle F, gv\rangle \qquad \text{for } v \in H_0^1(B^+).$$

Since

$$\int_{B^+} a^{ij} u_{x_i} (gw)_{x_j}\, dx = \int_{B^+} [a^{ij}(gu)_{x_i} w_{x_j} + a^{ij} u_{x_i} g_{x_j} w - a^{ij} u g_{x_i} w_{x_j}]\, dx$$

and

$$\langle F, gw\rangle = \int_{B^+} [(f^0 g + f^i g_{x_i}) w + f^i g w_{x_i}]\, dx$$

whenever $w \in H^1(B^+)$, we are in the situation considered in Remark 1 after Lemma 2.22 with u replaced by gu, f^0 by $f^0 g - a^{ij} u_{x_i} g_{x_j} + f^i g_{x_i} \in L^2(B^+)$, f^i by $f^i g + a^{ji} u g_{x_j} \in H^1(B^+)$. Hence $gu \in H^2(B^+)$, which implies $u|_{B_R{}^+} \in H^2(B_R{}^+)$, with

$$| u |^2_{H^2(B_R{}^+)} \leq | gu |^2_{H^2(B^+)} \leq C\varkappa(F; u).$$

Now let U be a bounded domain of $\mathbb{R}^N$ such that $U \cap \partial\Omega$ is straightened by a $C^{1,1}$ diffeomorphism $\Lambda: \bar{U} \to \bar{B}$. Then the function $u' \equiv (u \circ \Lambda^{-1})|_{B^+}$ solves a problem such as (4.75) and $u|_\omega$, where $\omega \equiv \Lambda^{-1}(B_R{}^+)$, belongs to $H^2(\omega)$ with

$$| u |^2_{H^2(\omega)} \leq C\varkappa(F; u).$$

Finally, we write Ω as $\bigcup_{j=0}^m \omega_j$, where $\omega_1, \ldots, \omega_m$ are chosen by the same criterion illustrated above for ω and $\omega_0 \subset\subset \Omega$: the full conclusion of the theorem is obtained by patching together the H^2 regularity results and estimates on $\omega_1, \ldots, \omega_m$ as well as on ω_0 (see Lemma 2.21). ☐

Returning to (4.42) we have the following theorem.

THEOREM 4.40. *Same assumptions about* $\partial\Omega = \Gamma$, a^{ij}, d^j *and* ζ *as in Theorem* 4.39. *Let* $\psi \in H^2(\Omega)$. *Then any solution* u *of* (4.42) *with* F *given by* (4.43) *belongs to* $H^2(\Omega)$ *with norm estimate* (4.74).

PROOF. Since, by Lemma 4.33, u satisfies (4.46) with f replaced by $Lu \in L^2(\Omega)$, we are led back to the situation investigated in Theorem 4.39. ☐

4.7. Regularity in Morrey and Campanato Spaces

4.7.1. The Case of Continuous Leading Coefficients

We are going to give sufficient conditions in order that all first derivatives of a solution u to (4.42) belong to $L^{2,\mu}(\Omega)$ whenever all first derivatives of the obstacle ψ do; this will lead to the membership of u in $C^{0,\delta}(\bar{\Omega})$, with $\delta = (\mu + 2 - N)/2$, if $N - 2 < \mu < N$, and even in $C^{1,\delta}(\bar{\Omega})$ with $\delta = (\mu - N)/2$, if $N < \mu < N + 2$ [see Theorems 1.17(ii) and 1.40].

THEOREM 4.41. *Assume $\partial\Omega$ of class C^1, Γ closed, $a^{ij} \in C^0(\bar{\Omega})$. Take μ in $]0, N[$ and let u solve* (4.42) *with*

$$\langle F, v\rangle = \int_\Omega (f^0 v + f^i v_{x_i})\, dx \qquad \text{for } v \in V,$$

where $f^0 \in L^{2,(\mu-2)^+}(\Omega)$, $f^1, \ldots, f^N \in L^{2,\mu}(\Omega)$, $\psi \in H^1(\Omega)$, $\psi|_{\partial\Omega\setminus\Gamma} \geq 0$, $\psi_{x_i} \in L^{2,\mu}(\Omega)$ for $i = 1, \ldots, N$. Then all first derivatives of u belong to $L^{2,\mu}(\Omega)$ with norm estimate

$$\begin{aligned} |\nabla u|_{2,\mu;\Omega} \leq C\Big(&|f^0|_{2,(\mu-2)^+;\Omega} + \sum_{i=1}^N |f^i|_{2,\mu;\Omega} \\ &+ |\psi|_{2;\Omega} + |\nabla\psi|_{2,\mu;\Omega} + |u|_{H^1(\Omega)}\Big), \end{aligned} \tag{4.76}$$

where C depends on the coefficients of $a(u, v)$ through the bound imposed on their $L^\infty(\Omega)$ norms, as well as through the constant α of uniform ellipticity and the modulus τ of uniform continuity of the a^{ij}'s.

PROOF. *Step* 1: *Preliminary reductions.* We need only prove the theorem under the additional assumption that $u \in L^{2,\mu}(\Omega)$ with $|u|_{2,\mu;\Omega}$ bounded by a quantity such as the right-hand side of (4.76). [This assumption is certainly satisfied for $\mu \leq 2$ by the mere membership of u in $H^1(\Omega)$: see Theorem 1.40.] For, if it is only known that $u \in L^{2,\mu'}(\Omega)$ with $\mu' < \mu$, then the theorem itself with μ replaced by μ' yields $u \in L^{2,\mu'+2}(\Omega)$ with norm estimate by Theorem 1.40, so that we can again arrive at the conclusion of the theorem with μ replaced by $\min(\mu, \mu' + 2)$, etc.

At this point it is not restrictive to assume $d^j = 0$, since $f^j - d^j u$ has the same regularity as f^j. Nor is it restrictive to assume that the bilinear form is coercive on V: in the general case, we need only replace the coefficient c by $c + \lambda$ and the free term f^0 by $f^0 + \lambda u$, with λ large enough. Finally, it is not restrictive to take $F = 0$, since we can always replace u

by $u - \hat{u}$ and ψ by $\psi - \hat{u}$, with $\hat{u}$ defined by

$$\hat{u} \in V, \qquad a(\hat{u}, v) = \langle F, v \rangle \qquad \text{for } v \in V \tag{4.77}$$

[the bilinear form being coercive; see Theorem 3.16(i) for what concerns the regularity of $\hat{u}$].

Step 2: Interior regularity. Let $\omega \subset\subset \Omega$, $x^0 \in \bar{\omega}$, $0 < r \leq d \equiv [\operatorname{dist}(\omega, \partial\Omega)] \wedge 1/2$, and solve

$$z \in H_0^1(B_r(x^0)),$$

$$\int_{B_r(x^0)} a_0^{ij} z_{x_i} v_{x_j}\, dx = \int_{B_r(x^0)} a_0^{ij} u_{x_i} v_{x_j}\, dx \qquad \text{for } v \in H_0^1(B_r(x^0)),$$

$a_0^{ij} \equiv a^{ij}(x^0)$, with the help of Theorem 2.1 and of the corollary of Theorem 1.43. Then (4.42) with $d^j = 0$, $F = 0$ yields

$$\begin{aligned}
&\int_{B_r(x^0)} a_0^{ij} z_{x_i} (v - u)_{x_j}\, dx \\
&\quad = \int_{B_r(x^0)} [a^{ij} u_{x_i} (v - u)_{x_j} + (a_0^{ij} - a^{ij}) u_{x_i} (v - u)_{x_j}]\, dx \\
&\quad \geq - \int_{B_r(x^0)} [(b_i u_{x_i} + cu)(v - u) + (a^{ij} - a_0^{ij}) u_{x_i} (v - u)_{x_j}]\, dx
\end{aligned}$$

whenever $v \in H^1(B_r(x^0))$ with $v \leq \psi$ in $B_r(x^0)$ and $v - u \in H_0^1(B_r(x^0))$: note that the trivial extension of $v - u$ to Ω is in V, so that v is the restriction to $B_r(x^0)$ of a function from V which equals u throughout $\Omega \setminus B_r(x^0)$.

We now set $w \equiv u - z$, so that $v - u = v - w - z$, and obtain

$$\begin{aligned}
\int_{B_r(x^0)} a_0^{ij} z_{x_i} z_{x_j}\, dx \leq \int_{B_r(x^0)} &\{(b^i u_{x_i} + cu)(v - w - z) \\
&+ (a^{ij} - a_0^{ij}) u_{x_i} [(v - w)_{x_j} - z_{x_j}] \\
&+ a_0^{ij} z_{x_i} (v - w)_{x_j}\}\, dx.
\end{aligned} \tag{4.78}$$

Poincaré's inequality applies to the functions $v - w$, $z \in H_0^1(B_r(x^0))$ (see the corollary of Theorem 1.43 again): for $\varepsilon > 0$ we have

$$\begin{aligned}
&\int_{B_r(x^0)} | (b^i u_{x_i} + cu)(v - w - z) |\, dx \\
&\quad \leq C(| \nabla u |_{2;x^0,r} + | u |_{2;x^0,r})(| v - w |_{2;x^0,r} + | z |_{2;x^0,r}) \\
&\quad \leq C(| \nabla u |_{2;x^0,r} + | u |_{2;x^0,r}) r (| \nabla (v - w) |_{2;x^0,r} + | \nabla z |_{2;x^0,r}) \\
&\quad \leq C(\varepsilon) r^2 (| \nabla u |_{2;x^0,r}^2 + | u |_{2;x^0,r}^2) + \varepsilon (| \nabla (v - w) |_{2;x^0,r}^2 + | \nabla z |_{2;x^0,r}^2).
\end{aligned}$$

The remaining terms on the right-hand side of (4.78) are bounded by a quantity

$$C(\varepsilon)[\tau^2(r) \mid \nabla u \mid_{2;x^0,r}^2 + \mid \nabla(v - w) \mid_{2;x^0,r}^2] + \varepsilon \mid \nabla z \mid_{2;x^0,r}^2.$$

By taking $\varepsilon = \alpha/4$ we can therefore deduce from (4.78) that

$$\begin{aligned} \mid \nabla z \mid_{2;x^0,r}^2 \leq C\{[r^2 + \tau^2(r)] \mid \nabla u \mid_{2;x^0,r}^2 \\ + \mid \nabla(v - w) \mid_{2;x^0,r}^2 + r^{\mu+2} \mid u \mid_{2,\mu;\Omega}^2\}. \end{aligned} \tag{4.79}$$

Let $0 < \varrho \leq r$. The function w belongs to $H^1(B_r(x^0))$ and satisfies

$$\int_{B_r(x^0)} a_0^{ij} w_{x_i} v_{x_j}\, dx = 0 \qquad \text{for } v \in H_0^1(B_r(x^0)),$$

so that Lemma 3.1 yields

$$\mid \nabla w \mid_{2;x^0,\varrho}^2 \leq C \frac{\varrho^N}{r^N} \mid \nabla w \mid_{2;x^0,r}^2 \leq C \frac{\varrho^N}{r^N} (\mid \nabla u \mid_{2;x^0,r}^2 + \mid \nabla z \mid_{2;x^0,r}^2) \tag{4.80}$$

and finally

$$\begin{aligned} \mid \nabla u \mid_{2;x^0,\varrho}^2 &\leq 2(\mid \nabla w \mid_{2;x^0,\varrho}^2 + \mid \nabla z \mid_{2;x^0,\varrho}^2) \\ &\leq C\Big(\frac{\varrho^N}{r^N} \mid \nabla u \mid_{2;x^0,r}^2 + \mid \nabla z \mid_{2;x^0,r}^2\Big) \\ &\leq C\Big\{\Big[\frac{\varrho^N}{r^N} + r^2 + \tau^2(r)\Big] \mid \nabla u \mid_{2;x^0,r}^2 \\ &\qquad + \mid \nabla(v - w) \mid_{2;x^0,r}^2 + r^{\mu+2} \mid u \mid_{2,\mu;\Omega}^2\Big\}. \end{aligned}$$

We now choose $v = w \wedge \psi$, which is admissible since $w \wedge \psi - u = (w - u) \wedge (\psi - u) = (-z) \wedge (\psi - u)$ belongs to $H_0^1(B_r(x^0))$.

The function $w \wedge \psi - w$ belongs to $H_0^1(B_r(x^0))$ and vanishes at all points of $B_r(x^0)$ where $w \leq \psi$, hence satisfies

$$\begin{aligned} &\int_{B_r(x^0)} a_0^{ij}(w \wedge \psi - w)_{x_i}(w \wedge \psi - w)_{x_j}\, dx \\ &\quad = \int_{B_r(x^0)} a_0^{ij}(w \wedge \psi)_{x_i}(w \wedge \psi - w)_{x_j}\, dx \\ &\quad = \int_{B_r(x^0)} a_0^{ij}\psi_{x_i}(w \wedge \psi - w)_{x_j}\, dx, \end{aligned}$$

so that

$$|\nabla(w \wedge \psi - w)|^2_{2;x^0,r} \leq C|\nabla\psi|^2_{2;x^0,r} \leq Cr^\mu |\nabla\psi|^2_{2,\mu;\Omega}.$$

Summing up,

$$|\nabla u|^2_{2;x^0,\varrho} \leq C\biggl\{\biggl[\frac{\varrho^N}{r^N} + r^2 + \tau^2(r)\biggr]|\nabla u|^2_{2;x^0,r} + r^\mu|\nabla\psi|^2_{2,\mu;\Omega} + r^{\mu+2}|u|_{2,\mu;\Omega}\biggr\}.$$

At this point we need only proceed as in Step 1 of the proof of Lemma 3.3 to conclude that $u_{x_1}|_\omega, \ldots, u_{x_N}|_\omega$ belong to $L^{2,\mu}(\omega)$ with norm estimate.

Step 3: Completion of the proof. Let

$$\Omega \cap B = B^+, \qquad \partial\Omega \cap B = (\partial\Omega\setminus\Gamma) \cap B = S^0.$$

Let $x^0 \in S_R{}^0$, $0 < R < 1$, and fix $r \in]0, (1 - R)/2[$. We solve

$$z \in H_0{}^1(B_r{}^+(x^0)),$$

$$\int_{B_r{}^+(x^0)} a_0{}^{ij} z_{x_i} v_{x_j}\,dx = \int_{B_r{}^+(x^0)} a_0{}^{ij} u_{x_i} v_{x_j}\,dx \qquad \text{for } v \in H_0{}^1(B_R{}^+(x^0)),$$

so that $w \equiv u - z$ belongs to $H_0{}^1(B_r{}^+(x^0) \cup S_r{}^+(x^0))$ and satisfies

$$\int_{B_r{}^+(x^0)} a_0{}^{ij} w_{x_i} v_{x_j}\,dx = 0 \qquad \text{for } v \in H_0{}^1(B_r{}^+(x^0)),$$

whereas $w \wedge \psi - u$ belongs to $H_0{}^1(B_r{}^+(x^0))$. We can estimate $|\nabla z|_{2;x^0,r,+}$ by proceeding as in Step 2 of the present proof: the only major difference is that here Poincaré's inequality in $H_0{}^1(B_r{}^+(x^0))$ must be utilized (see the corollary of Theorem 1.43). Next, we estimate $|\nabla w|_{2;x^0,\varrho,+}$ with the help of Lemma 3.9. At this point the same techniques employed in the first two steps of Lemma 3.11 yield $u_{x_1}|_{B_R{}^+}, \ldots, u_{x_N}|_{B_R{}^+} \in L^{2,\mu}(B_R{}^+)$ with norm estimate.

Now suppose that $\Omega \cap B = B^+$, $\partial\Omega \cap B = \Gamma \cap B = S^0$. After fixing $x^0 \in S_R{}^0$ for $0 < R < 1$ and $r \in]0, (1 - R)/2[$ we solve

$$z \in H_0{}^1(B_r{}^+(x^0) \cup S_r{}^0(x^0)),$$

$$\int_{B^+(x^0)} a_0{}^{ij} z_{x_i} v_{x_j}\,dx = \int_{B_r{}^+(x^0)} a_0{}^{ij} u_{x_i} v_{x_j}\,dx \qquad \text{for } v \in H_0{}^1(B_r{}^+(x^0) \cup S_r(x^0))$$

with the help of Poincaré's inequality in $H_0{}^1(B_r{}^+(x^0) \cup S_r{}^0(x^0))$ (see the corollary of Theorem 1.45); thus, $w \equiv u - z$ belongs to $H^1(B_r^+(x^0))$ and satisfies

$$\int_{B_r^+(x^0)} a_0^{ij} w_{x_i} v_{x_j}\, dx = 0 \qquad \text{for } v \in H_0^1(B_r^+(x^0) \cup S_r^0(x^0)),$$

whereas $w \wedge \psi - u$ belongs to $H_0^1(B_r^+(x^0) \cup S_r^0(x^0))$. We can again estimate $| \nabla z |_{2;x^0,r,+}$ by proceeding as in Step 1 of the present proof: here Poincaré's inequality in $H_0^1(B_r^+(x^0) \cup S_r^0(x^0))$ must be utilized again. After $| \nabla w |_{2;x^0,\varrho,+}$ has been estimated with the help of Lemma 3.9, the same techniques employed in Step 1 and 2 of the proof of Lemma 3.11 again yield $u_{x_1}|_{B_R^+}, \ldots, u_{x_N}|_{B_R^+} \in L^{2,\mu}(B_R^+)$ with norm estimate.

The final global result can at this point be obtained by locally straightening both $\partial\Omega \setminus \Gamma$ and Γ through C^1 diffeomorphisms, then patching together local regularity results in the interior of Ω as well as near $\partial\Omega \setminus \Gamma$ and Γ. □

4.7.2. The Case of Hölderian Leading Coefficients

THEOREM 4.42. *Theorem* 4.41 *remains valid if* μ *is taken in* $]N, N+2[$, *provided* $\partial\Omega$ *is assumed of class* $C^{1,\delta}$, $\Gamma = \emptyset$, *and* $a^{ij}, d^j \in C^{0,\delta}(\bar{\Omega})$, *where* $\delta = (\mu - N)/2$: C *in* (4.76) *then depends on the coefficients of* $a(u, v)$ *through the bound imposed on* $| a^{ij}, d^j |_{C^{0,\delta}(\bar{\Omega})}$ *and* $| b^i, c |_{\infty;\Omega}$ *as well as through* α.

PROOF. As in Step 1 of the proof of Theorem 4.41 it can be proven that it is not restrictive to assume $d^j = 0$ (note that $u \in C^{0,\delta}(\bar{\Omega})$, with norm estimate, by Theorem 4.41 itself), $F = 0$.

Let us study interior regularity. We repeat the same procedure as in Step 2 of the preceding proof, and arrive at

$$| \nabla z |_{2;x^0,r}^2 \le C[(r^2 + r^{2\delta}) | \nabla u |_{2;x^0,r}^2 + | \nabla(v - w) |_{2;x^0,r}^2 + r^{N+2} | u |_{\infty;\Omega}^2].$$

Since Lemma 3.1 yields

$$\begin{aligned} | \nabla w - (\nabla w)_{x^0,\varrho} |_{2;x^0,\varrho}^2 &\le C \frac{\varrho^{N+2}}{r^{N+2}} | \nabla w - (\nabla w)_{x^0,r} |_{2;x^0,r}^2 \\ &\le C \frac{\varrho^{N+2}}{r^{N+2}} | \nabla w - (\nabla u)_{x^0,r} |_{2;x^0,r}^2 \\ &\le C \frac{\varrho^{N+2}}{r^{N+2}} (| \nabla u - (\nabla u)_{x^0,r} |_{2;x^0,r}^2 + | \nabla z |_{2;x^0,r}^2). \end{aligned}$$

we have

$$
\begin{aligned}
| \nabla u - (\nabla u)_{x^0,\varrho} |_{2;x^0,\varrho}^2 &\le | \nabla u - (\nabla w)_{x^0,\varrho} |_{2;x^0,\varrho}^2 \\
&\le 2(| \nabla w - (\nabla w)_{x^0,\varrho} |_{2;x^0,\varrho}^2 + | \nabla z |_{2;x^0,\varrho}^2) \\
&\le C\Big[\frac{\varrho^{N+2}}{r^{N+2}} | \nabla u - (\nabla u)_{x^0,r} |_{2;x^0,r}^2 + (r^2 + r^{2\delta}) | \nabla u |_{2;x^0,r}^2 \\
&\quad + | \nabla (v - w) |_{2;x^0,r}^2 + r^{N+2} | u |_{\infty;\Omega}^2 \Big].
\end{aligned}
$$

In order to bound $| \nabla (v - w) |_{2;x^0,r}^2$ we again choose $v = w \wedge \psi$. This time we make use of the circumstance that since $w \wedge \psi - w \in H_0^1(B_r(x^0))$, we have

$$
a_0^{ij}(\psi_{x_i})_{x^0,r} \int_{B_r(x^0)} (w \wedge \psi - w)_{x_j} \, dx = 0
$$

by the divergence theorem, and obtain

$$
\begin{aligned}
\alpha | \nabla (w \wedge \psi - w) |_{2,x^0,r}^2 &\le \int_{B_r(x^0)} a_0^{ij}(w \wedge \psi - w)_{x_i}(w \wedge \psi - w)_{x_j} \, dx \\
&= \int_{B_r(x^0)} a_0^{ij}[\psi_{x_i} - (\psi_{x_i})_{x^0,r}](w \wedge \psi - w)_{x_j} \, dx \\
&\le \frac{\alpha}{2} | \nabla (w \wedge \psi - w) |_{2;x^0,r}^2 + C | \nabla \psi - (\nabla \psi)_{x^0,r} |_{2;x^0,r}^2 .
\end{aligned}
$$

Finally we remark that all first derivatives of u belong to $L^{2,N-\delta}(\Omega)$ by Theorem 4.41, so that

$$
\begin{aligned}
| \nabla u - (\nabla u)_{x^0,\varrho} |_{2;x^0,\varrho}^2 \le C\Big(&\frac{\varrho^{N+2}}{r^{N+2}} | \nabla u - (\nabla u)_{x^0,r} |_{2;x^0,r}^2 + r^{N+\delta} | \nabla u |_{2,N-\delta;\Omega}^2 \\
&+ r^{\mu} | \nabla \psi |_{2,\mu;\Omega}^2 + r^{N+2} | u |_{\infty;\Omega}^2 \Big).
\end{aligned}
$$

We can now proceed as in Step 4 of the proof of Lemma 3.3 and arrive at the membership of $u_{x_1}|_{\bar\omega}, \ldots, u_{x_N}|_{\bar\omega}$ in $C^{0,\delta/2}(\bar\omega)$ first, and then in $C^{0,\delta}(\bar\omega)$.

Regularity near $\partial\Omega$ presents no difference with respect to the above except that Lemma 3.9 must be used instead of Lemma 3.1. Note that using the same symbols as in Step 3 of the previous proof, we are only concerned with the case when the function $w \wedge \psi - w$ belongs to $H_0^1(B_r^+(x^0))$ because $\Gamma = \varnothing$.

Global regularity is at this point obvious. $\square$

REMARK. Let $\Gamma \neq \emptyset$. The example of Section 4.6.1 shows that Theorem 4.42 cannot remain valid, because u cannot be expected to belong to $C^{1,\delta}(\bar{\Omega})$ for $\delta > 1/2$. For $\delta \leq 1/2$, however, $C^{1,\delta}(\bar{\Omega})$ regularity can still be proven: see L. Caffarelli [25].

4.7.3. The Case of Discontinuous Leading Coefficients

We return to the setting of Theorem 4.41, except for $a^{ij} \in C^0(\bar{\Omega})$ weakened into $a^{ij} \in L^\infty(\Omega)$; we take $N \geq 3$. The proof of $L^{2,\mu}(\Omega)$ regularity of $u_{x_1}, \ldots, u_{x_N}$ must accordingly be modified as follows.

To begin with, for any solution $\hat{u}$ of (4.77) the only available regularity result $\hat{u}_{x_1}, \ldots, \hat{u}_{x_N} \in L^{2,\mu}(\Omega)$ (with norm estimate) concerns the range

$$0 < \mu < \mu_0 = N - 2 + 2\delta_0, \tag{4.81}$$

where δ_0 is the Hölder exponent of Theorem 2.14 (see Theorem 2.19). We shall therefore limit ourselves to the case (4.81).

It can again be proven that it is not restrictive to suppose that $u \in L^{2,\mu}(\Omega)$ with norm estimate, that $d^j = 0$, that the bilinear form is coercive, and that $F = 0$.

If z solves

$$z \in H_0^1(B_r(x^0)),$$

$$\int_{B_r(x^0)} a^{ij} z_{x_i} v_{x_j}\, dx = \int_{B_r(x^0)} a^{ij} u_{x_i} v_{x_j}\, dx \qquad \text{for } v \in H_0^1(B_r(x^0)),$$

where $x^0 \in \bar{\omega}$ with $\omega \subset\subset \Omega$, $0 < r \leq [\mathrm{dist}(\omega, \partial\Omega)] \wedge 1/2$, we arrive at an estimate such as

$$|\nabla z|^2_{2;x^0,r} \leq C(r^2 |\nabla u|^2_{2;x^0,r} + |\nabla(v - w)|^2_{2;x^0,r} + r^{\mu+2} |u|^2_{2,\mu;\Omega})$$

instead of (4.79). To the function $w = u - z$ we apply Lemma 2.17 instead of Lemma 3.1 and obtain

$$|\nabla w|^2_{2;x^0,\varrho} \leq C \frac{\varrho^{\mu_0}}{r^{\mu_0}} (|\nabla u|^2_{2;x^0,r} + |\nabla z|^2_{2;x^0,r}),$$

instead of (4.80), for $0 < \varrho \leq r$. We thus arrive at an inequality

$$|\nabla u|^2_{2;x^0,\varrho} \leq C\left[\left(\frac{\varrho^{\mu_0}}{r^{\mu_0}} + r^2\right) |\nabla u|^2_{2;x^0,r} + r^\mu |\nabla \psi|^2_{2,\mu;\Omega} + r^{\mu+2} |u|_{2,\mu;\Omega}\right],$$

which implies the regularity $u_{x_1}|_\omega, \ldots, u_{x_N}|_\omega \in L^{2,\mu}(\omega)$ and the corresponding norm estimate with the choice (4.81) of μ.

Regularity up to $\partial\Omega$ can be proven similarly: the analog of Lemma 2.17 for functions w satisfying either

$$w \in H_0^1(B_r^+(x^0) \cup S_r^+(x^0)),$$

$$\int_{B_r^+(x^0)} a^{ij} w_{x_i} v_{x_j}\, dx = 0 \qquad \text{for } v \in H_0^1(B_r^+(x^0))$$

or

$$w \in H^1(B_r^+(x^0)),$$

$$\int_{B_r^+(x^0)} a^{ij} w_{x_i} v_{x_j}\, dx = 0 \qquad \text{for } v \in H_0^1(B_r^+(x^0) \cup S_r^0(x^0))$$

can be deduced from Lemma 2.17 itself by proceeding as in the proof of Lemma 2.18.

Summing up, we have the following lemma.

LEMMA 4.43. *For μ satisfying* (4.81), *with δ_0 as in Theorem* 2.14, *Theorem* 4.41 *remains valid if the a^{ij}'s are merely taken from $L^\infty(\Omega)$; the constant C in* (4.76) *depends on the coefficients of the bilinear form only through the bound imposed on their respective $L^\infty(\Omega)$ norms and through α.*

For $N - 2 < \mu < N - 2 + 2\delta_0$ the previous lemma implies Hölder continuity of u throughout $\bar{\Omega}$, with Hölder exponent $\delta' = (\mu + 2 - N)/2$. We are now going to utilize Lemma 4.43 and prove that if the bilinear form is assumed coercive on V and ψ is simply required to be continuous, or Hölder continuous, then u is also continuous, or Hölder continuous.

First we take $\psi \in C^0(\bar{\Omega})$, $\psi \geq 0$ on $\partial\Omega \setminus \Gamma$, and construct a sequence $\{\psi_n\}$ of regular functions such that $\psi_n \geq \psi$ and $\psi_n \to \psi$ in $C^0(\bar{\Omega})$. If the bilinear form is coercive on V, there exists a unique solution u_n of (4.42) with ψ replaced by ψ_n; moreover,

$$| u_n - u_m |_{\infty;\Omega} \leq C \,| \psi_n - \psi_m |_{\infty;\Omega}$$

(see Lemma 4.29).

Assume $f^0 \in L^{2,(\mu-2)^+}(\Omega)$, $f^1, \ldots, f^N \in L^{2,\mu}(\Omega)$ with $N - 2 < \mu < N - 2 + 2\delta_0$: then, each u_n is continuous on $\bar{\Omega}$ by Lemma 4.43, and so is the limit u' of the Cauchy sequence $\{u_n\}$ in $C^0(\bar{\Omega})$. But, since $v \leq \psi$ implies $v \leq \psi_n$, $\{u_n\}$ is also bounded in V by coerciveness: hence, $u_n \rightharpoonup u'$ in V, and u' satisfies (4.42). We have thus proved the following theorem.

THEOREM 4.44. *Same assumptions as in Lemma* 4.43, *except for* $a(u, v)$ *being assumed coercive on* V *and* ψ *being simply taken from* $C^0(\bar{\Omega})$, *with* $\psi \geq 0$ *on* $\partial\Omega \setminus \Gamma$. *Then the solution of* (4.42) *belongs to* $C^0(\bar{\Omega})$.

Next, we have the following theorem.

THEOREM 4.45. *Same assumptions as in Lemma* 4.43, *except that* μ *is supposed* $> N - 2$, $a(u, v)$ *is assumed coercive on* V, *and* ψ *is taken from* $C^{0,\delta_1}(\bar{\Omega})$, $0 < \delta_1 < 1$, *with* $\psi \geq 0$ *on* $\partial\Omega \setminus \Gamma$. *Then the solution of* (4.42) *belongs to* $C^{0,\delta}(\bar{\Omega})$ *for some* $\delta \in]0, \delta_1[$.

PROOF. Once again it is not restrictive to assume $F = 0$; note that the solution of (4.77) is in $C^{0,\delta'}(\bar{\Omega})$ for $\delta' = (\mu + 2 - N)/2$. Let us first consider the case

$$\psi \geq \varepsilon > 0 \qquad \text{on } \partial\Omega \setminus \Gamma. \tag{4.82}$$

Denoting by $\tilde{\psi}$ a controlled C^{0,δ_1} extension of ψ to $\mathbb{R}^N$ (see Theorem 1.2), we set

$$\psi_n = (\varrho_n * \tilde{\psi})|_{\bar{\Omega}}.$$

Thus, $\psi_n > 0$ on $\partial\Omega \setminus \Gamma$ for n sufficiently large, and

$$| \psi_n - \psi |_{\infty;\Omega} \leq n^{-\delta_1} | \psi |_{C^{0,\delta_1}(\bar{\Omega})}$$

$$| \psi_n |_{C^1(\bar{\Omega})} \leq C n^{1-\delta_1} | \psi |_{C^{0,\delta_1}(\bar{\Omega})}$$

(see the remark following Lemma 1.8). Next we denote by u_n the solution of (4.42) with ψ replaced by ψ_n, $F = 0$: since $v = \psi_n \wedge 0$ is admissible in the v.i., we have

$$| u_n |_V \leq C | \psi_n |_{H^1(\Omega)}$$

by coerciveness, and from Lemma 4.43 we deduce

$$| u_n |_{C^{0,\delta'}(\bar{\Omega})} \leq C | \psi_n |_{C^1(\bar{\Omega})} \leq C n^{1-\delta_1} | \psi |_{C^{0,\delta_1}(\bar{\Omega})}.$$

At this point we utilize the fact that

$$| u_n - u |_{\infty;\Omega} \leq C | \psi_n - \psi |_{\infty;\Omega} \leq C n^{-\delta_1} | \psi |_{C^{0,\delta_1}(\bar{\Omega})}$$

(see Lemma 4.29) to obtain

$$\begin{aligned} | u(x) - u(y) | &\leq | u(x) - u_n(x) | + | u(y) - u_n(y) | + | u_n(y) - u_n(x) | \\ &\leq C | \psi |_{C^{0,\delta_1}(\bar{\Omega})} (n^{-\delta_1} + n^{1-\delta_1} | x - y |^{\delta'}) \end{aligned}$$

for $x, y \in \bar{\Omega}$, $0 < | x - y | \leq 1$. By choosing n between $| x - y |^{-\delta'}$ and $| x - y |^{-\delta'} + 1$ we arrive at

$$| u(x) - u(y) | \leq C | \psi |_{C^{0,\delta_1}(\bar{\Omega})} | x - y |^{\delta'\delta_1},$$

thus proving the theorem with $\delta = \delta'\delta_1$, under the additional assumption (4.82). The latter can finally be removed by first replacing ψ by $\psi + \varepsilon$ and solving the corresponding v.i., then letting $\varepsilon \to 0^+$. □

REMARK. In Theorems 4.44 and 4.45 the coerciveness assumption about $a(u, v)$ can be dispensed with if $f^0 \in L^q(\Omega)$ and $f^1, \ldots, f^N \in L^p(\Omega)$, where $p > N$ and $q = pN/(N + p)$. Then, indeed, the function $\hat{u} \equiv -u$, which satisfies

$$a\big(\hat{u}, (\hat{u} - k)^+\big) \leq \langle -F, (\hat{u} - k)^+ \rangle$$

for every $k \geq \max_{\bar{\Omega}} | \psi |$, is bounded not only from below, but also from above (see the remark following Lemma 2.8). This means that $f^0 + \lambda u$ satisfies the same assumptions as f^0, so that $a(u, v)$ can be replaced by $a(u, v) + \lambda \int_\Omega uv\, dx$.

4.8. Lipschitz Regularity by the Penalty Method

We again take $N \geq 3$.

THEOREM 4.46. *Assume $\partial\Omega$ of class C^2, $\Gamma = \emptyset$, $a^{ij} \in C^{0,1}(\bar{\Omega})$, $d^j \in C^{0,\delta}(\bar{\Omega})$ for some $\delta \in]0, 1[$. Let u solve* (4.42) *with*

$$\langle F, v \rangle = \int_\Omega (f^0 v + f^i v_{x_i})\, dx \qquad \text{for } v \in V,$$

where $f^0 \in L^{2,\mu-2}(\Omega)$, $f^1, \ldots, f^N \in L^{2,\mu}(\Omega)$ for some μ in $]N, N + 2[$, and $\psi \in C^{0,1}(\bar{\Omega})$, $\psi \geq 0$ on $\partial\Omega \setminus \Gamma$. Then $u \in C^{0,1}(\bar{\Omega})$ and

$$| u |_{C^{0,1}(\bar{\Omega})} \leq C\Big(| f^0 |_{2,\mu-2;\Omega} + \sum_{i=1}^{N} | f^i |_{2,\mu;\Omega} + | \psi |_{C^{0,1}(\bar{\Omega})} + | u |_{H^1(\Omega)}\Big),$$

where the constant C depends on the coefficients of the bilinear form through the bound imposed on their respective norms as well as through α.

PROOF. *Step* 1: *Preliminary reductions.* Under our present assumptions Theorem 4.41 yields $u_{x_1}, \ldots, u_{x_N} \in L^{2,\mu'}(\Omega)$ and therefore $u \in L^{2,\mu'+2}(\Omega)$

for any $\mu' < N$; thus, $f^0 - b^i u_{x_i} - cu \in L^{2,\mu-2}(\Omega)$ and $f^i - d^i u \in L^{2,\hat{\mu}}(\Omega)$ for $\hat{\mu} = \mu \wedge (N + 2\delta)$. By dint of Theorem 3.16(iii) a simple translation argument shows that it is not restrictive to assume $d^j = b^i = c = 0$, $F = 0$.

Step 2: *The penalized equation.* Set $\beta(w)(x) \equiv [w(x) - \psi(x)]^+$. It is easy to verify that β, as an operator $H^1(\Omega) \to V'$, is bounded, hemicontinuous, and T-monotone. Since $\beta(w)(x) = 0 \Leftrightarrow w(x) \leq \psi(x)$, β is a penalty operator associated with the convex set (4.27), and the solution of (4.42) (under the simplifications stipulated in Step 1) is the weak limit in V, as $\varepsilon \to 0^+$, of the sequence $\{u_\varepsilon\}$ defined by

$$u_\varepsilon \in V, \qquad Lu_\varepsilon + \frac{1}{\varepsilon}\beta(u_\varepsilon) = 0, \tag{4.83}$$

$L = -\partial(a^{ij}\partial/\partial x_i)/\partial x_j$ (see Theorem 4.19 and the remark following it). Note that an easy bootstrap argument based on Sobolev inequalities yields $u_\varepsilon \in H^{2,p}(\Omega)$ for any $p \in [2, \infty[$ [see Theorem 3.17(ii)].

Step 3: *Boundary estimates.* Since $\partial\Omega$ is of class C^2, there exists a positive number r such that, whenever $x^0 \in \partial\Omega$, $\overline{B_r(y)} \cap \bar{\Omega} = x^0$ for a suitable choice of $y = y(x^0)$ in $\mathbb{R}^N$ (a property that is usually expressed by saying that Ω satisfies a *uniform exterior sphere condition*; compare with the beginning of the proof of Lemma 3.26).

We now fix $x^0 \in \partial\Omega$, translate the origin of $\mathbb{R}^N$ in the center y of the corresponding exterior sphere, and introduce the smooth function

$$w^{(+)}(x) \equiv \zeta(r^{-\eta} - |x|^{-\eta}), \qquad x \in \mathbb{R}^N \setminus \{0\},$$

with ζ, $\eta > 0$ to be determined later. Of course, $w^{(+)}(x) \geq 0$ for $x \in \bar{\Omega}$; moreover,

$$w^{(+)}_{x_i}(x) = \zeta\eta x_i |x|^{-(\eta+2)},$$

$$w^{(+)}_{x_i x_j}(x) = \zeta\eta\delta^{ij} |x|^{-(\eta+2)} - \zeta\eta(\eta + 2)x_i x_j |x|^{-(\eta+4)}$$

with $\delta^{ij} = 0$ for $i \neq j$, $= 1$ for $i = j$.

Let M be the essential supremum over Ω of the function $x \mapsto |a^{ij}_{x_j}(x)x_i + a^{ij}(x)\delta^{ij}|$, and take η so large that

$$\begin{aligned} Lw^{(+)}(x) &= -a^{ij}_{x_j}(x)w^{(+)}_{x_i}(x) - a^{ij}w^{(+)}_{x_i x_j}(x) \\ &= \zeta\eta |x|^{-(\eta+2)}[a^{ij}(x)(\eta + 2)x_i x_j |x|^{-2} - a^{ij}_{x_j}(x)x_i - a^{ij}(x)\delta^{ij}] \\ &\geq \zeta\eta |x|^{-(\eta+2)}[\alpha(\eta + 2) - M] \geq 0, \end{aligned}$$

for $x \in \Omega$; *a fortiori* $Lw^{(+)} + (1/\varepsilon)\beta(w^{(+)}) \geq 0$ in Ω.

Next we denote by ϱ a positive real number such that $\Omega \subset B_\varrho$, and fix $\zeta = |\psi|_{C^{0,1}(\bar{\Omega})}\varrho^{\eta+1}/\eta$: hence,

$$|\nabla w^{(+)}(x)| = \zeta\eta\,|x|^{-(\eta+1)} \geq \zeta\eta\varrho^{-(\eta+1)} = |\psi|_{C^{0,1}(\bar{\Omega})}$$

for $x \in \bar{B}_\varrho \setminus B_r$.

If x is arbitrarily taken in Ω, we draw a line segment from x to 0 and denote by $\hat{x}$ the point on that segment that minimizes $\operatorname{dist}(x, \partial\Omega)$. Let $w^{(-)} \equiv -w^{(+)}$: since $w^{(-)}$ is a radial function,

$$\begin{aligned} w^{(-)}(x) - w^{(-)}(\hat{x}) &= -\,|w^{(-)}(x) - w^{(-)}(\hat{x})| \\ &\leq -\min_{\bar{B}_\varrho \setminus B_r} |\nabla w^{(-)}|\,|x - \hat{x}| \end{aligned}$$

by the mean value theorem, and

$$\begin{aligned} w^{(-)}(x) - \psi(x) &\leq w^{(-)}(x) - w^{(-)}(\hat{x}) - [\psi(x) - \psi(\hat{x})] \\ &\leq \min_{\bar{B}_\varrho \setminus B_r} |\nabla w^{(-)}|\,|x - \hat{x}| + |\psi|_{C^{0,1}(\bar{\Omega})}\,|x - \hat{x}| \leq 0. \end{aligned}$$

because $\psi(\hat{x}) \geq 0$. This shows that $\beta(w^{(-)}) = 0$ throughout Ω, hence that $Lw^{(-)} + (1/\varepsilon)\beta(w^{(-)}) = Lw^{(-)} = -Lw^{(+)} \leq 0$ in Ω.

Summing up, the functions $w^{(\pm)}$ have the properties

$$\begin{gathered} w^{(-)}(x^0) = 0 = w^{(+)}(x^0), \\ w^{(-)}(x) \leq 0 \leq w^{(+)}(x) \qquad \text{for } x \in \partial\Omega, \\ Lw^{(-)} + \frac{1}{\varepsilon}\,\beta(w^{(-)}) \leq 0 \leq Lw^{(+)} + \frac{1}{\varepsilon}\,\beta(w^{(+)}) \qquad \text{in } \Omega, \end{gathered}$$

which can be expressed by saying that $w^{(-)}(w^{(+)})$ is a *lower* (an *upper*) *barrier* in Ω relative to $L + (1/\varepsilon)\beta(\cdot)$ at x^0. Then, the T-monotonicity of $\beta(\cdot)$ yields

$$\begin{aligned} 0 &\geq \int_\Omega \{a^{ij}(u_\varepsilon - w^{(+)})_{x_i}[(u_\varepsilon - w^{(+)})^+]_{x_j} \\ &\qquad + \frac{1}{\varepsilon}\,[\beta(u_\varepsilon) - \beta(w^{(+)})](u_\varepsilon - w^{(+)})^+\}\,dx \\ &\geq \int_\Omega a^{ij}[(u_\varepsilon - w^{(+)})^+]_{x_i}[(u_\varepsilon - w^{(+)})^+]_{x_j}\,dx, \end{aligned}$$

hence $|\nabla[(u_\varepsilon - w^{(+)})^+]|_{2;\Omega} = 0$ by coerciveness, and finally $u_\varepsilon \leq w^{(+)}$ on $\bar{\Omega}$ by Poincaré's inequality in $H_0^1(\Omega)$. Analogously, $u_\varepsilon \geq w^{(-)}$ on $\bar{\Omega}$. Thus,

$$\frac{w^{(-)}(x) - w^{(-)}(x^0)}{|x - x^0|} \leq \frac{u_\varepsilon(x) - u_\varepsilon(x^0)}{|x - x^0|} \leq \frac{w^{(+)}(x) - w^{(+)}(x^0)}{|x - x^0|}$$

for $x \in \Omega$. We have obtained a uniform bound on

$$| u_\varepsilon(x) - u_\varepsilon(x^0) |/| x - x^0 | \qquad \text{for } x^0 \in \partial\Omega,\ x \in \Omega,\ \varepsilon > 0;$$

since $u_\varepsilon = 0$ on $\partial\Omega$, a passage to local coordinates shows that

$$| \nabla u_\varepsilon | \leq C \qquad \text{on } \partial\Omega \qquad \text{for } \varepsilon > 0.$$

Step 4: *Completion of the proof.* Fix $\varepsilon > 0$ and $k \in \{1, \ldots, N\}$. By (4.83) the function $z \equiv u_{\varepsilon x_k}$ [which belongs to $H^{1,p}(\Omega)$ for any $p < \infty$] satisfies

$$Lz + \frac{1}{\varepsilon} \chi_{\Omega^+}(z - \psi_{x_k}) = (a^{ij}_{x_k} u_{\varepsilon x_i})_{x_j} \tag{4.84}$$

in the sense of $\mathscr{D}'(\Omega)$, χ_{Ω^+} denoting the characteristic function of the set $\Omega^+ \subset \Omega$ where $u_\varepsilon > \psi$. We fix $\theta > | \nabla u_\varepsilon |_{\infty;\partial\Omega} \vee | \nabla\psi |_{\infty;\Omega}$ and set

$$\Omega_1 \equiv \{x \in \Omega \mid z_1(x) \equiv z(x) - \theta > 0\},$$
$$\Omega_2 \equiv \{x \in \Omega \mid z_2(x) \equiv z(x) + \theta < 0\}.$$

By our choice of θ, for $i = 1, 2$ we have $\Omega_i \subset\subset \Omega$ and $z_i \in H_0^1(\Omega_i)$ (see Problem 1.22); moreover, from (4.84) it follows that

$$Lz_1 \leq (a^{ij}_{x_k} u_{\varepsilon x_i})_{x_j} \qquad \text{in the sense of } H^{-1}(\Omega_1),$$
$$L(-z_2) \leq -(a^{ij}_{x_k} u_{\varepsilon x_i})_{x_j} \qquad \text{in the sense of } H^{-1}(\Omega_2).$$

Since $z_1 \colon \Omega_1 \to \mathbb{R}$ and $z_2 \colon \Omega_2 \to \mathbb{R}$ are nonnegative bounded functions, we arrive at a uniform bound

$$| z_i |_{2^*;\Omega_i} \leq C$$

(see Problem 2.5, with $p = 2$). We have thus found a bound on $| z |_{2^*;\Omega}$ since $\Omega = \Omega_1 \cup \Omega_2 \cup \{x \in \Omega \mid z(x) \leq \theta\}$, hence also on $| \nabla u_\varepsilon |_{2^*;\Omega}$ for $\varepsilon > 0$ by the arbitrariness of the index k.

By repeating the above procedure a convenient finite number of times we arrive at a uniform bound on $| \nabla u_\varepsilon |_{p;\Omega}$ for some $p > N$, so that

$$| z_i |_{\infty;\Omega_i} \leq C$$

by Lemma 2.8; hence,

$$| z |_{\infty;\Omega} \leq C,$$

and finally

$$| \nabla u_\varepsilon |_{\infty;\Omega} \leq C \qquad \text{for } \varepsilon > 0.$$

By the compactness of the imbedding $H^1(\Omega) \subsetneq L^2(\Omega)$, $\{u_\varepsilon\}$ converges to u in $L^2(\Omega)$, therefore (after extraction of a subsequence) a.e. in Ω. Lipschitz continuity follows from the inequalities

$$| u_\varepsilon(x) - u_\varepsilon(\xi) | \leq C | x - \xi | \qquad \text{for } x, \ \xi \in \bar{\Omega}, \qquad \varepsilon > 0;$$

for the norm estimate see Problem 4.14. □

REMARK. When (4.42) is replaced by (4.48), the choice of

$$\beta(w)(x) \equiv [w(x) - \varphi(x)]^- + [w(x) - \psi(x)]^+$$

yields a penalty operator which plays exactly the same role as the one of the preceding proof. Thus, Theorem 4.46 admits an obvious counterpart for the bilateral case.

4.9. Problems Involving Natural Growth of Nonlinear Terms

We momentarily go back to the minimum problem considered in the introduction to the present chapter and generalize it slightly as follows:

$$\begin{gathered} \text{minimize } \mathscr{J}(v) \equiv \frac{1}{2}\int_\Omega [a(v) \, | \nabla v |^2 + v^2] \, dx - \int_\Omega fv \, dx \\ \text{over a convex subset } \mathbb{K} \text{ of } H_0^1(\Omega \cup \Gamma), \end{gathered} \tag{4.85}$$

where $a(t)$ is a nonconstant bounded and smooth function ≥ 1 of $t \in \mathbb{R}$, $| a'(t) | \leq C$. The above functional fails to satisfy the requirement of being Gateaux differentiable at every $u \in H_0^1(\Omega \cup \Gamma)$: for $v \in H_0^1(\Omega \cup \Gamma) \cap L^\infty(\Omega)$ the function $\lambda \mapsto \mathscr{J}(u + \lambda v)$ does indeed admit a derivative at $\lambda = 0$, given by

$$\frac{d}{d\lambda} \mathscr{J}(u + \lambda v)|_{\lambda=0} = \int_\Omega \left[\frac{a'(u)}{2} | \nabla u |^2 v + a(u) u_{x_i} v_{x_i} + uv - fv\right] dx,$$

but the term

$$\int_\Omega \frac{a'(u)}{2} | \nabla u |^2 v \, dx$$

does not make sense if u and v are arbitrarily chosen in $H_0^1(\Omega \cup \Gamma)$ (unless $N = 1$). Thus, there can be no hope of tackling (4.85) in the light of Lemma 4.9.

If $\mathbb{K} \subset L^\infty(\Omega)$ we can still consider the v.i.

$$u \in \mathbb{K}, \qquad \langle A(u), v - u \rangle \geq \int_\Omega F(u, \nabla u)(v - u)\, dx \qquad \text{for } v \in \mathbb{K}, \tag{4.86}$$

where

$$\langle A(u), v \rangle \equiv \int_\Omega [a(u) u_{x_i} v_{x_i} + uv]\, dx, \qquad F(u, \nabla u) \equiv f - \frac{a'(u)}{2} | \nabla u |^2.$$

Note that the term $[a'(u)/2] \, | \nabla u |^2$ cannot be absorbed into the nonlinear functional $A(u) \in [H_0^1(\Omega \cup \Gamma)]'$ by conveniently redefining the latter; hence, (4.86) cannot be reduced to (4.17). If $\mathbb{K} \not\subset L^\infty(\Omega)$, (4.86) must be replaced by

$$u \in \mathbb{K}, \qquad \langle A(u), v - u \rangle \geq \int_\Omega F(u, \nabla u)(v - u)\, dx$$
$$\text{for } v \in \mathbb{K} \text{ such that } v - u \in L^\infty(\Omega),$$

which in the case $\mathbb{K} = H_0^1(\Omega \cup \Gamma)$ becomes

$$u \in H_0^1(\Omega \cup \Gamma), \qquad \langle A(u), v \rangle = \int_\Omega F(u, \nabla u) v\, dx$$
$$\text{for } v \in H_0^1(\Omega \cup \Gamma) \cap L^\infty(\Omega) \tag{4.87}$$

[the Euler–Lagrange equation of the minimum problem (4.85)]. It is intuitively clear that (4.86) must be somewhat easier to handle than (4.87). In this section we shall deal with rather general problems of the above types, starting with bilateral v.i.'s.

We fix p in $]1, \infty[$ and take $V = H_0^{1,p}(\Omega \cup \Gamma)$ with Γ of class C^1, assuming that the imbedding $V \subsetneq L^p(\Omega)$ is compact. Let A be the operator $H^{1,p}(\Omega) \to V'$ defined in (4.34) under assumptions (4.35), (4.38) with strict inequality sign when $\xi \neq \xi'$, (4.40) with $g \in L^{p'}(\Omega)$; we also assume A coercive with respect to the convex set defined by the bilateral constraints. Next, we introduce a Carathéodory function f of $x \in \Omega$ and $(\eta, \xi) \in \mathbb{R}^{1+N}$ such that given any $r \in]0, \infty[$,

$$| f(x, \eta, \xi) | \leq C \, | \xi |^p + f_0(x)$$
$$\text{for a.a. } x \in \Omega \qquad \text{and} \qquad \text{any } (\eta, \xi) \in \mathbb{R}^{1+N}, \qquad | \eta | \leq r, \tag{4.88}$$

where the constant $C > 0$ and the function $f_0 \geq 0$, $f_0 \in L^{p'}(\Omega)$, depend on r; (4.88) is called a *natural growth condition* (see our introductory remarks in the case $p = 2$).

We denote by $F(u, \nabla u)$, $u \in H^{1,p}(\Omega)$, the function $f(x, u(x), \nabla u(x))$ of $x \in \Omega$.

THEOREM 4.47. *Under the above assumptions about V, $u \mapsto A(u)$ and $u \mapsto F(u, \nabla u)$, there exists at least one solution to the bilateral v.i.*

$$\begin{gathered} u \in V, \qquad \varphi \leq u \leq \psi \qquad \text{in } \Omega \\ \langle A(u), v - u\rangle \geq \int_\Omega F(u, \nabla u)(v - u)\, dx \quad \text{for } v \in V, \quad \varphi \leq v \leq \psi \quad \text{in } \Omega \end{gathered} \tag{4.89}$$

provided φ, ψ belong to $L^\infty(\Omega)$ and there exists $v_0 \in V$ such that $\varphi \leq v_0 \leq \psi$.

PROOF. *Step 1: A class of auxiliary v.i.'s.* For $n \in \mathbb{N}$ we introduce the bounded function

$$f_n(x, \eta, \xi) \equiv \begin{cases} 0 & \text{for } f(x, \eta, \xi) = 0 \\ f(x, \eta, \xi) \dfrac{n \wedge |f(x, \eta, \xi)|}{|f(x, \eta, \xi)|} & \text{for } f(x, \eta, \xi) \neq 0 \end{cases}$$

and set $[F_n(u, \nabla u)](x) \equiv f_n(x, u(x), \nabla u(x))$. It is obvious that f_n is a Carathéodory function of $x \in \Omega$ and $(\eta, \xi) \in \mathbb{R}^{1+N}$; moreover, the function $a^0(x, \eta, \xi) - f_n(x, \eta, \xi)$ satisfies the same type of growth condition (4.35) as $a^0(x, \eta, \xi)$. By Theorem 4.21, therefore, the operator $B_n\colon V \to V'$ defined by

$$\langle B_n(u), v\rangle \equiv \langle A(u), v\rangle - \int_\Omega F_n(u, \nabla u) v\, dx$$

for $u, v \in V$ is of the Leray–Lions type, hence pseudomonotone. Since the coerciveness of A implies the coerciveness of B_n, from Theorem 4.17 (with A replaced by B_n, F by 0) we deduce that the bilateral v.i.

$$\begin{gathered} u_n \in V, \qquad \varphi \leq u_n \leq \psi \qquad \text{in } \Omega \\ \langle A(u_n), v - u_n\rangle - \int_\Omega F_n(u_n, \nabla u_n)(v - u_n)\, dx \geq 0 \\ \text{for } v \in V, \qquad \varphi \leq v \leq \psi \qquad \text{in } \Omega \end{gathered} \tag{4.90}$$

admits at least one solution.

Step 2: A uniform bound on $|u_n|_{H^{1,p}(\Omega)}$. For any $t > 0$ we can find a positive number $\delta(t)$ such that $\delta(t)\mathscr{E}(t)$, with $\mathscr{E}(t) \equiv e^{t(u_n - v_0)^2}$ (where v_0 is the function required in the statement of the theorem), is < 1 a.e. in Ω.

This means that the function

$$v_n \equiv [1 - \delta(t)\mathscr{E}(t)]u_n + \delta(t)\mathscr{E}(t)v_0$$

is an element of V which lies between φ and ψ (see Lemma 1.57). We insert $v = v_n$ in (4.90). Since

$$v_n - u_n = -\delta(t)(u_n - v_0)\mathscr{E}(t)$$

and

$$(v_n - u_n)_{x_i} = -\delta(t)(u_n - v_0)_{x_i}\mathscr{E}(t) - 2\delta(t)t(u_n - v_0)^2(u_n - v_0)_{x_i}\mathscr{E}(t),$$

we have

$$\begin{aligned}\int_\Omega & A^i(u_n, \nabla u_n)u_{nx_i}[1 + 2t(u_n - v_0)^2]\mathscr{E}(t)\,dx \\ &\leq \int_\Omega \{A^i(u_n, \nabla u_n)v_{0x_i}[1 + 2t(u_n - v_0)^2] \\ &\quad + [F_n(u_n, \nabla u_n) - A^0(u_n, \nabla u_n)](u_n - v_0)\}\mathscr{E}(t)\,dx.\end{aligned}$$

Since both u_n and v_0 lie between φ and ψ, from (4.35), (4.40), and (4.88) [which clearly holds also with f replaced by f_n, r being chosen $\geq \max(|\varphi|_{\infty;\Omega}, |\psi|_{\infty;\Omega})$] we deduce that

$$\begin{aligned}\alpha \int_\Omega & |\nabla u_n|^p[1 + 2t(u_n - v_0)^2]\mathscr{E}(t)\,dx \\ &\leq C_0 \int_\Omega [(1 + g)(1 + 2t) + (1 + |\nabla u_n|^{p-1} + h)|\nabla v_0|(1 + 2t) \\ &\quad + |\nabla u_n|^p |u_n - v_0| + f_0 + 1 + |\nabla u_n|^{p-1} + h]\mathscr{E}(t)\,dx.\end{aligned}$$

At this point we utilize the following estimates:

$$\begin{aligned}\int_\Omega & |\nabla u_n|^{p-1}|\nabla v_0|\mathscr{E}(t)\,dx \\ &\leq \left[\int_\Omega |\nabla u_n|^p\mathscr{E}(t)\,dx\right]^{1-1/p}\left[\int_\Omega |\nabla v_0|^p\mathscr{E}(t)\,dx\right]^{1/p},\end{aligned}$$

$$\begin{aligned}\int_\Omega & |\nabla u_n|^{p-1}\mathscr{E}(t)\,dx \\ &\leq \left[\int_\Omega |\nabla u_n|^p\mathscr{E}(t)\,dx\right]^{1-1/p}\left[\int_\Omega \mathscr{E}(t)\,dx\right]^{1/p},\end{aligned}$$

$$\begin{aligned}C_0 \int_\Omega & |\nabla u_n|^p |u_n - v_0|\,\mathscr{E}(t)\,dx \\ &\leq \frac{\alpha}{2}\int_\Omega |\nabla u_n|^p\mathscr{E}(t)\,dx + \frac{C_0^2}{2\alpha}\int_\Omega |\nabla u_n|^p(u_n - v_0)^2\mathscr{E}(t)\,dx.\end{aligned}$$

By choosing $t = C_0^2/4\alpha^2$ we arrive at the inequality

$$\frac{\alpha}{2}\int_\Omega |\nabla u_n|^p \mathscr{E}\left(\frac{C_0^2}{4\alpha^2}\right) dx \le C\left[1 + \int_\Omega |\nabla u_n|^p \mathscr{E}\left(\frac{C_0^2}{4\alpha^2}\right) dx\right]^{1-1/p},$$

which yields a uniform bound on $\int_\Omega |\nabla u_n|^p \mathscr{E}(C_0^2/4\alpha^2)\, dx$, hence on $|\nabla u_n|_{p;\Omega}$ because $\mathscr{E} \ge 1$ on $[0, \infty[$. Note that by (4.35) this yields, for each $j = 0, 1, \ldots, N$, a uniform bound on $|A^j(u_n, \nabla u_n)|_{p';\Omega}$ as well.

Step 3: *Completion of the proof.* By reflexivity, the conclusions of the preceding step yield

$$u_n \rightharpoonup u \text{ in } V, \qquad u_n \to u \text{ in } L^p(\Omega), \qquad u_n(x) \to u(x) \qquad \text{for a.a. } x \in \Omega,$$
$$A^j(u_n, \nabla u_n) \rightharpoonup h^j \qquad \text{in } L^{p'}(\Omega) \qquad \text{for } j = 0, 1, \ldots, N$$

as $n \to \infty$ (with the same symbol for a suitable subsequence of indices as for the original sequence). We are now going to show that with the notation (4.39),

$$\int_\Omega D_n\, dx \to 0 \qquad \text{as } n \to \infty. \tag{4.91}$$

For $t > 0$ set $\mathscr{F}(t) \equiv e^{t(u_n-u)^2}$, and let $\delta(t) > 0$ be such that $\delta(t)\mathscr{F}(t) < 1$ a.e. in Ω. Since $v = [1 - \delta(t)\mathscr{F}(t)]u_n + \delta(t)\mathscr{F}(t)$ is admissible in (4.90), we arrive at the inequality

$$\int_\Omega A^i(u_n, \nabla u_n)(u_n - u)_{x_i}[1 + 2t(u_n - u)^2]\mathscr{F}(t)\, dx$$
$$\le \int_\Omega [F_n(u_n, \nabla u_n) - A^0(u_n, \nabla u_n)](u_n - u)\mathscr{F}(t)\, dx,$$

whose right-hand side is majorized by a quantity

$$C\left[\int_\Omega (f_0 + 1 + |\nabla u_n|^{p-1} + h)\,|u_n - u|\mathscr{F}(t)\, dx\right.$$
$$\left. + \frac{\varepsilon}{2}\int_\Omega |\nabla u_n|^p \mathscr{F}(t)\, dx + \frac{1}{2\varepsilon}\int_\Omega |\nabla u_n|^p (u_n - u)^2 \mathscr{F}(t)\, dx\right]$$

for $\varepsilon > 0$ (see Step 2 above). Because of (4.40) we have

$$\alpha\int_\Omega |\nabla u_n|^p \mathscr{F}(t)\, dx \le \int_\Omega A^i(u_n, \nabla u_n)u_{nx_i}\mathscr{F}(t)\, dx + \int_\Omega (C + g)\mathscr{F}(t)\, dx,$$

$$\alpha\int_\Omega |\nabla u_n|^p (u_n - u)^2 \mathscr{F}(t)\, dx$$
$$\le \int_\Omega A^i(u_n, \nabla u_n)u_{nx_i}(u_n - u)^2 \mathscr{F}(t)\, dx + \int_\Omega (C + g)(u_n - u)^2 \mathscr{F}(t)\, dx.$$

We set

$$P_n(t) \equiv \int_\Omega (f_0 + 1 + |\nabla u_n|^{p-1} + h)\,|u_n - u|\,\mathscr{F}(t)\,dx,$$

$$Q(t) \equiv \int_\Omega (1+g)\mathscr{F}(t)\,dx,$$

$$R_n(t) \equiv \int_\Omega (1+g)(u_n - u)^2\mathscr{F}(t)\,dx + \int_\Omega A^i(u_n, \nabla u_n)u_{x_i}(u_n - u)^2\mathscr{F}(t)\,dx;$$

note that since u_n converges to u in $L^p(\Omega)$ and $|u_n - u|_{\infty;\Omega}$, $|\nabla u_n|_{p;\Omega}$, $|A^i(u_n, \nabla u_n)|_{p';\Omega}$ are uniformly bounded,

$$\lim_{n\to\infty} P_n(t) = \lim_{n\to\infty} R_n(t) = 0$$

for every fixed t. Thus,

$$\begin{aligned}\int_\Omega & A^i(u_n, \nabla u_n)(u_n - u)_{x_i}[1 + 2t(u_n - u)^2]\mathscr{F}(t)\,dx \\ &\le C_1\Big[P_n(t) + \frac{\varepsilon}{2\alpha}\int_\Omega A^i(u_n, \nabla u_n)(u_n - u)_{x_i}\mathscr{F}(t)\,dx \\ &\quad + \frac{\varepsilon}{2\alpha}\int_\Omega A^i(u_n, \nabla u_n)u_{x_i}\mathscr{F}(t)\,dx + \frac{\varepsilon}{2\alpha}Q(t) \\ &\quad + \frac{1}{2\varepsilon\alpha}\int_\Omega A^i(u_n, \nabla u_n)(u_n - u)_{x_i}(u_n - u)^2\mathscr{F}(t)\,dx \\ &\quad + \frac{1}{2\varepsilon\alpha}R_n(t)\Big].\end{aligned}$$

We set $t = K_\varepsilon \equiv C_1/4\varepsilon\alpha$ with $\varepsilon < 2\alpha/C_1$ and obtain

$$\begin{aligned}\Big(1 - \frac{C_1\varepsilon}{2\alpha}\Big)&\int_\Omega A^i(u_n, \nabla u_n)(u_n - u)_{x_i}\mathscr{F}(K_\varepsilon)\,dx \\ &\le C_1\Big[P_n(K_\varepsilon) + \frac{\varepsilon}{2\alpha}\int_\Omega A^i(u_n, \nabla u_n)u_{x_i}\mathscr{F}(K_\varepsilon)\,dx \\ &\quad + \frac{\varepsilon}{2\alpha}Q(K_\varepsilon) + \frac{1}{2\varepsilon\alpha}R_n(K_\varepsilon)\Big],\end{aligned}$$

hence

$$\left(1 - \frac{C_1\varepsilon}{2\alpha}\right) \int_\Omega D_n \mathscr{F}(K_\varepsilon)\, dx$$

$$\leq -\left(1 - \frac{C_1\varepsilon}{2\alpha}\right) \int_\Omega A^i(u_n, \nabla u)(u_n - u)_{x_i} \mathscr{F}(K_\varepsilon)\, dx$$

$$+ C_1\left[P_n(K_\varepsilon) + \frac{\varepsilon}{2\alpha} \int_\Omega A^i(u_n, \nabla u_n) u_{x_i} \mathscr{F}(K_\varepsilon)\, dx\right.$$

$$\left. + \frac{\varepsilon}{2\alpha} Q(K_\varepsilon) + \frac{1}{2\varepsilon\alpha} R_n(K_\varepsilon)\right].$$

Since $\mathscr{F}(K_\varepsilon) > 1$, we arrive at

$$\left(1 - \frac{C_1\varepsilon}{2\alpha}\right) \limsup_{n\to\infty} \int_\Omega D_n\, dx \leq C_1 \frac{\varepsilon}{2\alpha} \int_\Omega (h^i u_{x_i} + 1 + g)\, dx$$

because $A^i(u_n, \nabla u) \to A^i(u, \nabla u)$ in $L^{p'}(\Omega)$, $u_{nx_i} \rightharpoonup u_{x_i}$ in $L^p(\Omega)$, and (4.91) follows by letting $\varepsilon \to 0^+$ since the integral on the left-hand side above is ≥ 0.

At this point, since (4.40) holds, we can apply Lemma 4.22 and obtain

$$u_n \to u \qquad \text{in } V,$$

so that

$$A^j(u_n, \nabla u_n) \to A^j(u, \nabla u) \qquad \text{in } L^{p'}(\Omega) \qquad \text{for } j = 0, 1, \ldots, N$$

by Lemma 4.20. As a matter of fact, the proof of Lemma 4.20 can also be easily adapted to prove that

$$F_n(u_n, \nabla u_n) \to F(u, \nabla u) \qquad \text{in } L^1(\Omega);$$

after passing to a subsequence, we therefore have

$$F_n(u_n, \nabla u_n) \to F(u, \nabla u) \qquad \text{a.e. in } \Omega$$

with

$$| F_n(u_n, \nabla u_n) | \leq f^* \qquad \text{a.e. in } \Omega,$$

$f^* \in L^1(\Omega)$ (see Theorem 1.Q). Hence,

$$\int_\Omega F_n(u_n, \nabla u_n) u_n\, dx \to \int_\Omega F(u, \nabla u) u\, dx$$

by the dominated convergence theorem. We can prove that u solves (4.89), therefore, by a passage to the limit in (4.90). □

Theorem 4.47 will now be utilized to investigate the solvability of the equation

$$u \in V, \qquad \langle A(u), v\rangle = \int F(u, \nabla u)v\, dx \qquad \text{for } v \in V \cap L^\infty(\Omega); \tag{4.92}$$

note that (4.92) implies

$$-A^i(u, \nabla u)_{x_i} + A^0(u, \nabla u) = F(u, \nabla u) \qquad \text{in the sense of } \mathscr{D}'(\Omega).$$

We take V as in Theorem 4.47. For what concerns the operator A, we strengthen our previous assumptions by requiring also that it be strictly T-monotone and coercive. Finally, we again assume (4.88).

THEOREM 4.48. *In addition to the above assumptions, suppose that* $\varphi, \psi \in H^{1,pp'}(\Omega) \cap L^\infty(\Omega)$ *with* $\varphi \leq 0 \leq \psi$ *on* $\partial\Omega \setminus \Gamma$ *in the sense of* $H^{1,p}(\Omega)$, $A(\varphi) \leq F(\varphi, \nabla\varphi)$, $A(\psi) \geq F(\psi, \nabla\psi)$ *in the sense of* V'. *Then* (4.92) *admits at least one solution* u, *which lies between* φ *and* ψ.

PROOF. Consider the v.i. (4.90). Since $A(\varphi)$ and $F_n(u_n, \nabla u_n)$ [$A(\psi)$ and $F_n(u_n, \nabla u_n)$] admit an order upper bound $F(\varphi, \nabla\varphi) \vee F_n(u_n, \nabla u_n) \in L^{p'}(\Omega)$ [an order lower bound $F(\psi, \nabla\psi) \wedge F_n(u_n, \nabla u_n) \in L^{p'}(\Omega)$], we have

$$F(\psi, \nabla\psi) \wedge F_n(u_n, \nabla u_n) \leq A(u_n) \\ \leq F(\varphi, \nabla\varphi) \vee F_n(u_n, \nabla u_n) \qquad \text{in the sense of } V'. \tag{4.93}$$

(see Remark 1 at the end of Section 4.5).

The first consequence we draw from (4.93) is that the linear functional $A(u_n)$ can be continuously extended from V to $L^p(\Omega)$. In other words, there exists $f_n^* \in L^{p'}(\Omega)$ such that

$$\langle A(u_n), v\rangle = \int_\Omega f_n^* v\, dx \qquad \text{for } v \in V.$$

By (4.90),

$$\int_\Omega [f_n^* - F_n(u_n, \nabla u_n)](v - u_n)\, dx \geq 0 \qquad \text{for } v \in V, \qquad \varphi \leq v \leq \psi.$$

We can successively take $v = u_n + \chi_k(\varphi - u_n)$ and $v = u_n + \chi_k(\psi - u_n)$ in the above inequality, where $\{\chi_k\}_k \subset C_c^\infty(\Omega)$, $0 \leq \chi_k \leq 1$, $\chi_k \to \chi_I$ a.e.

in Ω, with $\chi_I \equiv$ characteristic function of a measurable subset I of Ω, $\operatorname{dist}(I, \partial\Omega) > 0$. We let $k \to \infty$ and obtain

$$\begin{aligned} &[f_n{}^* - F_n(u_n, \nabla u_n)](\varphi - u_n) \geq 0, \\ &[f_n{}^* - F_n(u_n, \nabla u_n)](\psi - u_n) \geq 0 \end{aligned} \tag{4.94}$$

a.e. in Ω, by the arbitrariness of I. Write $\Omega = \bigcup_{i=0}^{3} \Omega_i$, with

$$\begin{aligned} &\varphi < u_n < \psi \quad \text{in } \Omega_0, \qquad \varphi = u_n < \psi \quad \text{in } \Omega_1, \\ &\varphi < u_n = \psi \quad \text{in } \Omega_2, \qquad \varphi = u_n = \psi \quad \text{in } \Omega_3. \end{aligned}$$

By (4.94) and Theorem 1.56 we have

$$\begin{aligned} f_n{}^* = F_n(u_n, \nabla u_n) \qquad &\text{in } \Omega_0, \\ f_n{}^* \geq F_n(u_n, \nabla u_n), \qquad F(\varphi, \nabla\varphi) = F(u_n, \nabla u_n) \qquad &\text{in } \Omega_1, \\ f_n{}^* \leq F_n(u_n, \nabla u_n), \qquad F(\psi, \nabla\psi) = F(u_n, \nabla u_n) \qquad &\text{in } \Omega_2, \\ F(\varphi, \nabla\varphi) = F(\psi, \nabla\psi) = F(u_n, \nabla u_n) \qquad &\text{in } \Omega_3. \end{aligned}$$

From (4.93) we therefore deduce

$$F(u_n, \nabla u_n) \wedge F_n(u_n, \nabla u_n) \leq f_n{}^* \leq F(u_n, \nabla u_n) \vee F_n(u_n, \nabla u_n) \qquad \text{a.e. in } \Omega,$$

hence

$$\begin{aligned} \int_\Omega &[F(u_n, \nabla u_n) \wedge F_n(u_n, \nabla u_n)]v\, dx \\ &\leq \langle A(u_n), v\rangle \leq \int_\Omega [F(u_n, \nabla u_n) \vee F_n(u_n, \nabla u_n)]v\, dx \\ &\qquad\qquad \text{for } v \in V \cap L^\infty(\Omega), \qquad v \geq 0. \end{aligned}$$

Since $u_n \to u$ in V, $A(u_n) \rightharpoonup A(u)$ in V', $F(u_n, \nabla u_n) \wedge F_n(u_n, \nabla u_n) \to F(u, \nabla u)$ and $F(u_n, \nabla u_n) \vee F_n(u_n, \nabla u_n) \to F(u, \nabla u)$ in $L^1(\Omega)$ (for a subsequence of indices; see Step 3 of the preceding proof), u solves (4.83). ☐

Remark 1. Under the same assumptions about A as in Theorem 4.47, Theorem 4.48 can be given a (more difficult) proof that does not utilize the result about bilateral v.i.'s: see L. Boccardo, F. Murat, and J. P. Puel [16].

Remark 2. Under a natural growth assumption about the nonlinear function f, regularity results for solutions of equations or v.i.'s are extremely delicate. We refer the reader to S. Campanato [32], J. Frehse [52], J. Frehse and U. Mosco [53], M. Giaquinta and E. Giusti [66]. Let us also mention that some of the results of the next chapter will imply existence of regular solutions to problems such as (4.89) or (4.92), in the case of linear operators A. See also the remark following Theorem 5.14 below.

Problems

4.1. For the existence of a solution to (4.2), the proof of Lemma 4.2 utilizes the circumstance that Hilbert spaces are reflexive. An alternative method can be based on the identity

$$\left|\frac{u_m - u_n}{2}\right|_V^2 = \frac{1}{2}(d_m{}^2 + d_n{}^2) - \left|z - \frac{u_m + u_n}{2}\right|_V^2,$$

where $z \equiv \mathscr{I}F$ and $d_n \equiv |u_n - z|_V \to \inf_{v\in K} |v - z|_V$.

4.2. We identify $\mathbb{R}^N$ with its dual $(R^N)'$. If $\mathscr{J} \in C^1(\mathbb{R}^N)$ is (strictly) convex, then $A(x) \equiv \nabla\mathscr{J}(x)$ is (strictly) monotone on $\mathbb{R}^N$. On the other hand, an operator $R^2 \to R^2$ such as

$$A(x) \equiv (x_1, x_2 + \varphi(x_1)),$$

where φ is a nonconstant function from $C^1(\mathbb{R})$ with $|\varphi'| \leq 1$ on $\mathbb{R}$, is strictly monotone without being the gradient of a convex function.

4.3. For a counterexample to uniqueness under the assumptions of Theorem 4.17, take $q \in N$, $p = 2(q+1)$, $V = \mathbb{K} = H_0{}^{1,p}(\Omega)$ and

$$A(u) = -(|u_{x_i}|^{2q} u_{x_i})_{x_i} + (2q+1)z_{x_i x_i} |u_{x_i}|^{2q},$$

where z is a given function from $C_c{}^\infty(\Omega)$. (See J. A. Dubinskii [46].)

4.4. Let $\mathbb{K}$ be a convex subset of a reflexive Banach space V, and let A be a monotone and hemicontinuous operator $V \to V'$. Then (4.17) is equivalent to:

$$u \in \mathbb{K}, \ \langle A(v) - F, v - u\rangle \geq 0 \qquad \text{for } v \in \mathbb{K};$$

as a consequence, the set of solutions to (4.17) is convex. (See G. J. Minty [113].)

4.5. In addition to the assumptions of Problem 4.4, suppose that the equation

$$u \in V, \qquad A(u) = 0$$

can have at most one solution, and that $\mathbb{K}$ satisfies the strict convexity condition: $u, v \in \mathbb{K}$, $u \neq v$ and $0 < \lambda < 1 \Rightarrow \lambda u + (1 - \lambda)v \in$ interior of $\mathbb{K}$. Then (4.17) can have at most one solution.

4.6. For $h = 1, 2$ let K_h be a nonempty, closed and convex subset of a Hilbert space V; let $u = u_h$ solve (4.9) with $K = K_h$, $a(u, v)$ coercive, $F \in V'$; and let $w_h \in K_h$ be such that $w_1 + w_2 = u_1 + u_2$ and $a(w_1 - u_2, w_1 - u_1) = 0$. Then $w_1 = u_1$. In particular, consider (4.47) with $\varphi = \varphi_h$ and $\psi = \psi_h$ measurable in Ω: if $\varphi_1 \geq \varphi_2$ and $\psi_1 \geq \psi_2$, then $w_1 = u_1 \vee u_2$ and $w_2 = u_1 \wedge u_2$ are admissible, and $u_1 \geq u_2$. Consider also (4.43) with $\psi = \psi_h$ measurable in Ω, $\psi_1 \geq \psi_2$, and compare with the corollary to Lemma 4.26. (See Y. Haugazeau [74].)

4.7. Give the explicit expression of the solution to the v.i.

$$u \in K, \qquad \int_\Omega u(v - u)\,dx \geq \int_\Omega f(v - u)\,dx \qquad \text{for } v \in K,$$

where $K = \{v \in L^2(\Omega) \mid \varphi \leq v \leq \psi \text{ in } \Omega\}$ with f, φ, and ψ given in $L^2(\Omega)$.

4.8. A v.i. associated with a fourth-order operator is

$$u \in K, \qquad \int_\Omega \Delta u \Delta(v - u)\,dx \geq \langle F, v - u\rangle \qquad \text{for } v \in K,$$

where $K = \{v \in H_0^1(\Omega) \cap H^2(\Omega) \mid -1 \leq \Delta v \leq 1 \text{ in } \Omega\}$ and $F = f^i_{x_i}$ with $f^i \in L^p(\Omega)$, $2 \leq p < \infty$. Introduce the solution $\bar{u}$ to the Dirichlet problem $\bar{u} = 0$ on $\partial\Omega$, $-\Delta\bar{u} = F$, solve the v.i. for Δu and prove that, if $\partial\Omega$ is sufficiently regular, then $u \in H^{3,p}(\Omega)$. See H. Brézis and G. Stampacchia [23].

4.9. Let u_n solve (4.42) with ψ replaced by ψ_n, supposing that $\{\psi_n\}$ converges to ψ in V. Assume $a(u, v)$ coercive on V. Then $u_n \to u$ in V, where u is the solution of (4.42).

4.10. Utilize the method of translations to discuss the one-dimensional v.i.

$$u \in K, \qquad \int_0^1 u'(v' - u')\,dx \geq \int_0^1 f(v - u)\,dx \qquad \text{for } v \in K$$

where $K = \{v \in H_0^1(\Omega) \mid v \leq 0 \text{ in } \Omega\}$, $\Omega =]0, 1[$, and f is a first-order polynomial in x. Give numerical examples.

4.11. Let $f \equiv f^0 - f^i_{x_i}$, $f^j \in L^2(\Omega)$, satisfy $f \leq 0$ in the sense of $H^{-1}(\Omega)$. Then there exists $\{f_n\} \subset L^\infty(\Omega)$, $f_n \leq 0$ a.e. in Ω, such that $f_n \to f$ in $H^{-1}(\Omega)$. To see this, approximate f in $H^{-1}(\Omega)$ with $\{\tilde{f}_n\} \subset C_c^\infty(\Omega)$ and solve

$$\tilde{u}_n \in H_0^1(\Omega), \qquad \int_\Omega \tilde{u}_{nx_i} v_{x_i}\,dx = \int_\Omega \tilde{f}_n v\,dx \qquad \text{for } v \in H_0^1(\Omega),$$

then solve

$$u_n \in H_0^1(\Omega), \qquad u_n \leq \tilde{u}_n \qquad \text{in } \Omega,$$

$$\int_\Omega u_{nx_i}(v - u_n)_{x_i}\,dx \geq 0 \qquad \text{for } v \in H_0^1(\Omega), \qquad v \leq \tilde{u}_n \qquad \text{in } \Omega,$$

and finally pass to the limit: the functions $f_n \equiv -\Delta u_n$ have the required properties. (See G. M. Troianiello [146].)

4.12. Let $V = H_0^1(\Omega \cup \Gamma)$ with $\Gamma \neq \varnothing$ of class C^1. Assume $F \in V'$ and $\psi \in H^1(\Omega)$, $\psi \geq 0$ on $\partial\Omega \backslash \Gamma$, are such that there exists $(L\psi) \wedge f \in H^{-1}(\Omega)$, with L defined by (4.33) and $f \equiv$ restriction of F to $H_0^1(\Omega)$. Then a solution u of (4.42) satisfies

$$(L\psi) \wedge f \leq Lu \leq f \qquad [\text{in the sense of } H^{-1}(\Omega)].$$

(Compare with Theorem 4.32.) To see this, suppose that $a(u, v)$ is coercive on $H_0^1(\Omega)$, solve

$$\hat{u} - u \in H_0^1(\Omega), \qquad L\hat{u} = L\psi \qquad \text{in } \Omega,$$

and utilize the Lewy–Stampacchia inequality for the v.i. in $H_0^1(\Omega)$ with obstacle $\psi - \hat{u}$ and free term $f - L\psi$.

4.13. Give a local counterpart to Theorem 4.38 as follows. If ω is an open subset of Ω, take Ω' open with $\bar{\Omega}' \subset \bar{\Omega}$, $\partial\Omega'$ of class $C^{3,1}$ and $\text{dist}(\bar{\omega}, \partial\Omega' \cap \Omega) > 0$. Let $g \in C_c^\infty(\mathbb{R}^N)$ with $\text{dist}(\text{supp } g, \partial\Omega' \cap \Omega) > 0$, $g = 1$ on $\bar{\omega}$, and write the v.i. for gu in Ω' with obstacle $g\psi$: the $H^{2,\infty}$ regularity of $u|_\omega$ depends only on the suitable regularity of the data on $\bar{\Omega}'$.

4.14. Give the norm estimate of Theorem 4.46 by evaluating the various constants appearing in the proof.

5

Nonvariational Obstacle Problems

The first section in this chapter is based on the following considerations. Obstacle problems such as (4.44) and (4.48) can be formulated even when the operator L is of the nonvariational type; candidates as solutions are those functions u whose first and second derivatives are defined a.e. in Ω, so that Lu certainly makes sense. We can still avail ourselves of existence, uniqueness, and regularity results for v.i.'s if the leading coefficients of L are smooth. If not, we can approximate L by a sequence of operators to which variational tools do apply.

Now let the given functions f and ζ be replaced by functions $F(u, \nabla u)$ and $Z(u)$ that depend on the solution u itself [so that the linear operators

$$u \mapsto Lu, \qquad u \mapsto Bu$$

are replaced by the nonlinear ones

$$u \mapsto Lu - F(u, \nabla u), \qquad u \mapsto Bu - Z(u)].$$

We tackle the corresponding obstacle problems in Section 5.3. Our approach to existence results, centered around the Leray–Schauder fixed point theorem, utilizes the existence and uniqueness results of the first section in conjunction with *a priori* $H^{2,p}$ estimates on solutions. It is to the derivation of estimates of this sort that Section 5.2, in its turn, is devoted.

The last section deals with unilateral problems for the operator $u \mapsto Lu - F(u, \nabla u)$ (under Dirichlet boundary conditions) in cases when regularity assumptions about the obstacle ψ are too weak to guarantee the existence of a solution u in the previous, strong sense. We therefore introduce an appropriate substitute for a regular solution. This new notion enables us, in particular, to tackle problems where ψ depends on u itself.

5.1. Obstacle Problems for Linear Operators

Set

$$Lu \equiv -a^{ij}u_{x_ix_j} + a^iu_{x_i} + au, \qquad Bu \equiv \beta^iu_{x_i}|_\Gamma + \beta u|_\Gamma.$$

Throughout this chapter the following properties of regularity will be supposed to hold:

- Γ is closed in $\partial\Omega$, the latter being of class $C^{1,1}$;
- $a^{ij} \in C^0(\bar{\Omega})$ with a modulus of continuity τ, $a^{ij} = a^{ji}$, and

$$a^{ij}\xi_i\xi_j \geq a\,|\,\xi\,|^2 \quad \text{on } \bar{\Omega} \quad \text{for } \xi \in \mathbb{R}^N \quad (\alpha > 0);$$

- $\beta^1, \ldots, \beta^N \in C^{0,1}(\Gamma)$, and $\beta^i\nu^i > 0$ on Γ.

In the present section we shall also assume $a^1, \ldots, a^N, a \in L^\infty(\Omega)$, $a \geq 0$ in Ω, $\beta \in C^{0,1}(\Gamma)$, $\beta \geq 0$ on Γ, and

$$\operatorname*{ess\,sup}_{\Omega} a + \max_{\Gamma} \beta > 0 \qquad \text{if } \Gamma = \partial\Omega.$$

5.1.1. Bilateral Problems

We begin with the problem

$$\begin{aligned} &\varphi \leq u \leq \psi, \qquad (Lu - f)(u - \varphi) \leq 0 \\ &\quad \text{and} \quad (Lu - f)(u - \psi) \leq 0 \qquad \text{in } \Omega, \\ &u|_{\partial\Omega\setminus\Gamma} = 0, \qquad Bu = \zeta \qquad \text{on } \Gamma. \end{aligned} \tag{5.1}$$

This is a *nonvariational bilateral problem*, although of a special type: the condition on Γ is the same as in (2.19) or (3.36), not as in (4.48).

THEOREM 5.1. *For $2 \leq p < \infty$ assume $f \in L^p(\Omega)$, $\zeta \in H^{1/p',p}(\Gamma)$, $\varphi = \bigvee_{h=1}^m \varphi^h$ with $\varphi^h \in H^{2,p}(\Omega)$, $\varphi^h|_{\partial\Omega\setminus\Gamma} \leq 0$ and $B\varphi^h \leq \zeta$ on Γ, $\psi = \bigwedge_{h=1}^m \psi^h$ with $\psi^h \in H^{2,p}(\Omega)$, $\psi^h|_{\partial\Omega\setminus\Gamma} \geq 0$ and $B\psi^h \geq \zeta$ on Γ, $\varphi \leq \psi$ in Ω. Then (5.1) admits a unique solution $u \in H^{2,p}(\Omega)$, which satisfies the Lewy–Stampacchia inequalities*

$$\bigwedge_{h=1}^{m} (L\psi^h) \wedge f \leq Lu \leq \bigvee_{h=1}^{m} (L\varphi^h) \vee f \qquad \text{in } \Omega. \tag{5.2}$$

PROOF. *Step* 1: *Existence*. We consider the same functions a_n^{ij}, θ_n, operators L_n, and bilinear forms $\hat{a}_n(u, v)$ as in Step 1 of the proof of Lemma 3.21, so that $\theta_n B$ is the conormal derivative with respect to $\hat{a}_n(u, v)$.

Our present assumptions about a and β yield $\hat{a}_n(1, v) \geq 0$ whenever $v \in H_1^0(\Omega \cup \Gamma)$ is ≥ 0 in Ω, and also $\hat{a}_n(1, v) \neq 0$ for some $v \in H^1(\Omega)$ if $\Gamma = \partial\Omega$. This implies (Theorem 2.4) the validity of the weak maximum principle, so that the bilateral v.i.

$$\begin{gathered} u_n \in H_0^1(\Omega \cup \Gamma), \qquad \varphi \leq u_n \leq \psi \qquad \text{in } \Omega, \\ \hat{a}_n(u_n, v - u_n) \geq \int_\Omega f(v - u_n)\, dx + \int_\Gamma \theta_n \zeta (v - u_n)|_\Gamma \, d\sigma \qquad (5.3) \\ \text{for } v \in H_0^1(\Omega \cup \Gamma), \qquad \varphi \leq u_n \leq \psi \qquad \text{in } \Omega \end{gathered}$$

admits a unique solution by Theorem 4.30. Moreover, u_n belongs to $H^{2,p}(\Omega)$ and satisfies

$$\begin{gathered} \varphi \leq u_n \leq \psi, \qquad (L_n u_n - f)(u_n - \varphi) \leq 0 \\ \text{and} \quad (L_n u_n - f)(u_n - \psi) \leq 0 \qquad \text{in } \Omega, \qquad (5.4) \\ u_n|_{\partial\Omega \setminus \Gamma} = 0, \qquad Bu_n = \zeta \qquad \text{on } \Gamma, \end{gathered}$$

as well as

$$\bigwedge_{h=1}^{m} (L_n \psi^h) \wedge f \leq L_n u_n \leq \bigvee_{h=1}^{m} (L_n \varphi^h) \vee f \qquad \text{in } \Omega, \qquad (5.5)$$

by Lemmas 4.25, 4.36, and 4.37. But then, Theorem 3.28(i) yields a uniform bound

$$| u_n |_{H^{2,p}(\Omega)} \leq C \Big[\sum_{h=1}^{m} (| \varphi^h |_{H^{2,p}(\Omega)} + | \psi^h |_{H^{2,p}(\Omega)}) + | f |_{p;\Omega} + | \zeta |_{H^{1/p',p}(\Gamma)} \Big]:$$

by reflexivity, a subsequence of $\{u_n\}$, still denoted by the same symbol, converges weakly in $H^{2,p}(\Omega)$ [and strongly in $H^{1,p}(\Omega)$] toward some function u. We pass to the limit in (5.4) (as in the proof of Lemma 3.22 for what concerns Bu_n) and in (5.5), thus showing that u solves (5.1) and satisfies (5.2).

Step 2: *Uniqueness*. Assume that $u_1, u_2 \in H^{2,p}(\Omega)$ are two solutions to our problem. By Step 1, the problem

$$\begin{gathered} \varphi \leq v \leq u_1 \wedge u_2, \qquad (Lv - f)(v - \varphi) \leq 0 \\ \text{and} \quad (Lv - f)(v - u_1 \wedge u_2) \leq 0 \qquad \text{in } \Omega, \\ v|_{\partial\Omega \setminus \Gamma} = 0, \qquad Bv = \zeta \qquad \text{on } \Gamma \end{gathered}$$

admits a solution $v \in H^{2,p}(\Omega)$. Put

$$\Omega' \equiv \text{the subset of } \Omega \text{ where } v < u_1 \wedge u_2,$$
$$\Omega'' \equiv \text{the subset of } \Omega \text{ where } v = u_2 < u_1,$$
$$\Omega''' \equiv \text{the subset of } \Omega \text{ where } v = u_1.$$

In Ω' we have $Lv \geq f$ and $Lu_1 \leq f$ (since $u_1 > \varphi$); in Ω'' we have $Lv = Lu_2 \geq f$ (since $u_2 < \psi$) and $Lu_1 \leq f$ (since $u_1 > \varphi$); finally, in Ω''' we have $Lv = Lu_1$. [We have repeatedly exploited the fact that the first and second derivatives of a function $w \in H^{2,p}(\Omega)$ vanish a.e. in the subset of Ω where $w = 0$: see Theorem 1.56.] Thus,

$$Lv \geq Lu_1 \quad \text{in } \Omega,$$
$$(v - u_1)|_{\partial\Omega\setminus\Gamma} = 0, \qquad B(v - u_1) = 0 \quad \text{on } \Gamma,$$

so that Theorem 3.29 yields $v \geq u_1$ and therefore $v = u_1$. It can analogously be proven that $v = u_2$, whence uniqueness follows.

Note that, by uniqueness, the whole sequence $\{u_n\}$ of Step 1 converges weakly to u in $H^{2,p}(\Omega)$. □

5.1.2. Unilateral Problems

The unilateral counterpart of (5.1) is

$$\begin{aligned} &u \leq \psi, \qquad Lu \leq f \quad \text{and} \quad (Lu - f)(u - \psi) = 0 \qquad \text{in } \Omega, \\ &u|_{\partial\Omega\setminus\Gamma} = 0, \qquad Bu = \zeta \qquad \text{on } \Gamma. \end{aligned} \tag{5.6}$$

We have the following theorem.

THEOREM 5.2. *Under the same assumptions about f, ζ, and ψ as in Theorem* 5.1, (5.6) *admits a unique solution $u \in H^{2,p}(\Omega)$, which satisfies the Lewy–Stampacchia inequality*

$$Lu \geq \bigwedge_{h=1}^{m} (L\psi^h) \wedge f \qquad \text{in } \Omega. \tag{5.7}$$

PROOF. For existence and (5.7) we could again use an approximation procedure as in the proof of Theorem 5.1, this time with the aid of Theorem 4.27 and Lemmas 4.23, 4.33, 4.34. An alternative method utilizes instead Theorem 5.1 itself, as follows.

Let w be any function from $H^{2,p}(\Omega)$ satisfying

$$\begin{aligned} &w \le \psi \quad \text{and} \quad Lw \le f \qquad \text{in } \Omega, \\ &w|_{\partial\Omega\setminus\Gamma} \le 0, \qquad Bw \le \zeta \qquad \text{on } \Gamma \end{aligned} \tag{5.8}$$

[a possible choice being the solution of the b.v.p.

$$Lw = \bigwedge_{h=1}^{m} (L\psi^h) \wedge f \qquad \text{in } \Omega,$$

$$w|_{\partial\Omega\setminus\Gamma} = 0, \qquad Bw = \zeta \qquad \text{on } \Gamma;$$

Theorem 3.29 then yields

$$w \le \psi \qquad \text{in } \Omega].$$

By Theorem 5.1 the bilateral problem

$$\begin{aligned} &w \le u \le \psi, \qquad (Lu - f)(u - w) \le 0 \\ &\quad \text{and} \quad (Lu - f)(u - \psi) \le 0 \qquad \text{in } \Omega, \\ &u|_{\partial\Omega\setminus\Gamma} = 0, \qquad Bu = \zeta \qquad \text{on } \Gamma \end{aligned}$$

admits a unique solution $u \in H^{2,p}(\Omega)$, which verifies

$$\bigwedge_{h=1}^{m} (L\psi^h) \wedge f \le Lu \le (Lw) \vee f \le f \qquad \text{in } \Omega.$$

Then u satisfies (5.7), as well as the inequalities of the first line of (5.6); from them, the equality $(Lu - f)(u - \psi) = 0$ follows since $(Lu - f) \times (u - \psi) \le 0$.

Uniqueness can be proven as in Theorem 5.1, this time assuming $u_1, u_2 \in H^{2,p}(\Omega)$ to be solutions of (5.6) and solving: $v \in H^{2,p}(\Omega)$,

$$\begin{aligned} &v \le u_1 \wedge u_2, \qquad Lv \le f \quad \text{and} \quad (Lv - f)(v - u_1 \wedge u_2) = 0 \qquad \text{in } \Omega, \\ &v|_{\partial\Omega\setminus\Gamma} = 0, \qquad Bv = 0 \qquad \text{on } \Gamma, \end{aligned}$$

then showing $v = u_1$, $v = u_2$.

Observe that once uniqueness has been ascertained, the first part of this proof shows that the solution u is maximal among all functions $w \in H^{2,p}(\Omega)$ satisfying (5.8). ∎

For the case $1 < p < 2$ in Theorems 5.1 and 5.2 solve Problem 5.1.

If $\Gamma \neq \emptyset$, the most general formulation of a *nonvariational unilateral problem* is

$$\begin{gathered} u \leq \psi, \qquad Lu \leq f \quad \text{and} \quad (Lu - f)(u - \psi) = 0 \qquad \text{in } \Omega, \\ u|_{\partial\Omega \setminus \Gamma} = 0, \\ u|_\Gamma \leq \psi|_\Gamma, \qquad Bu \leq \zeta \quad \text{and} \quad (Bu - \zeta)(u - \psi)|_\Gamma = 0 \qquad \text{on } \Gamma \end{gathered} \tag{5.9}$$

[see (4.44)], which contains (5.6) as a special case. As the example of Section 4.6.2 shows, we cannot expect to solve (5.9) in $H^{2,p}(\Omega)$ for p much larger than 2, unless (5.9) reduces to (5.6). We take $p = 2$. By analogy with the definition introduced in Section 4.4.1, we call $w \in H^2(\Omega)$ a *subsolution* of (5.9) if it satisfies (5.8). When the assumptions about the leading coefficients of L are suitably strengthened, (5.9) can be tackled in the light of the variational theory. Indeed, we have the following lemma.

LEMMA 5.3. *Take the a^{ij}'s in $C^{0,1}(\bar{\Omega})$. Assume $f \in L^2(\Omega)$, $\zeta \in H^{1/2}(\Gamma)$ if $\Gamma \neq \emptyset$, and ψ measurable in Ω if $\Gamma = \emptyset$, $\psi \in H^1(\Omega)$ otherwise. A solution $u \in H^2(\Omega)$ of (5.9), if existing, is then maximal among all subsolutions (and therefore unique). Moreover, under the additional requirement that $\psi \in H^2(\Omega)$ with $\psi|_{\partial\Omega \setminus \Gamma} \geq 0$, such a solution does indeed exist and satisfies the Lewy–Stampacchia inequalities*

$$\begin{aligned} Lu &\geq (L\psi) \wedge f \qquad \text{in } \Omega, \\ Bu &\geq (B\psi) \wedge \zeta \qquad \text{on } \Gamma \end{aligned}$$

as well as a norm estimate

$$| u |_{H^2(\Omega)} \leq C(| f |_{2;\Omega} + | \zeta |_{H^{1/2}(\Gamma)} + | \psi |_{H^2(\Omega)}), \tag{5.10}$$

where C (independent of u, f, ζ, ψ) depends on the coefficients of L only through α and the bound on $| a^{ij} |_{C^{0,1}(\bar{\Omega})}$, $| a^i, a |_{\infty;\Omega}$.

PROOF. We introduce the same function θ and bilinear form $\hat{a}(u, v)$ as in Step 1 of the proof of Lemma 3.18, so that θBu is the conormal derivatives of u with respect to $\hat{a}(u, v)$. Any solution $u \in H^2(\Omega)$ of (5.9) is also a solution of the v.i.

$$\begin{gathered} u \in H_0^1(\Omega \cup \Gamma), \qquad u \leq \psi \qquad \text{in } \Omega, \\ \hat{a}(u, v - u) \geq \int_\Omega f(v - u)\, dx + \int_\Gamma \theta\zeta(v - u)|_\Gamma \, d\sigma \\ \text{for } v \in H_0^1(\Omega \cup \Gamma), \qquad v \leq \psi \qquad \text{in } \Omega \end{gathered} \tag{5.11}$$

(Lemma 4.23). Since $\hat{a}(1, v) \geq 0$ if $v \in H_0^1(\Omega \cup \Gamma)$ with $v \geq 0$, and in addition $\hat{a}(1, v) \neq 0$ for some $v \in H^1(\Omega)$ if $\Gamma = \partial\Omega$, the weak maximum principle holds and Theorem 4.27 applies.

The maximality property of u follows immediately, because any subsolution of (5.9) is a subsolution of (5.11) as well. Moreover, if $\psi \in H^2(\Omega)$ with $\psi|_{\partial\Omega \setminus \Gamma} \geq 0$, (5.11) admits a unique solution u, and the Lewy–Stampacchia inequalities follow from Lemma 4.33, the $H^2(\Omega)$ regularity with norm estimate

$$| u |_{H^2(\Omega)} \leq C(| f |_{2;\Omega} + | \zeta |_{H^{1/2}(\Gamma)} + | \psi |_{H^2(\Omega)} + | u |_{H^1(\Omega)}) \qquad (5.12)$$

from Theorem 4.40. By Lemma 4.23, u satisfies (5.9): notice that the pairing $\langle Bu - \zeta, (u - \psi)|_\Gamma \rangle$ equals the integral $\int_\Gamma (Bu - \zeta)(u - \psi)|_\Gamma \, d\sigma$, and the latter equals 0 if and only if the nonnegative function $(Bu - \zeta) \times (u - \psi)|_\Gamma$ vanishes a.e. $[N - 1]$ on Γ.

In order to pass from (5.12) to (5.10) we first of all apply the interpolation inequality (Lemma 1.37)

$$| u |_{H^1(\Omega)} \leq \varepsilon \, | u |_{H^2(\Omega)} + C(\varepsilon) \, | u |_{2;\Omega}, \qquad \varepsilon > 0,$$

which enables us to replace $| u |_{H^1(\Omega)}$ with $| u |_{2,\Omega}$ in the right-hand side of (5.12). Then we utilize Theorem 3.29 and solve: $w \in H^2(\Omega)$,

$$Lw = - | f | - | L\psi | \qquad \text{in } \Omega,$$

$$w|_{\partial\Omega \setminus \Gamma} = 0, \qquad Bw = - | \zeta | - | B\psi | \qquad \text{on } \Gamma$$

(where Theorem 1.61 has also been taken into account). Thus,

$$| w |_{H^2(\Omega)} \leq C(| f |_{2;\Omega} + | \zeta |_{H^{1/2}(\Gamma)} + | \psi |_{H^2(\Omega)})$$

by Theorem 3.28(i). Since w is a subsolution of (5.9), we have $u \geq w$ in addition to $u \leq \psi$, so that $| u |_{2;\Omega} \leq | \psi |_{2;\Omega} + | w |_{2;\Omega}$ and (5.10) follows. □

We now return to the general case $a^{ij} \in C^0(\bar{\Omega})$.

THEOREM 5.4. *Lemma* 5.3 *is still valid if the leading coefficients of L are merely taken in* $C^0(\bar{\Omega})$; *the constant in* (5.10) *now depends on them only through the bound on their* $L^\infty(\Omega)$ *norms as well as through* α *and* τ.

PROOF. *Step* 1: *Maximality among subsolutions.* Let $\{a_n^{ij}\} \subset C^\infty(\bar{\Omega})$ be the usual sequence of restrictions to $\bar{\Omega}$ of regularizations $\varrho_n * \tilde{a}^{ij}$ and set

$$L_n u \equiv -a_n^{ij} u_{x_i x_j} + a^i u_{x_i} + au.$$

A solution $u \in H^2(\Omega)$ of (5.8) is also a solution of its analog with L replaced by L_n and f by $f + (L_n - L)u$. Let w be any subsolution of (5.9) and consider the b.v.p.

$$L_n z_n = -|(L_n - L)u| - |(L_n - L)w| \qquad \text{in } \Omega,$$

$$z_n|_{\partial\Omega \setminus \Gamma} = 0, \qquad Bz_n = 0 \qquad \text{on } \Gamma.$$

By Theorem 3.29 such a problem admits a unique solution $z_n \in H^2(\Omega)$, which in addition is ≤ 0; by Theorem 3.28(i), $z_n \to 0$ as $n \to \infty$. The function $w_n \equiv w + z_n$ verifies

$$w_n \leq \psi \quad \text{and} \quad L_n w_n \leq f + (L_n - L)u \qquad \text{in } \Omega,$$

$$w_n|_{\partial\Omega \setminus \Gamma} \leq 0, \qquad Bw_n \leq \zeta \qquad \text{on } \Gamma,$$

so that $w_n \leq u$ by Lemma 5.3 and finally $w \leq u$.

Step 2: Proof of the norm estimate. We now follow a procedure analogous to the one utilized in the proof of Lemma 3.20.

Consider a point $x^0 \in \Gamma$ and fix any $r > 0$ such that $B_{2r}(x^0) \cap \partial\Omega \subset \Gamma$. Let $g \equiv g_{x^0,r} \in C^\infty(\mathbb{R}^N)$ with $0 \leq g \leq 1$, supp $g \subset B_{2r}(x^0)$ and $g = 1$ on $\overline{B_r(x^0)}$. The function $u' \equiv u'_{x^0,r} \equiv gu$ satisfies

$$u' \leq \psi', \qquad L_0 u' \leq f' \quad \text{and} \quad (L_0 u' - f')(u' - \psi') = 0 \qquad \text{in } \Omega,$$

$$u'|_{\partial\Omega \setminus \Gamma} = 0,$$

$$u'|_\Gamma \leq \psi'|_\Gamma, \qquad Bu' \leq \zeta' \quad \text{and} \quad (Bu' - \zeta')(u' - \psi')|_\Gamma = 0 \qquad \text{on } \Gamma,$$

where $L_0 \equiv -a^{ij}(x^0)\partial^2/\partial x_i \partial x_j + a^i \partial/\partial x_i + a$, and $\psi' \equiv g\psi$, $f' \equiv gf + (L_0 - L)u' - 2a^{ij}u_{x_i}g_{x_j} + uLg$, $\zeta' \equiv g|_\Gamma \zeta + \beta^i g_{x_i}|_\Gamma u|_\Gamma$: notice that

$$|f'|_{2;\Omega} \leq C[|f|_{2;\Omega} + \tau(2r)|u'|_{H^2(\Omega)} + |g|_{C^2(\overline{B_{2r}(x^0)})}|u|_{H^1(\Omega)}],$$

$$|\zeta'|_{H^{1/2}(\Gamma)} \leq C(|\zeta|_{H^{1/2}(\Gamma)} + |g|_{C^2(\overline{B_{2r}(x^0)})}|u|_{H^1(\Omega)})$$

with C independent of r. We can apply Lemma 5.3 with L, u, f, ζ, ψ, respectively, replaced by L_0, u', f', ζ', ψ', thus obtaining, for a small enough r, a bound

$$|u'|_{H^2(\Omega)} \leq C(|f|_{2;\Omega} + |\zeta|_{H^{1/2}(\Gamma)} + |\psi|_{H^2(\Omega)} + |u|_{H^1(\Omega)})$$

with $C = C(r)$ independent of u, f, ζ, ψ. Similar considerations can be repeated if $x^0 \in \partial\Omega \setminus \Gamma$ or $x^0 \in \Omega$.

At this point we utilize a straightforward compactness argument and arrive at (5.12), hence also at (5.10) (with the required type of dependence of C on the a^{ij}'s) by the same considerations as in the proof of Lemma 5.2.

Step 3: *Existence and Lewy–Stampacchia inequalities*. We again consider the sequence of operators $\{L_n\}$ introduced in Step 1, and find the unique solution u_n of (5.9) with L replaced by L_n thanks to Lemma 5.3. Because of the type of dependence the constant C in (5.10) now has on the leading coefficients, the norms $| u_n |_{H^2(\Omega)}$ are uniformly bounded. Hence, (a subsequence of) $\{u_n\}$ converges weakly in $H^2(\Omega)$ toward some function u, and $u_n \to u$ in $H^1(\Omega)$, $u_{nx_i}|_\Gamma \rightharpoonup u_{x_i}|_\Gamma$ in $H^{1/2}(\Gamma)$ (see Problem 1.27). A passage to the limit (with the aid of Lemma 1.60 for what concerns inequalities on Γ) shows that u satisfies (5.9) and the Lewy–Stampacchia inequalities. □

REMARK. The maximality property of solutions to (5.9) implies their monotonicity with respect to f, ζ, and ψ.

5.1.3. An Approximation Result

Under the assumptions of Theorem 5.1 about f, ζ, φ, and ψ we solved (5.1) by introducing $\{a_n{}^{ij}\} \subset C^\infty(\bar{\Omega})$ with $a_n{}^{ij} \to a^{ij}$ in $C^0(\bar{\Omega})$ and showing that u_n[solution of (5.3)] $\rightharpoonup u$[solution of (5.1)] in $H^{2,p}(\Omega)$. Since numerical tools such as the finite element method are available for the investigation of v.i.'s (see the notes to Chapter 4), the above approximation procedure is also useful in the numerical analysis of (5.1), provided the rate of convergence of u_n to u is estimated in some convenient norm. The result we have (with the above notations) is as follows.

THEOREM 5.5. *Let q be the number $Np/(N-2p)$ for $2p < N$, any real number for $2p = N$, ∞ for $2p > N$. Then,*

$$| u_n - u |_{q;\Omega} \leq C \max_{i,j=1,\dots,N} | a_n{}^{ij} - a^{ij} |_{\infty;\Omega} \Big[\sum_{h=1}^{m} (| \varphi^h |_{H^{2,p}(\Omega)} + | \psi^h |_{H^{2,p}(\Omega)}) + | f |_{p;\Omega} + | \zeta |_{H^{1/p',p}(\Gamma)} \Big], \tag{5.13}$$

where C (independent of u, f, ζ, φ and ψ) depends on the coefficients of L only through the bound on their $L^\infty(\Omega)$ norms, as well as through α and τ.

PROOF. Without loss of generality [since we can always operate the translations

$$\varphi \mapsto \varphi - \bar{u}, \qquad \psi \mapsto \psi - \bar{u}, \qquad u \mapsto u - \bar{u}$$

with $\bar{u} \in H^{2,p}(\Omega)$ solution of

$$L\bar{u} = f \quad \text{in } \Omega,$$
$$\bar{u}|_{\partial\Omega \setminus \Gamma} = 0, \qquad B\bar{u} = \zeta \qquad \text{on } \Gamma],$$

we assume $f = 0$, $\zeta = 0$. Then (5.3) becomes

$$\begin{gathered} u_n \in H_0^1(\Omega \cup \Gamma), \qquad \varphi \leq u_n \leq \psi \qquad \text{in } \Omega, \\ \hat{a}_n(u_n, v - u_n) \geq 0 \qquad \text{for } v \in H_0^1(\Omega \cup \Gamma), \quad \varphi \leq v \leq \psi \quad \text{in } \Omega, \end{gathered} \tag{5.14}$$

and its analog with n replaced by r can be rewritten as

$$\begin{gathered} u_r \in H_0^1(\Omega \cup \Gamma), \qquad \varphi \leq u_r \leq \psi \qquad \text{in } \Omega, \\ \hat{a}_n(u_r, v - u_r) \geq \int_\Omega [(L_n - L_r)u_r](v - u_r)\, dx \\ \text{for } v \in H_0^1(\Omega \cup \Gamma), \qquad \varphi \leq v \leq \psi \qquad \text{in } \Omega \end{gathered} \tag{5.15}$$

[because $u_r \in H^{2,p}(\Omega)$ with $Bu_r = 0$ on Γ]. Let $\hat{u} \equiv (u_r - u_n)^+$, $w_\varepsilon \equiv \hat{u}w/(\hat{u} + \varepsilon)$ with $\varepsilon > 0$ and $w \in C_c^1(\Omega \cup \Gamma)$, $0 \leq w \leq 1$, so that both functions $v' \equiv u_n + \varepsilon w_\varepsilon$ and $v'' \equiv u_r - \varepsilon w_\varepsilon$ lie between φ and ψ. We insert $v = v'$ in (5.14), $v = v''$ in (5.15), and obtain

$$\int_\Omega g_{nr} w_\varepsilon\, dx \geq \hat{a}_n(u_r - u_n, w_\varepsilon) = \hat{a}_n(\hat{u}, w_\varepsilon) \tag{5.16}$$

with $g_{nr} \equiv (L_n - L_r)u_r$ where $u_r > u_n$, $g_{nr} \equiv 0$ elsewhere. By proceeding as in Step 3 of the proof of Theorem 4.27, we decompose $\hat{a}_n(\hat{u}, w_\varepsilon)$ as a sum $I_1(\varepsilon) + \varepsilon I_2(\varepsilon)$, with $I_1(\varepsilon) \to \hat{a}_n(\hat{u}, w)$ as $\varepsilon \to 0^+$. If

$$\hat{a}_n(\hat{u}, w) > \int_\Omega g_{nr} w\, dx \tag{5.17}$$

for some function w as above, we arrive at

$$\limsup_{\varepsilon \to 0^+} \varepsilon I_2(\varepsilon) < 0 \tag{5.18}$$

[see (4.52)] since

$$\int_\Omega g_{nr} w_\varepsilon\, dx \to \int_\Omega g_{nr} w\, dx.$$

But (5.18) is self-contradictory, and (5.17) must be false. This shows that

$$\hat{a}_n(\hat{u}, w) \le \int_\Omega g_{nr} w \, dx \qquad \text{for } w \in H_0^1(\Omega \cup \Gamma), \qquad w \ge 0;$$

by the weak maximum principle, $(u_r - u_n)^+ \le z_{nr}$ in Ω, where

$$z_{nr} \in H_0^1(\Omega \cup \Gamma), \qquad \hat{a}_n(z_{nr}, v) = \int_\Omega g_{nr} v \, dx \qquad \text{for } v \in H_0^1(\Omega \cup \Gamma).$$

But z_{nr} belongs to $H^{2,p}(\Omega)$, with

$$\begin{aligned} |z_{nr}|_{H^{2,p}(\Omega)} &\le C |g_{nr}|_{p;\Omega} \\ &\le C |(L_n - L_r)u_r|_{p;\Omega} \\ &\le C \max_{i,j=1,\ldots,N} |a_n^{ij} - a_r^{ij}|_{\infty;\Omega} |u_r|_{H^{2,p}(\Omega)}. \end{aligned}$$

At this point we recall that

$$|z_{nr}|_{q;\Omega} \le |z_{nr}|_{H^{2,p}(\Omega)}$$

by Theorems 1.33 and 1.41, and that

$$|u_r|_{H^{2,p}(\Omega)} \le C \sum_{h=1}^{m} (|\varphi^h|_{H^{2,p}(\Omega)} + |\psi^h|_{H^{2,p}(\Omega)})$$

with C independent of r. Since the roles of r and n can be interchanged, we have proved that (5.13) (with $f = 0$, $\zeta = 0$) holds if u is replaced by u_r, a^{ij} by a_r^{ij}, and the sought-for conclusion follows as $r \to \infty$. ▯

5.1.4. Systems of Unilateral Problems

In the proof of Theorem 5.2 we solved a unilateral problem by reducing it to a bilateral one. We now take $\Gamma = \emptyset$ (for simplicity's sake) and consider two unilateral problems which are coupled through their respective obstacles as follows:

$$\begin{gathered} v^1 \le v^2 - \varphi, \qquad v^2 \le v^1 + \psi, \\ Lv^1 \le f^1, \qquad Lv^2 \le f^2, \\ (Lv^1 - f^1)(v^1 - v^2 + \varphi) = (Lv^2 - f^2)(v^2 - v^1 - \psi) = 0 \qquad \text{in } \Omega, \\ v^1|_{\partial\Omega} = v^2|_{\partial\Omega} = 0, \end{gathered} \tag{5.19}$$

where f^1, f^2, φ, and ψ are measurable functions on Ω. If $(v^1, v^2) \in [H^2(\Omega)]^2$ satisfies (5.19) the function $u \equiv v^2 - v^1$ is an element of $H^2(\Omega)$ which vanishes on $\partial\Omega$ and verifies

$$\varphi \leq u \leq \psi$$

as well as

$$\begin{aligned}(Lu - f)(u - \varphi) &= (Lv^2 - f^2)(u - \varphi) + (-Lv^1 + f^1)(u - \varphi) \\ &\leq (-Lv^1 + f^1)(v^2 - v^1 - \varphi) = 0\end{aligned}$$

and

$$\begin{aligned}(Lu - f)(u - \psi) &= (Lv^2 - f^2)(u - \psi) + (-Lv^1 + f^1)(u - \psi) \\ &\leq (Lv^2 - f^2)(v^2 - v^1 - \psi) = 0\end{aligned}$$

in Ω, with $f \equiv f_2 - f_1$. This means that a solution u of the bilateral problem (5.1) with $\Gamma = \varnothing$ can be obtained from a solution (v^1, v^2) of the system of unilateral problems (5.19). Vice versa we have the following theorem.

THEOREM 5.6. *Take φ, ψ as in Theorem* 5.1 $(p = 2)$, $f = f^2 - f^1$ *with f^1, $f^2 \in L^2(\Omega)$. Suppose that $u \in H^2(\Omega)$ satisfies* (5.1) *with $\Gamma = \varnothing$ and let u^1, $u^2 \in H^2(\Omega)$ solve the b.v.p.'s*

$$Lu^1 = f^1 - (Lu - f)^+, \qquad Lu^2 = f^2 - (Lu - f)^- \qquad \text{in } \Omega,$$

$$u^1|_{\partial\Omega} = u^2|_{\partial\Omega} = 0,$$

so that $u^2 - u^1 = u$. Then (u^1, u^2) is the maximal solution of (5.19); *uniqueness holds if and only if the set*

$$S \equiv \{x \in \Omega \mid \varphi(x) = \psi(x)\}$$

has measure 0.

PROOF. From $\varphi \leq u \leq \psi$ we deduce

$$u^1 \leq u^2 - \varphi, \qquad u^2 \leq u^1 + \psi.$$

Besides, the inequalities

$$Lu^1 \leq f^1, \qquad Lu^2 \leq f^2$$

are obvious. On the other hand,

$$(Lu^1 - f^1)(u^1 - u^2 + \varphi) = (Lu - f)^+(u - \varphi) = 0,$$

the latter equality consequent on the fact that $Lu \leq f$ in any measurable subset of Ω where $u > \varphi$, and analogously

$$(Lu^2 - f^2)(u^2 - u^1 - \psi) = (Lu - f)^-(\psi - u) = 0.$$

This suffices to prove that (u^1, u^2) is a solution of (5.19).

Now let $(v^1, v^2) \in [H^2(\Omega)]^2$ be any other such solution. By our previous considerations, uniqueness yields $v^2 - v^1 = u$ and therefore

$$Lv^2 - f^2 - (Lv^1 - f^1) = Lu - f = (Lu - f)^+ - (Lu - f)^-.$$

This implies the existence of a function $g \in L^2(\Omega)$, $g \leq 0$, such that the nonpositive functions $Lv^1 - f^1$ and $Lv^2 - f^2$ satisfy

$$\begin{aligned} Lv^1 - f^1 &= -(Lu - f)^+ + g, \\ Lv^2 - f^2 &= -(Lu - f)^- + g \qquad \text{in } \Omega. \end{aligned} \tag{5.20}$$

By Theorem 3.29,

$$v^1 \leq u^1 \quad \text{and} \quad v^2 \leq u^2 \qquad \text{in } \Omega,$$

thus the maximality of (u^1, u^2). On the other hand, we can find a necessary and sufficient condition on a function g as above in order that the corresponding pair (v^1, v^2) in (5.20) satisfy (5.19). Indeed, the identity

$$0 = (Lv^1 - f^1)(v^1 - v^2 + \varphi) = (Lu - f)^+(u - \varphi) - g(u - \varphi)$$

is valid in Ω if and only if, in any measurable subset of Ω where $u - \varphi > 0$ and therefore $(Lu - f)^+ = 0$, the identity $g(u - \varphi) = 0$ is valid and therefore $g = 0$. By the same token, the identity

$$0 = (Lv^2 - f^2)(v^2 - v^1 - \psi) = -(Lu - f)^-(u - \psi) + g(u - \psi)$$

is valid in Ω if and only if $g = 0$ in any measurable subset of Ω where $u - \psi < 0$. Summing up, any solution $(v^1, v^2) \in [H^2(\Omega)]^2$ to (5.19) must satisfy (5.20) with

$$g = 0 \qquad \text{in } \Omega \setminus S.$$

If $|S| = 0$, then $g = 0$ a.e. in Ω, and $(v^1, v^2) = (u^1, u^2)$. If instead $|S| > 0$, then the solution $(v^1, v^2) \in [H^2(\Omega)]^2$ to (5.20) with $g \equiv$ characteristic function of S satisfies (5.19) and differs from (u^1, u^2). ∎

System (5.19) is a particular case of the following system of unilateral problems:

$$v^k \le M^k(v), \qquad Lv^k \le f^k, \qquad (Lv^k - f^k)[v^k - M^k(v)] = 0 \qquad \text{in } \Omega,$$
$$v^k|_{\partial\Omega} = 0 \qquad \text{for } k = 1, \ldots, m, \tag{5.21}$$

where $v \equiv (v^1, \ldots, v^m)$ and

$$M^k(v) \equiv \bigwedge_{j \ne k} v^j + \psi^k.$$

For $m > 2$ we cannot solve (5.21) by reducing it to a bilateral problem. We proceed instead by putting some more restrictions on the ψ^k's.

THEOREM 5.7. *For $k = 1, \ldots, m$ let $f^k \in L^2(\Omega)$, $\psi^k \in H^2(\Omega)$ with $\psi^k|_{\partial\Omega} \ge 0$ and $L(\psi^j + \psi^k) \ge 0$ if $j \ne k$. Then* (5.21) *admits a maximal solution $u \equiv (u^1, \ldots, u^m) \in [H^2(\Omega)]^m$.*

PROOF. We set

$$\alpha^k \equiv \frac{L\psi^k}{2} - \bigvee_{i=1}^{m} \left(\frac{L\psi^i}{2} - f^i \right)$$

(obviously, $\alpha^k \le f^k$) and

$$\mathbb{D} \equiv \{v \equiv (v^1, \ldots, v^m) \in [H^2(\Omega)]^m \mid$$
$$\alpha^k \le Lv^k \le f^k \text{ in } \Omega,\ v^k|_{\partial\Omega} = 0 \text{ for } k = 1, \ldots, m\}.$$

$\mathbb{D}$ is convex and closed; moreover, it is also bounded in $[H^2(\Omega)]^m$ by Theorem 3.28(i). Let $v \in \mathbb{D}$. For each k Theorem 5.2 yields the existence of a unique function $z^k \in H^2(\Omega)$ such that

$$z^k \le M^k(v), \qquad Lz^k \le f^k, \qquad (Lz^k - f^k)[z^k - M^k(v)] = 0 \qquad \text{in } \Omega,$$
$$z^k|_{\partial\Omega} = 0; \tag{5.22}$$

moreover,

$$Lz^k \ge \left(\bigwedge_{j \ne k} Lv^j + L\psi^k \right) \wedge f^k \qquad \text{in } \Omega. \tag{5.23}$$

Since $v \in \mathbb{D}$ implies

$$\begin{aligned} \bigwedge_{j \ne k} Lv^j + L\psi^k &\ge \bigwedge_{j \ne k} \alpha^j + L\psi^k \\ &= \bigwedge_{j \ne k} \left[\frac{L\psi^j}{2} - \bigvee_{i=1}^{m} \left(\frac{L\psi^i}{2} - f^i \right) \right] + L\psi^k \\ &= \bigwedge_{j \ne k} \left(\alpha^k + \frac{L\psi^j}{2} + \frac{L\psi^k}{2} \right) = \alpha^k + \frac{1}{2} \bigwedge_{j \ne k} L(\psi^j + \psi^k), \end{aligned}$$

(5.23) yields $Lz^k \geq \alpha^k$ by the restriction imposed on the ψ^k's. Thus, $z \equiv (z^1, \ldots, z^m) \in \mathbb{D}$. Denote by σ the mapping $v \mapsto z$: we have proved that $\sigma(\mathbb{D}) \subseteq \mathbb{D}$.

Now let $u_0{}^k \in H^2(\Omega)$ solve

$$Lu_0{}^k = f^k \qquad \text{in } \Omega,$$
$$u_0{}^k|_{\partial\Omega} = 0.$$

Of course, $u_0 \equiv (u_0{}^1, \ldots, u_0{}^m)$ belongs to $\mathbb{D}$, and for each component of $u_1 \equiv \sigma(u_0) \in \mathbb{D}$ Theorem 3.29 yields

$$u_1{}^k \leq u_0{}^k \qquad \text{in } \Omega.$$

Let $u_{n+1} \equiv \sigma(u_n)$ for $n \in \mathbb{N}$. If $u_n{}^k \leq u_{n-1}^k$ for each k, then $M^k(u_n) \leq M^k(u_{n-1})$, and therefore $u_{n+1}^k \leq u_n{}^k$ by monotonicity with respect to obstacles (see the remark after Theorem 5.4). Thus each sequence $\{u_n{}^k\}_n \subset \mathbb{D}$, being nonincreasing as well as bounded in $H^2(\Omega)$, converges weakly in that space and strongly in $H^1(\Omega)$. We can pass to the limit as $n \to \infty$ in the system of unilateral problems (5.22) written for $v = u_n$, $z = u_{n+1}$. Let $u^k \equiv \lim_{n\to\infty} u_n{}^k$, $u \equiv (u^1, \ldots, u^m)$. Then not only $u \in \mathbb{D}$, but also

$$u^k \leq M^k(u), \qquad Lu^k \leq f^k, \qquad (Lu^k - f^k)[u^k - M^k(u)] = 0 \qquad \text{in } \Omega,$$
$$u^k|_{\partial\Omega} = 0 \qquad \text{for } k = 1, \ldots, m.$$

This shows that u is a solution to (5.21). As for maximality: if $v \in [H^2(\Omega)]^m$ is any other solution, Theorem 3.29 applied to each component v^k of v yields $v^k \leq u_0{}^k$, hence $v^k \leq u_n{}^k$ for $n \in \mathbb{N}$ by recurrence, and finally $v^k \leq u^k$. ∎

5.2. Differential Inequalities

From now on we suppose that the coefficients $a^1, \ldots, a^N, a$ (from the definition of L) and β (from the definition of B) vanish identically.

In this section we provide global and local bounds on functions u satisfying a differential inequality

$$\begin{gathered} |Lu| \leq K|\nabla u|^2 + f_0 \qquad \text{in } \Omega \\ \text{with } K > 0 \qquad \text{and} \qquad f_0 \in L^p(\Omega), \qquad f_0 \geq 0. \end{gathered} \tag{5.24}$$

As a preliminary, we give an interpolation inequality in $H^{2,p}(\Omega)$ which will be needed in the sequel.

5.2.1. Interpolation Results

LEMMA 5.8. *Let* $1 \leq p < \infty$, $1 \leq q \leq \infty$, *and* $2/r = 1/p + 1/q \leq 2$ (*where* $1/\infty$ *stands for* 0). *There exists a constant* C *such that*

$$| \nabla u |^2_{r;\mathbb{R}^N} \leq C \sum_{i,j=1}^{N} | u_{x_i x_j} |_{p;\mathbb{R}^N} \, | u |_{q;\mathbb{R}^N} \tag{5.25}$$

whenever $u \in C_c^\infty(\mathbb{R}^N)$.

PROOF. We begin with the case $N = 1$, $1 < p < \infty$, $1 \leq q < \infty$.

Let $u \in C_c^\infty(\mathbb{R})$. We claim that, whenever I is a bounded interval of lenght λ,

$$\int_I | u' |^r \, dx \leq \hat{C}\left[\lambda^{1+r-r/p} \left(\int_I | u'' |^p \, dx \right)^{r/p} + \lambda^{-(1+r-r/p)} \left(\int_I | u |^q \, dx \right)^{r/q} \right] \tag{5.26}$$

with $\hat{C}$ independent of u as well as of I. To prove (5.26) we set $]a, b[= I$ and apply the mean value theorem in any interval $[\xi, \eta]$ with $a < \xi < a + \lambda/4$, $b - \lambda/4 < \eta < b$:

$$\frac{u(\eta) - u(\xi)}{\eta - \xi} = u'(\zeta)$$

for a suitable ζ between ξ and η, hence

$$| u'(x) | = \left| \int_\zeta^x u''(t) \, dt + u'(\zeta) \right| \leq \int_I | u''(t) | \, dt + 2 \frac{| u(\xi) | + | u(\eta) |}{\lambda}$$

for $x \in I$. After integrating with respect to ξ from a to $a + \lambda/4$ and with respect to η from $b - \lambda/4$ to b, we find

$$\begin{aligned} | u'(x) | &\leq \int_I | u''(t) | \, dt + \frac{C}{\lambda^2} \int_I | u(t) | \, dt \\ &\leq \lambda^{1-1/p} \left(\int_I | u'' |^p \, dt \right)^{1/p} + C\lambda^{-1-1/q} \left(\int_I | u |^q \, dt \right)^{1/q} \end{aligned}$$

by Hölder's inequality; therefore,

$$\int_I | u' |^r \, dx \leq \hat{C}\left[\lambda^{1+r-r/p} \left(\int_I | u'' |^p \, dx \right)^{r/p} + \lambda^{1-r-r/q} \left(\int_I | u |^q \, dx \right)^{r/q} \right],$$

the last inequality amounting to (5.26) because $1 - r/q = r/p - 1$.

We shall now prove

$$\int_I |u'|^r\,dx \le 2\hat{C}\left(\int_{-\infty}^{\infty} |u''|^p\,dx\right)^{r/2p}\left(\int_{-\infty}^{\infty} |u|^q\,dx\right)^{r/2q} \tag{5.27}$$

which is nothing but (5.25) in the case at hand by the arbitrariness of I.

Without loss of generality we assume $\operatorname{supp} u \subset]0, \infty[$ and restrict ourselves to intervals $I =]0, \lambda[$. Let $\lambda_0 \equiv \lambda/k$ with $k \in N$ arbitrarily fixed, and consider (5.26) with I replaced by $I_0 \equiv]0, \lambda_0[$. If

$$\lambda_0^{1+r-r/p}\left(\int_{I_0} |u''|^p\,dx\right)^{r/p} \ge \lambda_0^{-(1+r-r/p)}\left(\int_{I_0} |u|^q\,dx\right)^{r/q} \tag{5.28}$$

we set $I_1 \equiv I_0$ and obtain

$$\int_{I_1} |u'|^r\,dx \le 2\hat{C}\left(\frac{\lambda}{k}\right)^{1+r-r/p}\left(\int_{-\infty}^{\infty} |u''|^p\,dx\right)^{r/p}. \tag{5.29}$$

Suppose that (5.28) does not hold: we then denote by I_1 the bounded interval of length λ_1 obtained by increasing the right endpoint of I_0 until we reach the equality sign in (5.28) with I_0 replaced by I_1 and λ_0 by λ_1. (Note that u'' cannot vanish identically unless u does.) We obtain

$$\int_{I_1} |u'|^r\,dx \le 2\hat{C}\left(\int_{I_1} |u''|^p\,dx\right)^{r/2p}\left(\int_{I_1} |u|^q\,dx\right)^{r/2q}. \tag{5.30}$$

Starting at the right endpoint of I_1 we repeat the above procedure, choosing an interval I_2 of length λ_2, and so on until I is covered. This requires k steps at most.

We now sum our estimates (5.29) and (5.30) with I_1 replaced by I_j, λ_1 by λ_j, and arrive at

$$\int_I |u'|^r\,dx \le 2\hat{C}\left(\frac{\lambda}{k}\right)^{1+r-r/p} k\left(\int_{-\infty}^{\infty} |u''|^p\,dx\right)^{r/p}$$
$$+ 2\hat{C}\left(\int_{-\infty}^{\infty} |u''|^p\,dx\right)^{r/2p}\left(\int_{-\infty}^{\infty} |u|^q\,dx\right)^{r/2q}$$

with the aid of Hölder's inequality. Now let $k \to \infty$: since $p > 1$, (5.27) follows.

If $N > 1$ we apply the preceding one-dimensional result to each function $x_i \mapsto u(x_1, \ldots, x_i, \ldots, x_N)$ treating the variables x_j for $j \ne i$ as parameters, and obtain (5.25) by utilizing Hölder's inequality in $N-1$ dimensions.

Finally, we cover the cases $q = \infty$ and $p = 1$ through easy passages to the limit. □

Lemma 5.8 is instrumental in proving the following theorem.

THEOREM 5.9. *Take p, q, r as in Lemma 5.8. If* $u \in H^{2,p}(\Omega) \cap L^q(\Omega)$ *then* $u_{x_i}, \ldots, u_{x_N} \in L^r(\Omega)$ *and there exists a constant C (independent of u) such that*

$$| \nabla u |^2_{r;\Omega} \leq C \, | u |_{H^{2,p}(\Omega)} \, | u |_{q;\Omega} . \tag{5.31}$$

PROOF. By Theorem 1.30 u admits an extension $\tilde{u} \in H^{2,p}(\mathbb{R}^N)$ with compact support and

$$| \tilde{u} |_{H^{2,p}(\mathbb{R}^N)} \leq C \, | u |_{H^{2,p}(\Omega)} .$$

Moreover, the construction of $\tilde{u}$ shows that the latter belongs to $L^q(\mathbb{R}^N)$ with

$$| \tilde{u} |_{q;\mathbb{R}^N} \leq C \, | u |_{q;\Omega} .$$

The proof of the theorem is achieved by showing that $\tilde{u}_{x_1}, \ldots, \tilde{u}_{x_N} \in L^r(\mathbb{R}^N)$ and that (5.25) holds with u replaced by $\tilde{u}$. To do this we approximate $\tilde{u}$, both in $H^{2,p}(\mathbb{R}^N)$ and in $L^q(\mathbb{R}^N)$, with the regularizing sequence $\{\varrho_n * \tilde{u}\} \subset C_c^\infty(\mathbb{R}^N)$: Lemma 5.8 shows that $\{(\varrho_n * \tilde{u})_{x_i}\}$, $i = 1, \ldots, N$, is a Cauchy sequence in $L^r(\mathbb{R}^N)$, and the conclusion follows from a passage to the limit. □

REMARK. It is clear that the above result remains valid if the regularity assumption about $\partial\Omega$ is weakened into the requirement that Ω has the extension property $(2, p)$. In particular consider cubes, denoted by Q_R, such as $]-R, R[^N$ or $]-R, R[^{N-1} \times]0, 2R[$ (see Problem 1.17). If $u \in H^{2,p}(Q_R) \cap L^q(Q_R)$ and $R' > R$ we first estimate the $L^r(Q_{R'})$ norms of first derivatives of the function $x \mapsto u(Rx/R')$, $x \in Q_{R'}$, then perform the change of variables $x \mapsto R'x/R$ inside the integrals, and finally obtain

$$| \nabla u |^2_{r;Q_R} \leq C\Big(| u |_{p;Q_R}/R^2 + | \nabla u |_{p;Q_R}/R + \sum_{i,j=1}^{N} | u_{x_i x_j} |_{p;Q_R} \Big) | u |_{q;Q_R}$$

with $C = C(R')$ independent of R.

5.2.2. A Global Bound

From now on we assume $p > N$ [so that $H^{2,p}(\Omega)$ injects compactly in $C^1(\bar{\Omega})$].

LEMMA 5.10. *Let* $u \in H^{2,p}(\Omega)$ *satisfy* (5.24) *as well as* $u|_{\partial\Omega \setminus \Gamma} = 0$, *and let* $M \in]0, \infty[$ *be* $\geq |u|_{\infty;\Omega} + |Bu|_{C^0(\Gamma)}$. *Then there exists a constant* $\bar{\varkappa}$, *depending on* u *only through* M, *such that*

$$|u|_{H^{2,p}(\Omega)} \leq \bar{\varkappa}(1 + |Bu|_{H^{1/p',p}(\Gamma)}). \tag{5.32}$$

PROOF. *Step* 1: *The case* $f_0 \in L^\infty(\Omega)$. If f_0 belongs to $L^\infty(\Omega)$ the absolute value of the function

$$g(x) \equiv [Lu(x) + u(x)]/[|\nabla u(x)|^2 + 1]$$

is bounded a.e. in Ω by $K + |f_0|_{\infty;\Omega} + |u|_{\infty;\Omega}$. For $t, \sigma \in [0, 1]$ we introduce the b.v.p.

$$\begin{aligned} (L + 1)z^{t\sigma} &= \sigma g(|\nabla z^{t\sigma}|^2 + t) \qquad \text{in } \Omega, \\ z^{t\sigma}|_{\partial\Omega \setminus \Gamma} = 0, \qquad Bz^{t\sigma} + z^{t\sigma}|_\Gamma &= t\sigma(Bu + u|_\Gamma) \qquad \text{on } \Gamma. \end{aligned} \tag{5.33}$$

If $z^{t\sigma} \in H^{2,p}(\Omega)$ solves (5.33) and $z^{s\sigma} \in H^{2,p}(\Omega)$ its analog with t replaced by s, we set $w \equiv z^{s\sigma} - z^{t\sigma}$, $M' \equiv |s - t|(M + |g|_{\infty;\Omega})$. The function $M' - w$ satisfies

$$\begin{aligned} (L + 1)(M' - w) &- \sigma g(z^{s\sigma} + z^{t\sigma})_{x_i}(M' - w)_{x_i} \\ &= M' - \sigma g(s - t) \geq 0 \qquad \text{in } \Omega \end{aligned}$$

as well as

$$(M' - w)|_{\partial\Omega \setminus \Gamma} \geq 0,$$

$$B(M' - w) + (M' - w)|_\Gamma = M' - (s - t)\sigma(Bu + u|_\Gamma) \geq 0 \qquad \text{on } \Gamma.$$

We can apply Theorem 3.29 to the operator

$$-a^{ij}\partial^2/\partial x_i\, \partial x_j - \sigma g(z^{s\sigma} + z^{t\sigma})_{x_i}\partial/\partial x_i + 1$$

and deduce that $M' - w \geq 0$. Analogously, $M' + w \geq 0$. We have thus proved that

$$|w|_{\infty;\Omega} \leq |s - t|(M + |g|_{\infty;\Omega}). \tag{5.34}$$

Hence a solution of (5.33), if existing, is unique; moreover, the choice $t = 0$ yields

$$|z^{s\sigma}|_{\infty;\Omega} \leq M + |g|_{\infty;\Omega}. \tag{5.35}$$

On the other hand, w satisfies

$$(L+1)w = \sigma g(|\nabla z^{s\sigma}|^2 - |\nabla(w - z^{s\sigma})|^2) + \sigma g(s-t) \qquad \text{in } \Omega,$$

$$w|_{\partial\Omega\setminus\Gamma} = 0,$$

$$Bw + w|_\Gamma = (s-t)\sigma(Bu + u|_\Gamma) \qquad \text{on } \Gamma,$$

and Theorem 3.28(i) yields a bound

$$|w|_{H^{2,p}(\Omega)} \leq C[|g|_{\infty;\Omega}(\big|\,|\nabla z^{s\sigma}|^2\big|_{p;\Omega} + \big|\,|\nabla w|^2\big|_{p;\Omega} + 1) + \big|Bu + u|_\Gamma\big|_{H^{1/p',p}(\Gamma)}]. \tag{5.36}$$

To estimate $\big|\,|\nabla w|^2\big|_{p;\Omega}$ we apply (5.31) with $q = \infty$ and $r = 2p$, so that

$$\big|\,|\nabla w|^2\big|_{p;\Omega} \leq C|w|_{H^{2,p}(\Omega)}|w|_{\infty;\Omega}. \tag{5.37}$$

We proceed analogously for what concerns $\big|\,|\nabla z^{s\sigma}|^2\big|_{p;\Omega}$, and deduce from (5.34)–(5.37) an estimate

$$|z^{s\sigma} - z^{t\sigma}|_{H^{2,p}(\Omega)} \leq C\{|g|_{\infty;\Omega}[(M + |g|_{\infty;\Omega})(|z^{s\sigma}|_{H^{2,p}(\Omega)} + |s-t|\,|z^{s\sigma} - z^{t\sigma}|_{H^{2,p}(\Omega)}) + 1] + \big|Bu + u|_\Gamma\big|_{H^{1/p',p}(\Gamma)}\}.$$

In particular, if $|s-t| \leq m^{-1}$ with $m \in N$ large enough we obtain

$$|z^{t\sigma}|_{H^{2,p}(\Omega)} \leq |z^{s\sigma}|_{H^{2,p}(\Omega)} + C\{|g|_{\infty;\Omega}[(M + |g|_{\infty;\Omega})|z^{s\sigma}|_{H^{2,p}(\Omega)} + 1] + \big|Bu + u|_\Gamma\big|_{H^{1/p',p}(\Gamma)}\}. \tag{5.38}$$

For $i = 1, \ldots, m$ we now denote by u_i the solution (whenever it exists) of (5.33) with $t = i/m$ and $\sigma = 1$. We shall prove that $u_1, \ldots, u_m$ do indeed exist and satisfy

$$|u_i|_{H^{2,p}(\Omega)} \leq \varkappa_i(1 + |Bu + u|_\Gamma|_{H^{1/p',p}(\Gamma)}), \tag{5.39}$$

where each $\varkappa_i$ depends on the same parameters as $\bar{\varkappa}$ in (5.32). This will accomplish the proof for $f_0 \in L^\infty(\Omega)$, since $u = u_m$ and

$$|Bu + u|_\Gamma|_{H^{1/p',p}(\Gamma)} \leq |Bu|_{H^{1/p',p}(\Gamma)} + |u|_{H^{1,p}(\Omega)} \leq |Bu|_{H^{1/p',p}(\Gamma)} + \varepsilon|u|_{H^{2,p}(\Omega)} + C(\varepsilon)M|\Omega|^{1/p}$$

for all $\varepsilon > 0$ (see Lemma 1.37).

If u_1 exist, the corresponding estimate (5.39) follows immediately from (5.38) with $s = 0$, $t = 1/m$, $\sigma = 1$. To prove existence we apply the Leray–Schauder fixed point theorem. Let z be arbitrarily given in $C^1(\bar{\Omega})$, σ in $[0, 1]$, and denote by $\mathscr{C}(\sigma, z)$ the solution $v \in H^{2,p}(\Omega)$ of the linear b.v.p.

$$(L+1)v = \sigma g(|\nabla z|^2 + 1/m) \qquad \text{in } \Omega,$$

$$v|_{\partial\Omega\setminus\Gamma} = 0, \qquad Bv + v|_\Gamma = \frac{1}{m}\sigma(Bu + u|_\Gamma) \qquad \text{on } \Gamma$$

[see Theorem 3.29(i)]. By taking Theorem 3.28(i) into account it is easy to ascertain that $\mathscr{C}$ is a compact mapping from $[0, 1]\times C^1(\bar{\Omega})$ into $C^1(\bar{\Omega})$. If $v = \mathscr{C}(\sigma, v)$ (and therefore $v = z^{(1/m)\sigma}$) for some $\sigma \in [0, 1]$, a uniform bound on $|v|_{C^1(\bar{\Omega})}$ follows from (5.38) with $t = 1/m$ and $s = 0$. Finally, $\mathscr{C}(0, z) = 0$. The existence of the fixed point u_1 of the mapping $\mathscr{C}(1, \cdot)$ is now consequent on Theorem 1.K.

By the same token, (5.33) with $t = 1/m$ is uniquely solvable (with norm estimate) also for $0 < \sigma < 1$. This means that for $0 \le \sigma \le 1$ a uniform $H^{2,p}$ bound on $z^{(2/m)\sigma}$, whenever the latter exists, is provided by (5.38) with $s = 1/m$, $t = 2/m$. We can therefore apply Theorem 1.K again, and arrive at the existence of u_2 [with the corresponding norm estimate (5.39)]. The final conclusion follows by repeating the above procedure a convenient number of times.

Step 2: *The general case.* If f_0 does not belong to $L^\infty(\Omega)$, we introduce the bounded function

$$h(x) \equiv [Lu(x) + u(x)]/[|\nabla u(x)|^2 + f_0(x) + 1]$$

and solve the linear b.v.p.

$$(L+1)v = h(1 + f_0) \qquad \text{in } \Omega,$$

$$v|_{\partial\Omega\setminus\Gamma} = 0, \qquad Bv = 0 \qquad \text{on } \Gamma.$$

The function $w \equiv u - v$ satisfies

$$(L+1)w = g'(|\nabla w|^2 + 1),$$

$$w|_{\partial\Omega\setminus\Gamma} = 0, \qquad Bw = Bu \qquad \text{on } \Gamma$$

where $g' \equiv h\,|\nabla u|^2/(|\nabla w|^2 + 1)$. Since

$$|g'| \le |h|\,\frac{2\,|\nabla v|^2 + 2\,|\nabla w|^2}{|\nabla w|^2 + 1} \qquad \text{in } \Omega,$$

and

$$| v |_{H^{2,p}(\Omega)} \leq C(K + M + 1)(1 + | f_0 |_{p;\Omega}),$$

by Theorem 3.28(i), (5.32) follows from Step 1 with u replaced by w. □

5.2.3. A Local Bound

LEMMA 5.11. *Let $u \in H^{2,p}(\Omega)$ vanish on $\partial\Omega \cap \Omega'$ and satisfy* (5.24) *with Ω replaced by $\Omega'' \equiv \Omega \cap \Omega'$, where Ω' is some open subset of $\mathbb{R}^N$. Whenever ω is an open subset of Ω whose closure lies in Ω', there exists a constant $\bar{\varkappa}$, depending on u only through its modulus of uniform continuity in $\bar{\Omega}''$ and the bound on $| u |_{\infty;\Omega''}$, such that*

$$| u |_{H^{2,p}(\omega)} \leq \bar{\varkappa}. \tag{5.40}$$

PROOF. *Step 1: A family of cubes.* We suppose that $\bar{\omega}$ contains the origin 0 of $\mathbb{R}^N$.

If $0 \in \Omega$ we denote by $Q_r{}^a$ the cubes $]-r(1-a), r(1-a)[^N$ for $0 \leq a < 1$ and $0 < r \leq R$, where R is so small that $Q_R{}^0 \subset \Omega''$.

If instead $0 \in \partial\Omega$ we suppose that a relatively open portion of $\partial\Omega$ near 0 lies on the hyperplane $x_N = 0$, and denote by $Q_r{}^a$ the cubes $]-r(1-a), r(1-a)[^{N-1} \times]0, 2r(1-a)[$ for $0 \leq a < 1$ and $0 < r \leq R$, where R is so small that $Q_R{}^0 \subset \Omega''$ and

$$]-R, R[^{N-1} \times \{0\} \subset \partial\Omega \cap \Omega'.$$

In both cases above elementary considerations show that we can construct cutoff functions $g_r{}^a \in C_c^\infty(\mathbb{R}^N)$ with

$$0 \leq g_r{}^a \leq 1, \qquad g_r{}^a = 1 \quad \text{on } Q_r{}^a, \qquad g_r{}^a = 0 \quad \text{in } \Omega \diagdown Q_r{}^{a/2},$$
$$| g^a_{rx_i} | \leq C/ar \quad \text{and} \quad | g^a_{rx_ix_j} | \leq C/(ar)^2 \ \ (i, j = 1, \ldots, N)$$

whenever $0 < a < 1$ and $0 < r \leq R$.

Step 2: Estimates over cubes. We let a vary in $]0, 1/2]$ and set $w_r{}^a \equiv g_r{}^a u$. From (5.24) we deduce

$$\begin{aligned} | Lw_r{}^a |_{p;\Omega} &= | Lw_r{}^a |_{p;Q_r{}^{a/2}} \\ &= | g_r{}^a Lu + uLg_r{}^a + 2a^{ij} u_{x_i} g^a_{rx_j} |_{p;Q_r{}^{a/2}} \\ &\leq K | \, | \nabla u |^2 |_{p;Q_r{}^{a/2}} + | f_0 |_{p;Q_r{}^{a/2}} \\ &\quad + \frac{C}{ar} | \nabla u |_{p;Q_r{}^{a/2}} + \frac{C}{(ar)^2} | u |_{p;Q_r{}^{a/2}}. \end{aligned} \tag{5.41}$$

To estimate $\big|\,|\nabla u|^2\big|_{p;Q_r^{a/2}}$ we introduce the function $z \equiv (u - u_0)\times (g_r^{a/2})^{2p}$, with $u_0 \equiv u(0)$ and therefore $u_0 = 0$ if $0 \in \partial\Omega$. Since $z \in H_0^1(Q_r^{a/4})$, we have

$$\begin{aligned}
I^{\pm} &\equiv \int_{Q_r^{a/4}} \{[(u_{x_i})^{\pm}]^{2p-1}\}_{x_i} z\,dx \\
&= -\int_{Q_r^{a/4}} [(u_{x_i})^{\pm}]^{2p-1}\{u_{x_i}(g_r^{a/2})^{2p} + (u-u_0)[(g_r^{a/2})^{2p}]_{x_i}\}\,dx \\
&= \mp \sum_{i=1}^{N} \int_{Q_r^{a/4}} [(u_{x_i})^{\pm}]^{2p}(g_r^{a/2})^{2p}\,dx \\
&\quad - 2p\int_{Q_r^{a/4}} [(u_{x_i})^{\pm}]^{2p-1}(u-u_0)(g_r^{a/2})^{2p-1} g_{rx_i}^{a/2}\,dx
\end{aligned} \tag{5.42}$$

as well as

$$I^{\pm} = (2p-1)\int_{Q_r^{a/4}} [(u_{x_i})^{\pm}]^{2p-2} u_{x_i x_i}(u-u_0)(g_r^{a/2})^{2p}\,dx \tag{5.43}$$

by Theorem 1.56 and Lemma 1.57. We utilize (5.42) and (5.43) to majorize $\big|\,|\nabla u|^2\big|_{p;Q_r^{a/2}}^p$ with

$$\begin{aligned}
\int_{Q_r^{a/4}} |\nabla u|^{2p}(g_r^{a/2})^{2p}\,dx \le C \operatorname*{osc}_r u\bigg(&\int_{Q_r^{a/4}} |\nabla u|^{2(p-1)} \sum_{i,j=1}^{N} |u_{x_i x_j}|\,dx \\
&+ |\nabla g_r^{a/2}|_{\infty;\mathbb{R}^N} \int_{Q_r^{a/4}} |\nabla u|^{2p-1}\,dx\bigg),
\end{aligned}$$

where

$$\operatorname*{osc}_r u \equiv \max_{\overline{Q_r^0}} u - \min_{\overline{Q_r^0}} u.$$

Finally, since

$$|st| \le \frac{\delta^p |s|^p}{p} + \frac{|t|^{p'}}{\delta^{p'} p'}$$

for $s, t \in \mathbb{R}$ and $\delta > 0$, Hölder's inequality yields

$$\begin{aligned}
\big|\,|\nabla u|^2\big|_{p;Q_r^{a/2}}^p &\le C \operatorname*{osc}_r u\bigg(\int_{Q_r^{a/4}} |\nabla u|^{2p}\,dx\bigg)^{1/p'} (|u|_{H^{2,p}(Q_r^{a/4})} \\
&\quad + |\nabla g_r^{a/2}|_{\infty;\mathbb{R}^N} |\nabla u|_{p;Q_r^{a/4}}) \\
&\le C(\varepsilon)(\operatorname*{osc}_r u)^{p'} \int_{Q_r^{a/4}} |\nabla u|^{2p}\,dx \\
&\quad + \varepsilon^p(|u|_{H^{2,p}(Q_r^{a/4})}^p + |\nabla g_r^{a/2}|_{\infty;\mathbb{R}^N}^p |\nabla u|_{p;Q_r^{a/4}}^p)
\end{aligned} \tag{5.44}$$

for $\varepsilon > 0$, with $C(\varepsilon)$ independent of a and r.

Next, we majorize $\int_{Q_r^{a/4}} |\nabla u|^{2p}\,dx$ with

$$C_0 |u|_{\infty;Q_r^{a/4}}^p \left\{ \sum_{i,j=1}^N |u_{x_ix_j}|_{p;Q_r^{a/4}}^p + [(1-a/4)r]^{-p} |\nabla u|_{p;Q_r^{a/4}}^p + [(1-a/4)r]^{-2p} |u|_{p;Q_r^{a/4}}^p \right\},$$

where $C_0 = C_0(R)$ is independent of a and r (see the remark after Theorem 5.9). Let $r = r(\varepsilon) > 0$ be so small that

$$C(\varepsilon)(\operatorname*{osc}_r u)^{p'} C_0 |u|_{\infty;\Omega''}^p < \varepsilon^p:$$

since $(1-a/4)r \geq ar$, from (5.44) we deduce (if $\varepsilon \leq 1$) an estimate

$$\big| |\nabla u|^2 \big|_{p;Q_r^{a/2}}^p \leq 2\varepsilon^p |u|_{H^{2,p}(Q_r^{a/4})}^p + C[(ar)^{-p} |\nabla u|_{p;Q_r^{a/4}}^p + (ar)^{-2p} |u|_{p;Q_r^{a/4}}^p]$$

with C independent of a and r.

At this point we go back to (5.41): by Theorem 3.28(i) we have

$$\begin{aligned} |u|_{H^{2,p}(Q_r^a)}^p &\leq |w_r^a|_{H^{2,p}(\Omega)}^p \leq C |Lw_r^a|_{p;\Omega}^p \\ &\leq C_1[\varepsilon^p |u|_{H^{2,p}(Q_r^{a/4})}^p + |f_0|_{p;Q_r^{a/4}}^p + (ar)^{-p} |\nabla u|_{p;Q_r^{a/4}}^p \\ &\qquad + (ar)^{-2p} |u|_{p;Q_r^{a/4}}^p]. \end{aligned} \tag{5.45}$$

with C_1 independent of ε, a, and r. Let ε be so small that $C_1\varepsilon^p \leq 1/(2\times 4^{2p+1})$, and let r be fixed correspondingly. As in Problem 1.18 an estimate

$$|\nabla u|_{p;Q_r^{a/4}}^p \leq \delta |u|_{H^{2,p}(Q_r^{a/4})}^p + \frac{C_2}{\delta} |u|_{p;Q_r^{a/4}}^p,$$

with C_2 independent of δ as well as of a and r, is valid whenever $\delta > 0$ is sufficiently small, say $\delta \leq \hat{C}(1-a/4)^p r^p$. We take $\delta = (\eta a r)^p$, where $\eta > 0$ satisfies $(\eta r/2)^p \leq \hat{C}(7r/8)^p$ and $C_1\eta^p \leq 1/(2\times 4^{2p+1})$. With the above choice of ε, r, and η (5.45) yields

$$\begin{aligned} |u|_{H^{2,p}(Q_r^a)}^p &\leq \frac{1}{4^{2p+1}} |u|_{H^{2,p}(Q_r^{a/4})}^p \\ &\qquad + C_1[|f_0|_{p;Q_r^{a/4}}^p + (1 + C_2\eta^{-p})(ar)^{-2p} |u|_{p;Q_r^{a/4}}^p] \end{aligned} \tag{5.46}$$

whenever $0 < a \leq 1/2$. Set $\Phi(a) \equiv a^{2p} |u|_{H^{2,p}(Q_r^a)}^p$,

$$H(a) \equiv C_1[a^{2p} |f_0|_{p;\Omega''}^p + (1 + C_2\eta^{-p}) r^{-2p} |u|_{p;\Omega''}^p].$$

From (5.46) it follows that

$$\Phi(a) \leq \frac{1}{4}\,\Phi\left(\frac{a}{4}\right) + H(a),$$

and by recurrence

$$\Phi(a) \leq \frac{1}{4^n}\,\Phi\left(\frac{a}{4^n}\right) + \sum_{i=0}^{n-1} \frac{1}{4^i}\, H\left(\frac{a}{4^i}\right), \qquad n \in \mathbb{N}.$$

Letting $n \to \infty$ we obtain

$$\Phi(a) \leq \sum_{i=0}^{\infty} \frac{1}{4^i}\, H\left(\frac{a}{4^i}\right) \leq H(a) \sum_{i=0}^{\infty} \frac{1}{4^i}$$

because Φ is bounded on $]0, 1/2]$ and H is increasing. This provides a bound on, say, $|\, u\,|_{H^{2,p}(Q_r{}^{1/2})}$.

Step 3: *Completion of the proof.* Let x be arbitrarily fixed in $\bar{\omega}$.

If x lies in Ω a bound of the required type on $|\, u\,|_{H^{2,p}(Q(x))}$, where $Q(x) \subset \Omega$ is a suitable open cube centered at x, is obtained from Step 2 through a translation of x into the origin 0 of $\mathbb{R}^N$.

If instead x is a boundary point we straighten a relatively open portion of $\Omega' \cap \partial\Omega$ near x, say $U(x) \cap \partial\Omega$, through a $C^{1,1}$ diffeomorphism Λ: $\overline{U(x)} \to \bar{B}$ with $\Lambda(x) = 0$ [$U(x)$ being a suitable bounded domain of $\mathbb{R}^N$]. Since (5.24), with Ω replaced by $U(x) \cap \Omega$, is transformed by Λ into a similar inequality in B^+, Step 2 again yields the desired $H^{2,p}$ bound in a suitable cube $]-\varrho, \varrho[^{N-1} \times]0, 2\varrho[\subset B^+$, hence also in its image under Λ^{-1}. We set

$$Q'(x) \equiv \Lambda^{-1}(]-\varrho, \varrho[^{N-1} \times]-2\varrho, 2\varrho[).$$

Since $\bar{\omega}$ is covered by the family of all open sets $Q(x)$, $x \in \bar{\omega} \cap \Omega$, and $Q'(x)$, $x \in \bar{\omega} \cap \partial\Omega$, the sought-for $H^{2,p}$ bound in ω follows from the compactness of $\bar{\omega}$. □

REMARK 1. Both in (5.32) and in (5.40) $\varkappa$ depends on the a^{ij}'s only through the bound on their $L^\infty(\Omega)$ norms, α and τ.

REMARK 2. A major difficulty in the proof of Lemma 5.11 comes from the type of dependence the constant $\bar{\varkappa}$ is required to have on u. The reader may want to give a simpler proof of (5.40) with $\bar{\varkappa}$ depending on u through $|\, u\,|_{H^{1,p}(\Omega'')}$. See J. Frehse [51]. □

5.3. Obstacle Problems for Nonlinear Operators

We generalize (5.1) as follows:

$$\begin{aligned} &\varphi \leq u \leq \psi, \qquad [Lu - F(u, \nabla u)](u - \varphi) \leq 0 \\ &\qquad \text{and} \quad [Lu - F(u, \nabla u)](u - \psi) \leq 0 \qquad \text{in } \Omega, \\ &u|_{\partial\Omega\smallsetminus\Gamma} = 0, \qquad Bu = Z(u) \qquad \text{on } \Gamma. \end{aligned} \tag{5.47}$$

Here and throughout $F(u, \nabla u)$ denotes the function $x \mapsto f(x, u(x), \nabla u(x))$, $x \in \Omega, f(x, \eta, \xi)$ being a Carathéodory function of $x \in \Omega$ and $(\eta, \xi) \in \mathbb{R}^{1+N}$, whereas $Z(u)$ denotes the function $x \mapsto \zeta(x, u(x))$, $x \in \Gamma$, where ζ belongs to $C^{0,1}(\Gamma \times \mathbb{R})$. Of course, (5.47) amounts to (5.1) when $F(u, \nabla u) = -a^i u_{x_i} - au + f$ with $f = f(x)$, and $Z(u) = -\beta u|_\Gamma + \zeta$ with $\zeta = \zeta(x)$.

We shall first investigate the solvability of (5.47), then apply the results thus obtained to the nonlinear generalization of (5.6), that is,

$$\begin{aligned} &u \leq \psi, \qquad Lu \leq F(u, \nabla u) \quad \text{and} \quad [Lu - F(u, \nabla u)](u - \psi) = 0 \qquad \text{in } \Omega, \\ &\qquad u|_{\partial\Omega\smallsetminus\Gamma} = 0, \qquad Bu = Z(u) \qquad \text{on } \Gamma, \end{aligned} \tag{5.48}$$

as well as to the unconstrained nonlinear b.v.p.

$$\begin{aligned} &Lu = F(u, \nabla u) \qquad \text{in } \Omega, \\ &u|_{\partial\Omega\smallsetminus\Gamma} = 0, \qquad Bu = Z(u) \qquad \text{on } \Gamma. \end{aligned} \tag{5.49}$$

[Note the difference with the linear case: for (5.49) we have not proved yet existence and uniqueness results corresponding to Theorem 3.29.]

5.3.1. Existence

On the function $f(x, \eta, \xi)$ we now impose the following natural growth condition: given any $r \in \,]0, \infty[$,

$$\begin{aligned} &| f(x, \eta, \xi) | \leq K | \xi |^2 + f_0(x) \\ &\text{for a.a. } x \in \Omega \qquad \text{and} \qquad \text{any } (\eta, \xi) \in \mathbb{R}^{1+N}, \qquad | \eta | \leq r \end{aligned} \tag{5.50}$$

where the constant $K > 0$ and the function $f_0 \geq 0, f_0 \in L^p(\Omega)$, depend on r. [Compare with (4.88).]

In the proof of the next result a crucial role is played by Lemma 5.10.

THEOREM 5.12. *Assume* (5.50). *Let* $\varphi = \bigvee_{h=1}^{m} \varphi^h$ *with* $\varphi^h \in H^{2,p}(\Omega)$, $\varphi^h|_{\partial\Omega\setminus\Gamma} \leq 0$ *and* $B\varphi^h \leq Z(\varphi^h)$ *on* Γ, $\psi = \bigwedge_{h=1}^{m} \psi^h$ *with* $\psi^h \in H^{2,p}(\Omega)$, $\psi^h|_{\partial\Omega\setminus\Gamma} \geq 0$ *and* $B\psi^h \geq Z(\psi^h)$ *on* Γ. *If* $\varphi \leq \psi$ *on* $\bar{\Omega}$, (5.47) *admits a maximal and a minimal solution in* $H^{2,p}(\Omega)$.

PROOF. *Step* 1: *An intermediate existence result.* For $x \in \Gamma$ the function $\zeta_\lambda(x, \eta) \equiv \zeta(x, \eta) + \lambda\eta$, where λ is any positive number $\geq [\zeta]_{1;\Gamma\times R}$, is nondecreasing. We denote by $Z_\lambda(w)$ the function $x \mapsto \zeta_\lambda(x, w(x))$, and set $\hat{Z}_\lambda(w) \equiv Z_\lambda(\varphi \vee w \wedge \psi)$. It is easy to verify that

$$| \hat{Z}_\lambda(w) |_{H^{1/p',p}(\Gamma)} \leq C(1 + | w |_{C^{0,1}(\bar{\Omega})})$$

if w belongs to $C^{0,1}(\bar{\Omega})$, or in particular (see the remark at the end of Section 1.2.2) to $C^1(\bar{\Omega})$. Moreover,

$$B\varphi^h + \lambda\varphi^h|_\Gamma \leq \hat{Z}_\lambda(w) \leq B\psi^h + \lambda\psi^h|_\Gamma \qquad \text{on } \Gamma.$$

Let w be arbitrarily fixed in $C^1(\bar{\Omega})$, σ in [0, 1]. By Theorem 5.1 the bilateral problem

$$\sigma\varphi \leq v \leq \sigma\psi, \qquad [Lv - F(w, \nabla w)](v - \sigma\varphi) \leq 0$$
$$\text{and} \quad [Lv - F(w, \nabla w)](v - \sigma\psi) \leq 0 \qquad \text{in } \Omega,$$
$$v|_{\partial\Omega\setminus\Gamma} = 0, \qquad Bv + \lambda v|_\Gamma = \sigma\hat{Z}_\lambda(w) \qquad \text{on } \Gamma$$

admits a unique solution $v \in H^{2,p}(\Omega)$; moreover,

$$\bigwedge_{h=1}^{m} (\sigma L\psi^h) \wedge F(w, \nabla w) \leq Lv \leq \bigvee_{h=1}^{m} (\sigma L\varphi^h) \vee F(w, \nabla w) \qquad \text{in } \Omega. \tag{5.51}$$

We denote by $\mathscr{C}$ the operator $(\sigma, w) \mapsto v$. Note that $\mathscr{C}(0, w) = 0$ for all w.

Let w vary in a bounded subset of $C^1(\bar{\Omega})$. Then $F(w, \nabla w)$ remains bounded in $L^p(\Omega)$ and $\hat{Z}_\lambda(w)$ in $H^{1/p',p}(\Gamma)$; by (5.51), $\mathscr{C}(\sigma, w)$ remains bounded, independently of σ, in $H^{2,p}(\Omega)$ [Theorem 3.28(i)]. As $F(w_n, \nabla w_n) \to F(w, \nabla w)$ in $L^p(\Omega)$ and $\hat{Z}_\lambda(w_n) \to \hat{Z}_\lambda(w)$ in $C^0(\Gamma)$ whenever $w_n \to w$ in $C^1(\bar{\Omega})$, it is easy to conclude that $\mathscr{C}$ is a compact operator $[0, 1] \times C^1(\bar{\Omega}) \to C^1(\bar{\Omega})$. Theorem 1.K can be applied to the mapping $\mathscr{C}(1, \cdot)$, and the existence of a solution to (5.47) can thus be proven, if an *a priori* bound on $| v |_{C^1(\bar{\Omega})}$ is provided for all possible fixed points $v = \mathscr{C}(\sigma, v)$, $0 \leq \sigma \leq 1$. But any such function v satisfies (5.51) with $w = v$, hence

$$| Lv | \leq | F(v, \nabla v) | + \sum_{h=1}^{m} (| L\varphi^h | + | L\psi^h |) \qquad \text{in } \Omega.$$

Fix some $r \geq \max(|\varphi|_{\infty;\Omega}, |\psi|_{\infty;\Omega})$: (5.50) yields (5.24) with u replaced by v, f_0 by $f_0 + \sum_{h=1}^{m} (|L\varphi^h| + |L\psi^h|)$. By Lemma 5.10,

$$|v|_{H^{2,p}(\Omega)} \leq \bar{\varkappa}(1 + |\sigma Z_\lambda(v) - \lambda v|_\Gamma|_{H^{1/p',p}(\Gamma)})$$

with $\bar{\varkappa}$ independent of v since the quantity $|v|_{\infty;\Omega} + |Bv|_{C^0(\Gamma)}$ is uniformly bounded. It is easy to majorize the right-hand-side term above with a quantity $C(1 + |v|_{C^1(\bar{\Omega})})$. But, exactly as in Lemma 1.37, it can be proven that

$$|v|_{C^1(\bar{\Omega})} \leq \varepsilon |v|_{H^{2,p}(\Omega)} + C(\varepsilon) |v|_{L^p(\Omega)}$$

for $\varepsilon > 0$. This enables us to arrive at an *a priori* bound on $|v|_{H^{2,p}(\Omega)}$, hence on $|v|_{C^1(\bar{\Omega})}$.

Step 2: *Maximal and minimal solutions.* We define

$$u_{\max}(x) \equiv \sup\{u(x) \mid u \text{ solves } (5.47)\},$$

$$u_{\min}(x) \equiv \inf\{u(x) \mid u \text{ solves } (5.47)\},$$

and proceed to prove that both $u_{\max}$ and $u_{\min}$ are solutions.

Denoting by $\{x^k\}_k$ a countable dense subset of $\bar{\Omega}$, we construct for each k a sequence $\{u^{k,n}\}_n \subset H^{2,p}(\Omega)$ of solutions to (5.47) such that

$$u_{\max}(x^k) = \lim_{n\to\infty} u^{k,n}(x^k).$$

Then we consider (5.47) with φ replaced by $u^{1,1}$ and correspondingly find a solution $U^1 \in H^{2,p}(\Omega)$. The open subset $\{U^1 > \varphi\}$ of Ω where U^1 is strictly larger than φ can be decomposed as $\{U^1 > u^{1,1}\} \cup \{U^1 = u^{1,1} > \varphi\}$. In $\{U^1 > u^{1,1}\}$ the inequality

$$[LU^1 - F(U^1, \nabla U^1)](U^1 - u^{1,1}) \leq 0$$

yields

$$LU^1 \leq F(U^1, \nabla U^1),$$

and therefore also

$$[LU^1 - F(U^1, \nabla U^1)](U^1 - \varphi) \leq 0. \tag{5.52}$$

In $\{U^1 = u^{1,1} > \varphi\}$, $u^{1,1}$ verifies

$$Lu^{1,1} \leq F(u^{1,1}, \nabla u^{1,1})$$

as well as (by Theorem 1.56)

$$Lu^{1,1} = LU^1$$

and

$$F(u^{1,1}, \nabla u^{1,1}) - \Gamma(U^1, \nabla U^1),$$

so that (5.52) is again satisfied. This shows that (5.52) is valid throughout Ω, hence that U^1 is still a solution of the original problem (5.47).

We can inductively define a nondecreasing sequence $\{U^j\} \subset H^{2,p}(\Omega)$, where U^j solves (5.47) with φ replaced, as it is admissible, by $\bigvee_{i=1}^{j} u^{i,j} \vee U^{j-1}$. Since

$$\begin{aligned}\{U^j > \varphi\} = &\left\{U^j > \bigvee_{i=1}^{j} u^{i,j} \vee U^{j-1}\right\}\\ &\cup \{U^j = u^{1,j} > \varphi\} \cup \cdots \cup \{U^j = u^{j,j} > \varphi\}\\ &\cup \{U^j = U^{j-1} > \varphi\},\end{aligned}$$

an analysis as above shows that, if U^{j-1} solves (5.47), then U^j satisfies

$$[LU^j - F(U^j, \nabla U^j)](U^j - \varphi) \leq 0$$

in Ω and is therefore a solution of (5.47).

We utilize the final estimate of the previous step for all functions $U^j = \mathscr{C}(1, U^j)$ and see that the U^j's are uniformly bounded in $H^{2,p}(\Omega)$. By monotonicity, the whole sequence converges weakly in $H^{2,p}(\Omega)$ toward a function U; a passage to the limit in (5.47) written for $u = U^j$ shows that U is a solution of the same obstacle problem. Since

$$\lim_{j\to\infty} U^j(x^k) = u_{\max}(x^k),$$

U satisfies

$$U(x^k) \geq u(x^k)$$

whatever $k \in \mathbb{N}$, and therefore by density

$$U(x) \geq u(x)$$

whatever $x \in \bar{\Omega}$, if u is a solution to our problem. This means that throughout $\bar{\Omega}$, U equals $u_{\max}$, and the latter is a solution. The proof concerning $u_{\min}$ is analogous. □

We can now easily move on to the study of (5.48) and (5.49).

THEOREM 5.13. *In addition to the assumptions of Theorem* 5.12, *suppose that each function* φ^h *satisfies* $L\varphi^h \leq F(\varphi^h, \nabla\varphi^h)$ *in* Ω. *Then the set of all solutions to* (5.48) *which lie above* φ *coincides with the set of all solutions to* (5.47); *therefore, it is not empty and admits a maximal and a minimal element.*

PROOF. It suffices to notice that any solution of (5.47) satisfies

$$Lu - F(u, \nabla u) = L\varphi^h - F(\varphi^h, \nabla\varphi^h) \leq 0$$

in the subset of Ω where $u = \varphi^h$ for some h. □

Call a function $v \in H^{2,p}(\Omega)$ a *subsolution* of (5.48) if

$$v \leq \psi, \qquad Lv \leq F(v, \nabla v) \qquad \text{in } \Omega,$$

$$v_{\partial\Omega\setminus\Gamma} \leq 0, \qquad Bv \leq Z(v) \qquad \text{on } \Gamma:$$

Theorem 5.13 admits the following corollary.

COROLLARY. *Same assumptions about* f *and* ψ *as in Theorem* 5.12. *If* (5.48) *admits a subsolution, then it admits also a solution from* $H^{2,p}(\Omega)$ *which is maximal among all subsolutions* (*in particular, among all solutions*).

Next, we have the following theorem.

THEOREM 5.14. *In addition to the assumptions of Theorem* 5.13, *suppose that each function* ψ^h *satisfies* $L\psi^h \geq F(\psi^h, \nabla\psi^h)$ *in* Ω. *Then the set of all solutions to* (5.49) *which lie between* φ *and* ψ *coincides with the set of all solutions to* (5.47); *therefore, it is not empty and admits a maximal and a minimal element.*

(Compare with Theorem 4.48.)

REMARK. If the a^{ij}'s are in $C^{0,1}(\bar{\Omega})$ the operator L can be put into the divergence form

$$-\frac{\partial}{\partial x_j}\left(a^{ij}\frac{\partial}{\partial x_i}\right) + \frac{\partial a^{ij}}{\partial x_j}\frac{\partial}{\partial x_i}\colon H^1(\Omega) \to H^{-1}(\Omega),$$

so that the assumptions of Theorems 5.12–5.14 about φ and ψ can be somewhat weakened. Consider for instance the Dirichlet case $\Gamma = \varnothing$ in Theorem 5.12: it suffices to assume $\varphi, \psi \in H^1(\Omega)$ with $\varphi \leq \psi$ a.e. in Ω,

$\varphi|_{\partial\Omega} \le 0 \le \psi|_{\partial\Omega}$ and $L\varphi \le h'$, $L\psi \ge h''$ in the sense of $H^{-1}(\Omega)$, where $h', h'' \in L^p(\Omega)$.

5.3.2. Uniqueness

We now make the following assumption about the behavior of $f(x, \eta, \xi)$ in η and ξ:

$$\begin{aligned} &f(x, \eta', \xi') - f(x, \eta'', \xi'') \le \sigma_r(\eta' - \eta'') + \tau_r(|\xi' - \xi''|) \\ &\qquad\qquad\qquad \text{for a.a. } x \in \Omega, \\ &\text{any } \eta', \eta'' \in \mathbb{R} \qquad \text{with } \eta' \ge \eta'' \quad \text{and} \quad |\eta'|, |\eta''| \le r, \\ &\text{any } \xi', \xi'' \in \mathbb{R}^N \qquad \text{with } |\xi'|, |\xi''| \le r \end{aligned} \tag{5.53}$$

$(0 < r < \infty)$; here, $\sigma_r(t)$ is continuous and < 0 for $t > 0$, $\tau_r(t) \to 0$ as $t \to 0^+$. Note that (5.53) is certainly satisfied when f is continuous on $\bar{\Omega} \times \mathbb{R} \times \mathbb{R}^N$ and decreasing with respect to η. {Take

$$\sigma_r(t) \equiv \max_{x \in \bar{\Omega}, |\eta| \le r, |\xi| \le r} [f(x, \eta + t, \xi) - f(x, \eta, \xi)],$$

$\tau_r \equiv$ modulus of uniform continuity of f on $\bar{\Omega} \times [-r, r] \times B_r$.}

As for $\zeta(x, \eta)$, we require that if $\Gamma \ne \varnothing$,

$$\eta \mapsto \zeta(x, \eta) \text{ is decreasing for } x \in \Gamma. \tag{5.54}$$

THEOREM 5.15. *Assume* (5.53) *and* (5.54). *Then a solution* $u \in H^{2,p}(\Omega)$ *of* (5.47) *with* φ *and* ψ *measurable in* Ω, *if existing, is unique.*

PROOF. Let $v \in H^{2,p}(\Omega)$ be another solution of (5.47).

The function $v - u$ cannot attain a positive maximum at a point $x^0 \in \Gamma$, since then the necessary inequality

$$\beta^i(x^0)[v_{x_i}(x^0) - u_{x_i}(x^0)] \ge 0$$

would contradict the other inequality

$$\zeta(x^0, v(x^0)) < \zeta(x^0, u(x^0))$$

consequent on (5.54). Since $v - u$ vanishes on $\partial\Omega \setminus \Gamma$, it can attain a positive maximum only at a point $x^0 \in \Omega$. If such a situation occurs there exists an open neighborhood $U \subset \Omega$ of x^0 such that

$$\varphi \leq u < v \leq \psi$$

and therefore

$$Lu \geq F(u, \nabla u), \qquad Lv \leq F(v, \nabla v)$$

a.e. in U. Since (5.53) implies

$$F(v, \nabla v)(x) - F(u, \nabla u)(x) \leq \sigma_r(v(x) - u(x)) + \tau_r(|\nabla v(x) - \nabla u(x)|)$$

for a.a. $x \in U$ if r is large enough, we arrive at

$$\operatorname*{ess\,lim\,sup}_{x \to x^0} L(v - u)(x) \leq \sigma_r[v(x^0) - u(x^0)] < 0 \tag{5.55}$$

because $\nabla v(x^0) = \nabla u(x^0)$. But (5.55) contradicts Bony's maximum principle (Lemma 3.24) and therefore $v \leq u$. By the same token $v \geq u$, and uniqueness follows. □

Theorem 5.15 clearly contains a uniqueness result for (5.48) and (5.49) as well, since the latter problems can obviously be interpreted as bilateral ones. More specifically, the same procedure as above yields the following maximality property for solutions of (5.48).

LEMMA 5.16. *Under assumptions* (5.53) *and* (5.54) *a solution* $u \in H^{2,p}(\Omega)$ *of* (5.48) *with* ψ *measurable in* Ω, *if existing, is maximal among all subsolutions.*

Passing to (5.49), we call $v \in H^{2,p}(\Omega)$ a *subsolution* if

$$Lv \leq F(v, \nabla v) \qquad \text{in } \Omega,$$

$$v_{\partial\Omega \setminus \Gamma} \leq 0, \qquad Bv \leq Z(v) \qquad \text{on } \Gamma,$$

a *supersolution* if the above inequalities hold with reversed signs; then we have the following lemma.

LEMMA 5.17. *Under assumptions* (5.53) *and* (5.54) *a solution* $u \in H^{2,p}(\Omega)$ *of* (5.49), *if existing, is maximal among all subsolutions and minimal among all supersolutions.*

The requirement that f be nonincreasing (instead of decreasing) with respect to η, even if $f \in C^0(\bar{\Omega} \times \mathbb{R} \times \mathbb{R}^N)$, is not sufficient to guarantee uniqueness:

EXAMPLE. Let Ω be the annulus defined by the inequalities $1 < |x| < 3$ and set $L = -\Delta$,

$$f(x, \eta, \xi) = -(N-1)x_i\xi_i/|x|^2 - |\xi|^{1/2}.$$

The unconstrained Dirichlet problem

$$Lu = F(u, \nabla u) \qquad \text{in } \Omega,$$
$$u|_{\partial\Omega} = 0$$

is solved by the functions $u_0(x) \equiv 0$ and $u_1(x) \equiv (||x| - 2|^3 - 1)/12$, as well as by all functions

$$u_t(x) \equiv \begin{cases} [(t+1-|x|)^3 - t^3]/12 & \text{for } 1 \leq |x| \leq 1+t, \\ 0 \quad \text{for } 1+t < |x| < 3-t, \\ [(|x| - 3 + t)^3 - t^3]/12 & \text{for } 3-t < |x| \leq 3 \end{cases}$$

with $0 < t < 1$.

5.4. Generalized Solutions and Implicit Unilateral Problems

5.4.1. Generalized Solutions

Up until now our approach to nonvariational obstacle problems has required that all derivatives up to the second order of a function u belong to some Lebesgue space for (Lu to make sense and) u to be admissible as a solution. In the present section we relinquish such a requirement in the case of problem (5.48) with $\Gamma = \varnothing$, that is,

$$u \leq \psi, \qquad Lu \leq F(u, \nabla u) \ \text{ and } \ [Lu - F(u, \nabla u)](u - \psi) = 0 \qquad \text{in } \Omega,$$
$$u|_{\partial\Omega} = 0. \tag{5.56}$$

We denote by $\Sigma(\psi)$ the family of all subsolutions $v \in H^{2,p}(\Omega)$ of problem (5.56) with $\psi \in L^\infty(\Omega)$, by $\sigma(\psi)$ the supremum in $L^\infty(\Omega)$ of $\Sigma(\psi)$ if the latter is nonvoid (see Lemma 1.54). Of course, $\sigma(\psi) \leq \psi$ in Ω. Basic properties of the mapping $\psi \mapsto \sigma(\psi)$ are listed here below.

LEMMA 5.18. *Let $\psi \in L^\infty(\Omega)$ with $\Sigma(\psi) \neq \varnothing$. Then*

(i) $\sigma(\sigma(\psi))$ *exists and equals* $\sigma(\psi)$;

(ii) *$\sigma(\psi')$ exists and verifies the inequality*

$$\sigma(\psi') \geq \sigma(\psi) \qquad \text{in } \Omega$$

if $\psi' \in L^\infty(\Omega)$ with $\psi' \geq \psi$ in Ω;

(iii) *$\sigma(\psi')$ exists and verifies the norm estimate*

$$| \sigma(\psi') - \sigma(\psi) |_{\infty;\Omega} \leq | \psi' - \psi |_{\infty;\Omega}$$

for all $\psi' \in L^\infty(\Omega)$, provided the function $\eta \mapsto f(x, \eta, \xi)$ is nonincreasing in $\mathbb{R}$ for a.a. $x \in \Omega$ and any $\xi \in \mathbb{R}^N$;

(iv) *for $0 \leq \lambda \leq 1$, $\sigma(\lambda\psi + (1 - \lambda)\psi')$ exists and verifies the inequality*

$$\sigma(\lambda\psi + (1 - \lambda)\psi') \geq \lambda\sigma(\psi) + (1 - \lambda)\sigma(\psi') \qquad \text{in } \Omega$$

if $\psi' \in L^\infty(\Omega)$ with $\Sigma(\psi') \neq \varnothing$, provided the function $(\eta, \xi) \mapsto f(x, \eta, \xi)$ is concave in $\mathbb{R}^{1+N}$ for a.a. $x \in \Omega$.

Proof. The obvious identity $\Sigma(\sigma(\psi)) = \Sigma(\psi)$ yields (i). As for (ii), it follows from the inclusion $\Sigma(\psi) \subseteq \Sigma(\psi')$. Let us pass to (iii) and set $k \equiv | \psi - \psi' |_{\infty;\Omega}$. We know that $\sigma(\psi) \leq \sigma(\psi' + k)$. But the monotonicity hypothesis implies $v - k \in \Sigma(\psi')$ and therefore $v \leq \sigma(\psi') + k$ whenever $v \in \Sigma(\psi' + k)$, so that $\sigma(\psi' + k) \leq \sigma(\psi') + k$, and finally $\sigma(\psi) - \sigma(\psi') \leq k$; the roles of ψ and ψ' can obviously be interchanged. To conclude we tackle (iv). If $v \in \Sigma(\psi)$ and $v' \in \Sigma(\psi')$, then $w \equiv \lambda v + (1 - \lambda)v'$ satisfies

$$w \leq \lambda\psi + (1 - \lambda)\psi'$$

as well as, by concavity,

$$\begin{aligned} Lw &= \lambda Lv + (1 - \lambda)Lv' \\ &\leq \lambda F(v, \nabla v) + (1 - \lambda)F(v', \nabla v') \leq F(w, \nabla w) \end{aligned}$$

in Ω, hence $w \in \Sigma(\lambda\psi + (1 - \lambda)\psi')$. This means that

$$\lambda\Sigma(\psi) + (1 - \lambda)\Sigma(\psi') \subseteq \Sigma(\lambda\psi + (1 - \lambda)\psi'),$$

and (iv) follows. □

By the corollary to Theorem 5.13, $\sigma(\psi)$ is the maximal solution from $H^{2,p}(\Omega)$ of (5.56) if f and ψ satisfy the same assumptions as in Theorem 5.12, with $\Sigma(\psi) \neq \varnothing$.

When ψ is merely required to be continuous on $\bar{\Omega}$ and ≥ 0 on $\partial\Omega$, $\sigma(\psi)$ still resembles a regular solution of (5.56) as closely as can be expected. Indeed we have the following theorem.

THEOREM 5.19. *Suppose that the function $\eta \mapsto f(x, \eta, \xi)$ is nonincreasing in $\mathbb{R}$ for a.a. $x \in \Omega$ and any $\xi \in \mathbb{R}^N$, that (5.50) holds, and that $\psi \in C^0(\bar{\Omega})$ with $\psi|_{\partial\Omega} \geq 0$, $\Sigma(\psi) \neq \varnothing$. Then $u \equiv \sigma(\psi)$ belongs to $C^0(\bar{\Omega})$ and $u|_\omega$ to $H^{2,p}(\omega)$ whenever ω is open with $\bar{\omega} \subset D^\psi \equiv \{x \in \bar{\Omega} \mid u(x) < \psi(x)\}$. Moreover, u satisfies*

$$u \leq \psi \qquad \text{in } \Omega, \qquad Lu = F(u, \nabla u) \qquad \text{in } \Omega \cap D^\psi,$$

$$u|_{\partial\Omega} = 0.$$

PROOF. Let $\{\psi_n\} \subset H^{2,p}(\Omega)$, with $\psi_n|_{\partial\Omega} \geq 0$, converge toward ψ in $C^0(\bar{\Omega})$ as $n \to \infty$. Since $\Sigma(\psi_n) \neq \varnothing$, the corollary to Theorem 5.13 yields the existence of the maximal solution $u_n = \sigma(\psi_n) \in H^{2,p}(\Omega)$ to (5.56) with ψ replaced by ψ_n. Then $u_n \to u$ in $C^0(\bar{\Omega})$ [Lemma 5.18(iii)], so that u is continuous on $\bar{\Omega}$ and vanishes on $\partial\Omega$. Now let Ω' be an open subset of $\mathbb{R}^N$ such that $\bar{\omega} \subset \Omega'$ and $\overline{\Omega \cap \Omega'} \subset D^\psi$. For all n sufficiently large we have $u_n < \psi_n$ on $\overline{\Omega \cap \Omega'}$ and therefore

$$Lu_n = F(u_n, \nabla u_n) \qquad \text{in } \Omega \cap \Omega'.$$

By (5.50) (with $r \geq \sup_n |u_n|_{\infty;\Omega}$) we can apply Lemma 5.11 and obtain a bound on $|u_n|_{H^{2,p}(\omega)}$, which is independent of n because (Theorem 1.M) the u_n's are uniformly bounded throughout $\bar{\Omega}$ and share a common modulus of uniform continuity there. At this point standard arguments yield the full conclusion of the lemma. □

If $\sigma(\psi)$ is known to be regular, the last statement of Theorem 5.19 is strengthened as follows.

LEMMA 5.20. *In addition to the hypotheses of Theorem 5.19 suppose that $\sigma(\psi)$ belongs to $H^{2,p}(\Omega)$. Then (5.56) is solvable in $H^{2,p}(\Omega)$, and $\sigma(\psi)$ is its maximal solution.*

PROOF. Since $\Sigma(\sigma(\psi)) \neq \varnothing$ and $\sigma(\psi) = 0$ on $\partial\Omega$, (5.56) with ψ replaced by $\sigma(\psi)$ admits a maximal solution $u = \sigma(\sigma(\psi)) \in H^{2,p}(\Omega)$. But then $\sigma(\psi) = u$ by Lemma 5.18(i), and the conclusion follows from the inequality

$$Lu \leq F(u, \nabla u) \qquad \text{in } \Omega.$$

□

[Of course, $\sigma(\psi)$ need not belong to $H^{2,p}(\Omega)$, even if f vanishes identically: see the beginning of Section 4.5.]

The considerations of this subsection motivate our calling $\sigma(\psi)$ the *generalized maximal solution* of (5.56).

REMARK. In the case of bilateral problems (for linear operators) a satisfactory notion of generalized solutions is provided by a rather elaborate approach based on Theorem 5.6: see M. G. Garroni and M. A. Vivaldi [59].

5.4.2. Implicit Unilateral Problems

We now want to deal with the situation arising when the obstacle ψ in (5.56), instead of being kept fixed, "varies with the solution u." More precisely, we consider an *implicit unilateral problem* such as

$$\begin{gathered} u \leq M(u), \qquad Lu \leq F(u, \nabla u) \\ \text{and} \quad [Lu - F(u, \nabla u)][u - M(u)] = 0 \qquad \text{in } \Omega, \\ u|_{\partial\Omega} = 0, \end{gathered} \tag{5.57}$$

where M is a mapping between functions spaces.

We cannot expect to find a solution $u \in H^{2,p}(\Omega)$ of (5.57) in the case of an arbitrary mapping M. We can, however, look for a function u that equals the generalized maximal solution of (5.56) with $\psi = M(u)$, that is, for a fixed point of the mapping $S \equiv \sigma \circ M$. This is the approach we follow. About f we assume:

- that the function $\eta \mapsto f(x, \eta, \xi)$ is nonincreasing in $\mathbb{R}$ for a.a. $x \in \Omega$ and any $\xi \in \mathbb{R}^N$;
- that the function $(\eta, \xi) \mapsto f(x, \eta, \xi)$ is concave in $\mathbb{R}^{1+N}$ for a.a. $x \in \Omega$;
- that (5.50) holds;
- that the unconstrained Dirichlet problem

$$L\bar{u} = F(\bar{u}, \nabla\bar{u}) \qquad \text{in } \Omega,$$

$$\bar{u}|_{\partial\Omega} = 0$$

admits a solution $\bar{u} \in H^{2,p}(\Omega)$ maximizing all subsolutions. [So that $\sigma(\psi)$ exists and satisfies $\sigma(\psi) \leq \bar{u}$ in Ω whatever $\psi \in L^\infty(\Omega)$].

As for M, we suppose that it is a continuous, nondecreasing and concave mapping from the normed space of functions $u \in C^0(\bar{\Omega})$ vanishing on $\partial\Omega$ into $C^0(\bar{\Omega})$, so that the same is true of S by Lemma 5.18(ii)–(iv); moreover, we suppose that there exist $\underline{u} \in C^0(\bar{\Omega})$ with $\underline{u}|_{\partial\Omega} = 0$ and $t \in]0, 1[$ such that

$$\begin{aligned} &\underline{u} \leq \bar{u} \quad \text{on } \bar{\Omega}, \qquad M(\underline{u}) \geq 0 \quad \text{on } \partial\Omega, \\ &u_t \equiv (1-t)\underline{u} + t\bar{u} \leq S(\underline{u}) \quad \text{on } \bar{\Omega}. \end{aligned} \tag{5.58}$$

If the nonpositive continuous function $\underline{u} \equiv \sigma(0)$ is > -1 on $\bar{\Omega}$, the above requirements are met by

$$[M(u)](x) \equiv 1 + \bigwedge_{\substack{y \in \bar{\Omega} \\ y \geq x}} u(y) = 1 + \bigwedge_{\substack{y \in \mathbb{R}^N \\ y \geq x}} \tilde{u}(y), \tag{5.59}$$

where $y \geq x$ means $y_1 \geq x_1, \ldots, y_N \geq x_N$ and $\tilde{u}$ denotes the trivial extension of u to $\mathbb{R}^N$, if $u \in C^0(\bar{\Omega})$ with $u|_{\partial\Omega} = 0$. [Note that $u_t \leq M(\underline{u})$ for $t > 0$ small enough, so that $u_t \in \Sigma(M(u))$ by the concavity assumption about f.] The implicit unilateral problem corresponding to the choice of the operator (5.59) plays a fundamental role in the theory of stochastic impulse control: see A. Bensoussan and J. L. Lions [13].

THEOREM 5.21. *Under the above assumptions about f and M there exists a unique fixed point $u \in C^0(\bar{\Omega})$, with $u|_{\partial\Omega} = 0$, of the mapping S; moreover, u is the limit in $C^0(\bar{\Omega})$ of the sequence $\{S^n(u^0)\}$, with rate of convergence*

$$| S^n(u^0) - u |_{\infty;\Omega} \leq (1-t)^n \, | \bar{u} - \underline{u} |_{\infty;\Omega}, \tag{5.60}$$

whenever $u^0 \in C^0(\bar{\Omega})$ satisfies $\underline{u} \leq u^0 \leq \bar{u}$ on $\bar{\Omega}$.

PROOF. Fix any u^0 as above. The continuous function $M(u^0)$ is $\geq M(\underline{u})$ on $\bar{\Omega}$ and in particular ≥ 0 on $\partial\Omega$; $S(u^0)$ is continuous on $\bar{\Omega}$, vanishes on $\partial\Omega$, and satisfies

$$\underline{u} \leq u_t \leq S(\underline{u}) \leq S(u^0) \quad \text{on } \bar{\Omega}$$

[see (5.58)]. By induction, $\underline{u} \leq S^n(u^0)$ on $\bar{\Omega}$ for every $n \in \mathbb{N}$.

Next, let v, w belong to $C^0(\bar{\Omega})$ with $\underline{u} \leq v$, $w \leq \bar{u}$, and denote by λ a number from $]0, 1]$ satisfying

$$\lambda(\underline{u} - w) \leq v - w \leq \lambda(v - \underline{u}).$$

We claim that

$$(1-t)\lambda[\underline{u}-S(w)] \leq S(v)-S(w) \leq (1-t)\lambda[S(v)-\underline{u}]. \tag{5.61}$$

To prove the right-hand-side inequality we observe that

$$S(w) \geq (1-\lambda)S(v)+\lambda S(\underline{u}),$$

since S is concave and w verifies

$$w \geq (1-\lambda)v+\lambda\underline{u};$$

therefore,

$$\begin{aligned} S(w) &\geq (1-\lambda)S(v)+\lambda[(1-t)\underline{u}+t\bar{u}] \\ &\geq [1-\lambda(1-t)]S(v)+\lambda(1-t)\underline{u} \end{aligned}$$

because $S(\underline{u}) \geq u_t$ and $S(v) \leq \bar{u}$. The left-hand-side inequality is proven analogously.

At this point we choose $v = S(u^0)$, $w = u^0$, $\lambda = 1$, and deduce that

$$(1-t)[\underline{u}-S(u^0)] \leq S^2(u^0)-S(u^0) \leq (1-t)[S^2(u^0)-\underline{u}]$$

by (5.61). Then we choose $v = S^2(u^0)$, $w = S(u^0)$, $\lambda = 1-t$, and arrive at

$$(1-t)^2[\underline{u}-S^2(u^0)] \leq S^3(u^0)-S^2(u^0) \leq (1-t)^2[S^3(u^0)-\underline{u}].$$

Thus proceeding we prove that

$$(1-t)^{n-1}[\underline{u}-S^{n-1}(u^0)] \leq S^n(u^0)-S^{n-1}(u^0) \leq (1-t)^{n-1}[S^n(u^0)-\underline{u}],$$

hence that

$$| S^n(u^0)-S^{n-1}(u^0)|_{\infty;\Omega} \leq (1-t)^{n-1} | \bar{u}-\underline{u} |_{\infty;\Omega}, \tag{5.62}$$

for every $n \in N$.

By (5.62) the series $\sum_{n=1}^{\infty} [S^n(u^0)-S^{n-1}(u^0)]$ converges uniformly, and so does the sequence $\{S^n(u^0)\}$, toward a function $u \in C^0(\bar{\Omega})$ with $u|_{\partial\Omega} = 0$. By the continuity of S,

$$S(S^n(u^0)) \to S(u) \qquad \text{in } C^0(\bar{\Omega}),$$

so that $u = S(u)$. Both the uniqueness of the fixed point u and the estimate (5.60) follow from another iteration of (5.61), starting this time with $v = u^0$, $w = u$, $\lambda = 1$. □

5.4.3. The Implicit Unilateral Problem of Stochastic Impulse Control

The continuous function $u = S(u)$ whose existence and uniqueness are guaranteed by Theorem 5.21 "solves" (5.57) only in a weak sense, namely, that of Theorem 5.19 for $\psi = M(u)$. But of course, under certain specific choices of the operator M sufficient information can be obtained for results of further regularity. This is in particular the case of the operator (5.59): indeed we have the following theorem.

THEOREM 5.22. *For some* $\delta \in \,]0, 1[$ *let* $\partial\Omega$ *be of class* $C^{2,\delta}$ *and* $a^{ij} \in C^{0,\delta}(\bar{\Omega})$, $f \in C^{0,\delta}(\bar{\Omega}\times\mathbb{R}\times\mathbb{R}^N)$. *Suppose the function* $\eta \mapsto f(x, \eta, \xi)$ *non-increasing in* $\mathbb{R}$ *for* $(x, \xi) \in \bar{\Omega}\times\mathbb{R}^N$, *and assume* (5.50) *with* f_0 *constant, say* $f_0 = K$. *For* M *given by* (5.59) *a fixed point* $u \in C^0(\bar{\Omega})$ *with* $u|_{\partial\Omega} = 0$ *of the mapping* S, *if existing, belongs to* $H^{2,p}(\Omega)$ *and solves* (5.57) *in the usual sense.*

The proof of this theorem relies on the circumstance that throughout $\bar{\Omega}$, $M(u)$ inherits some regularity from the regularity u has in the set $D_u \equiv \{x \in \bar{\Omega} \mid u(x) < [M(u)](x)\}$, as is illustrated by the next two lemmas.

LEMMA 5.23. *Let* $u \in C^0(\bar{\Omega})$ *with* $u|_{\partial\Omega} = 0$. *For every* $x^0 \in \bar{\Omega}$ *there exist an open subset* $V(x^0)$ *of* $\mathbb{R}^N$, *a sphere* $B_r(x^0)$ *with* $r = r(x^0) > 0$, *and a family* $T(x^0)$ *of vectors* $\xi \geq 0$ *from* $\mathbb{R}^N$ *such that*

$$\overline{V(x^0) \cap \Omega} \subset D_u, \tag{5.63}$$

$$B_r(x^0) + T(x^0) \subset V(x^0) \tag{5.64}$$

and

$$[M(u)](x) = 1 + \bigwedge_{\xi \in T(x^0)} \tilde{u}(x + \xi) \qquad \textit{for } x \in \overline{B_r(x^0) \cap \Omega}. \tag{5.65}$$

PROOF. For $x \in \bar{\Omega}$ let $\Delta(x)$ denote the set of all vectors $y \in \bar{\Omega}$ such that $y \geq x$ and $u(y) = [I(u)](x) \equiv [M(u)](x) - 1$. If $y \in \Delta(x)$, then

$$1 + u(y) = [M(u)](x) \leq [M(u)](y)$$

and therefore $u(y) < [M(u)](y)$. This shows that $\Delta(x) \subset D_u$. Fix $x = x^0$. Since $\Delta(x^0)$ is compact, there exists some $\hat{\varrho} > 0$ such that $V(x^0) \equiv \Delta(x^0) + B_{\hat{\varrho}}$ satisfies (5.63). On the other hand, any sequence $\{y^n\}$ such that $y^n \in \Delta(x^n)$ with $\{x^n\} \subset \bar{\Omega}$ and $x^n \to x^0$ admits a subsequence which converges toward some $y^0 \in \Delta(x^0)$. This means that given any $\varrho > 0$, the inclusion

$$\Delta(x) \subset \Delta(x^0) + B_\varrho$$

holds for every $x \in \bar{\Omega}$ with $|x - x^0| < \hat{r}$ provided $\hat{r} > 0$ is small enough. We fix $\varrho < \hat{\varrho}$ and choose $r < \hat{r}$ with $2r + \varrho < \hat{\varrho}$: setting

$$\Delta'(x) \equiv \{\xi \in \mathbb{R}^N \mid x + \xi \in \Delta(x)\},$$

for every $x \in \overline{B_r(x^0) \cap \Omega}$ we have

$$\Delta'(x) \subset \Delta'(x^0) + B_{r+\varrho}$$

and hence

$$\Delta'(x) + B_r(x^0) \subset \Delta(x^0) + B_{2r+\varrho} \subset V(x^0).$$

At this point (5.64) and (5.65) are immediately ascertained for

$$T(x^0) \equiv \bigcup_{x \in \overline{B_r(x^0) \cap \Omega}} \Delta'(x). \qquad \square$$

LEMMA 5.24. *Let $u \in C^0(\bar{\Omega})$ with $u|_{\partial\Omega} = 0$ and $u|_{\bar{\omega}} \in C^{0,1}(\bar{\omega})$ whenever ω is open with $\bar{\omega} \subset D_u$. Then $M(u) \in C^{0,1}(\bar{\Omega})$.*

PROOF. Fix $x^0 \in \bar{\Omega}$. By (5.63) $u|_{\overline{V(x^0) \cap \Omega}}$ is Lipschitzian and so is obviously $\tilde{u}|_{\overline{V(x^0)}}$. From (5.64) and (5.65) it is easy to deduce that

$$|[M(u)](x') - [M(u)](x'')| \le [\tilde{u}]_{1;\overline{V(x^0)}} |x' - x''|$$

for $x', x'' \in \overline{B_r(x^0) \cap \Omega}$, hence that $M(u) \in C^{0,1}(\bar{\Omega})$ because $\bar{\Omega}$ can be covered by a finite number of spheres such as $B_r(x^0)$. $\square$

PROOF OF THEOREM 5.22. Thanks to Theorem 5.19 (see also Problem 3.10), our present assumptions about $\partial\Omega$, a^{ij}, and f yield $u|_{\bar{\omega}} \in C^{2,\delta}(\bar{\omega})$ whenever ω is open with $\bar{\omega} \subset D_u$, and therefore $M(u) \in C^{0,1}(\bar{\Omega})$ by the preceding lemma.

For $i, j = 1, \ldots, N$ let $\{a_n^{ij}\} \subset C^\infty(\bar{\Omega})$ converge to a^{ij} in $C^0(\bar{\Omega})$ with $[a_n^{ij}]_{\delta;\bar{\Omega}}$ bounded independently of $n \in \mathbb{N}$ (see Lemma 1.8 and the remark following Lemma 1.9). Each operator

$$L_n \equiv -a_n^{ij} \frac{\partial^2}{\partial x_i \, \partial x_j}$$

can be put into divergence form and applied to $M(u)$. We can show that

$$L_n M(u) \geq -\hat{K} \qquad \text{in the sense of } H^{-1}(\Omega), \tag{5.66}$$

for some positive constant $\hat{K}$. To wit, we begin by fixing any $x^0 \in \Omega$ and introducing $V(x^0)$, $B_r(x^0)$, $T(x^0)$ as in Lemma 5.23. Let

$$\Omega'_{\xi,\varepsilon} \equiv \{x \in B_r(x^0) \cap \Omega \mid u(x+\xi) < -\varepsilon/2\},$$
$$\Omega''_{\xi,\varepsilon} \equiv \{x \in B_r(x^0) \cap \Omega \mid \tilde{u}(x+\xi) > -\varepsilon\}$$

for $\xi \in T(x^0)$ and $\varepsilon > 0$. Both $\Omega'_{\xi,\varepsilon}$ and $\Omega''_{\xi,\varepsilon}$ are open, and their union equals $B_r(x^0) \cap \Omega$. As x varies in $\Omega'_{\xi,\varepsilon}$, $x + \xi$ varies in Ω as well as [by (5.64)] in $V(x^0)$. But then [by (5.63)] the function $u^{(\xi)}: x \mapsto \tilde{u}(x+\xi)$ and all its first and second derivatives admit, when restricted to $\Omega'_{\xi,\varepsilon}$, an L^∞ bound independent of ξ and ε. Therefore,

$$L_n u^{(\xi)} \geq -K^0 \qquad \text{in } \Omega'_{\xi,\varepsilon} \tag{5.67}$$

with $K^0 \geq 0$ independent of n as well. Through a partition of unity every function $v \in C_c^\infty(B_r(x^0) \cap \Omega)$, $v \geq 0$, can be decomposed as the sum of two nonnegative smooth functions v' and v'' with $\operatorname{supp} v' \subset \Omega'_{\xi,\varepsilon}$, $\operatorname{supp} v'' \subset \Omega''_{\xi,\varepsilon}$. By Lemma 4.28, therefore, (5.67) suffices for the validity of the inequality

$$L_n w^{(\xi,\varepsilon)} \geq -K^0 \qquad \text{in the sense of } H^{-1}(B_r(x^0) \cap \Omega)$$

with $w^{(\xi,\varepsilon)} \equiv u^{(\xi)}|_{B_r(x^0)\cap\Omega} \wedge (-\varepsilon)$ (note that $w^{(\xi,\varepsilon)} = -\varepsilon$ in $\Omega''_{\xi,\varepsilon}$), hence also of the inequality

$$L_n w^{(\xi)} \geq -K^0 \qquad \text{in the sense of } H^{-1}(B_r(x^0) \cap \Omega) \tag{5.68}$$

with $w^{(\xi)} \equiv w^{(\xi,0)}$ after a passage to the limit as $\varepsilon \to 0^+$. We let ξ vary in $T(x^0)$: the same proof as in the case of a finite number of the ξ's (Lemma 4.28) shows that (5.68) implies

$$L_n\Big(\bigwedge_{\xi \in T(x^0)} w^{(\xi)}\Big) \geq -K^0 \qquad \text{in the sense of } H^{-1}(B_r(x^0) \cap \Omega),$$

and (5.66) follows easily since

$$M(u)|_{B_r(x^0)\cap\Omega} = 1 + \bigwedge_{\xi \in T(x^0)} w^{(\xi)}.$$

We now avail ourselves of the remark at the end of Section 5.3.1. We arbitrarily fix $\varphi = \bigvee_{h=1}^{m} \varphi^h$ with $\varphi^h \in \Sigma(M(u))$ and find a solution $v_n \in H^{2,p}(\Omega)$ to the bilateral problem

$$\begin{gathered}\varphi \leq v_n \leq M(u), \qquad [L_n v_n - F(v_n, \nabla v_n)](v_n - \varphi) \leq 0 \\ \text{and} \quad [L_n v_n - F(v_n, \nabla v_n)][v_n - M(u)] \leq 0 \qquad \text{in } \Omega, \\ v_n|_{\partial\Omega} = 0,\end{gathered}$$

with

$$(-\hat{K}) \wedge F(v_n, \nabla v_n) \leq L_n v_n \leq \bigvee_{h=1}^{m} (L_n \varphi^h) \vee F(v_n, \nabla v_n) \qquad \text{in } \Omega, \tag{5.69}$$

and therefore also

$$| L_n v_n | \leq K(| \nabla v_n |^2 + 1) + \hat{K} + \sum_{h=1}^{m} | L_n \varphi^h | \qquad \text{in } \Omega$$

for a suitable choice of r from (5.50). But then Lemma 5.10 (see also Remark 1 at the end of Section 5.2.3) provides a uniform bound

$$| v_n |_{H^{2,p}(\Omega)} \leq C \qquad \text{for } n \in N.$$

This means that a subsequence of $\{v_n\}$ converges weakly in $H^{2,p}(\Omega)$ and strongly in $C^1(\bar{\Omega})$ toward a function $v \geq \varphi$, which satisfies

$$\begin{gathered}v \leq M(u), \quad Lv \leq F(v, \nabla v) \\ \text{and} \quad [Lv - F(v, \nabla v)][v - M(u)] = 0 \qquad \text{in } \Omega, \\ v|_{\partial\Omega} = 0\end{gathered} \tag{5.70}$$

since φ^h is a subsolution of the above unilateral problem (see Theorem 5.13). Moreover, the left-hand-side inequality of (5.69) becomes

$$Lv \geq F(v, \nabla v) \wedge (-\hat{K}) \qquad \text{in } \Omega,$$

so that

$$| Lv | \leq K(| \nabla v |^2 + 1) + \hat{K} \qquad \text{in } \Omega.$$

This means that $| v |_{H^{2,p}(\Omega)}$ is bounded independently of the choice of φ. We can therefore proceed as in Step 2 of the proof of Theorem 5.12 and prove that (5.70) admits a solution $v_{\max} \in H^{2,p}(\Omega)$ which majorizes all subsolutions. But then $v_{\max} = \sigma(M(u)) = u$, and the desired regularity of u is proven. ◻

Problems

5.1. Theorems 5.1 and 5.2 remain valid for $1 < p < 2$.

5.2. Same assumptions as in Theorem 5.1. If $w \in H^{2,p}(\Omega)$ satisfies (5.8), the solution u of (5.1) satisfies $u \geq w$ in Ω. Compare with Step 1 of the proof of Theorem 4.30.

5.3. For $n = 0, 1, \ldots$ denote by u_n the solution to (5.6) with $\psi = 0$, $f = f_n$, and $\zeta = \zeta_n$, where $f_n \in L^p(\Omega)$ and $\zeta_n \in H^{1/p',p}(\Gamma)$ $(1 < p < \infty)$, $\zeta_n \leq 0$ on Γ. If $f_n \to f_0$ in $L^p(\Omega)$ and $\zeta_n \to \zeta_0$ in $H^{1/p',p}(\Gamma)$, then $u_n \to u_0$ in $H^{2,p}(\Omega)$. To see this, note that $Lu_n = f_n\chi_n$ with $\chi_n \equiv$ characteristic function of the subset Ω_n of Ω where $u_n < 0$. Passing to suitable subsequences, still indexed by n, one sees that $\chi_n \to 1$ a.e. in Ω_0, so that

$$\int_{\Omega_0} | f_n\chi_n - f_0 |^p \, dx \to 0;$$

moreover, since $Lu_n \rightharpoonup Lu$ in $L^p(\Omega)$,

$$\int_{\Omega\setminus\Omega_0} | Lu_n |^p \, dx = \int_\Omega (Lu_n)g_n \, dx \to 0,$$

where $g_n \equiv | f_n |^{p-2} f_n(1 - \chi_0)$ if $f_n \neq 0$, $g_n \equiv 0$ otherwise.

5.4. An interesting consequence of Theorem 5.4 is that a function $\psi \in H^2(\Omega)$ satisfying $L\psi \geq 0$, $\psi|_{\partial\Omega\setminus\Gamma} \geq 0$, $(B\psi)\psi|_\Gamma \leq 0$ on Γ is ≥ 0 in Ω: indeed, the solution u of (5.6) with $f = 0$, $\zeta = 0$ vanishes identically. [Compare with Theorem 3.29 for the case $\Gamma = \varnothing$, and with Lemmas 3.25, 3.26 for the case $\psi \in H^{2,p}(\Omega)$, $p > N$.]

5.5. Theorems 5.12–5.14 remain valid, for $\Gamma = \varnothing$, if the linear operator $u \mapsto -a^{ij}(x)u_{x_ix_j}$ is replaced by the nonlinear one $u \mapsto -a^{ij}(x, u)u_{x_ix_j}$ with $a^{ij} \in C^0(\bar{\Omega} \times \mathbb{R})$,

$$\alpha \, | \xi |^2 \leq a^{ij}(x, \eta)\xi_i\xi_j \leq \alpha^{-1} \, | \xi |^2 \qquad \text{for } \xi \in \mathbb{R}^N$$

whatever $(x, \eta) \in \bar{\Omega} \times \mathbb{R}$ $(0 < \alpha < 1)$. To see this, apply a very general result of N. S. Trudinger [153], O. A. Ladyzhenskaya and N. N. Ural'tseva [95], which in particular provides two constants $\delta \in]0, 1[$ and $H > 0$, dependent only on α, K, f_0, and $| u |_{\infty;\Omega}$, such that $[u]_{\delta;\bar{\Omega}} \leq H$ whenever u is a function from $H^{2,N}(\Omega)$ satisfying

$$| a^{ij}(x, u)u_{x_ix_j} | \leq K \, | \nabla u |^2 + f_0 \qquad \text{in } \Omega$$

with $f_0 \in L^N(\Omega)$, $f_0 \geq 0$. (See G. M. Troianiello [150].)

5.6. In (5.56) take $a^{ij} \in C^{0,1}(\bar{\Omega})$, $F(u, \nabla u) = -a^i u_{x_i} - au - f$ with $a^1, \ldots, a^N$, $a \in L^\infty(\Omega)$, $a \geq 0$ in Ω, and $f \in L^p(\Omega)$. If $\psi \in C^0(\bar{\Omega})$, $\psi|_{\partial\Omega} \geq 0$, is such that the set of functions $v \in H_0^1(\Omega)$ satisfying $v \leq \psi$ in Ω is nonvoid, then $\sigma(\psi)$

coincides with the solution of the corresponding variational inequality. [Note that this is obviously true when the continuity of ψ throughout $\bar{\Omega}$ is strengthened into the requirement $\psi \in H^{2,p}(\Omega)$.] If the v.i. does not make sense (as in the example of Section 4.3.1), we can still avail ourselves of the notion of a generalized solution: the estimate of Problem 2.6 can be utilized to prove that, in addition to all the properties stated in Theorem 5.19, $\sigma(\psi)$ belongs to $H^1_{\mathrm{loc}}(\Omega)$ and satisfies $\tilde{L}\sigma(\psi) \leq f|_\omega$ in the sense of $H^{-1}(\omega)$ whenever $\omega \subset\subset \Omega$, with $\tilde{L}u \equiv -(a^{ij}u_{x_i})_{x_j} + (a^{ij}_{x_j} + a^i)u_{x_i} + au$.

5.7. What is the role of the exponent p in the proof of Theorem 5.22? And what additional assumption guarantees the membership of the fixed point u in $H^{2,q}(\Omega)$ for any finite q?

Bibliographical Notes

Chapter 1

The contents of Section 1.2 are certainly familiar to most readers. We only mention that the example in Section 1.2.1 is taken from D. Gilbarg and N. S. Trudinger [67], and the proof of Theorem 1.2 from A. Kufner, O. John, and S. Fučik [92].

In Section 1.3 the only nonstandard topics are those of the last subsection. Their presentation is largely based on A. Kufner, O. John, and S. Fučik [92], with some modifications in the proof of Theorem 1.12.

Almost all results of Section 1.4 are the contribution of S. Campanato [27, 28] (see also N. Meyers [110] for what concerns Theorem 1.17); functions of bounded mean oscillation were introduced by F. John and L. Nirenberg [81].

The theory of Sobolev spaces stems from the works of several authors: S. L. Sobolev [137, 138], of course, but also, e.g., B. Levi [100], L. Tonelli [145], C. B. Morrey, Jr. [116], J. Deny and J. L. Lions [45]. The equivalence of Levi's and Sobolev's definitions can be obtained as a consequence of Theorem 1.20 (whose proof as adopted here follows J. Nečas [127]). For what concerns density results we mention N. Meyers and J. Serrin [111] (Theorem 1.26) and S. Agmon [2] (Theorem 1.27).

Sobolev inequalities (Sections 1.6.1 and 1.6.3) are due to S. L. Sobolev [138], L. Nirenberg [129], and E. Gagliardo [58] for $kp < N$, to C. B. Morrey, Jr. [117] for $kp > N$. The proof of Theorem 1.34 (see F. Rellich [132] and V. I. Kondrachov [89]) is taken from H. Brézis [19]; the other results of Section 1.6.2 are based on C. B. Morrey, Jr. [118].

Both Sections 1.7 and 1.8 utilize a more or less standard approach to their arguments {although in the literature the intrinsic definition of $H^{1/p',p}(\Gamma)$ is usually preferred to its more rapid construction via the quotient space technique adopted here: e.g., see A. Kufner, O. John, and S. Fučik [92]}. Section 1.8.1 is based on H. H. Schaefer [134]; Theorem 1.55 is due to R. Klee [88]. The core of Section 1.8.2 can be considered to be Theorem 1.56 (G. Stampacchia [143]), whose proof here is taken, for its first part, from D. Gilbarg and N. S. Trudinger [67]. (The proof of the last statement, in our Step 2, is simpler than the one suggested by D. Kinderlehrer and G. Stampacchia [87].)

Chapter 2

Many authors, starting with K. O. Friedrichs [57], have developed the variational approach to elliptic b.v.p.'s in the last 40 years. In these notes we shall confine ourselves to the sources of the results proven in the Chapter 2.

Theorem 2.1 is due to P. D. Lax and A. Milgram [98]. The Fredholm alternative for elliptic b.v.p.'s was developed by O. A. Ladyzhenskaya and N. N. Ural'tseva [94] and G. Stampacchia [141]. The weak maximum principle of Theorem 2.4 is due to M. Chicco [36] and N. S. Trudinger [152]; the proof in the text is Trudinger's, except for some minor modifications.

The results of Section 2.3 are due to G. Stampacchia [139]; for the proofs we followed C. Miranda [114].

The core of Section 2.4 is the result of E. De Giorgi [44] and J. Nash [126] on Hölder continuity of solutions to equation (2.36). Our approach to the whole topic has been based partly on E. Giusti [68] (Section 2.4.1), partly on J. Moser [122] and C. B. Morrey, Jr. [118] (Section 2.4.2), and finally on S. Campanato [31] (Section 2.4.3).

The results of Section 2.5 are special cases of contributions by L. Nirenberg [128].

Section 2.6 is based on E. Giusti [68].

Chapter 3

Most of the fundamental results of the nonvariational $C^{k,\delta}$ theory for the Dirichlet problem were obtained in the 1930s by E. Hopf [77], J. Schauder [135, 136], and R. Caccioppoli [24]. These authors introduced far-reaching techniques of *a priori* estimates, starting with majorization formulas for constant coefficient operators, then considering variable coefficient operators as perturbations of the former ones (a device originally perfected by A. Korn [90]). Existence theorems were finally deduced from the above-mentioned estimates through simple methods of functional analysis.

In more recent years R. Fiorenza [49] extended the $C^{k,\delta}$ theory to the regular oblique derivative problem.

$H^{k,p}$ a priori estimates were obtained, again through a preliminary treatment of the constant coefficient case, by D. Greco [70] and A. E. Koželev [91] for the Dirichlet problem, by S. Agmon, A. Douglis, and L. Nirenberg [3] for the regular oblique derivative problem, (the latter being a special case of the general boundary value problems investigated by these authors for elliptic operators of arbitrary order). Existence criteria, inevitably more complex than in the $C^{k,\delta}$ theory, were then deduced by M. Chicco [37, 38].

Thorough presentations of the $C^{k,\delta}$ case can be found in the books by O. A. Ladyzhenskaya and N. N. Ural'tseva [94] and C. Miranda [115]; for Dirichlet boundary conditions the book by C. B. Morrey Jr. [118] illustrates also the $H^{k,p}$ estimates, that of D. Gilbarg and N. S. Trudinger [67] existence and uniqueness of solutions.

In 1965 S. Campanato [28] initiated a series of innovative contributions to the methodology of the whole theory under consideration. He approached the subject of regularity and *a priori* estimates in Hölder function spaces by utilizing the general machinery of $L^{2,\mu}$ spaces. Then S. Campanato and G. Stampacchia [33] showed that the regularity theory in $H^{k,p}$ could be deduced from the above through an interpolation argument. Campanato's method has been in recent years extended to topics such as parabolic problems or nonlinear elliptic systems (e.g., we refer to S. Campanato [29, 32]).

It is on S. Campanato [28] and S. Campanato and G. Stampacchia [33] that the first five sections of the present chapter are largely based, with contributions from the dissertation of F. Vespri [154] for what concerns the study of Neumann boundary conditions. (See also J. Peetre [130] and M. Giaquinta [65].) Lemma 3.6 is due to G. Stampacchia [142]; its proof here is taken from S. Campanato [30]. The example of Section 3.2 is taken from Chapter IV of V. P. Mikhailov [112], that of Section 3.3 from N. G. Meyers [109], that of Section 3.5 from M. Chicco [40].

Many of the procedures utilized in various proofs of Sections 3.6 and 3.7 can by now be considered standard in the field, though their rearrangement is partly new. Existence and uniqueness in $H^{2,p}$ are treated here more simply than in M. Chicco [37, 38], at the price, however, of less generality in the assumptions about the lower-order coefficients of the operators. Lemma 3.24 is taken from J. M. Bony [17]. Lemmas 3.27 and 3.26 are straightforward generalizations of classical results by E. Hopf [76, 78]. The example of Section 3.7 is taken from the Introduction of O. A. Ladyzhenskaya and N. N. Ural'tseva [94]. Theorem 3.30 is a special case of a general result whose proof can be found in A. Zygmund [155]. The (complete) proof of Lemma 3.31 can be found in F. John and L. Nirenberg [81].

Chapter 4

The theory of v.i.'s originated in Italy from the independent works of G. Fichera [48] and G. Stampacchia [140] in the early 1960s. The intense research that flourished internationally since can be roughly viewed as consisting of three strands:

- abstract existence results (culminating in the unifying approach of H. Brézis [18] to pseudomonotone operators);
- regularity results in more "concrete" cases involving partial differential operators, still the main source of difficulties;
- applications of v.i.'s in such diverse fields as elasticity theory, control theory, hydraulics, etc.

Existing monographs on v.i.'s usually find their motivations in the third strand above: e.g., see J. L. Lions [104], G. Duvaut and J. L. Lions [47], C. Baiocchi and A. Capelo [8], A. Bensoussan and J. L. Lions [12]. More attention to regularity questions is devoted by D. Kinderlehrer and G. Stampacchia [87], A. Friedman [56], and M. Chipot [43].

For the material of our Sections 4.1–4.3 the main reference is J. L. Lions [103]. The proof of Stampacchia's Theorem 4.4 is taken from J. L. Lions and G. Stampacchia [105], that of Fichera's Theorem 4.7 from P. Hess [75]. Theorem 4.21 is a fundamental result of J. Leray and J. L. Lions [99], generalized slightly by dint of a device, due to R. Landes [97], in the proof of Lemma 4.22; the second part of Lemma 4.22 is taken from L. Boccardo, F. Murat, and J. P. Puel [16].

The results of Section 4.4 are due to the present author; in more particular cases Theorem 4.27 was previously proven by M. Chicco [39] and P. L. Lions [107] with completely different methods.

Lewy–Stampacchia inequalities are named after the paper by H. Lewy and G. Stampacchia [102], dealing with a potential-theoretic approach to a minimum problem of the type illustrated in the Introduction. The passage to a variational setting with applications to regularity of solutions is due to U. Mosco and G. M. Troianiello [121]. For more general results see B. Hanouzet and J. L. Joly [72], O. Nakoulima [125], and U. Mosco [120]; the latter article provides the simple arguments of the proof of Theorem 4.32. Regularity results of the same type as Lemma 4.34 were first obtained, with different techniques, by H. Lewy and G. Stampacchia [101] and H. Brézis and G. Stampacchia [22].

Interior $H^{2,\infty}$ regularity was proved by H. Brézis and D. Kinderlehrer [20] and C. Gerhardt [62]. Global $H^{2,\infty}$ regularity was first proved by R. Jensen [80] who, however, used a norm estimate (Lemma 4.4 in A. Friedman [56]) that is not quite correct: compare with M. Chipot [43]. The proof of Theorem 4.38 is based on C. Gerhardt [63]. The example of Section 4.6.2 is attributed to E. Shamir by H. Brézis and G. Stampacchia [22]; the proof of Theorem 4.39 is basically due to J. L. Lions [103] (see also D. Kinderlehrer [86]).

The techniques of Section 4.7 were introduced (for the study of interior regularity) by M. Giaquinta [64]; the proof of Theorem 4.45, however, is essentially that of M. Biroli [15]. For a different approach see J. Frehse [52].

Theorem 4.46 is due to M. Chipot [41].

Except for some minor changes, the proof of Theorem 4.47 comes from L. Boccardo, F. Murat, and J. P. Puel [16]. The proof of Theorem 4.48 is ours (but see the remark following it); the idea of reducing a nonlinear equation to a v.i. was first utilized by J. P. Puel [131].

By no means does our treatment of (elliptic) v.i.'s do justice to the richness of existing results. Among our omissions we could mention numerical aspects (see R. Glowinski, J. L. Lions, and R. Trémolières [69]), regularity of the free boundary (see A. Friedman [56]), and v.i.'s that are not of the obstacle type {see H. Brézis and M. Sibony [21] and P. L. Lions [108] for what concerns the convex set (4.31)}.

Chapter 5

Nonvariational obstacle problems were introduced by A. Friedman [55] and A. Bensoussan and J. L. Lions [12] as auxiliary tools in the theory of stochastic control, for the case when the dynamic system at hand is governed by a merely continuous diffusion term. Among the subsequent contributions to the subject we mention the papers by G. M. Troianiello [147–149], P. L. Lions [106], and M. G. Garroni and M. A. Vivaldi [59, 60], all dealing with linear operators, and the papers by G. M. Troianiello [150, 151] and M. G. Garroni and M. A. Vivaldi [61], where nonlinear operators are taken up; for a class of degenerate problems see I. Capuzzo Dolcetta and M. G. Garroni [34].

The presentation provided here is largely taken from the author's articles. In particular, the notion of a generalized solution and its applications to the study of implicit unilateral problems are based on G. M. Troianiello [148].

Theorems 5.6 and 5.7 are based on their variational counterparts, respectively studied by O. Nakoulima [125] and J. L. Joly and U. Mosco [82]. Theorem 5.8 extends a result previously proven, with different techniques, by M. G. Garroni and M. A. Vivaldi [61].

The results of Section 5.2.1 are due to L. Nirenberg [129]. Lemma 5.10 is based on H. Amman [5] (see also H. Amman and M. G. Crandall [6] and K. Inkmann [79]). The proof of Lemma 5.11, due to the present author, makes a crucial use of some techniques by O. A. Ladyzhenskaya, V. Solonnikov, and N. N. Ural'tseva [96] as well as of some by J. Frehse [51].

Step 2 of the proof of Theorem 5.12 utilizes an idea in an article by K. Akô [4], which also contains the example of Section 5.3.2. Theorem 5.14 extends previous results of H. Amann and M. G. Crandall [6] and J. L. Kazdan and R. J. Kramer [85].

In a variational setting implicit unilateral problems enter the theory of *quasivariational inequalities*, introduced by A. Bensoussan and J. L. Lions [11]: see A. Bensoussan and J. L. Lions [13], J. L. Joly and U. Mosco [82], C. Baiocchi and A. Capelo [8] as well as, for what concerns in particular the impulse control problem, J. L. Joly, U. Mosco, and G. M. Troianiello [83], I. Capuzzo Dolcetta and M. A. Vivaldi [35], B. Hanouzet and J. L. Joly [71, 73], L. Caffarelli and A. Friedman [26], U. Mosco [120], and A. Bensoussan, J. Frehse, and U. Mosco [14].

References

1. Adams, R. A. *Sobolev Spaces*, Academic, New York (1975).
2. Agmon, S. *Lectures on Elliptic Boundary Value Problems*, Van Nostrand, Princeton, New Jersey (1965).
3. Agmon, S., Douglis, A., and Nirenberg, L. Estimates near the boundary for solutions of elliptic partial differential equations satisfying general boundary conditions, I, *Commun. Pure Appl. Math.* **12**, 623–727 (1959).
4. Akô, K. On the Dirichlet problem for quasi-linear elliptic differential equations of second order, *J. Math. Soc. Japn.* **13**, 45–52 (1961).
5. Amann, H. Existence and multiplicity theorems for semi-linear elliptic boundary value problems, *Math. Z.* **150**, 281–295 (1976).
6. Amann, H., and Crandall, M. G. On some existence theorems for semilinear elliptic equations, *Indiana Univ. Math. J.* **27**, 779–790 (1978).
7. Babuška, I., Error bounds for the finite element method, *Numer. Math.* **16**, 322–333 (1971).
8. Baiocchi, C., and Capelo, A. *Variational and Quasivariational Inequalities. Application to Free Boundary Problems*, Wiley, New York (1984).
9. Baiocchi, C., Gastaldi, F., and Tomarelli, F. Inéquations variationnelles non coercives, *C. R. Acad. Sci. Paris Sér. I Math.* **299**, 647–650 (1984).
10. Baiocchi, C., Gastaldi, F., and Tomarelli, F. Some existence results on non-coercive variational inequalities, *Pubblicazioni I.A.N. Pavia* (1984).
11. Bensoussan, A., and Lions, J.-L. Nouvelle formulation de problèmes de contrôle impulsionnel et applications, *C. R. Acad. Sci. Paris Sér. A-B* **276**, A1189-A1192 (1973).
12. Bensoussan, A., and Lions, J.-L. *Applications des inéquations variationnelles en contrôle stochastique*, Dunod, Paris (1978); Engl. transl. *Applications of Variational Inequalities in Stochastic Control*, North-Holland, Amsterdam (1982).
13. Bensoussan, A., and Lions, J.-L. *Contrôle impulsionnel et inéquations quasi-variationnelles*, Dunod, Paris (1980); Engl. transl. *Impulse Control and Quasi-Variational Inequalities*, Gauthiers-Villars, Paris (1984).
14. Bensoussan, A., Frehse, J., and Mosco, U. A stochastic impulse control problem with quadratic growth Hamiltonian and the corresponding quasi variational inequality, *J. Reine Angew. Math.* **331**, 124–145 (1982).

15. Biroli, M. A De Giorgi–Nash–Moser result for a variational inequality, *Boll. Un. Mat. Ital. A* (5) **16**, 598–605 (1979).

16. Boccardo, L., Murat, F., and Puel, J.-P. Résultats d'existence pour certains problèmes elliptiques quasilinéaires, *Ann. Sc. Norm. Sup. Pisa Cl. Sci.* (4) **11**, 213–235 (1984).

17. Bony, J.-M. Principe du maximum dans les espaces de Sobolev, *C. R. Acad. Sci. Paris Sér. A-B* **265**, A333–A336 (1967).

18. Brézis, H. Équations et inéquations nonlinéaires dans les espaces vectoriels en dualité, *Ann. Inst. Fourier (Grenoble)* **18**, 115–175 (1968).

19. Brézis, H. *Analyse fonctionnelle. Théorie et applications*, Masson, Paris (1983).

20. Brézis, H., and Kinderlehrer, D. The smoothness of solutions to nonlinear variational inequalities, *Indiana Univ. Math. J.* **23**, 831–844 (1974).

21. Brézis, H., and Sibony, M. Equivalence de deux inéquations variationnelles et applications, *Arch. Rational Mech. Anal.* **41**, 254–265 (1971).

22. Brézis, H., and Stampacchia, G. Sur la régularité de la solution d'inéquations elliptiques, *Bull. Soc. Math. France* **96**, 153–180 (1968).

23. Brézis, H., and Stampacchia, G. Remarks on some fourth order variational inequalities, *Ann. Scuola Norm. Sup. Pisa Cl. Sci.* (4) **4**, 363–371 (1977).

24. Caccioppoli, R. Sulle equazioni ellittiche a derivate parziali con n variabili indipendenti, *Rend. Accad. Lincei* **19**, 83–89 (1934).

25. Caffarelli, L. Further regularity for the Signorini problem, *Commun. Partial Differential Equations* **4**, 1067–1075 (1979).

26. Caffarelli, L. A., and Friedman, A. Regularity of the solution of the quasi-variational inequality for the impulse control problem, *Commun. Partial Differential Equations* **3**, 745–753 (1978).

27. Campanato, S. Proprietà di Hölderianità di alcune classi di funzioni, *Ann. Scuola Norm. Sup. Pisa Sci. Fis. Mat.* (3) **17**, 175–188 (1963).

28. Campanato, S. Equazioni ellittiche del secondo ordine e spazi $\mathscr{L}^{2,\lambda}$, *Ann. Mat. Pura Appl.* (4) **69**, 321–382 (1965).

29. Campanato, S. Equazioni paraboliche del secondo ordine e spazi $\mathscr{L}^{2,\theta}(\Omega, \delta)$, *Ann. Mat. Pura Appl.* (4) **73**, 55–102 (1966).

30. Campanato S. Su un teorema di interpolazione di G. Stampacchia, *Ann. Scuola Norm. Sup. Pisa Sci. Fis. Mat.* (3) **20**, 649–652 (1966).

31. Campanato, S. Alcune osservazioni relative alle soluzioni di equazioni ellittiche di ordine $2m$, in *Atti del Convegno su le equazioni alle derivate parziali*, Oderisi, Gubbio, 1967.

32. Campanato, S. *Sistemi ellittici in forma divergenza. Regolarità all'interno*, Quaderni Scuola Norm. Sup. Pisa, 1980.

33. Campanato, S., and Stampacchia, G. Sulle maggiorazioni L^p nella teoria delle equazioni ellittiche, *Boll. Un. Mat. Ital.* **20**, 393–399 (1965).

34. Capuzzo Dolcetta, I., and Garroni, M. G. Oblique derivative problems and invariant measures, *Ann. Scuola Norm. Sup. Pisa Cl. Sci.* (4), to appear.

35. Capuzzo Dolcetta, I., and Vivaldi, M. A. Existence of a regular solution of a quasivariational inequality in an unbounded domain, *Comm. Partial Differential Equations* **3**, 443–470 (1978).

36. Chicco, M. Principio di massimo forte per sottosoluzioni di equazioni ellittiche di tipo variazionale, *Boll. Un. Mat. Ital.* **22**, 368–372 (1967).

37. Chicco, M. Solvability of the Dirichlet problem in $H^{2,p}(\Omega)$ for a class of linear second order elliptic partial differential equations, *Boll. Un. Mat. Ital.* **4**, 374–387 (1971).
38. Chicco, M. Third boundary value problem in $H^{2,p}(\Omega)$ for a class of linear second order elliptic partial differential equations, *Rend. Ist. Mat. Univ. Trieste* **4**, 85–94 (1972).
39. Chicco, M. Esistenza ed unicità della soluzione di una disequazione variazionale associata a un operatore ellittico del secondo ordine a coefficienti discontinui, *Rend. Sem. Mat. Univ. Padova* **57**, 17–37 (1977).
40. Chicco, M. Appartenenza ad $H^{1,p}(\Omega)$ $(2 \leq p < +\infty)$ delle soluzioni di una classe di disequazioni variazionali ellittiche. *Boll. Un. Mat. Ital. B* (4) **3**, 137–148 (1984).
41. Chipot, M. Sur la regularité Lipschitzienne de la solution d'inéquations elliptiques, *J. Math. Pures Appl.* (9) **57**, 69–76 (1978).
42. Chipot, M. Regularity for the two obstacle problems, in *Free Boundary Problems, Vol. II* (*Pavia*, 1979), Ist. Naz. Alta Mat. Francesco Severi, Rome (1980).
43. Chipot, M. *Variational Inequalities and Flow in Porous Media*, Appl. Math. Sciences, Vol. 52, Springer-Verlag, Berlin (1984).
44. De Giorgi, E. Sulla differenziabilità e l'analiticità delle estremali degli integrali multipli regolari, *Mem. Accad. Sci. Torino Cl. Sci. Fis. Mat. Natur.* **3**, 25–43 (1957).
45. Deny, J., and Lions, J.-L. Les espaces du type Beppo Levi, *Ann. Inst. Fourier* (*Grenoble*) **5**, 305–370 (1955).
46. Dubinskii, J. A. Quasilinear elliptic and parabolic equations of arbitrary order (in Russian), *Usp. Mat. Nauk* **23**, 45–90 (1968).
47. Duvaut, G., and Lions, J.-L. *Inequalities in Mechanics and Physics*, Springer-Verlag, Berlin (1976).
48. Fichera, G. Problemi elastostatici con vincoli unilaterali: il problema di Signorini con ambigue condizioni al contorno, *Atti Accad. Naz. Lincei Mem. Cl. Sci. Fis. Mat. Natur. Sez. Ia* (8) **7**, 91–140 (1963–64).
49. Fiorenza, R. Sui problemi di derivata obliqua per le equazioni ellittiche, *Ricerche Mat.* **8**, 83–110 (1959).
50. Frehse, J. On the regularity of the solution of a second order variational inequality, *Boll. Un. Mat. Ital.* **6**, 312–315 (1972).
51. Frehse, J. On the regularity of solutions to elliptic differential inequalities, in *Mathematical Techniques of Optimization, Control and Decision*, J.-P. Aubin, A. Bensoussan, and I. Ekeland, editors, Birkhäuser, Basel (1981).
52. Frehse, J. On the smoothness of solutions of variational inequalities with obstacles, in *Partial Differential Equations*, B. Bojarski editor, Banach Center Publications Vol. 10, Polish Scientific Publishers, Warsaw (1983).
53. Frehse, J., and Mosco, U. Irregular obstacles and quasi-variational inequalities of stochastic impulse control, *Ann. Scuola Norm. Sup. Pisa Cl. Sci.* (4) **9**, 105–157 (1982).
54. Friedman, A. *Partial Differential Equations*, Holt, Rinehart and Winston, New York (1969).
55. Friedman, A. *Stochastic Differential Equations and Applications*, Vol. 2, Academic, New York (1976).
56. Friedman, A. *Variational Principles and Free-Boundary Problems*, Wiley, New York (1982).
57. Friedrichs, K. O. The identity of weak and strong extensions of differential operators, *Trans. Am. Math. Soc.* **55**, 132–151 (1944).

58. Gagliardo, E. Proprietà di alcune classi di funzioni in più variabili, *Ricerche Mat.* **7**, 102–137 (1958).
59. Garroni, M. G., and Vivaldi, M. A. Bilateral inequalities and implicit unilateral systems of the non variational type, *Manuscripta Math.* **33**, 177–215 (1980).
60. Garroni, M. G., and Vivaldi, M. A. Bilateral evolution problems of non-variational type: existence, uniqueness, Hölder-regularity and approximation of solutions, *Manuscripta Math.* **48**, 39–69 (1984).
61. Garroni, M. G., and Vivaldi, M. A. Approximation results for non linear evolution problems of non-variational type, *Nonlinear Anal.* **8**, 301–312 (1984).
62. Gerhardt, C. Regularity of solutions of nonlinear variational inequalities, *Arch. Rational Mech. Anal.* **52**, 389–393 (1973).
63. Gerhardt, C. Global $C^{1,1}$-regularity for solutions of quasilinear variational inequalities, *Arch. Rational Mech. Anal.* **89**, 83–92 (1985).
64. Giaquinta, M. Remarks on the regularity of weak solutions to some variational inequalities, *Math. Z.* **177**, 15–31 (1981).
65. Giaquinta, M. *Multiple Integrals in the Calculus of Variations and Nonlinear Elliptic Systems*, Princeton University Press, Princeton, New Jersey (1983).
66. Giaquinta, M., and Giusti, E. Global $C^{1,\alpha}$- regularity for second order quasilinear elliptic equations in divergence form, *J. Reine Angew. Math.* **351**, 55–65 (1984).
67. Gilbarg, D., and Trudinger, N. S. *Elliptic Partial Differential Equations of Second Order. Second Edition*, Springer-Verlag, Berlin (1983).
68. Giusti, E. *Equazioni ellittiche del secondo ordine*, Quaderni dell'Unione Matematica Italiana, Pitagora Editrice, Bologna (1978).
69. Glowinski, R., Lions, J.-L., and Trémolières, R. *Numerical Analysis of Variational Inequalities*, North-Holland, Amsterdam (1981).
70. Greco, D. Nuove formole integrali di maggiorazione per le soluzıoni di un'equazione lineare di tipo ellittico ed applicazioni alla teoria del potenziale, *Ricerche Mat.* **5**, 126–149 (1956).
71. Hanouzet, B., and Joly, J.-L. Un résultat de régularité pour une inéquation quasi-variationnelle du type de Neumann intervenant dans un problème de contrôle impulsionnel, *J. Math. Pures Appl.* (9) **56**, 327–337 (1977).
72. Hanouzet, B., and Joly, J.-L. Méthodes d'ordre dans l'interprétation de certaines inéquations variationnelles et applications, *J. Funct. Anal.* **34**, 217–249 (1979).
73. Hanouzet, B., and Joly, J.-L. Convergence uniforme des interés definissant la solution d'inéquations quasi-variationnelles et application à la regularité, *Numer. Funct. Anal. Optim.* **1**, 399–414 (1979).
74. Haugazeau, Y. Sur des inéquations variationnelles, *C. R. Acad. Sci. Paris Sér. A-B* **265**, A95–A98 (1967).
75. Hess, P. On semi-coercive nonlinear problems, *Indiana Univ. Math. J.* **23**, 645–654 (1974).
76. Hopf, E. Elementare Bemerkungen über die Lösungen partieller Differentialgleichungen zweiter Ordnung vom elliptischen Typus, *Sitzungsber. Berliner Akad. Wiss., Math.-Phys. Kl.* **19**, 147–152 (1927).
77. Hopf, E. Über den funktionalen, insbesondere den analytischen Charakter der Lösungen elliptischer Differentialgleichungen zweiter Ordnung, *Math. Z.* **34**, 194–233 (1932).
78. Hopf, E. A remark on linear elliptic differential equations of second order, *Proc. Am. Math. Soc.* **3**, 791–793 (1952).

79. Inkmann, F. Existence and multiplicity theorems for semilinear elliptic equations with nonlinear boundary conditions, *Indiana Univ. Math. J.* **31**, 213–221 (1982).
80. Jensen, R. Boundary regularity for variational inequalities, *Indiana Univ. Math. J.* **29**, 495–504 (1980).
81. John, F., and Nirenberg, L. On functions of bounded mean oscillation, *Commun. Pure Appl. Math.* **14**, 415–426 (1961).
82. Joly, J.-L., and Mosco, U. À propos de l'existence et de la régularité des solutions de certaines inéquations quasi-variationnelles, *J. Funct. Anal.* **34**, 107–137 (1979).
83. Joly, J.-L., Mosco, U., and Troianiello, G. M. On the regular solution of a quasi-variational inequality connected to a problem of stochastic impulse control, *J. Math. Anal. Appl.* **61**, 357–369 (1977).
84. Kadlec, J., and Nečas, J. Sulla regolarità delle soluzioni di equazioni ellittiche negli spazi $H^{k,\lambda}$, *Ann. Scuola Norm. Sup. Pisa Sci. Fis. Mat.* (3) **21**, 527–545 (1967).
85. Kazdan, J. L., and Kramer, R. J. Invariant criteria for existence of solutions to second-order quasilinear elliptic equations. *Commun. Pure Appl. Math.* **31**, 619–645 (1978).
86. Kinderlehrer, D. Remarks about Signorini's problem in linear elasticity, *Ann. Scuola Norm. Sup. Pisa Cl. Sci.* (4) **8**, 605–645 (1981).
87. Kinderlehrer, D., and Stampacchia, G. *An Introduction to Variational Inequalities and their Applications*, Academic, New York (1980).
88. Klee, V. L. Jr. Boundedness and continuity of linear functionals, *Duke Math. J.* **22**, 263–270 (1955).
89. Kondrachov, V. I. Certain properties of functions in the spaces L^p (in Russian), *Dokl. Akad. Nauk SSSR* **48**, 535–538 (1945).
90. Korn, A. Zwei Anwendungen der Methode der sukzessiven Annäherungen, in *Schwarz-Festschrift*, Berlin (1915).
91. Koželev, A. E. On boundedness in L^p of derivatives of solutions of elliptic differential equations (in Russian), *Mat. Sb.* **38**, 359–372 (1956).
92. Kufner, A., John, O., and Fučik, S. *Function Spaces*, Noordhoff, Leyden (1977).
93. Kuratowski, K. *Topology*, Vol. I, Academic, New York (1966).
94. Ladyzhenskaya, O. A., and Ural'tseva, N. N. *Linear and Quasilinear Elliptic Equations*, Izdat. "Nauka", Moscow (1964) (in Russian); Engl. transl., Academic, New York (1968).
95. Ladyzhenskaya, O. A., and Ural'tseva, N. N. Estimate of Hölder norm for solutions of second order quasilinear elliptic equations of general form (in Russian), *Usp. Mat. Nauk* **35**, 144–145 (1980).
96. Ladyzhenskaya, O. A., Solonnikov, V. A., and Ural'tseva, N. N. *Linear and Quasilinear Equations of Parabolic type*, Translations of Mathematical Monographs Vol. 23, American Mathematical Society, Providence, Rhode Island (1968).
97. Landes, R. On Galerkin's method in the existence theory of quasilinear elliptic equations, *J. Funct. Anal.* **39**, 123–148 (1980).
98. Lax, P. D., and Milgram, A. M. Parabolic equations, in *Contributions to the Theory of Partial Differential Equations*, L. Bers, S. Bochner, and F. John, editors, Princeton University Press, Princeton, New Jersey (1954).
99. Leray, J., and Lions, J.-L. Quelques résultats de Višik sur les problèmes elliptiques nonlinéaires par les méthodes de Minthy–Browder, *Bull. Soc. Math. France* **93**, 97–107 (1965).
100. Levi, B. Sul principio di Dirichlet, *Rend. Circ. Mat. Palermo* **22**, 293–359 (1906).

101. Lewy, H., and Stampacchia, G. On the regularity of the solution of a variational inequality, *Commun. Pure Appl. Math.* **22**, 153–188 (1969).

102. Lewy, H., and Stampacchia G. On the smoothness of superharmonics which solve a minimum problem, *J. Analyse Math.* **23**, 227–236 (1970).

103. Lions, J.-L. *Quelques méthodes de résolution des problèmes aux limites non linéaires*, Dunod, Paris (1969).

104. Lions, J.-L. *Optimal Control of Systems Governed by Partial Differential Equations*, Springer-Verlag, New York (1971).

105. Lions, J.-L., and Stampacchia, G. Variational inequalities, *Commun. Pure Appl. Math.* **20**, 493–519 (1967).

106. Lions, P.-L. Problèmes elliptiques du 2ème ordre non sous forme divergence, *Proc. R. Soc. Edinburgh Sect. A* **84**, 263–271 (1979).

107. Lions, P.-L. A remark on some elliptic second order problems, *Boll. Un. Mat. Ital. A* **17**, 267–270 (1980).

108. Lions, P.-L. An estimate of the Lipschitz norm of solutions of variational inequalities and applications, in *Variational Inequalities and Complementarity Problems. Theory and Applications*, R. W. Cottle, F. Giannessi, J.-L. Lions, editors, Wiley, New York (1980).

109. Meyers, N. G. An L^p-estimate for the gradient of solutions of second order elliptic divergence equation, *Ann. Scuola Norm. Sup. Pisa Sci. Fis. Mat.* **17**, 189–206 (1963).

110. Meyers, N. G. Mean oscillation over cubes and Hölder continuity, *Proc. Am. Math. Soc.* **15**, 717–721 (1964).

111. Meyers, N. G., and Serrin, J. $H = W$, *Proc. Natl. Acad. Sci. U.S.A.* **51**, 1055–1056 (1964).

112. Mikhailov, V. P. *Partial Differential Equations*, MIR, Moscow (1978).

113. Minty, G. J. Monotone (non linear) operators in Hilbert space, *Duke Math. J.* **29**, 341–346 (1962).

114. Miranda, C. Alcune osservazioni sulla maggiorazione in L^ν delle soluzioni deboli delle equazioni ellittiche del secondo ordine, *Ann. Mat. Pura Appl.* (4) **61**, 151–169 (1963).

115. Miranda, C. *Partial Differential Equations of Elliptic Type. Second Edition*, Springer-Verlag, Berlin (1970).

116. Morrey, C. B., Jr. On the solutions of quasilinear elliptic partial differential equations, *Trans. Am. Math. Soc.* **43**, 126–166 (1938).

117. Morrey, C. B., Jr. Functions of several variables and absolute continuity, II, *Duke Math. J.* **6**, 187–215 (1940).

118. Morrey, C. B., Jr. *Multiple Integrals in the Calculus of Variations*, Springer-Verlag, Berlin (1966).

119. Mosco, U. Convergence of convex sets and of solutions of variational inequalities, *Adv. Math.* **3**, 510–585 (1969).

120. Mosco, U. Implicit Variational problems and quasi variational inequalities, in *Nonlinear Operators and Calculus of Variations*, J.-P. Gossez, E. J. Lami Dozo, J. Mawhin, and L. Waelbroeck editors, Lect. Notes in Math. 543, Springer-Verlag, Berlin (1979).

121. Mosco, U., and Troianiello, G. M. On the smoothness of solutions of unilateral Dirichlet problems, *Boll. Un. Mat. Ital.* **8**, 57–67 (1973).

122. Moser, J. A new proof of de Giorgi's theorem concerning the regularity problem for elliptic differential equations, *Commun. Pure Appl. Math.* **13**, 457–468 (1960).

123. Moser, J. On Harnack's theorem for elliptic differential equations, *Commun. Pure Appl. Math.* **14**, 577–591 (1961).
124. Naimark, M. A. *Normed Rings*, P. Noordhoff, Groningen, The Netherlands (1964).
125. Nakoulima, O. Sur une notion de solution faible pour les inéquations variationnelles d'évolution à deux obstacles, *C. R. Acad. Sci. Paris Sér. A-B* **284**, A1037–A1040 (1977).
126. Nash, J. Continuity of solutions of parabolic and elliptic equations, *Am. J. Math.* **80**, 931–954 (1958).
127. Nečas, J. *Les méthodes directes en théorie des équations elliptiques*, Masson, Paris (1967).
128. Nirenberg, L. Remarks on strongly elliptic partial differential equations, *Commun. Pure Appl. Math.* **8**, 649–675 (1955).
129. Nirenberg, L. On elliptic partial differential equations, *Ann. Scuola Norm. Sup. Pisa Sci. Fis. Mat.* (3) **13**, 115–162 (1959).
130. Peetre, J. On the theory of $\mathscr{L}_{p,\lambda}$ spaces, *J. Funct. Anal.* **4**, 71–87 (1969).
131. Puel, J.-P. Existence, comportement à l'infini et stabilité dans certains problèmes quasilinéaires elliptiques et paraboliques d'ordre 2, *Ann. Scuola Norm. Sup. Pisa Cl. Sci.* (4) **3**, 89–119 (1976).
132. Rellich, R. Ein Satz uber mittlere Konvergenz, *Nachr. Akad. Wiss. Göttingen Math.-Phys. Kl. II*, 30–35 (1930).
133. Rudin, W. *Real and Complex Analysis*, McGraw-Hill, New York (1966).
134. Schaefer, H. H. *Topological Vector Spaces*, Springer-Verlag, New York (1971).
135. Schauder, J. Über lineare elliptische Differentialgleichungen zweiter Ordnung, *Math. Z.* **38**, 257–282 (1934).
136. Schauder, J. Numerische Abschätzungen in elliptischen linearen Differentialgleichungen, *Studia Math.* **5**, 34–42 (1934).
137. Sobolev, S. L. On a theorem of functional analysis (in Russian), *Math. Sb.* **46**, 471–496 (1938); Engl. transl.: *Amer. Math. Soc. Transl.* **34**, 39–68 (1963).
138. Sobolev, S. L. *Applications of Functional Analysis in Mathematical Physics*, Transl. Math. Mon. Vol. 7, American Mathematical Society, Providence, Rhode Island (1963).
139. Stampacchia, G. Équations elliptiques à données discontinues, *Seminaire Schwartz, 1960–61 5e année: Équations aux dérivées partielles et interpolation*, 4.01–4.16.
140. Stampacchia G. Formes bilinéaires coercitives sur les ensembles convexes, *C. R. Acad. Sci. Paris* **258**, 4413–4416 (1964).
141. Stampacchia G. Le problème de Dirichlet pour les équations elliptiques du second ordre à coefficients discontinus, *Ann. Inst. Fourier* (*Grenoble*) **15**, 189–258 (1965).
142. Stampacchia, G. The spaces $\mathscr{L}^{(p,\lambda)}$, $N^{(p,\lambda)}$ and interpolation, *Ann. Scuola Norm. Sup. Pisa* **19**, 443–462 (1965).
143. Stampacchia G. *Équations elliptiques du second ordre à coefficients discontinus*, Les Presses de l'Université de Montréal, Montréal (1966).
144. Taylor, A. E., and Lay, D. C. *Introduction to Functional Analysis. Second Edition*, Wiley, New York (1980).
145. Tonelli, L. Sulla quadratura delle superficie. *Atti Reale Accad. Lincei* **3**, 633–638 (1926).
146. Troianiello, G. M. On the regularity of solutions of unilateral variational problems, *Rend. Accad. Sci. Fis. Mat. Napoli* **62**, 200–214 (1975).

147. Troianiello, G. M. Unilateral Dirichlet problems of the nonvariational type, *Ann. Mat. Pura Appl.* (4) **122**, 365–381 (1979).
148. Troianiello, G. M. On a class of unilateral evolution problems, *Manuscripta Math.* **29**, 353–384 (1979).
149. Troianiello, G. M. Some unilateral problems of the nonvariational type, in *Mathematical Techniques of Optimization, Control and Decision*, J.-P. Aubin, A. Bensoussan, and I. Ekeland, editors, Birkhäuser, Basel (1981).
150. Troianiello, G. M. Maximal and minimal solutions to a class of elliptic quasilinear problems, *Proc. Am. Math. Soc.* **91**, 95–101 (1984).
151. Troianiello, G. M. Regular solutions to nonvariational obstacle problems for parabolic operators, *Math. Nachr.* **124**, 133–169 (1985).
152. Trudinger, N. S. Linear elliptic operators with measurable coefficients, *Ann. Scuola Norm. Sup. Pisa Sci. Fis. Mat.* (3) **27**, 265–308 (1973).
153. Trudinger, N. S. Local estimates for subsolutions and supersolutions of general second order elliptic quasilinear equations, *Invent. Math.* **61**, 67–79 (1980).
154. Vespri, F. Regolarità negli spazi $\mathscr{L}^{2,\lambda}$ delle soluzioni di sistemi ellittici lineari del II ordine. Tesi di laurea, Univ. di Pisa (1982).
155. Zygmund, A. *Trigonometric Series*, Vol. II, Cambridge, Cambridge University Press (1959).

Index of Special Symbols and Abbreviations

Index